International Scientific Conference of Young Scientists: Advanced Materials in Construction and Engineering 2014

IOP Conference Series: Materials Science and Engineering
Volume 71

Tomsk, Russia
15 - 17 October 2014

ISBN: 978-1-63439-882-4
ISSN: 1757-8981

Printed from e-media with permission by:

Curran Associates, Inc.
57 Morehouse Lane
Red Hook, NY 12571

Some format issues inherent in the e-media version may also appear in this print version.

Printed by Curran Associates, Inc. (2015)

For permission requests, please contact the Institute of Physics
at the address below.

Institute of Physics
Dirac House, Temple Back
Bristol BS1 6BE UK

Phone: 44 1 17 929 7481
Fax: 44 1 17 920 0979

techtracking@iop.org

Additional copies of this publication are available from:

Curran Associates, Inc.
57 Morehouse Lane
Red Hook, NY 12571 USA
Phone: 845-758-0400
Fax: 845-758-2634
Email: curran@proceedings.com
Web: www.proceedings.com

Table of contents

Volume 71

International Scientific Conference of Young Scientists: Advanced Materials in Construction and Engineering
15–17 October 2014, Tomsk, Russia

Accepted papers received: 24 November 2014
Published online: 20 January 2015

Technical regulation and quality control in construction

Properties, structure and phase transition of the structural metal materials under different energy deposition, structure and properties of ultra-fine grained and nano materials

IOP Conf. Series: Materials Science and Engineering **71** (2015) 011001 doi:10.1088/1757-899X/71/1/011001

PREFACE

In October 15-17, 2014 *International Scientific Conference of Young Scientists: Advanced Materials in Construction and Engineering* (TSUAB2014) took place at Tomsk State University of Architecture and Building (Tomsk, Russia).

The Conference became a discussion platform for researchers in the fields of studying structure and properties of advanced building materials and included open lectures of leading scientists and oral presentations of master, postgraduate and doctoral students. A special session was devoted to reports of school children who further plan on starting a research career.

The Conference included an industrial exhibition where companies displayed the products and services they supply. The companies also gave presentations of their products within the Conference sessions.

KEY SPEAKERS

We would like to express our sincere gratitude to the key speakers who gave oral presentations at the plenary sessions:

K. Kovler, Prof. Dr. (Technion – Israel Institute of Technology, Haifa, Israel)
V. Khozin, Prof. Dr. (Kazan State University of Architecture and Engineering, Kazan, Russia)
A. Pichugin, Prof. Dr. (Novosibirsk State Agricultural University, Novosibirsk, Russia)
V. Vereschagin, Prof. Dr. (National Research Tomsk Polytechnic University, Tomsk, Russia)
A. Kudiakov, Prof. Dr. (Tomsk State University of Architecture and Building, Tomsk, Russia)
V. Starenchenko, Prof. Dr. (Tomsk State University of Architecture and Building, Tomsk, Russia)

Their contribution to the Conference was tremendous and we highly appreciate their presence, participation in discussions and being a good example for young scientists on their early-stage research careers.

CONFERENCE SESSIONS

Oral presentations were given within the following research directions (conference sessions):

- Modern ways of managing structure and properties of building materials
- Innovative technologies in building materials production
- Technical regulation and quality control in construction
- Properties, structure and phase transition of the structural metal materials under different energy deposition, structure and properties of ultra-fine grained and nano materials
- Modeling of plastic flow and other processes in the materials
- Structure and properties of ultrafine grained- and nano-materials
- Advanced materials: students' and school children opinion

The results presented at the Conference included not only fundamental research but also the applied one in the fields of nanomaterials, nanotechnologies, formation of structure and properties of materials under impacts of different nature, modeling of porous materials, phase state and phase change in metals and alloys.

ACKNOWLEDGEMENTS

Organizing committee of the Conference express their sincere gratitude to colleagues from National Research Tomsk Polytechnic University, Department of Architecture and Construction of Tomsk Region, the Union of Builders of Tomsk Region, "Kopylovskii Brick" Company
Our sponsors are
- Tomsk Building Company, LLC (Tomsk, Russia);
- Kopylovskii Brick, LLC (Tomsk, Russia);
- Polyalpan Factory, LLC (Tomsk, Russia).

ORGANIZING COMMITTEE

Chairman:
Vasilii A. Klimenov – TSUAB vice-rector for research, Dr. Eng. Sc., Prof.

Co-chairmen:
Natalia O. Kopanitsa – Head of the Research Institute of Building Materials, Dr. Eng. Sc., Prof.
Alexander I. Kudyakov – Head of Building Materials Department of TSUAB, Dr. Eng. Sc., Prof.
Vladimir A. Starenchenko – Head of the General Faculty of TSUAB, Dr. Phys.-Math. Sc., Prof.

Organizing Committee
The key speaker of the Conference was the leading scientist – professor K. Kovler of Technion – Israel Institute of Technology, who gave two oral presentations.

PROGRAM COMMITTEE

Chairman:
Viktor A. Vlasov – TSUAB rector, Dr. Phys.-Math. Sc., Prof.

Co-chairmen:
Konstantin Kovler – Head of Building Materials and Technology Department, National Building Research Institute, Technion – Israel Institute of Technology, Dr. Sc. in Structural Engineering, Prof.
Anatoly P. Malinovsky – Head of the Civil Engineering Faculty of TSUAB, PhD, Prof.

Members of Program Committee
Bazhenov Y. M., Moscow, Russia
Lyakhovich L.S., Tomsk, Russia
Baitasov T.M., Astana, Kazakhstan
Vereshchagin V.I., Tomsk, Russia
ChernyshevE.M., Voronezh, Russia
Vatin N.I., St. Petersburg, Russia
Yumasheva E. I., Moscow, Russia
Oliferuk V., Warsaw, Poland
Panin V.E., Tomsk, Russia
Potekaev A.I., Tomsk, Russia
Pustovgar A.P., Moscow, Russia
Sharkeev Y.P., Tomsk, Russia
Pichugin A.P., Novosibirsk, Russia
Ovcharenko G.I., Barnaul, Russia
Mashkin N.A., Novosibirsk, Russia
Kudyakov A.I., Tomsk, Russia
Kopanitsa N.O., Tomsk, Russia
Kuznetsova T.V., Moscow, Russia

IOP Conf. Series: Materials Science and Engineering **71** (2015) 011001 doi:10.1088/1757-899X/71/1/011001

Fisher H., Weimar, Germany
Mori T., Sapporo, Japan
Mescheryakov Y.G., St. Petersburg, Russia
KozlovE.V., Tomsk, Russia
NazirovR.A., Krasnoyarsk, Russia
PaninS.V., Tomsk, Russia
Shpeter A.K.,Tomsk, Russia
Hasanov O.L., Tomsk, Russia

RESULTS

A total amount of *180 participants* from Russian and foreign institutions attended the Conference, including researchers from: L.N.Gumilyov Eurasian National University (Kazakhstan), Kazakh National Technical University After K.I. Satpayev (Kazakhstan), Technion – Israel Institute of Technology (Israel), Bauhaus-Universität Weimar: Universität (Germany), Kyrgyz Russian Slavic University named after the First President of Russia B.N. Yeltsin (the Kyrgyz Republic), State Scientific Institution "Institute for Single Crystals" of National Academy of Sciences of Ukraine (Ukraine), Institute for Scintillation Materials (Ukraine), Moscow State University of Civil Engineering (Russia) and many others.

All the institutions deal with studying structure and properties of structural and building materials, ultra-fine grained and nanomaterials also under energy depositions and the problems connected with them.

Within the conference framework school children and students had a chance to talk with experienced researchers within conference sessions and round tables, ask questions, and discuss their projects. The unique character of the conference is in combining different ways of interaction between the representatives of research community, young scientists and industry representatives (part of the program included excursions to the construction companies). The result of Conference is new projects for cooperation with Russian and foreign scientists.

The Conference was held with the financial support of Russian Foundation for Basic Research, project No. 14-38-10074\14 мол_г.

Prof. Dr. Natalia O. Kopanitsa,
Head of Building Materials Institute
Co-chairman of Organizing Committee, TSUAB2014

IOP Conf. Series: Materials Science and Engineering **71** (2015) 011001 doi:10.1088/1757-899X/71/1/011001

Conference Opening, October 15, 2014
prof. Dr. V.A. Vlasov, rector of TSUAB

Key speaker, October 15, 2014
prof. Dr. K. Kovler

Plenary session, October 15, 2014

Plenary session, October 15, 2014

Master-class within the Conference

Exhibition within the Conference

IOP Conf. Series: Materials Science and Engineering **71** (2015) 011002 doi:10.1088/1757-899X/71/1/011002

Peer review statement

All papers published in this volume of *Journal of Physics: Conference Series* have been peer reviewed through processes administered by the proceedings Editors. Reviews were conducted by expert referees to the professional and scientific standards expected of a proceedings journal published by IOP Publishing.

IOP Conf. Series: Materials Science and Engineering **71** (2015) 012001 doi:10.1088/1757-899X/71/1/012001

Long-term strength properties of HVFA concretes

M Špak[1] and R Bašková[2]
[1] Technical University of Košice, Institute of Construction Technology and Management, Faculty of Civil Engineering, Košice, 04200, Slovakia
[2] Technical University of Košice, Institute of Construction Technology and Management, Faculty of Civil Engineering, Košice, 04200, Slovakia

E-mail: matej.spak@tuke.sk

Abstract. Fly ash from coal burning is used as active addition for concrete in Middle-Europe region for several decades. The intensity of its utilization increases still. In the role of supplementary cement addition it serves as binder, whereby it helps to reduce final price of concrete as well as improves both the rheological properties of fresh concrete and several characteristics of hardened concrete. Fly ash presents the co-product of energetic industry. Its production increases together with growth of energy consumption. These factors bring the opportunity and requirement of production of concretes with high volume of fly ash based addition. Thus, significant economic, environmental, technological and technical benefits can be achieved by using of high amount of fly ash for concrete production.

1. Introduction

Cement binder together with both aggregate and mixing water present elementary components needed for concrete production. Cement composes only 15 % by wt. of concrete. Even though, it presents most expensive part of concrete. Thus, significant decreasing of material costs can be achieved by relatively small saving of used cement. Cement production is very expensive process, because of high requirement on energy consumption. To produce the clinker is needed to burn several raw materials (limestone, clay, silicon sand, iron ore etc.) at high temperature, approximately at 1400 °C. Clinker products create during the burning process as well as huge amount of greenhouse gases, especially CO_2. This gas is generated both from fuel burning and thermic decomposition of initial raw materials. World-wide cement production was about 4000 million tons (3300 million tons of clinker capacity) in 2013 [1]. Cement industry shares of total CO_2 production with 5 % [2]. Mentioned facts give us relevant arguments to increase utilization of energetic by-products for building materials production.

1.1. Fly ash

One of these alternative materials is fly ash from coal burning. Fly ash is by-produced by both brown and black coal burning in power plant or heating plant. During the process of coal burning, almost all of organic components burn out. Fly ash as a product of burning contains nearly 100 % of mineral-only components, alumina-silicates especially. A part of molten fine-size mineral particles flows together with exhaust gas and during rapid cooling glassy-amorphous particles (fly ash) create. These particles have spherical shape, mostly. Consequently, the particles are filtered form exhaust gas by set of multiple filters and electrostatic separators [3, 4].

Fly ash produced in power plants and heating plants presents industrial waste and require to be stored or disposed. Storage of fly ash on damp is usually expensive. Moreover, fly ash storage can cause the environmental loading. World-wide, it is produces approximately 550 million tons of fly ash annually. Currently, over 30 % of fly ash produced in Europe (EU 15) is utilized as addition for both concrete and cements as well as for production of other building materials. The rest of that is deposed on damp.

Fly ash is ranked among type II additions, thus active additions according to EN 206-1 [5]. In term of reactive mechanism, fly ash is hydraulic active pozzolanic addition. It means that fly ash doesn't set and harden itself as well as in not latent hydraulic. It consists amorphous SiO_2 which reacts with $Ca(OH)_2$ formed by reaction of cement with water. Result of this pozzolanic reaction is C-S-H gel. Chemical composition is determining for pozzolanic properties of fly ash but its mineralogical composition as well as granulometric properties is much more important [6]. Commonly, both chemical and mineralogical composition of fly ash is quite non-homogenous. Even though, fly ash can be utilized for cement and concrete production very well. In the case of commercial usage of fly ash it has to satisfy requirements of standards. These requirements are divided into chemical, physical and others (durability, dangerous components content, radioactivity etc.) according to EN 450-1 [7].

1.2. High volume fly ash concrete

According to Malhotra definition, HVFA (High Volume Fly Ash) concrete is material in which the fly ash presents over than 50 % wt. of total binder amount [6]. This type of concrete is also able to satisfy the requirements on high-performance concrete. There, the common dosage of fly ash for concrete is around 15 to 25 % by wt. of cement binder, in Slovakia. In this point of view, concrete with fly ash content above 30 % can be classed as HVFA concrete.

2. Materials and methods

2.1. Materials

Fly ash addition is usually used in combination with OPC. It is well known that the combination of fly ash with another types of cement bring very good results in term of technological, technical, environmental and economic aspects. Especially, CEM III type cement together with added fly ash creates binder which possesses ternary behaviour. Ternary binder system is based on synergic effect of clinker, blast-furnace slug and fly ash [8].

Cement type CEM II/A-S 42.5 R according to EN 197-1 was used as a binder. This cement came from cement plant Holcim Turňa nad Bodvou, Slovakia. Mineral addition presented fly ash from brown coal burning from power plant Bukocel Hencovce. Slovakia. This fly ash was certified as a type II active addition for concrete according to EN 450-1. Chemical composition of used fly ash is shown in the table 1.

Table 1. Chemical composition of fly ash.

Component	SiO_2	Al_2O_3	Fe_2O_3	CaO	MgO	Alkalis	SO_3	LOI
Content [%]	52.56	22.45	9.02	6.74	1.80	1.31	0.98	2.86

Aggregate fractions 0/4, 4/8 and 8/16 according to EN 12620 were used as filler. Particular fractions came from the quarry Holcim Geča, Slovakia. Aggregate from Geča location is particular crushed gravel. Plasticizing admixture was used for both the improving the consistence of fresh concrete and decreasing of mixing water. Superplasticizer Berament® HT121 based on PCE modified by naphthalene was used. Mixing water according to EN 1008 was used.

2.2. Methods

Described experiment is aimed on evaluation of short-term mechanical properties of hardened concrete according to requirements of standard EN 206-1 (compressive strength after 28 days of hardening) as well as long-term mechanical properties (after 360 days of hardening). Several experimental mixes were prepared. Particular mixes differed by fly ash content. The content of fly ash is presented by binder ratio (fa/c = fly ash/cement).

Recipes of particular mixes were designed according to previous experiences. Nine mixes were prepared which varied by binder ratio and/or water/total binder ratio (mixes marked FA0 to FA5). Table 2 contains characteristics of designed mixes. Part of cement was supplemented by fly ash with content started from 23 % wt. up to 50 % wt. Mix FA0 contains cement only. Mixes FA3 and FA4 were prepared with two different water/total binder ratios. Both pure water/binder ratio as well as water/binder ratio following the k-value concept is shown in the Table 2. Fly ash addition k-value in combination with CEM II/A-S type cement is set to 0.2. Consistence of all prepared fresh concretes was set to S3 class according to EN 206-1, thus 100-150 mm measured by Slump test.

Table 2. Characteristics of experimental mixes.

Mix mark	CSA* [%]	fa/c ratio	w/(c+fa) ratio	w/(c+k.fa) ratio
FA0-0	0	0.00	0.53	0.53
FA1-0	50	1.00	0.46	0.87
FA1-1	50	1.00	0.43	0.80
FA2-1	43	0.75	0.45	0.75
FA3-1	37	0.58	0.47	0.69
FA4-1	30	0.43	0.47	0.63
FA5-1	23	0.30	0.48	0.59
FA4-2	30	0.43	0.48	0.64
FA3-5	37	0.58	0.50	0.75

*CSA – proportion of cement supplementary addition (fly ash)

All mixes were prepared at laboratory ambient condition. Cube samples with dimension 150x150x150 mm were made after mixing and consistence measuring. Two cube samples for compressive strength test were made from every mix. Consequently, samples stayed in the cast for next 24 hours. Released samples were cured in water condition for up to 350 days. Compressive strength of particular samples was tested after 2, 7, 28, 56, 120 and 350 days.

3. Results

Six mixes with different binder (fly ash/cement) ratio were prepared (mixes FA1 to FA5 listed in the Table 2), including reference mix. Compressive strength and its in-time development of hardened concretes during first 28 days are shown on the Figure 1. The concrete with lowest fly ash content achieved highest strength (except non-fly ash content concrete). It is caused by higher clinker content compared to more-fly ash content concretes.

Process of cement hydration is faster than pozzolanic reaction of fly ash in concrete. Therefore, higher content of fly ash in concrete causes more gentle strength development. Concrete with non-fly ash content (0% fa mix on Figure 1) achieved highest compressive strength after 28 days of hardening and its strength development was faster, obviously.

Strength development of reference concrete decelerated after 56 days of hardening, significantly. On the contrary, strength development of fly ash-containing concretes was growing continuously, due to pozzolanic reactions of fly ash. Compressive strengths within 350 days setting and of hardening are shown on Figure 2. It is clear, that the compressive strengths of FA4 and FA5 concretes are higher than that of FA0 – reference concrete. FA4 and FA5 concretes achieved about 3.7 and 7 % higher compressive strength than FA0 concrete, respectively. In these cases, the synergic effect of ternary binder system has proved. Curves of compressive strength development shown on Figures 1 and 2, correspondent with fly ash content very well. Comparison of curves shows that compressive strength of reference concrete grows fast during first 28 days and then slows down. At the other side, compressive strength of fly ash-containing concretes grows slow at the start of hardening period but continues during next several months.

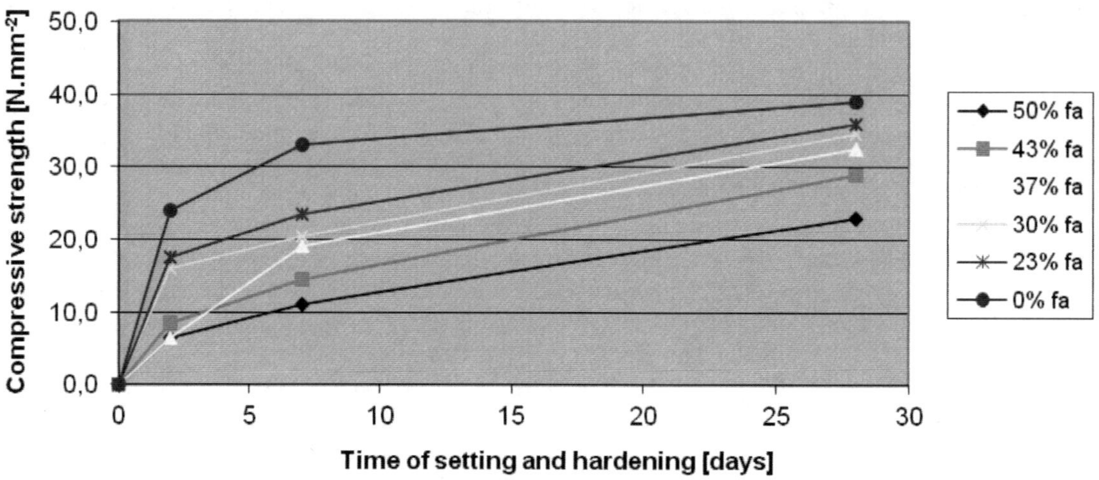

Figure 1. Development of compressive strength within first 28 days.

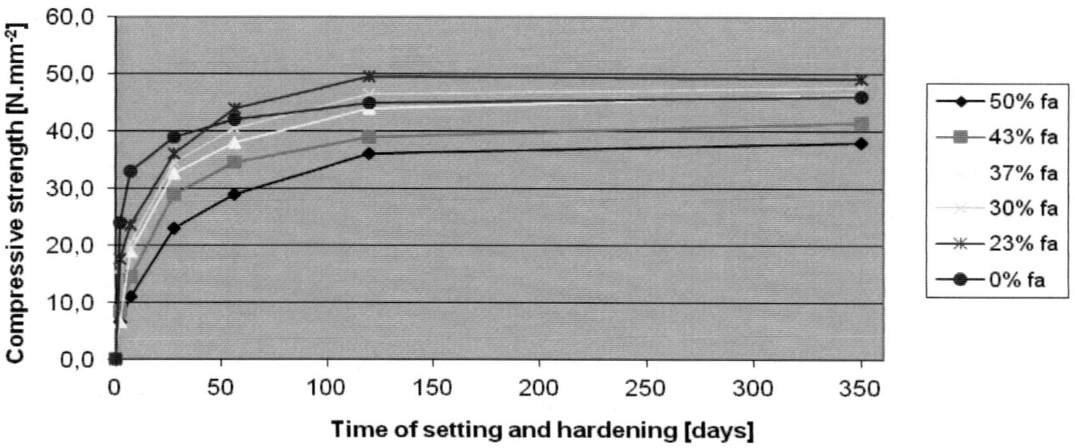

Figure 2. Development of compressive strength within 350 days.

Compressive strength of reference concrete after 350 days of hardening is only about 18 % higher than after 28 days. In the case of HVFA concrete is the increase from 37 to 65 % for FA5 down to FA1, respectively. In this point of view, FA1 mix concrete is optimal, in the case of requirement on higher long-time compressive strength.

Percentage increase of compressive strength within whole time of setting and hardening of concrete is shown on the Figure 3. Monitored time is divided into 6 periods. First period started from preparing of fresh. Columns on the Figure 3 describe the strength development behaviour of particular concretes. It is clear that reference concrete achieved more than 50 % of total compressive strength within first period 0 – 2 days. Its strength increase is no significant within period 121 – 350 days. On the other hand, strength increase of high-fly ash content concretes is more significant within period 8 - 28 days. This result confirms the pozzolanic behaviour of fly ash in HVFA concrete, too.

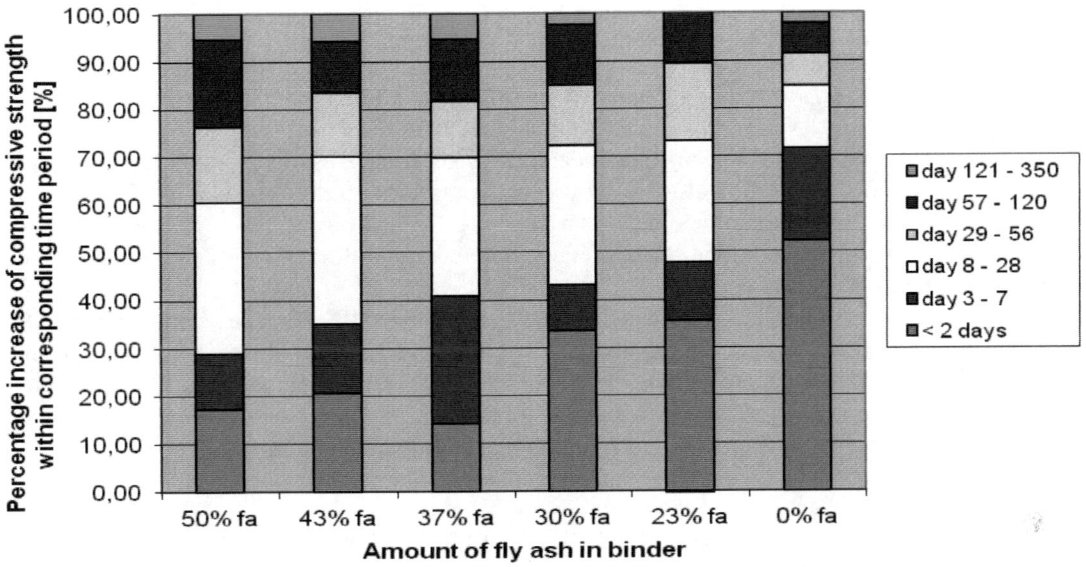

Figure 3. Compressive strength growth gradients of mixes.

It is well known that water/binder ratio influences on strength properties of hardened concrete. Moreover, binder ratio (fly ash/cement ratio) affects the concrete strength characteristics. Correlation between binder ratio and compressive strength of concrete after both 28 days and 350 day is shown on the Figure 4. Influence of the ratio on compressive strength after 28 days of hardening is more significant compared to influence on strength after 350 days. Reference concrete has got fly ash/cement ratio 0.00. Generally, increase of fly ash content in total binder causes decrease of compressive strength of hardened concrete. Following the trend growth from Figure 4, influence of binder ration on compressive strength after 350 days is approximately 2 times lower than the influence after 28 days. On the basis of measured results it is apparent that the long-term compressive strength of HVFA concrete achieves high values.

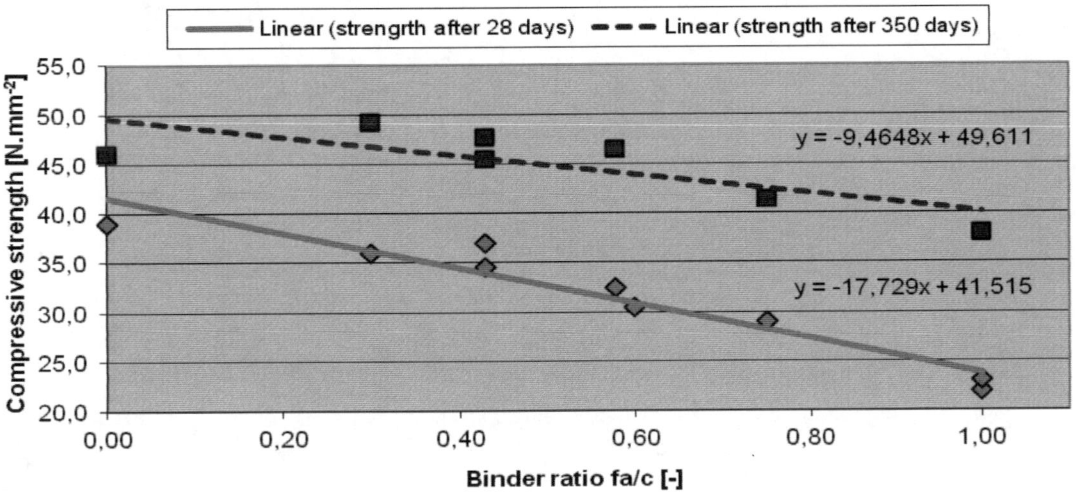

Figure 4. Correlation between compressive strength after both 28 and 350 days and binder ratio.

4. Conclusion

High volume fly ash concrete (HVFAC) presents interesting alternative for common only cement-based concrete. Besides improving of workability of fresh concrete by using of fly ash, it brings a lot of useful properties as a waterproof increasing, higher resistance of hardened concrete against aggressive environment etc. Despite of partial decreasing of strength grow rate it exists many cases when the high volume fly ash concrete is well useful. Long-term strength properties of HVFAC are often better comparing to common cement-based concrete. This behaviour is very useful for special massive structure due to low hydration heat development as well as for structure where gently hydration rate is required.

It is very important to keep low water/binder ratio, therefore high efficient plasticizing admixture should be used. Influence of added mixing water on compressive strength of hardened concrete should evaluate by use of k-value, according to EN 206-1. This value is specified as 0.2 for all cements expect the CEM I type cement. Thus, high amount of fly ash added to the concrete as a binder can cause that the type concrete wouldn't meet the requirements of present standards.

References

[1] Mineral Commodity Summaries 2014 (U.S. Department of the Interior U.S. Geological Survey United States Government Printing Office Washington)

[2] Worrnell E, Price L, Martin N, Hendriks Ch and Ozawa M 2011 Carbon Dioxide Emissions From The Global Cement Industry. *Annual Review of Energy and the Environment* **26** 303-29

[3] Lutze D and Berg W 2009 Príručka. *Popolček v betóne. Základy výroby a použitia* (Bratislava BetónRacio, s.r.o.)

[4] Špak M, Halaša I, Šuster M and Vojtechovský O 2012 *Informácie o používaní popolčeka do betónu* (Bratislava BetónRacio, s.r.o.)

[5] EN 206-1: 2000 Concrete. Part 1. Specification, properties, production and conformity

[6] Malhotra VM and Metha PK 2005 *High-performance, high-volume fly ash concrete* (Ottawa Supplementary Cementing Materials for Sustainable Development Inc.)

[7] EN 450-1:2012 Fly ash for concrete. Part 1. Definition, specifications and conformity criteria

[8] Špak M 2010 Betóny s vysokým obsahom popolčeka *Stavebné hmoty* **6** 26-29

Expanded polystyrene as the bearing building material of low energy construction

P Mesaros[1], M Spisakova[1], L Kyjakova[1] and T Mandicak[1]
[1]Technical University of Kosice, Civil Engineering Faculty, Kosice, Slovakia

E-mail: tomas.mandicak@tuke.sk

Abstract. Sustainability of buildings is a really important issue for the construction industry. Sustainable buildings are characterized by the lower construction costs for energy consumption and operations, they are environmentally friendly, able to save natural resources and they are comfortable and healthy for their users. The European Union supports this trend through its Strategy 2020, respectively with document Energy Roadmap 2020. The strategy 2020 sets greenhouse gas emissions 20% lower than 1990, 20% of energy from renewable and 20% increase in energy efficiency. It manifests itself in introduction of modern technologies of house building. One potential for the energy saving is construction of low-energy buildings using modern materials. This paper focuses on the analysis of the low-energy buildings made by expanded polystyrene as the bearing building material. The paper analyzes their design and describes the benefits of this modern but unusual type of construction technology for houses. The examples from abroad clearly indicate that this technology has potential in modern architecture. The success and exploration of this technology potential in the conditions of Slovak construction sector is closely related to interest of investors and users of further sustainable houses which are design according the Strategy 2020 conditions.

1. Introduction

The increasing environmental impact from the construction becomes a serious problem that can cause significant damage, not only to ecosystems but also to the health and wellbeing of field workers and nearby residents of building sites [1]. Therefore, it is necessary to approach and continuous effort within the industry in order to achieve the objectives sustainable construction and reducing of the environmental impacts of construction.

The authors Shen and Zhang [2] suggest that the impacts of construction activities on the environment include: (i) competition for land with other activities such as agriculture; (ii) adverse effect on the plots of land which are developed, and their environment, such as changing their ecological characteristics; (iii) consumption of substantial volumes of physical resources, both renewable and non-renewable; (iv) production of substantial volumes of wastes; (v) consumption of large amounts of energy during the processing of materials, the construction process and in the use of constructed items; (vi) contribution to air pollution from the dust and substances, including some toxic ones, which are released during the production and transportation of materials, and in some construction operations; (vii) disruption of the lives of the people living in the vicinity of the project through traffic diversions, noise pollution and others.

Considering this, the construction industry has a significant environmental footprint, especially in terms of greenhouse gas emissions and energy consumption. The European Commission estimates that 35% of all of Europe's greenhouse emissions come from buildings. On the other hand, the climate change and energy sustainability presents one of the five targets for the European Union in 2020. The construction sector provides many opportunities for achieving of strategy 2020 targets. One of them is optimize of the design and management of constructions based on modern methods of construction using the unusually and innovative environmentally friendly construction materials - expanded polystyrene as the bearing building material.

2. Sustainability in affordable housing

Housing is one of the most basic human needs and is a key competent in the sustainable development of a community. Sustainability has been defined as economic growth that meets the current generation compromising the opportunity and the potential for future generation needs [3]. In general, sustainable development is divided into three interacting sectors: economic, environmental, and social [4] which resent three pillars of sustainable development [5].

Sustainable construction can influence the short and long term economic targets. The cost reduction is possible thought the using of local suitable materials (lower transportation costs); the energy efficiency of sustainable construction (lower economic demands of realization and operation of construction), etc.

Social sustainability focuses on the people using the building. Their current and future needs influence the design, which aims to create a highly-flexible plan that allows the building to be easily re-purposed as needs change by the design. A flexible design means the structure can be used longer, preventing the negative impact involved in tearing down an old building and rebuilding a new one.

Measures to address environmental sustainability can include increasing a building's energy efficiency, finding ways for the building to produce its own energy or heat, installing water-reduction measures, finding ways to reduce waste and using of “green” and local building materials. Even the location and orientation of a building can make it more environmentally friendly.

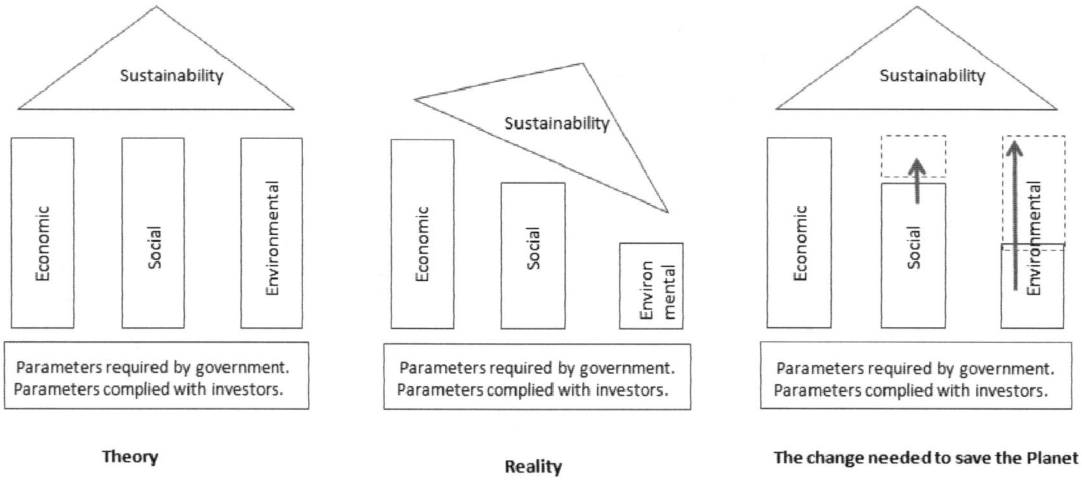

Figure 1. Three pillars of sustainable development [5].

Environmentally preferable products have less negative effects on human health and the environment when compared with competing products that serve the same purpose. In the design phase of construction is necessary to consider the selection of construction technology and building materials and products and prioritize environmentally-friendly forms. It is necessary to think about the theoretical and real sustainable development in the practice [5]. As seen in Figure 1, the individual pillars of sustainable development have different height pillars (influence) at present; therefore, it is necessary to do particular steps to balance their perception. The sustainability must be supported by the strong and the same high pillars in all three areas by the selection appropriate method of construction and construction material within the sustainable design and management of construction.

IOP Conf. Series: Materials Science and Engineering **71** (2015) 012002 doi:10.1088/1757-899X/71/1/012002

One of the possible ways of sustainable construction is the application of modern methods of construction (MMC).

3. Modern methods of construction

The majority of constructions are still constructed using traditional technology (masonry, concreting). On the other hand, the last few years it has been increased use of modern methods of construction (MMC) for housing, driven by a range of factors including demands for faster construction, skills shortages and sustainability of construction.

The modern methods of construction primarily involves the manufacture of constructions in factories, with potential benefits such as faster construction, fewer housing defects, reductions in energy use and waste [6] and offer significant potential to minimise construction waste [7] and construction safety risks [8]. The MMC is reasonably common in building construction but are less often used in civil engineering. For civil engineering projects, MMC can include use to pre-cast (pre-fabricated) components or preassembled structure. MMC can result in changes to on-site practice and may require different teams of specialist contractors; for example, compare operational needs on site to place large precast concrete components (handling areas, cranes and operators) with the demands of cast in – situ concrete (erection and striking of shuttering, temperature monitoring/control of large pours). The authors of study [7] divided the MMC products into these categories: (i) volumetric modular; (ii) timber frame and steel frame flat packs; (iii) prefabricated kitchen and bathroom pods; (iv) composite/insulated/sandwich (non load-bearing) panels; (v) light steel frame systems; (vi) pre-cast structural panels; (vii) pre-cast hollow-core flooring; (viii) structural insulated panels; (ix) pre-cast concrete cladding; (x) tunnel form construction; (xi) insulating concrete formwork. The wide range of MMC products has various applications in construction practice.

Many of the benefits of using MMC for housing are as yet unproven or contentious. On the other hand, the advantages and disadvantages are closely related to drivers and barriers of prefabrication use [9]. We can divide the main advantages of MMC [6] according the three pillars of sustainable development:

- economic – MMC houses typically have fewer defects and can be built more quickly, components provide a better quality and standards, the construction process can be speeded up by the mass production of prefab components in the factories,
- social – there may be fewer accidents and less impact on local residents during construction, reduces labour intensive activities and provides a safer working environment, designers from different disciplines can work closely together in the early design state to help to reduce abortive work,
- environmental – the houses can be more energy efficient, may involve less transport of materials, and produce less waste.

Typically MMC involves the manufacture of construction parts offsite in a specially designed factory. The two main products of MMC are:

- panels – including ready-made walls, floors and roofs. These are transported to the site and assembled quickly, often within a day. Some panels have wiring and plumbing already inside them, making construction even faster,
- modules – ready-made rooms or their parts, which can be pieced together to make a whole house or flat but are used most frequently for bathrooms or kitchens, where all the fittings are added in the factory. Also known as 'pods'.

A range of materials is used for MMC, the most common being wood, steel and concrete. One of the latest trends in the field of MMC is application of expanded polystyrene as the bearing building material.

4. Expanded polystyrene house

The Japanese company Japan Dome House Co. inspired by the development of its model Inuit tribe [10], which throughout the ages build their homes out of ice blocks. This fact was the basis of the concept of production similar dwellings, except that the ice blocks replaced polystyrene.

The concept of polystyrene was associated mainly with the production of disposable beverage cups, trays and containers for food from fast food. Even using polystyrene insulating boom in Japan acted as massive as in our country. Therefore, were and still are new houses made of such material – expanded polystyrene at the center of attention investors and users.

In Europe, in 2008, at the peak of the real estate market boom and later started the trend "fashion" of energy efficient buildings, engineers Pacura and Rogula [10] searched cheaper construction methods, that would be energy efficient and they came up with the idea of using expanded polystyrene as the bearing building material for the whole house.

Meantime, the use of foamed polystyrene in building industry has been limited to thermal insulations, filling technological spaces and protection of construction products during transportation. Thus, only compressive forces are present in the structure of the building that allow distribution of internal loads connected with the objects operational use and external loads connected with snow and wind [10].

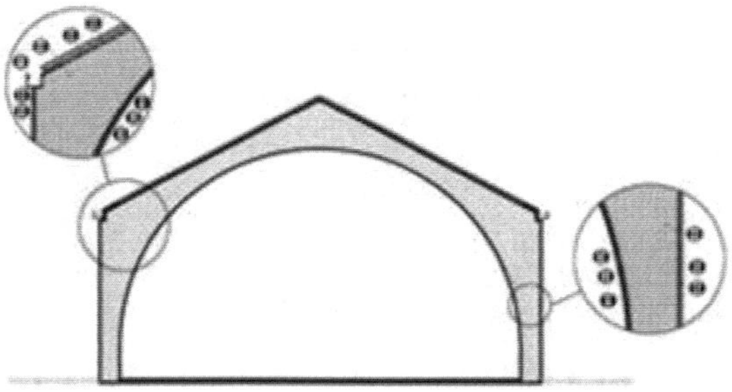

Figure 2. Basic shape the house of expanded polystyrene [11].

4.1. Technology and specifications of expanded polystyrene house

Based on the research [10], it was developed an innovative house design in the modern methods of construction – the expanded polystyrene in the form of prefabricated elements, which is supporting structure and thermal insulation of walls, as well as. Low specific gravity of material facilitates transportation, unloading and joining individual elements into the structure. All structural elements are delivered to the building site in the form of prefabricates that are joined together with the use of simple technological operations. An important thing is the innovative use of the heat-insulating material (foamed polystyrene) that is characterised by very low thermal conductivity, eliminating heat leakage bridges and any timber roof is not used. Roofing materials, for example roofing sheets, are fastened directly to the structure of the building. No expensive foundations are necessary, either. Due to low weight of the material the building can be erected on reinforced concrete slab with thermal and damp insulation. Moreover, low building cost is the advantage of the solution. The result is an energy efficient house, where energy use index stays within 50-60 kWh/ (m^2year), whereas the index in for buildings built by the traditional methods of construction exceeds 100-120 kWh/ (m^2year).

Figure 3. Expanded polystyrene houses [11].

Walls of a building, made of foamed polystyrene, accumulate heat, and at the same time – through recuperation – maintain optimal ventilation of rooms ensuring constant inflow of fresh air. Polystyrene used to building houses is an environmentally friendly material. During the production stage is used pure energy. During the using stage, the polystyrene does not decompose and is resistant to aging. These are the facts [11] which are confirmed by independent appraisers and research institutes based on many years of observation of behaviour polystyrene in all possible applications in the construction industry. In general, the aging becomes visible when the material is suitable for use, or in the subsequent decomposition [11].

This technology has some advantages, which are: short construction duration (2 - 3 months), high structural strength, very good heat insulation characteristics. The presented study will consider to a comparison of three variants of the constructions – modular construction, masonry construction and polystyrene construction on the basis of economic and time parameters.

4.2. Comparison of economic and time parameters of construction

The base for comparative analysis of time parameters are schedules prepared for three variants of construction. The time duration was selected with approximately the same building volume 428 m^3 considering the construction schedule. The results of schedule for all 3 construction technology are presented on the Figure 4. Modular construction is characterized by the significant shorter construction time; construction time of polystyrene construction is 90 days ad tradition methods of construction (masonry) presents 188 days. The construction duration of polystyrene is shorter than half that masonry construction, but longer than modular construction.

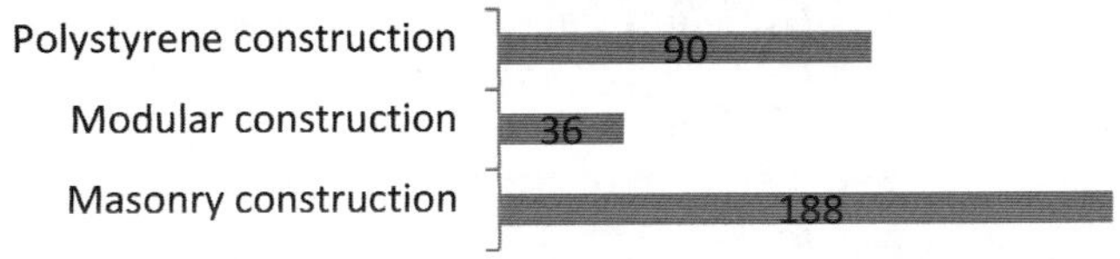

Figure 4. Comparison of overall construction duration for selected construction variants (in days).

The background for comparative analysis of economic parameters is prepared budgets of these three variants. Total costs were divided into the following groups: foundation, bearing structural system, windows and doors, interior partitions, heating, staircase and roof. The Figure 5 shows only the total construction costs. The graph presents the clear confirmation of efficiency of modular construction in economic indicators. The total cost of the house, realized by modular construction opposed to masonry construction, are lower by 20 650 EUR. The graph shows clearly the effectiveness of modular construction in economic indicators. The total cost of the house, realized by modular design unlike masonry structures are below 20 650 EUR. Polystyrene construction is one of the latest building technologies, the cost of this design compared to masonry construction and modular construction are higher in 19%. In spite of increased cost have polystyrene construction advantages. Among the best include short construction cycle (from 2 to 3 months), high structural strength and very good heat insulation characteristics what should reduce the operating cost for construction in future.

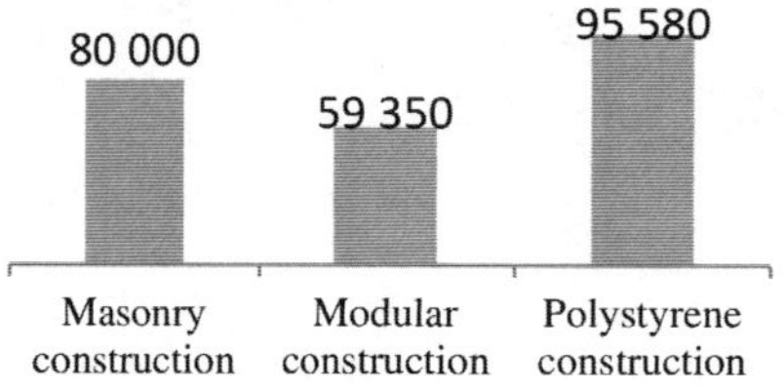

Figure 5. Comparison of total cost of selected construction variant (in EUR).

5. Conclusion

The results of a comparative analysis of polystyrene, modular and masonry construction shown that modular construction provide the better economic and time parameters. In spite of increased cost have polystyrene construction advantages. The examples from abroad [11] clearly indicate that this technology has potential in modern architecture. The success and exploration of this technology potential in the conditions of Slovak construction sector is closely related to interest of investors and users of further sustainable houses which are design according the Strategy 2020 conditions.

Acknowledgements

The article presents a partial research result of project VEGA - 1/0677/14 „Research of construction efficiency improvement through MMC technologies".

References

[1] Li X, Zhu Y and Zhang Z 2009 An LCA-based environmental impact assessment model for construction processe *Building and Environment* **45**

[2] Shen L and Zhang Z 1999 *ISO 14 000: The process towards sustainable construction - working paper* RICS Research Foundation

[3] WCED (World Commission on Environment and Development) 1987 Our common future (Oxford University Press: Oxford)

[4] Keiner M 2005 *Link-History, definition(s) and models of sustainable development*

[5] Tažikova A 2014 *Sustainability of existing buidings obtained by reconstruction. Integrated approaches to the desing and management of buildings reconstruction*

[6] Modern methods of house building 2003 Postnote

[7] Current practices and future potential in modern methods of construction 2007 Waste and resource programme *WAS003-001: Full Final Report*

[8] Strukova Z 2009 *Specific perception of safety risks of construction projects. Innovative approach to medelling of intelligent structural elemetns in construction*

[9] Lovel H and Smith S J 2008 *Agencement in housing markets: The case of the UK construction indurstry. Geogorum*

[10] Kapiszewski J 2014 M3 System: Budowlana rewolucja, czyli styropianowe domy *Gazeta prawna.pl*

[11] Plaziak M M3 System [online source] Available at: www.m3system.pl

[12] Kaleja P and Kozlovská M 2014 Comparative analysis of selected technology of modular construction. *Proc. of Student Conference of Civil Engineering and Architecture - Young Scientist* (Košice : TU) 1-7

Modular mobility investigation of polymer binder bitumen

D Ayupov[1], D Makarov[1], A Murafa[1], V Khozin[1], Y Khakimullin[2], V Sundukov[3], A Khakimov[3] and B Gizatullin[4]
[1]Kazan State University of Architecture and Engineering, Department of Technology of Building Materials and Construction, Kazan, 420043, Russia
[2]Kazan National Research Technological University, Department of Chemistry and Processing Technology of Elastomers, Kazan, 420015, Russia
[3]Kazan State University of Architecture and Engineering, Physics Department, Kazan, 420043, Russia
[4]Kazan Federal University, Department of Physics of Molecular Systems, Kazan, 420000, Russia

E-mail: ayupov_damir@rambler.ru

Abstract. This paper is aimed to obtain polymer binder bitumen for the road applications with improved properties. Objectives included studying the degradation of crumb rubber during devulcanization in melted petroleum bitumen for its modification and resulted structural properties of bitumen. Using equilibrium swelling technique reduced density of the polymer chains was observed and analysis of sol-gel fractions showed a significant decrease in gel fraction. Under selected method of devulcanization 64% of the backbone rubber remained. With The H^1 NMR relaxation method the reduction of bitumen molecular mobility was observed due to thickening of its light fractions. The effectiveness of devulcanization was optimized using a new agent in a powder form and vacuum application. The developed binder has an improved spectrum of physical and technical properties such as softening point temperature, hardness, elasticity, frost resistance, low temperature characteristics.

1. Introduction

Modification of bitumen polymers is an effective way to obtain high performance technology of building materials on the basis of organic binder. High relative cost of thermoplastic elastomers and rubbers hinders their widespread application for bitumen properties improvements.

At the same time the problem of used car tyre recycling is well known all over the world. Today the main usage of crumb rubber, produced by grinding of old tyres is to mix it with thermoplastics and thermosets [1-2] to reduce the cost of raw polymer blends, but crumb rubber degrades such mixtures' properties to some extent.

Direct modification of bitumen crumb rubber (CR) is considered ineffective, as it acts in bitumen as filler, causing swelling and forming elasticity centers, but does not create a polymer grid. "The dry" additional procedure method of crumb rubber in asphalt concrete mixture is considered unsatisfactory [3]. Therefore, to implement the properties of polymer in bitumen the rubber crumb must be devulcanized, making cross-linked polymer into linear (or branched) and therefore soluble.

We have developed a chemical method of rubber devulcanization for bitumen modification, where devulcanization is performed directly in bitumen. It should be noted that unlike devulcanization in solvents used for rubber regeneration [4], in this case there is no need to allocate a dissolved rubber from the solvent, since the purpose of crumb rubber devulcanization in bitumen is to obtain bitumen-polymer binder.

The composition included bitumen BND 90/130 (100 wt. parts), crumb rubber (20 wt. parts), devulcanizing agent (0.06 – 0.7 wt. parts). The technology of producing bitumen – polymer composition was addressed. Preparation of the bitumen-polymer compositions was performed in the following way: bitumen was heated in the reactor to a temperature 90°C with constant stirring, rubber crumb and devulcanizing agent were introduced sequentially, after vacuum was applied in the reactor with temperature rise to 220 °C. The stirring time was 1.5 hours.

In this mode bitumen was modified with the rubber crumb. We used Neozon D (naphthyl-2) - a secondary aromatic amine (Figure1) - as a devulcanizing agent.

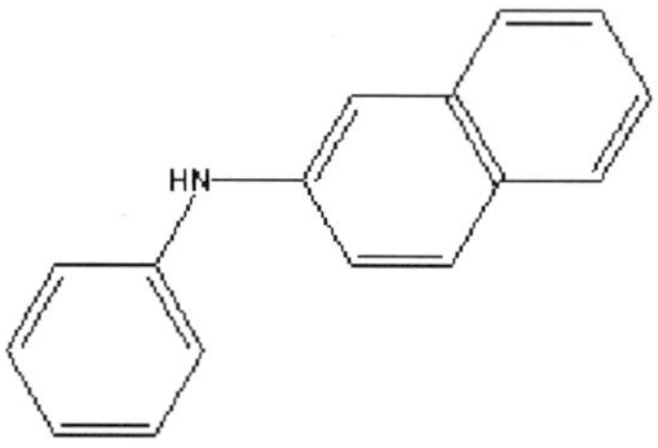

Figure 1. Neozone D structural formula.

Our experiments demonstrated that neozone D is an effective agent for devulcanization; it reduced a gel fraction to 35 % and grid circuit density by 5.5 fold. The resulted molecular weight of rubber in asphalt is more than 60 % of the original weight; therefore rubber degradation presumably occurs mainly by sulfur links.

The developed devulcanization technology of crumb rubber in asphalt environment allows obtaining a composite polymeric bitumen binder with enhanced technical properties. However the implementation of the composition obtained is hampered by the lack of opportunity to provide vacuum processing in manufacturing. So, we have produced the similar composition using the same technology in rotor – pulsation apparatus.

According to GOST 39 – 79 “Naphtam – 2. Specification” neozon D can be manufactured both in powder form and in form of crumbling flakes. Experiments indicate that the form of neozone D affects the efficiency of vulcanization.

The main properties of polymer modified bitumen (PMB) are represented in Table 1.

Table 1. The main properties of the developed bitumen-polymer composition (BPC).

Composition	Softening temperature, °C	Penetration n_{25}, 0,1 mm	Penetration n_0, 0,1 mm	Ductility $_{25}$, cm	Ductility $_0$, cm	Elasticity $_{25}$, %	Elasticity $_0$, %	Flexibility , °C	Penetration index
Bitumen 90/130	44	97	50	95	0	13	-	5	-1.2
Bitumen 90/130+20% CR	58	56	15	14	0,5	55	65	-5	-1.2
Bitumen 90/130+20% CR+0,1% powder neozone, vacuum	76	45	36	9,7	5	83	73	-25	3.5
BND 90/130+20%CR+ 0,1% powder neozone, RPA	68	55	40	10	4	78	72	-20	2.8
BND 90/130+20%CR+ 0,1% flakes neozone, RPA	67	58	44	13	3.3	74	73	-20	2.7

IOP Conf. Series: Materials Science and Engineering **71** (2015) 012003 doi:10.1088/1757-899X/71/1/012003

As seen from the Table 1, the best performance composition was obtained in vacuum. Powder neozon D was the most effective devulcanizing agent.

2. Nuclear magnetic resonance study of free induction decay relaxation

The measurements were carried out at the Physics of Molecular Systems Department in Kazan Federal University using NMR analyser "Chromatec Proton -20M". Resonance operating frequency on proton was 20 GHz, the detailed device specifications can be found online at (http://www.chromatec.ru/products/main/nmr/proton20m/).

Maltene fractions contribute strongly to H^1 NMR signal because of their significant presence and bitumen and higher hydrogen content (compared with asphaltenes and resins). H^1 NMR experiments provide qualitative information about molecular mobility of hydrogen containing molecules and structures due to complex composition of bitumen.

The asphaltene fraction of bitumen in the NMR experiment is characterized by a complex form of free inducation decay (FID), consisting of "solid" Gaussian component and Lorenzian recession component [5].

Resins and n-paraffins crystalline fraction, found in heavy crude oil (bitumen) may also contribute into Gaussian component [6].

Low molecular uncrystallized fraction of bitumen contributed to Lorentz component of FID. The decomposition into three components was used in processing of the experimental data: two exponentially decaying with time (Lorentzian components) with large T_a, T_b relaxation times and solid-state Gaussian component with a quadratic dependence on time inside the exponential function with smaller relaxation time T_c.

Thus, the experimental FID were fitted with formula:

$$A(t) = A(0)\left(p_c \exp\left(-\frac{t^2}{T_{2c}^2}\right) + p_b \exp\left(-\frac{t}{T_{2b}}\right) + p_a \exp\left(-\frac{t}{T_{2a}}\right)\right) \quad (1)$$

Where P_a , P_b , P_c – fractions, and T_a , T_b , T_c , - corresponding fractions of slowly and rapidly relaxing Lorentzian component and "Gaussian" solid component respectively.

Because of short time constant T_c is close to dead time the apparent proportion of solid component in the NMR signal is substantially less than the true proton proportion of this component. However, in terms of technological properties of bitumen, it is much more interesting to obtain information about the "fluid" fraction and therefore the analysis of Gaussian signal component is beyond the scope of this paper.

Because of a lower molecular weight and higher mobility, the slowly relaxing FID components are to a great extent responsible for the contribution in proton signal of maltene fraction.

Neozone affects the bitumen – polymer binders' structure as shown in Figure 2.

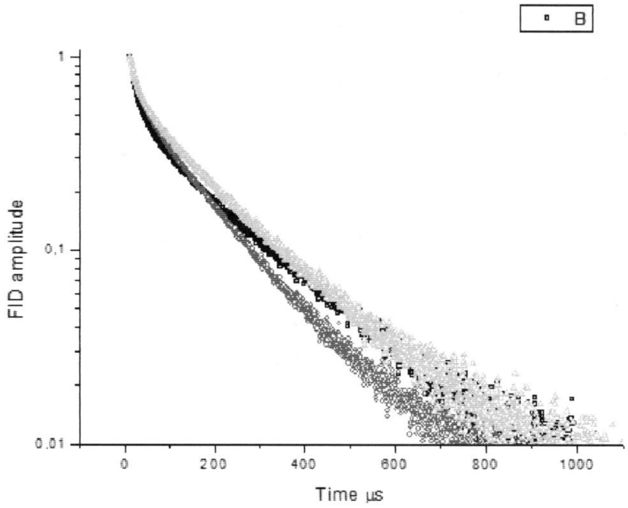

Figure 2. FID curves of free induction of the developed systems over time.

Black – BND 90/130
Red – the optimal composition of flake neozone in RPA
Green – the optimal composition of powder neozone in RPA

FID for bitumen with RK (green triangles) differs little from the FID of pure bitumen (black squares). In the presence of neozone (red circles) the slope of the "tail" of free induction decay is markedly increased, indicating a decrease in molecular mobility of uncrystallized low molecular weight components of bitumen.

By introducing powder form of neozone relaxation time of slowly relaxing component "a" decreased from 255 ms to 220 ms, and also its fraction was diminished from 30 % to 22 %. Relaxation times for "b and "c" components remained constant, but their fractions increased from 34 % to 38 % and from 36 % to 40 % respectively.

Presence of devulcanizing agent leads to reduced molecular mobility due to better swelling of devulcanization product. Figure 3 illustrates the greater efficiency of powder form neozone over flake one.

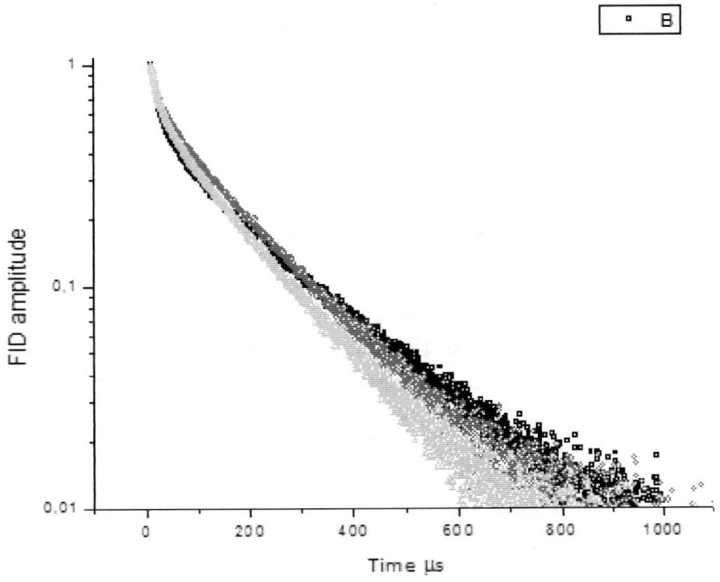

Figure 3. Free induction decay curves of the developed compounds.

Black – BND 90/130
Red – the optimal composition of flake neozone in RPA
Green – the optimal composition of powder neozone in RPA

In the presence of powder form of neozone the "tail" steepness of FID decay increases to a greater extent in comparison to flake form of neozone.

Vacuum effect is observed in Figure 4.

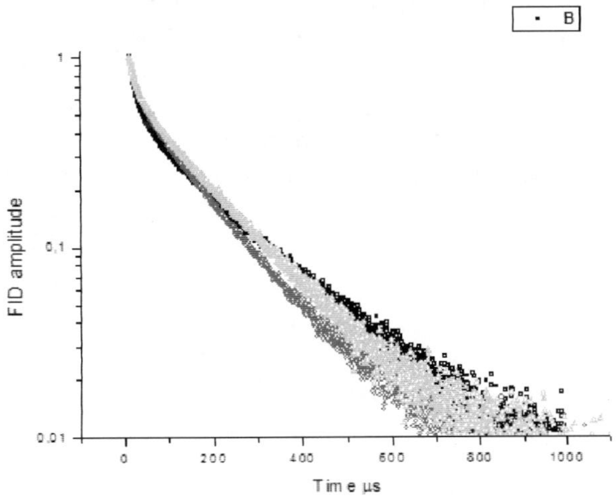

Figure 4. Effect of vacuum on FID curves of developed systems.

Black – BND 90/130
Green – optimal composition with powder neozone prepared in RPA
Red – optimal composition with powder neozone prepared in vacuum

For a product with vacuum application the contribution of slowly relaxing components “a” is reduced to 14 % comparing to 24 % at ADP of “b” component increases from 41 % to 50 %. Relaxation times and the solid component’s fractions are not sensitive to the method of processing. Thus, bitumen-polymer binder with the best performance was obtained in vacuum with the application of powder-like neozone.

3. Conclusions

The experimental results suggest the following conclusions:

1. To implement the properties of polymer in bitumen the rubber crumb must be devulcanized, making cross-linked polymer into linear (or branched) and therefore soluble.
2. Because of a lower molecular weight and higher mobility, the slowly relaxing FID components are to a great extent responsible for the contribution in proton signal of maltene fraction.
3. Degradation of sulfur cross-links increases when neozone is added. Presence of devulcanizing agent leads to reduced molecular mobility due to better swelling of devulcanization product
4. Resulting bitumen-compositions possess improved properties such as softening point, hardness, elasticity, resistance to frost, low temperature characteristics.
5. Devulcanization by powder form of devulcanizing agent is more effective than by flake form.
6. Using vacuum improves rubber crumb devulcanization in melted bitumen.

References

[1] Mujal-Rosas K, Orrit-Prat R and Marin-Genesca M. 2013 Caracteristicas dielectricas, termicas y mecanicas de la poliamida (PA) reforzada con neumaticos fuera de uso (GTR) *J. Afinidad* **LXX 561** 14-24

[2] Mujal-Rosas R, Orrit-Prat J, Ramis-Juan X, Marin-Genesca M and Rahhali A. 2012 Study on Dielectric, Mechanical and Thermal Properties of Polypropylene (PP) Composites With Ground Tyre Rubber (GTR) *J Polym Polym Compos* **20 9** 797-808

[3] Duhovny G S and Sachkova A V 2013 Increased crack resistance of asphalt concrete pavements *World transport* **1** 60-63

[4] Rooj S, Basak G C, Maji P K, Bhowmick A K 2011 New route for devulcanization of natural rubber and the properties of devulcanized rubber *J Polym Environ* **19** 382-390

[5] Shkalikov N V, Ganeeva Yu M, Yusupova T N and Skirda V D 2008 The characterization of asphaltenes by H^1 NMR relaxation method: microsecond range of spin-spin relaxation times. *Magnetic Resonance in Solids Electronic Journal* **10** 1 11-19

[6] Shkalikov N V 2010 Investigation of heavy oils and their components by NMR *Diss.*

IOP Conf. Series: Materials Science and Engineering **71** (2015) 012004 doi:10.1088/1757-899X/71/1/012004

Thermal insulation of pipelines by foamed glass-ceramic

A S Apkaryan[1,2] and A I Kudyakov[3]
[1]Institute of Strength Physics and Materials Science, Tomsk, 634021, Russia
[2]Tomsk State University of Control Systems and Radio Electronics, Tomsk, 634055, Russia
[3]Tomsk State University of Architecture and Building, Tomsk, 634003, Russia

E-mail: asaktc@ispms.tsc.ru

Abstract. Based on broken glass, clay and organic additives granular insulating glass crystalline material and technology of its receipt are developed. The regularities of the effect of composition and firing temperature on the properties of the granules are specified. The resulting granular thermally insulating material is produced with a bulk density of 260-280 kg/m^3, pellet strength - 1.74 MPa , thermal conductivity - 0.075 W/m °C , water absorption - 2.6 % by weight. The effect of the basic physical characteristics of the components of the charge on the process of pore formation is studied. According to the research results, basic parameters affecting the sustainability of the swelling glass are specified. Rational charge composition, thermal and gas synthesis mode are chosen so that the partial pressure of gases is below the surface tension of the melt. This enables the formation of granules with small closed pores and vitrified surface. The article is the result of studies on the application of materials for pipe insulation of heating mains with foamed glass ceramics.

1. Introduction

One of the main issues of the economy is preservation of fuel and energy resources. Priority role in the solution of this issue is given to energy saving in construction and industry, namely, the effective thermal protection of buildings and structures, heat insulation of industrial installations and pipelines of large diameters. Great attention in the protection of pipelines of large diameters is paid to the quality of the applied heat-insulating materials, which should have low thermal conductivity and corrosion resistance for providing stability to mechanical influence. Now for insulation of pipes mats, semi-cylinders, shells on the basis of staple fiber (URSA, ISOVER, Tisma, KNAUF, and so on), basalt fibers (Teparak, Rockpipe, Rockwool, Paroc, Nobasil), rubber (K-Flex, Armaflex) and blast furnace slag are used [1,2].

The purpose of work is development of technology of glass-ceramic insulation pellets of glass waste, raw clay, organic additives and shells on their basis for insulation of pipes of large diameters.

Object of research are glass-crystalline granules and products for thermal insulation of pipes.

In order to achieve the goal stated tasks were the following: 1. To develop the composition of the charge and to investigate the processes of the pores formation and structure formation of granules of glass-ceramic material. 2. To make calculations and analysis of thermal protection and efficient use of coverings of products on the basis of granulated glass-ceramic material.

2. Materials and methods

When conducting research different wastes of glass, clay, coke and sawdust were used for glass-ceramic pellets.

2.1. Waste glass

For the production of foam glass, glasses of following chemical composition are used: SiO_2 - 60...72,5 %, Fe_2O_3 - 0...2,5 %, CaO - 4...6,0 %, MgO - 1,5...2,5 %, Na_2O - 12,5...15,0 %, Al_2O_3 - 1,2...1,57 % [1]. The most similar to the structure is the window glass, glass bottles and lamp glass (Table 1). During the production of granulated foam glass ceramics it is necessary to carry out the correction of the functional composition of charge taking into account every type of glass used.

Chemical compositions of windowpanes are the closest to this. Their distinctive characteristic is low level of Al_2O_3 compared to other types of glass.

Table 1. Chemical composition of the crushed glass used.

Type of glass	Glass chemical composition, %										
	SiO_2	Al_2O_3	Fe_2O_3	CaO	MgO	R_2O	SO_3	K_2O	Na_2O	Cr_2O_3	BaO
Window glass	73.0	0.9	0.12	8.7	3.6	13.6	0.53	-	-	-	-
Bottle glass	70.7	4.0	0.65	6.5	3.5	-	-	-	14.5	0.15	-
Lamp glass SL96-1	71.9	1.5	0.1	5.5	3.5	-	-	0.15	16.1	-	2.0

2.2. Clay raw materials

Clay was introduced in the charge to increase the mechanical strength of interporous partitions, the surface of granules and reduce water absorption. For the studies clay rocks of Tomsk region were used. Clay minerals are presented by kaolinite, hydromica and montmorillonite. In addition, clay contain quartz, feldspars, carbonates, oxides and organic impurities. Kandinsky field meets the requirements for the production of pellets clay to the great extend. The best results were obtained using clay that is classified as pulverous loam, with a domination of SiO_2 – 70.73%; Al_2O_3 – 17.12% and Fe_2O_3 – 5.68% in its chemical composition. Good foaming is ensured by the presence in the clay of hydromica, montmorillonite, compounds of iron and feldspar.

2.3. Coke

Coke is used as the blowing agent. In the justification of the choice of blowing agent the coincidence of temperature intervals when appeared melt of the desired viscosity and formed the greatest partial pressure of gaseous products when burning of organic matter are taken into account.

2.4. Sawdust

Sustainable gassing is introduced to reduce the viscosity of the melt and security together with coke. In the compositions of "glass powder - clay - coke – water", it is better to use coniferous breeds of wood. Sawdust has been introduced in the charge by joint grinding with broken glass that ensured they were chopping up of particles less than 560 mm with a high value of the coefficient of the form of wood fiber (the ratio of length to thickness equal to 4).

In the course of the study the following charge of composition was used (% by weight): crushed glass 67 – 84; clay 8 – 25; carbonite 5 and sawdust – 3. The charge degree of fineness 0.45 – 0.15 mm.

The granules were made in the following way: crushed glass was crushed in a grinder to 1.5 – 3.0 mm, and then in a ball mill combined with sawdust to the particle specific surface of 300 – 350% m^2/kg. In such a way the sawdust was added to the charge during the moment of glass grinding, which enabled their pulverizing to a wood powder together with obtaining of a high shape factor of wood fiber (length-to-thickness ratio) equal to 4 and the size of particles no more than 560 microns. A grinding process of carbonite and clay also occurs in the ball mill. The obtained glass-wool and carbonite-clay powders were dosed and loaded in a rod vibrating mill where they were jointly mechanoactivated to a specific surface of 400 m^2/kg. The obtained charge was gaged with water (10-

15% of the charge mass) for gaining the required plasticity, and then was formed into granules of 3-8 mm in size.

The experiments were carried out with a permanent carbonite value of 5% and organic additives (sawdust) of 3%.

Granule burning was carried out at temperature of 830-850 °C within a sintering range of foaming mixtures equal to 4 – 6 minutes.

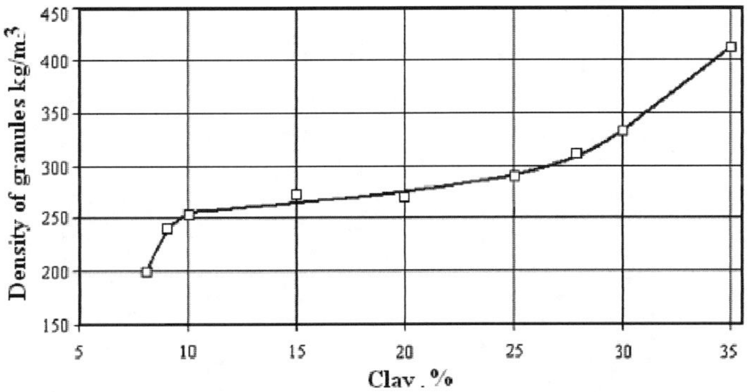

Figure 1. Impact content of clay in the charge on the density of granules.

Pore formation and sustainability of porous structure depend on surface energy on the phase boundary "gas – melt". In Figure 1, the results of studies of the influence of content of clay on the density of granules are shown. When the clay content is in the charge from 8 to 25 % favorable conditions for the formation of pores and produce pellets with a density of 200 to 300 kg/m^3 are created. If you increase the clay content in the charge of more than 25%, pores to unite, the gas will be released into the atmosphere - foam settles.

In order to solve this problem it was decided to use sawdust as an organic additive in the charge. It was experimentally estimated that adding up to 3% of sawdust with a simultaneous increase of the foaming temperature up to 830 °C and the time of pore formation up to 5 minutes the amount of gases and their partial pressure would increase. Furthermore, the viscosity of the hot melt begins to decrease and the particle sintering commences, leading to the formation of glass ceramic mass.

Clay as a fusible additive plays the role of stabilizer in the rational composition of a charge. The stabilizing impact of clay enables the formation of so-called "energy barrier" which increases the mechanical strength of partitions with the narrowest thickness. In the process of foaming, pores are mutually combined under slow speed. The gases in the pores work against the energy caused by the surface tension. The presence of clay in a hot melt significantly increases the energy of the surface tension and the gas pressure in pores decreases as the pores diminish. As the result, there are up to 92% of closed pores of 0.2 – 0.4 mm in size with the thickness of the dividing partition within the bounds of 0.07 – 0.1 micron up to 12 microns. The granule density is 260 kg/m^3.

A thermal shock in the zone of high temperatures (830 °C) in a reductive conditions with 3.0% CO containment is characteristic for granular thermal processing. Furthermore, the mechanical strength comes to the maximum while granules are in the high temperature zone for 5-6 minutes. Thermal cycle "foaming→ pore formation→ annealing" provides an optimal time for granules in the foaming zone, regulating the size and the quantity of pores and prevents the process of bubble coalescence (increasing of pore size), for which greater time in the high temperature zone is required. Research resulted in RF patent № 2374191was received. After completion of the synthesis in a rotary kiln granules size of 8 - 10 mm are able to pyroplastic status. There was a proposal on the possibility of the formation of granules of various geometric shapes of plates, cylinders and other segments of the shells for the insulation of pipes.

Annealing is necessary to fix the porous structure and to relieve temperature stresses in the pores. Thus, technological rules of manufacturing heat-insulating products made of glass-ceramic granules consist of three stages: formation and expansion, manufacture of building products, annealing products. For the production of heat-insulating materials from granules foamed glass-ceramic can also be used by the technology of gluing of granules. Technical and economic efficiency of heat-insulating coating of pipes of foamed glass ceramic has been defined for pipelines located in the open air at a temperature of 110 °C (Figure 2).

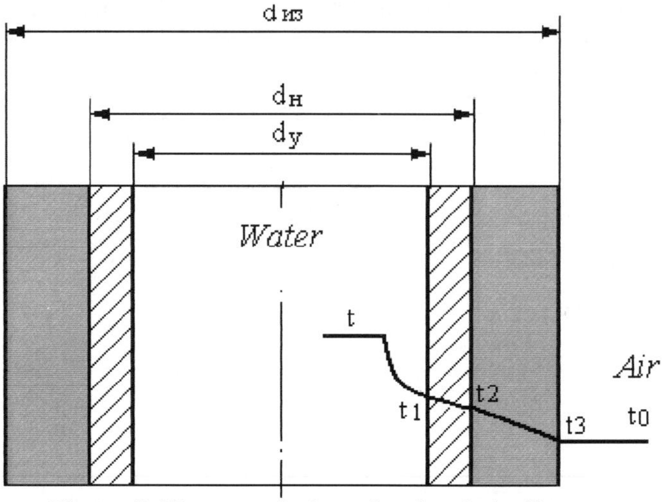

Figure 2. Heat transfer in an insulated pipeline.

Heat losses through the isolated surface with insulation thickness $d_{из}$, m shall be determined according to the formula 1, and the thermal resistance - by the formula 2. The temperature of the inner and outer environments $t_в$ and $t_н$ accept the calculated heat carrier temperature and the ambient air.

Mathematical formulation of the problem takes the following form [3-5]:

$$q_l = \frac{\pi (t - t_0)}{R_l} K, \quad (1)$$

where:

$$R_l = \frac{1}{\alpha_1 d_1} + \frac{1}{2\lambda_2} \ln \frac{d_н}{d_у} + \frac{1}{2\lambda_{из1}} \ln \frac{d_{из1}}{d_н} + \frac{1}{2\lambda_{из2}} \ln \frac{d_{из2}}{d_{из1}} + \frac{1}{\alpha_2 d_{из2}} \quad (2)$$

q_l - the density of a thermal stream flowing through patterns, W/m^2;
t - the coolant temperature, °C;
t_0 - the ambient temperature to be considered equal to the average temperature of the heating period (t_0 = - 5,2 °C);
d_y - conditional diameter of the pipeline, m;
$d_н$ - the outer diameter of the pipe, m;
$d_{из}$ - outer diameter of the heat-insulating shell, m;
α_1 - heat transfer coefficient from the coolant pipe inner surface, W/(m^2·°C);
α_2 - coefficient of heat transfer from the outer surface of the insulation W/(m^2 °C), taken equal to 29 W/(m^2 ·°C) [4].
λ, $\lambda_{из}$ - thermal conductivity of the pipe material and the heat insulation of the shell, respectively, W/(m ·°C).
K - coefficient of additional losses, taking into account the conductive heat loss through the inclusion of thermal insulation structures caused by the presence of fasteners and bearings.
K = 1.2 [4].

For comparison, the heat losses through the insulation use articles having an average density close to the maximum design density of foamed glass ceramic [1-5]. Characteristics of pipes and heating medium is shown in Table 2 and comparative thermal characteristics of the products in Table 3.

Table 2. Physical and technical characteristics of the pipe.

Name	Conditional diameter, d_y, m	The outer diameter, $d_н$,m	Temperature, ° C.	Thermal conductivity, W/m °C
Mild steel (plain pipe)	0,3	0,325	110	50.0

Table 3. Thermo-physical characteristics of heat-insulating materials.

№	Heat-insulating materials	Density, kg/m³	Coefficient of thermal conductivity, W/m °C	Temperature operation, °C
1	Vermiculite shell	230	$0{,}0814+0{,}000233\ t_m$	450
2	Mineral wool segments	350-400	$0{,}0814+0{,}003\ t_m$	840
3	Sovelite shell grade 500	200	$0{,}107+0{,}00116\ t_m$	400
4	Nyuvele shell	350	$0{,}0808 + 0{,}000097\ t_m$	375
5	Foamed glass-ceramic	200	$0{,}059+0{,}00016t_m$	- 50 up to 720

When using a shell made of foamed glass-ceramic heat loss is 1,37 times less than when using vermiculite shell, 1.45 times less than application of segments of mineral wool; 2,47 times less than at application of sovelite shell marks 500 and 1,25 times less than nyuvele shell.

Figure 3 graphically shows dependence of thermo-physical characteristics for mineral wool segments, shell of granular foamed glass-ceramic, and vermiculite shells.

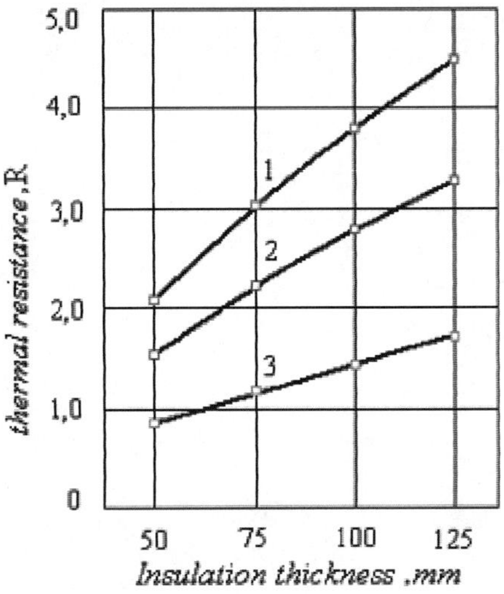

Figure 3. Dependence of the thermal resistance of the insulation materials the thickness of the insulation layer.
1 - shell of granular foamed glass-ceramic, 2 - mineral wool segments, 3 - vermiculite shell

IOP Conf. Series: Materials Science and Engineering **71** (2015) 012004 doi:10.1088/1757-899X/71/1/012004

3. Conclusion

Application of granulated foamed glass-ceramic significantly reduces heat loss, and as a consequence saves fuel and energy resources.

Application of granular materials from foamed glass enables to decrease the thickness of the insulating layer of pipelines, as thermal resistance depends on the thickness and heat conductivity of insulation materials. The thickness of the insulating layer is calculated from the conditions of minimal heat loss of the environment, the difference of the temperature and the thermal properties of heat-insulating materials.

Taking into account the dignity of the other insulation materials, as well as taking into account the physical and technical characteristics of materials foamed glass-ceramics, one can say that foamed glass-ceramics has its place in the use of insulation of buildings and energy constructions.

References

[1] Апкарьян А и Кудяков А 2013 *Теплоизоляционные материалы. Пеностеклокерамика. Оптимизация теплофизических процессов* (Томск: Изд-во ТУСУР) с 216
[2] Kudyakov A and Apkaryan A 2013 Granulated insulation material *Bulletin of science* **4** 10-217
[3] Блази В 2005 *Справочник проектировщика. Строительная физика* (Москва: Техносфера) 536 с
[4] СНиП 41-03-2003 2004 *Тепловая изоляция оборудования трубопроводов* (Москва: Стройиздат)
[5] Исаченко В, Осипова В и Сукомел А 1975 *Теплопередача. Учебник для вузов* (Москва: Энергия) 488 с

Influence of electrified surface of cementitious materials on structure formation of hardened cement paste

A Alekseev[1] and A Gusakov[2]
[1]Tomsk State University of Architecture and Building, Department of Construction Engineering Technology, Tomsk, 634003, Russia
[2]Tomsk State University of Architecture and Building, Department of Economics and Construction Organization, Tomsk, 634003, Russia

E-mail: alekseev10@yandex.ru

Abstract. To provide high strength and durability of concrete it is necessary to study the influence of physical and chemical and mechanical principles of dispersed cementitious systems. The experimental bench was developed to study the influence of electrified surface of cementitious materials on structure formation of hardened cement paste. The test bench allows accelerating the processes of dissolution of cementing materials in water due to influence of electric discharge on their surface. Cement activation with high-voltage corona discharge when AC current is applied allows increasing the ultimate compressive strength of hardened cement paste by 46% at the age of one day and by 20% at the age of 28 days.

1. Introduction

High-voltage discharge as a mean of obtaining plasma state of matter is widely used in science and engineering. High-voltage discharge is considered to be the one occurring in gas under the action of AC or DC current applied to electrodes.

Construction engineering allows not only to manufacture products but also to study physical and chemical, mechanical principles of dispersed cementitious systems in order to get the required material properties, find the most effective and economic production processes of obtaining mortar and concrete.

Composition and product type of cement hydration can be controlled at nano-level by composing the structure inside the molecules. The system classifying chemical reactions is built on transfer of electrons, electron pair, proton and atomic molecular particles. It is possible to obtain new chemical compounds of different and necessary basic capacity with different content of hydrated water, of different size and form of crystals thus providing the required strength, frost resistance and water resistance of dispersed cementitious systems.

The main strength of cement paste is provided by crystals and crystal joints that form the hydrated new growths, their sizes being within the range of 10…17 nm. The hydration products less than 10 nm in size are placed in the spaces between crystals, thus combining all the new growths in a single solid body.

P.A. Rebinder [10] noted that all the ways to enhance the production processes in condensed dispersed systems are based on controlling the properties of structure that was formed by the system particles. The basic parameters describing structure of system is total potential energy of hydration defined as:

$$E=E_0+E_{pol}+E_{disp}+E_{end}+E_{rep}+E_{solv}+E_{ext}\ , \qquad (1)$$

where E_0 – ion-dipole interaction energy; E_{pol} – polarization energy; E_{disp} – dispersion-interaction energy; E_{end} – endo-effect; E_{rep} – dipole-ion repulsive energy; E_{solv} – solvent polarization energy; E_{ext}– interaction energy of solvent molecules (distant hydration) and the surrounding water.

According to Sychev's [11] classification, the hydration process is divided into a number of simple processes: adsorption at the active centers of water molecules, excitation of active centers, dissociation of the water molecule, formation of active groups, surface bond breaking, release of structural components (Ca^{2+}, SiO^{4+}), polymerization of [SiO^{4+}] groups into dimers, binding of Ca^{2+} ions with dimer, trimeric anionic grouping, formation of hydrate nucleus with high-surface area, adsorption of water on the surface of hydrates, excitation of centers with adsorption energy, water molecules dissociation and formation of H^+ and OH^- active groups, transfer of protons and OH-groups through the layer of new growths into reaction zone, catalyzation of the process of breaking bond Si–O–Ca on a new surface of minerals under the layer of new-growths by means of sharp increase in active groups concentration as a result of formation of fine-grained hydrates and increase in adsorption surface where the water molecules dissociate into protons and OH-groups; formation of nucleus of hydrating phases in Taylor layer (Figure 1).

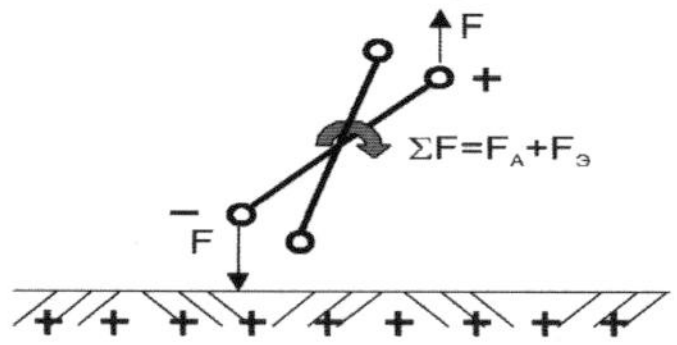

Figure 1. Polarization of dielectrics in the external electric field of the cementitious material surface.

The connection between strength, the average adhesive power in the contact and the number of contacts is defined by the known correlation by Rebinder and Schukin:

$$P_m = a_1\overline{F}_c n^{2/3} = a_2\overline{F}_c \varphi S^2 = a_3\overline{F}_c \varphi R^{-2} \qquad (2)$$

where Pm – yield value; $\overline{F}_c$ – the average adhesive power in contact; n – number of contacts per unit volume; φ – dispersed system; S – specific surface area with the size R; a_1, a_2, a_3 – constants.

2. Research design and results

At the Department "Construction Engineering Technology" of Tomsk State University of Architecture and Building (TSUAB) the research is held to establish the influence patterns of high-voltage discharge both on the cement and on cement suspension [1–7]. First of all high-voltage discharge allows to accelerate the processes of cementing agents dissolution in water by means of influence of electric discharges on their surfaces, i.e. the more intensive dissolution of cement particles in water occurs.

$$\frac{RT}{V_m}\ln\alpha = \frac{2\sigma}{r} - \frac{q^4}{8\pi\varepsilon}\cdot\frac{1}{r^4}\ , \qquad (3)$$

where R – universal gas constant; T – absolute temperature; $Vm = M/\gamma$ – mole extent; M – molecular weight; γ – relative weight of the dissolving phase; $\alpha = c/c\infty$ – oversaturation degree; σ – the specific free surface energy; r – radius of equilibrium particles (it is suggested that the particles have a round shape); q – electric discharge of particle with radius r; ε – dielectric constant of solvent.

Implementation of ozone technologies requires developing ozone synthesis systems, one of the ways being barrier-surface discharge. Barrier-surface discharge differs from barrier discharge by dynamic capacity due to treatment of dielectric with plasma discharge.

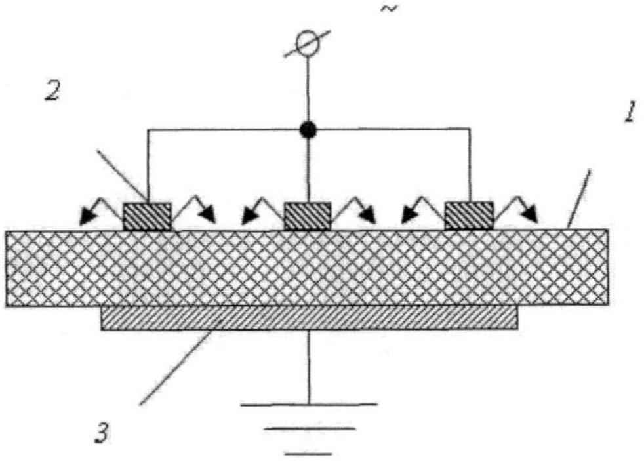

Figure 2. Barrier-surface discharge layout: 1-dielectric; 2-corona electrode; 3-inductive electrode.

Barrier-surface discharge works as follows: in case of placing wire or metal strip 2 on one side of dielectric plate 1 (Figure 2), and covering the other side of dielectric with conductive material 3 and provide AC or pulsing voltage between the electrodes, the current will flow in curcuit. At low voltages it is insignificant displacement current as the dielectric is a condenser. With increase of definite voltage the discharge spreads over the surface of dielectric from the corona electrode.

To study the influence of electrified surface of cementitious material on structure formation of hardened cement paste the experimental bench was developed. The test bench consists of: autotransformer (T1); step-up transformer (T2); discharge node (Figure3).

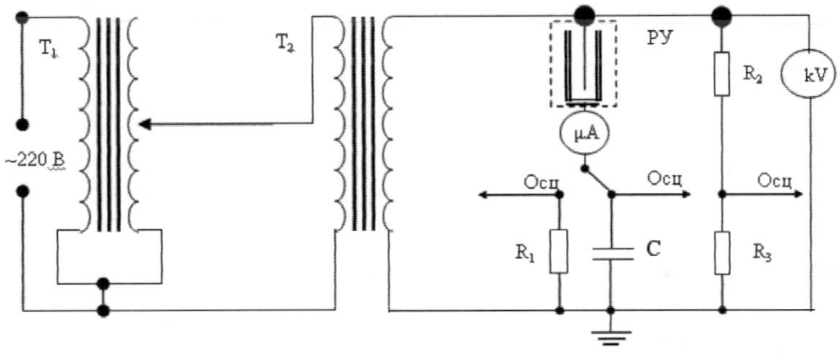

Figure 3. Test bench layout.

Hardened cement paste prepared on mineral aggregates electrified in ionized air environment under appropriate process conditions has better physical and mechanical properties, which was confirmed by the performed studies. Therefore cement paste has better performance properties as compared with the one prepared by traditional technology that was also confirmed by a number of studies.

The samples under study were as follows: water-cement ratio from 0,3 to 0,4; cement mark ПЦ400. Cement was subjected to activation in high-voltage corona discharge during 5, 10, 15, 20, 25 seconds. The samples were incubated during 28 days in normal hardening conditions.

Resulting from the experimental studies on the influence of electrified surface of cementitious materials on the structure formation of hardened cement paste the following data was obtained (Figure 4).

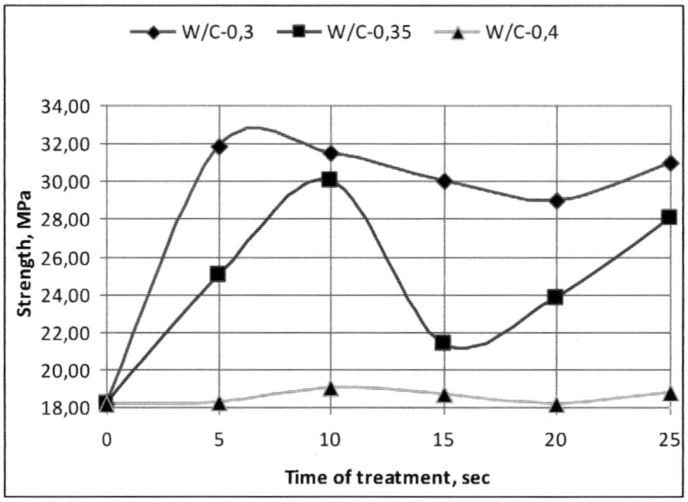

Figure 4. Compressive strength of hardened cement paste

3. Conclusions

Cement activation with high-voltage corona discharge when AC current is applied allows increasing ultimate compressive strength of hardened cement paste per 46% at the age of one day, and per 20% at the age of 28 days.

The universal nature of the developed test bench includes: application as a part of almost any production technology of any items and structures with the use of cement; low energy consumption; possibility of combined use with other cement activation technologies; using for activation of any cementitious materials; faster activation time compared to other activation technologies; no defects while cement activation.

References

[1] Alekseev A and Novostroinii A 2013 *Proc.1st All-Russian Scientific Conf. of young scientists with international participation* (Tomsk: Publishing Office of TSUAB) pp 350-353

[2] Alekseev A, Khonin D and Kuzugashev A 2010 Influence of electro-physical activation of the surface of dry building mixtures *Vestnik of TSUAB* **1** 123–29

[3] Gorshkov N and Khonin D 2011 Activation of cement dispersed systems in the field of high voltage category *Proc. 8th Int. Conf. of Students and Young Scientists "Prospects of Fundamental Sciences Development"* (Tomsk) p 529

[4] Alekseev A and Novostroinii A 2013 Study of structural disperse cement systems electrified surface with mineral binders *Proc. 10th Int. Conf. of Students and Young Scientists "Prospects of Fundamental Sciences Development*" (Tomsk) p 728

[5] Алексеев А А, Гусаков А М, Хонин Д В и Кузугашев А Н 2009 Активация портландцемента в поле высоковольтного барьерного разряда «*Строительство 2009»: материалы юбилейной Международной научно-практической конференции* 152–153

[6] Alekseev A, Gusakov A, Khonin D and Kuzugashev A 2009 Activation of portland cement in high-voltage barrier discharge field *Construction-2009 Proc. of the Anniversary Int. Research and Practice Conference* p 152–53

[7] Недавний О И 2000 Некоторые аспекты активации минерального заполнителя в среде ионизированного воздуха *Вестник ТГАСУ* **1** 120-128

IOP Conf. Series: Materials Science and Engineering **71** (2015) 012005 doi:10.1088/1757-899X/71/1/012005

[8] Лебовка Н И и Манк В В 1986 Об ориентационном упорядочении молекул мультиполярной жидкости во внешнем электрическом поле вблизи раздела фаз *Коллоидный журнал* **XLVIII** 483-486
[9] Дерягин Б В и Поповский Ю.М. 1982 Жидкокристаллическое состояние граничных слоев некоторых полярных жидкостей *Коллоидный журнал* **XLIV** 863-866
[10] Ребиндер П П, Щукин Е Д и Марголис Л Я 1964 О механической прочности пористых дисперсных тел. *Доклады Академии наук СССР* **3** 695-698
[11] Сычев М С 1984 Современные представления о механизме гидратации цементов *Цементная промышленность* **3** 53

IOP Conf. Series: Materials Science and Engineering **71** (2015) 012006 doi:10.1088/1757-899X/71/1/012006

Improving the reliability of road materials based on micronized sulfur composites

K K Abdrakhmanova[1]
[1]Kazakh National Technical University, Department of Technology of Building Materials and Construction, Almaty, 050013, Kazakhstan

E-mail:klara.abd@gmail.com

Abstract. The work contains the results of a nano-structural modification of sulfur that prevents polymorphic transformations from influencing the properties of sulfur composites where sulfur is present in a thermodynamic stable condition that precludes destruction when operated. It has been established that the properties of sulfur-based composite materials can be significantly improved by modifying sulfur and structuring sulfur binder by nano-dispersed fiber particles and ultra-dispersed state filler. The paper shows the possibility of modifying Tengiz sulfur by its' fragmenting which ensures that the structured sulfur is structurally changed and stabilized through reinforcement by ultra-dispersed fiber particles allowing the phase contact area to be multiplied. Interaction between nano-dispersed fibers of chrysotile asbestos and sulfur ensures the implementation of the mechanical properties of chrysotile asbestos tubes in reinforced composite and its integrity provided that the surface of chrysotile asbestos tubes are highly moistened with molten sulfur and there is high adhesion between the tubes and the matrix that, in addition to sulfur, contains limestone microparticles. Ability to apply materials in severe operation conditions and possibility of exposure in both aggressive medium and mechanical loads makes produced sulfur composites required by the road construction industry.

1. Introduction

The potential approach to the improvement of sulfur composites is based on synergetic interaction effects at the phase boundary. In reinforced, frame and matrix composites and materials of interpenetrating structure types the greatest potential for positive change in their properties is based on the modification of binding agents themselves [1].

Negative properties of sulfur such as fragility and low hardness are revealed as a result of polymorphic transformation as it is cooled down from melting. As polymorphic transformation takes place - structural changes at which sulfur abruptly changes its properties happen. The strength of sulfur-containing materials is determined by the allotropic form of the sulfur modification, the structure's variety - evidently or latently - crystal, sizes of sulfur particles. It is known that at - temperature lower than 95,5 ^{0}C -monoclinic sulfur is transformed into a rhombic shape thus it is re-crystallized resulting in integrity damage. That is the major obstacle to wide application of sulfur as binding element.

Necessary condition allowing the expenditure of the field of application of sulfur-based materials is the decrease of the scale level of sulfur heterogeneity down to nano-level where the concept of allotropic sulfur modifications loses its significance.

Taking it into consideration, study of the possibility of production of nano-composites through modifying the sulfur by nano-dispersed fiber-type particles in order to eliminate any internal stresses and embrittlement of sulfur seems promising. The solution is focused not only on reduction of size of the particle, but also on the special properties that are inherent to nano-layers, nano-crystals and nano-particles and related to so-called size effects [2,3]

2. Sulfur composites

2.1. Materials

Sulfur is a by-product of the refining of the Tengiz field's associated oil gas. Sulfur density: 1.92-2.07, g/cm^3 ; melting temperature: 103-119°C; specific surface: 1.04-2.61m^2/g; hydrogen sulfide content: 8.2-45.6 mg/kg; sulfur dioxide content: 4.5-5.2 g/kg.

Chrysotile asbestos has been applied as a reinforcing agent. The developed technology uses chrysotile asbestos's unique property of splitting into very thin elastic, durable tubular fibers with cross dimensions of less than one micrometer. Asbestos density: 2.4-2.6 g/cm^3; porosity: 5-6 %; water absorption: 3-4%; hygroscopic moisture removal temperature: 100-120°C; tensile strength: 800 MPa; specific surface: 20m^2/g; module of elasticity: 80-100 GPa.

Limestone has been applied as mineral filler. Limestone density: 1.4 -1.8 g/cm^3; density: 2.41-2.61g/cm^3; porosity: 11.5-40%; water absorption: 8-20%; compressive strength in dry state: 54-98.3 kg/cm^2 ; compressive strength in water-saturated state: 40.5-78.5 kg/cm^2 ; strength loss after freezing tests: 10.8-46.9%.

2.2. Optimization of sulfur composites

In order to develop sulfur micronization methods composites with the sulfur/limestone ratio varied in Range 2 : 3, asbestos content: from 3-9% were studied. Properties of reinforced composites significantly depend on the properties of the components' composition and relative positions, peculiarities of their interaction at interphase boundary, diffusion of the matrix and fibers' components, which means that a certain ratio shall be kept between fibers and matrix and their selection shall be well-reasoned. When developing the methods of reinforcing the micronized sulfur with fiber materials there has been studied the influence of the extent to which the composite is filled with the reinforcing component to form a microstructure in which each asbestos tube is surrounded with the binder matrix layer to ensure the highest adhesion. The results of the study of quantity of the binding component in sulfur composite showed the optimum content of asbestos to be 7% (Figure 1).

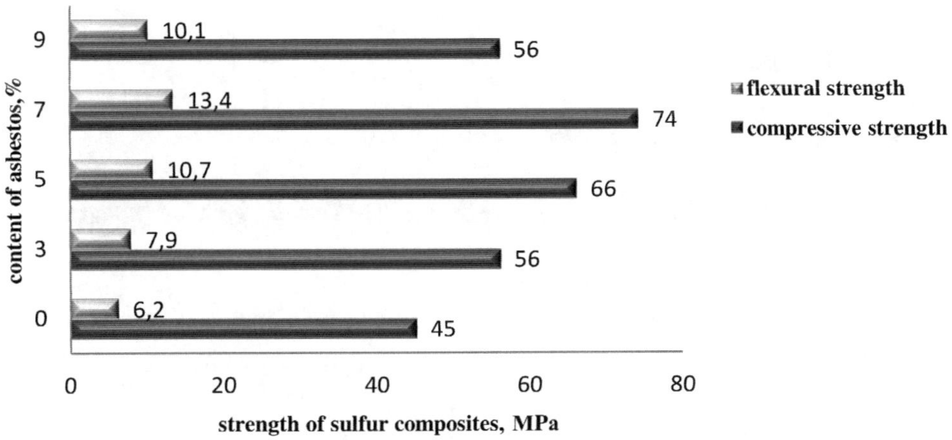

Figure 1. Dependence of the strength of sulfur composites from asbestos content.

3. Optimization of technology of sulfur composites processing

Sulfur micronization implies breaking down the particles into nano-size ones, involving high power consumption to increase the boundary surface. The micronization can be carried out by: restoring the sulfur from polysulfide solutions; ultrasound dispergation; mechanical dispergation. When sulfur is restored from polysulfide solutions it is difficult to produce dimensional samples due to very thin sulfur layers on the surfaces. Sulfur in ultrasound dispergator shall be dispergated in water medium and then dried, which significantly complicates the process. Mechanical dispergation of pure sulfur: it cannot be ground in any breaker due to high electrostatic charging and its aggregation.

Sulfur was grounded in vibrating mill in two ways - together with filler and binding component and separately. When grounded separately sulfur was found to be sticking to the mill walls and the grinding balls. It has been established that sulfur dispergation requires application of desintegrators or planetary mills as well as modern ultra-fine grinding equipment with mechanic-chemical activation and deagglomeration effects. Solution is in development of the method of sulfur micronization through grinding it in dry state together with the composite components being limestone and asbestos.

The optimum sulfur composite heat processing parameters have been determined based on changes in their strength characteristics. Molten sulfur was produced at 130-140^{0}C, but attempts to introduce reinforcing fibers into the molten sulfur were unsuccessful as significant viscosity of the molten sulfur does not allow distributing fibers even with nano-dimensions. As the result, components were ground together, which allowed them to be distributed evenly. The sulfur composite properties investigation results are shown in Figure 2.

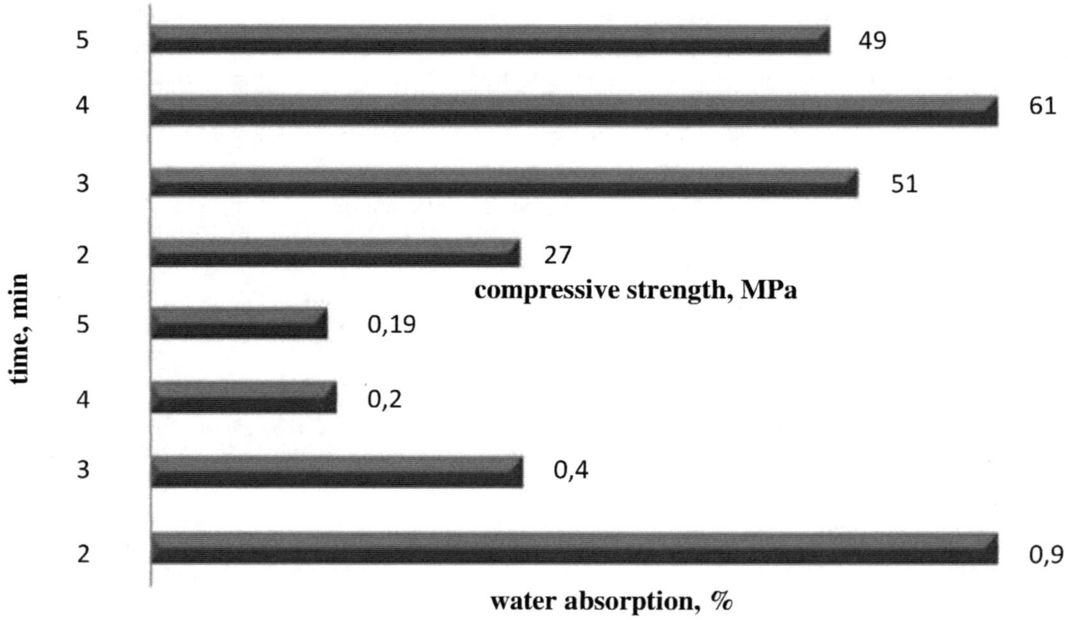

Figure 2. Dependence of the sulfur composites properties from heat-treatment duration.

The results of comparative tests of sulfur composites and cement concretes have shown that by physical and mechanical properties the sulfur composite is not inferior to the traditional concrete and by some indexes it is superior. For instance, sulfur composites' strength characteristics are 20-30% higher, water absorption is 3 times lower, frost resistance is twice as high and wearability is 30% lower.

Stable structure of sulfur composites with binary micro-fillers (mineral and fiber) is characterized by improved operational strength and deformation parameters. Durability of nanostructured sulfur

IOP Conf. Series: Materials Science and Engineering **71** (2015) 012006 doi:10.1088/1757-899X/71/1/012006

composites is determined by the character and value of forces of interactions of relations occurring on the modified binding filler boundary surface.

4. Discussion of results

Elimination of electronic insufficiency through introducing stabilizers is required in order to stabilize polymer sulfur molecules that have unsaturated electronic structure resulting in instability. Regulation of their quantity and melting temperature mode allows production of sulfur binding agent with stable physical-and-mechanical properties [4].

Appling filler for the sulfur composite in quantities comparable to the sulfur content creates prerequisites for heterogeneous primary nucleation when crystallized on hard particles [5]. As the result, filler has structure-forming effect on the sulfur matrix. When applying limestone as filler in the sulfur composite matrix calcite microcrystals are found. They have perfect cleavage extracted from limestone when ground and make the sulfur composite matrix highly durable (Figure 3a). Filling the molten substance with the fiber filler results in forming of foliated-and-flaky sulfur composite structure and the tubular structure of chrysotile asbestos, when using a capillary suction method, retains the molten sulfur in a form of ultra-thin particles. Joint grinding sulfur with fillers allows to form an optimal fine-grained homogeneous dense structure that ensures high properties of sulfur composites. Microstructure of sulfur composite represented by tightly bound particles fillers and filling the pore space of plasticized sulfur, which is a single matrix with numerous jumpers (Figure 3 b).

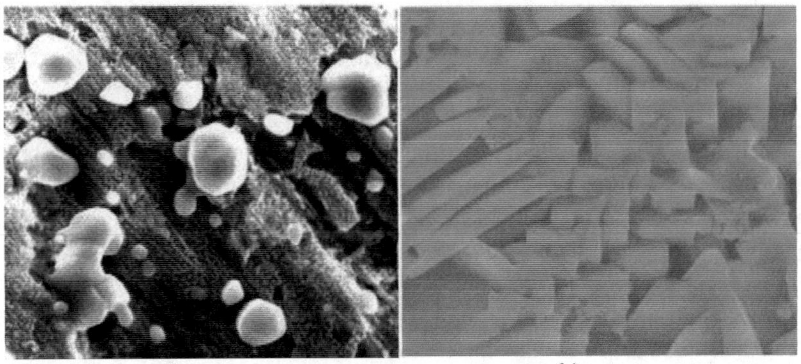

a) b)

Figure 3. Microstructure of sulfur composites: a) sulfur composites filled with milled limestone; b) sulfur composites filled with milled limestone and chrysotile asbestos.

During process of create of sulfur composites formed the homogeneous nanostructure of limestone and nano-tubular structure of chrysotile asbestos. In the result occur fragmentation of melt sulfur in the form of drip inhomogeneous nano-particles of possible amorphous structure from which then be forming ultrafine sulfur crystals. Distributed within a carbonate-chrysotile matrix sulfur crystals pass and the matrix in turn takes on deformation of the crystal lattice at polymorphic restructuring of the metastable monoclinic sulfur in a stable orthorhombic modification.

Theoretically dense packing is not possible as each fiber has to be surrounded by a layer of the matrix to ensure the highest adhesion of fibers to transmit mechanical forces between the reinforcing fibers and the matrix and for the achievement maximum monolithic composite. In addition, the high adhesion of the composite components shall be maintained in operational conditions under active external influences on sulfur composites including moisture, corrosive fluids, high mechanical loads, thermal processes [6, 7]. Minimum change in the properties of the reinforcing fibers under the influence of matrix components or operating conditions, which is typical for asbestos in the studied parameters of operation, is of fundamental importance.

The size of the microparticles of the original components affect the strength and elasticity of sulfur composites as the seal the composite occur in the initial stage as the result of rapprochement of sulfur particles and fillers, and then creation of a monolithic structure as the result of the penetration of particulate sulfur in the micropores of porous natural-fibrous fillers. The sulfur is stabilized and immobilized in the microporous cavities carbonate-asbestos matrix, providing permanence of properties.

Creation of fragmented structure of the sulfur composite allows explaining the improvement of his properties in relation to the characteristics of the original components. Such method of micronization original particles sulfur composites refers to the so-called matrix isolation nano-particles of sulfur in the massive body of the nanocomposite [8]. Feature of the synthesized matrix is its porous-fibrous microstructure. A uniform distribution of sulfur in the space of the matrix occurs in during co-grinding components, at the heat treatment the sulfur does not form massive layers and sulfur particles fragmented in the chrysotile-carbonate matrix.

The sulfur in the molten state has a very low surface tension. The maximum reduction in viscosity of sulfur allows to perform capillary effect and the walls of filled tubes in turn will prevent crystallization of sulfur in the form of large crystals that reduces the brittleness of the composites and allows plasticity which is not peculiar to traditional sulfur-containing materials. Manifestation of this effect in the sulfur composites promotes synergy of its physical and mechanical properties.

The improved properties are obtained not by chemical interaction of the components of sulfur composites, but by creation of three-dimensional nanostructured systems. Superdispersed mineral fibers form the bulk structure with enormous internal surface area. High adhesion properties of the fibers to nanostructured sulfur makes the framework almost inseparable from the sulfur binder, which excludes destructurization even in the case of sufficiently strong and corrosive effects on the surface of composites

At the transition on nanotechnology with decreasing grain size crystalline sulfur increasing the volume fraction of intergranular material and the surface of grain boundaries prevent to the dislocation motion. For small nanoscale grains the increase of strength occurs due to the low density of existing dislocations and the difficulty of formation of new dislocation, i.e. the formation of nanosized structures based on sulfur allows obtaining materials with high properties

Thus, the nanostructuring of the sulfur particles provides uniform microstructure with insulation particles of sulfur by the nanofibres chrysotile asbestos and ultrafine particles of carbonate. Uniform distribution of the components in the spatial structure during the heat treatment prevents the formation of massive layers of sulfur, which is fragmented in carbonate-chrysotile matrix, compensating the internal stresses, and provides the high properties of sulfur composites.

5. Conclusions

Modification of properties sulfur compositions is caused not by the chemical interaction of the composite components, but by creation of three-dimensional nano-structured system where synergistic effect is manifested. Special role in the formation of the microstructure sulfur compositions belongs to the fillers and reinforcing components, triggering the phenomenon of capillary suction, which has stabilizing effect on the physical and mechanical properties sulfur compositions.

References

[1] Shpak A and Gorbyk P 2009 *Nanomaterials and Supramolecular Structures. Physics Chemistry and Applications* (Nederlands: Springer) p 425
[2] Poole Ch and Owens F 2011 *Introduction to Nanotechnology* (New Jersey: Wiley Interscience) p 392
[3] Cao G and Wang Y 2003 *Nanostructured & Nanomaterials. Synthesis, Properties & Applications* (Singapore: World Scientific Puplishing Co) p 581
[4] Urakaev F and Bazarov L 1999 Kinetics of homogeneous nucleation of monodisperse spherical sulphur and anatase particles in water-acid systems *Journal of Crystal Growth*

205 223-232
[5] Massalimov I 2001 On the possible mechanism of energy transmission by mechanical impact. In: *Proceedings of FBMT* (Novosibirsk: SB RAN) p 59
[6] Marchiukaitis G 1994 Properties of concrete with wastes from sulphur production. In: *Proceedings of 12 International Baustofftag* (Weimar: IBAUSI) pp 249-255
[7] Massalimov I and Chuikin F 2013 Protection of building constractions with sulfur impregnating solution. *Study of Civil Engineering and Architecture SCEA* pp 219-24
[8] Hirth J and Lothe J 1972 *Theory of dislocations* (New York: McGraw-Hill Company) p 214

Influence of crystallized glyoxal on properties of gypsum construction mixes

L Anikanova[1], A Kudyakov[1], O Nikitina[1] and D Tolstov[1]
[1]Tomsk State University of Architecture and Building, Building Materials Department, Tomsk, 634003, Russia

E-mail: v.olga.nikitina@gmail.com

Abstract. The article presents the results of research of crystallized glyoxal influence on properties of gypsum binding and dry construction mixes. The article describes construction mixes with crystallized glyoxal formulation, processing technology of equal distribution of crystallized glyoxal in construction mixes, structure formation of construction mix with varied proportions of glyoxal and technical and performance characteristics of hardened construction mixes. It is discovered that maximum strength of construction mortar made of dry construction mixes is achieved by 0,1% of glyoxal content in construction gypsum.

1. Introduction

Various modifying additives are used in modern production of dry construction mixes. Process of structure formation of construction mortar made of dry construction mixes is accomplished within "mineral binder – filler material – water" system. It is possible to modulate properties of mixes and make construction mortars with specified qualities for various operating conditions by means of introduction of different organic and mineral modifying additives.

Introduction of modifying additives in dry construction mixes facilitates management of technological and mechanical properties of mixes and mortars, reduction of power consumption, which allows utilizing modern technology and increases competitive advantage of produces in terms of its quality and applicability. Development and production of liquid and powdered multi-purpose modifying additives for construction mixes is essential today [1, 2, 3]. Surface active substances, water-soluble polymers, water dispersions of polymers and electrolyte additives are utilized as modifying additives in construction mixes nowadays[4, 5, 6].

Crystallized glyoxal can be named among new additives widely used in various branches of industry [7]. In production of construction materials this new additive is still rarely used due to lack of research on structure formation processes, lack of scientifically approved formulation and production techniques, reliable data on objective technical and performance characteristics of final construction product and specification documents.

2. Materials and methods

2.1. Characteristic of initial materials

As initial materials in the research of gypsum mixes and dry construction mixes we used gypsum plaster, crystallized glyoxal, methylcelluloseof Bermocoll type, sand, tartaric acid and hydrated lime. The enumerated substances have the following characteristics:

IOP Conf. Series: Materials Science and Engineering **71** (2015) 012007 doi:10.1088/1757-899X/71/1/012007

- gypsum plaster, G5 type, standard consistency 55%, beginning and end of solidification process is in 6 and 30 minutes accordingly. Gypsum plaster complies with requirements of All Union State Standard (GOST) 125-79;
- mortar river sand, size modulus 1,96, real and apparent density is 2500 and 1430 kg/m3 accordingly, concentration of pulverescent and argillaceous admixtures is 1,5%. Sand complies with requirements of All Union State Standard (GOST) 8736-93;
- hydrated lime, concentration of CaO + MgO is 69%, water content is 3%, 0,08 mm sieve residue percentage is 12%. Hydrated lime complies with requirements of All Union State Standard (GOST) 9179-77;
- crystallized glyoxal is water-solublelarge grain powder. Structural formula of crystallized glyoxal is given below (Figure 1).

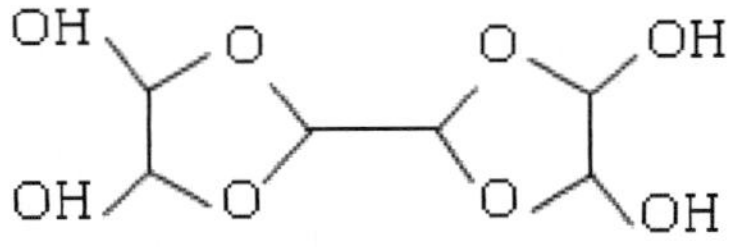

Figure 1. Structural formula of crystallized glyoxal.

- Concentration of crystallized glyoxal in additive is 84,4%, type A. Crystallized glyoxal is manufactured by Tomsk LLC "Novochem" and complies with requirements of TOR2633-004-67017122-2011 [7]. Crystallized glyoxal with market-gradeparticle size is badly combined with dry construction mixes and is not dissolved in water under traditional methods of production of mortar mixes. On solidifying of gypsum block non-dissolved particles of glyoxal can be found throughout the whole body that is why in the research we used glyoxal powdered to 200 m2/kg specific surface.

Technical characteristics of methylcellulose and tartaric acid are given below (Tables 1, 2).

Table 1. Technical characteristics methylcellulose of Bermocoll type.

Bermocollis produced in form of white, non-toxic substance without scent. Typical parameters:	
Apparent density, kg/m3	400…600
Product density, gr/cm3	1,33
Water content (in package), %	about 4
Salt content(NaCl), %	about 4
Water solution reaction	Neutral / faintly acid
Coating properties under 20 C and relative humidity under stretching, MPa,	45…55
%	25…35

Table 2. Physical and chemical parameters of tartaric acid.

Parameter	Standard
1. Weight percentage of tartaric acid ($C_4H_6O_6$), %, not less than	99,5
2. Weight percentage of water-insoluble substances, %, not more than	0,010
3. Weight percentage of residue on ignition (in form of sulphates), %, not less than	0,03
4. Weight percentage of sulphates(SO_4),%, not more than	0,010
5. Weight percentage of phosphates(PO_4), %, not more than	0,0020
6. Weight percentage of chlorides (Cl), %, not more than	0,0020
7. Weight percentage of iron (Fe), %, not more than	0,0010
8. Weight percentage of calcium(Ca), %, not more than	0,002

2.2. Research procedure

For research of gypsum binding mixes a standard procedure described in All Union State Standard125-79 was used; for research of gypsum dry mixes standard procedures described in All Union State Standard31386-2008 and 31387-2008were used.

Ready components were thoroughly mixed in blending tank for 5 minutes. Then the dry mix was poured into water and was mixed in forced mixer for 5 – 7 minutes. From final blend sample 70x70x70 cubes were molded. Technological tests on watered mix were performed immediately after its preparation and physical and mechanical tests were performed on naturally hardened 7 days old sample cubes. In tests amount of glyoxal was 0,1; 0,5 and 1,0% upon 0,4 water-gypsum ratio.

3. Results and analysis

Water requirement (standard consistency with given plasticity) and kinetics of structure formation (period of hardening and strength of samples at different times) were chosen as main criteria of building gypsum mixes with glyoxal additive quality evaluation.

Gypsum mixes standard consistency research results are given in Figure 2. Standard consistency of gypsum paste under introduction of crystallized glyoxal modifying additive insignificantly rises from 55% (control sample) to 60% (1% of glyoxal).

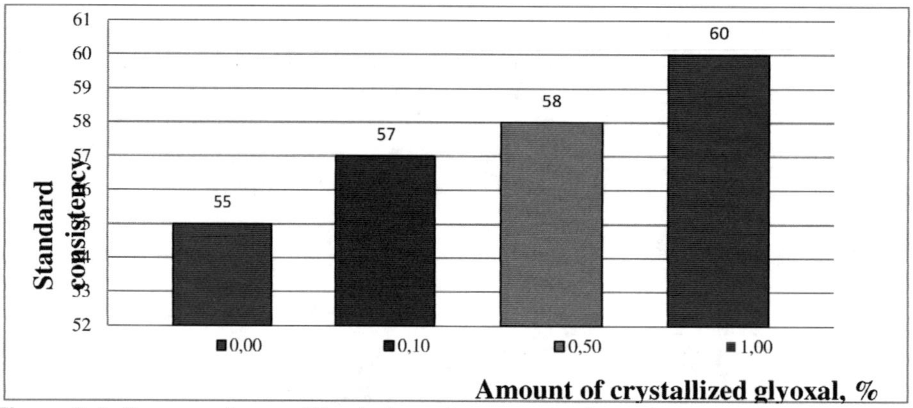

Figure2. Influence of crystallized glyoxal on standard consistency of gypsum paste.

In gypsum mixes with crystallized glyoxal acceleration of initial structure formation is observed. In addition, introduction of 1% of glyoxal leads to drastic reduction of gypsum paste hardening period: beginning of hardening process is reduced from 20 (without glyoxal adding) to 10 minutes and termination of hardening process is reduced from 15 to 5 minutes accordingly (Figure 3). In gypsum mixes introduction of 1% of glyoxal leads to sufficient raise in heat emission under hydration.

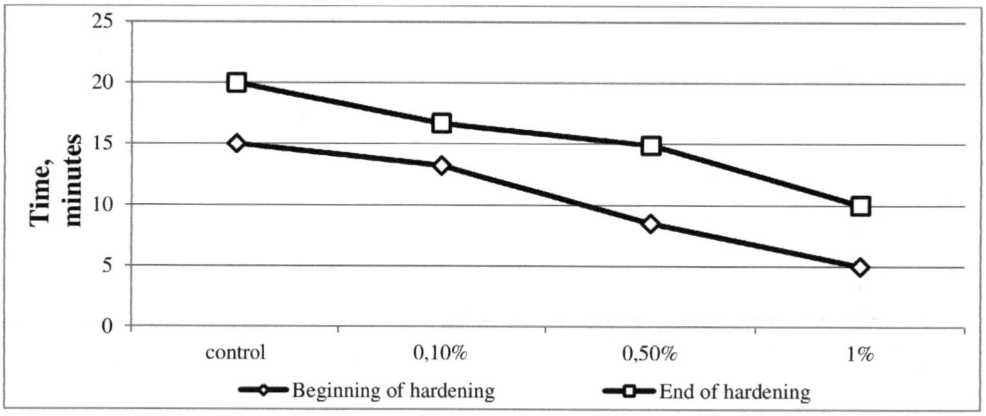

Figure 3. Influence of crystallized glyoxal on period of hardening of gypsum paste.

Research results of crystallized glyoxal impact on strength parameters of 7 days old hardened gypsum mixes are presented in Figure 4. It is proved that introduction of 0,1% of glyoxal leads to significant improvement of compressive strength from 8 to 16 MPa (100%). In case amount of glyoxal is raised over 0,5% strength characteristics of hardened gypsum mixes do not change.

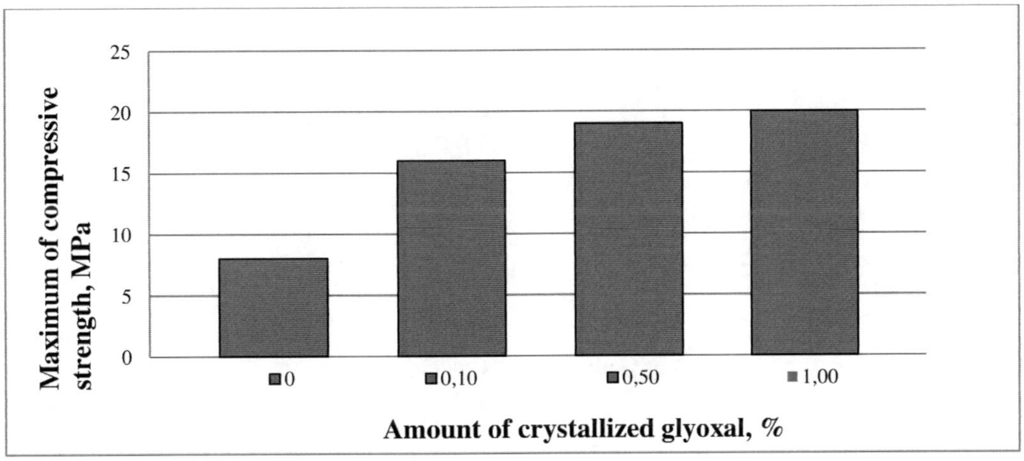

Figure 4. Influence of crystallized glyoxal on strength of gypsum mixes.

In the research of influence of glyoxal on characteristics of dry construction mixes, the content of dry construction mix produced in Tomsk by LLC "Bogatyr" was accepted as a basic one. Maximum strength gain of solutions made of dry construction mixes (80%) was achieved with introduction of 0,1% of glyoxal. Research results are given in Figure 5. Increase in glyoxal amount over 0,1% leads to decrease in strength characteristics of solutions.

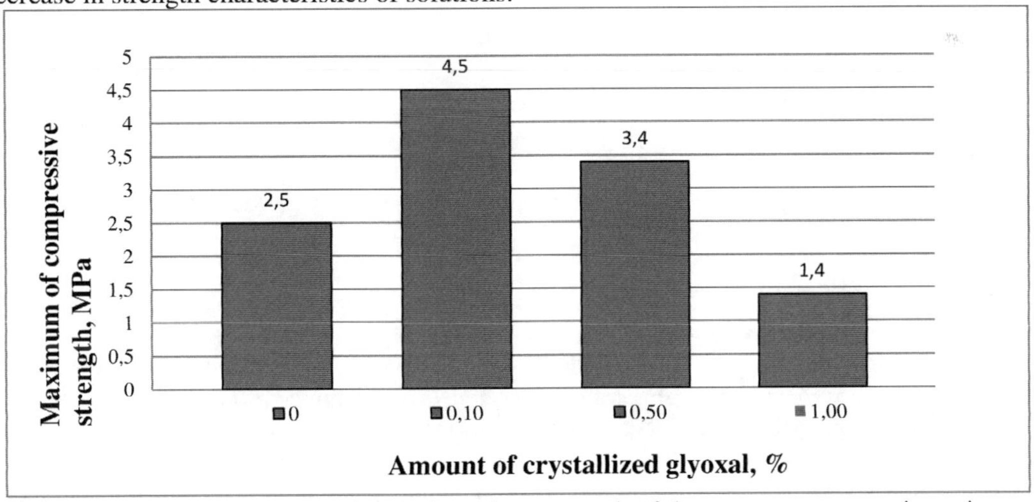

Figure 5. Influence of crystallized glyoxal on strength of dry gypsum construction mixes.

Optimized composition of dry construction mixes with crystallized glyoxal additive recommended for industrial utilization is given in Table 3.

Table 3. Compositions of dry construction mixes with crystallized glyoxal.

No	Type	Material consumption, kg, per 1 ton of mix					
		construction gypsum	sand	methylcellulose	hydrated lime	tartaric acid	glyoxal
1	25	766	200	1,9	30,0	1,0	0,8

Comparative physical and mechanical characteristics of control and developed mixes with crystallized glyoxal are given in Table 4. In solutions made of 0,1% crystallized glyoxal-added mixes adhesion (bond strength) of solution increases by 50%. It is suggested that strength and adhesion increase becomes possible due to formation of cross-linking polymeric structures of glyoxal with other additives of construction mixes. This hypothesis is being further physically and chemically tested.

Table 4. Physical and mechanical characteristics of dry construction mixes.

Parameter	Physical and mechanical characteristics of mixes	
	with standard additives in accordance with TOR 5745-002-61212785-2012	with crystallized glyoxal additive
Dry mix Water content, %, not more than	0,3	0,3
1,25 sieve residue percentage, %, not more than	5	5
Binding mix Cone slump, cm	8-12	12-14
Water holding capacity, %, not less than	95	95
Retaining of initial mobility, min, not less than	45	50
Solution Compressive strength at 7 days old, MPa	2,5	4,5
Average density, kg/m3, not more than	1200	1200
Adhesion, MPa	0,4	0,6

3. Technology of production of crystallized glyoxal-based dry construction mixes

Utilization of glyoxal in the form of fine powder (surface area=200 м^2/kg) into production of mixes does not add complexity to existing technology of dry construction mixes production. Introduction of glyoxal in recommended quantities along with other polymer additives is performed by measuring device directly into concrete mixing drum. Glyoxal is added after all other additives after which dry construction mix is stirred for 2 – 3 minutes more. Production technology manual on production of dry gypsum mixes by LLC "Bogatyr" in Tomsk, Russia was developed.

4. Conclusion

The research allowed discovering peculiarities of influence of crystallized glyoxal with surface area=200 м^2/kg on properties of gypsum and dry construction mixes, developing composition of dry construction mixes and technology of their preparation, working out production technology manualon production of dry mixes in construction.The reported study was supported by RFBR, research project No. 14-03-31922/14 *мол*_a. ("My first grant") and by LLC "Novochem" (Tomsk, Russia).

Acknowledgement
The research performed with the financial support of Russian Foundation for Basic Research, Competition of research projects performed by young scientists (My First Grant) No. 14-03-31922 “Studying composite acid fluoride-based binders of enhanced water resistance to obtain dry building mixes”.

References
[1] Kudyakov A, Anikanova L, Redlikh V and Sarkisov Yu 2012 The effect of sulfate and sodium sulfite on the processes of structure formation acid fluorides compositions *Building materials* **10** 50
[2] Modifying additives to dry building mixtures 2009 *Construction materials, equipment, technologies of XXI century. Dry building mixtures* **5-6** 38
[3] Lysenko E, Kotlayrova L, Tkachenko G, Trichenko I and Yundin A 2003 *Modern finishing and facing materials* (Rostov n/D: “Feniks”)
[4] Mikul’skij V, Sakharov G and others 2007 *Building materials* (Moscow: Publisher Association building universities)
[5] US 20090186963 A1. Eduardo, Jose A, Mark G and Philip J 2012 *Additive composition for mortars, cements and joint compounds and cementitious compositions made therefrom* (United States Patent)
[6] US 3365319 A Michael J 1968 *High strength inorganic cement compositions* (United States Patent)
[7] TR 2633-004-67017122-2011 Crystallizedglyoxal [Electronic resource] http://novochem.ru/products/crystalline%20glyoxal Date of access 23.06.20141
[8] Anikanova L, Kudyakov A, Nikitina O, Derghanskaya E and Fisher H-B 2013 Composite binders for dry building mixtures *International collection of proceedings* 258

IOP Conf. Series: Materials Science and Engineering **71** (2015) 012008 doi:10.1088/1757-899X/71/1/012008

Regulation of confining liquid for cement systems properties by means of electromagnetic fields

N P Gorlenko[1], B I Laptev[2], Ju S Sarkisov[1], G N Sidorenko[2], A K Kulchenko[3] and T S Minakova[4]
[1]Tomsk State University of Architecture and Building, Tomsk, 634003, Russia
[2]Nove Tehnologije d.o.o., Ljubljana, Slovenija
[3]Surgut State University, Surgut, 628418, Russia
[4]National Research Tomsk State University, Tomsk, 634050, Russia

E-mail: gorlen52@mail.ru

Abstract. The paper studies theoretical aspects of methods for electrochemical, electromagnetic, magnetic activation of water as a confining liquid for cement systems. The ideas about processes of water structural organization are shown to be the basis for one of possible mechanisms of activating impact. Experimental data for electric conductivity, electric capacity on the example of distilled water treatment by means of magnetic field which confirm the change of water structure have been presented.

1. Introduction

The creation of resource-energy-saving technologies for obtaining the materials of different technological dedication possessing high production properties is considered to be one of the most important tasks of modern materials science. The application of external actions of different nature in the course of which an active state of water-containing dispersed system can be reached is thought to be the most technologically available, economically and ecologically efficient method. However it is quite evident that without identification of thermodynamic and kinetic regularities taking place at this phenomenon it is impossible to control effectively the creation of the materials with target properties. Despite the numerous investigations made by both national and foreign scientists [1-3], most aspects remain unclarified, and the theory of activation processes has been worked out at the phenomenological level predominantly. This fact doesn't serve as an obstacle for application of the methods of external actions in practical purposes effectively, but it limits the possibilities of directed control over final properties of compounds based on binding agents. Taking into consideration the variety of methods for confining liquid activation, some definite types of low energy external actions on the cement systems are studied in this paper. From the authors' view they have practical value in the technologies of products manufacture.

2. Theoretical ideas about low energy water activation

This paper is aimed at systematization and development of the ideas about physic-chemical entity of definite low energy external actions, possessing electromagnetic nature, and at experimental substantiation of theoretical aspects of a magnetic field and thermal impact influence on the processes of structure formation in water as confining liquid for cement systems.

The transformation of liquids from their inert state into chemically active one and the control over discrete structures of the system by means of slight changes in their structure and properties are thought to be the low energy activation of liquids. It also includes the control over kinetics of the process with the help of impact on the systems of the source with energy which equals to the thermal energy or a little less (*kT*). The physic-chemical system can be considered as an active one if the internal potential energy store doesn't temporarily correspond to its thermodynamic equilibrium values being the result of some external impact. In case it occurs, this impact is called as an activating one. It is believed that maximum possible values of energy for activating impact equal to 10 J/mol. In this case low energetic impacts can be effective because they influence the process of diffusely-developed intermediate structures in the system, in contrast to high energy impacts in which the transformation from one discrete state into another occurs as a "shock". In the basis of this statement the regularities of experimentally observed facts can be found. For example, the phenomenon of disproportional dependency of energy-consuming processes development in the system of energetic equivalent inserted from outside.

Nowadays the following widespread methods of different media activation can be distinguished: magnetic, electromagnetic, electrochemical, hypersonic, thermal, mechanical, hydrodynamic, acoustic, plasmatic, disruptive-impulsive and chemical ones. They can be relatively divided into three groups [4]: physical (reagentless); chemical (reagent); combined. We will consider some definite methods of low energy activation of cement systems.

2.1 Electrochemical activation

In the course of water or water-salt solutions activation made by means of electrodes injection polarized with direct current, the active metastable compounds appear. Being used practically they quickly disappear, acting as a catalyst, an initiator and reagents in different chemical reactions, which occur predominantly on the interface "electrode - solution". The formation of a double electrical layer at the surface of electrode with high potential (10^6 -10^8V/sm) results in the occurrence of new metastable compounds which are capable to exist independently during a long period of time in the target thermodynamic conditions.

The peculiarity of this electrochemical activation is the formation of active microbubbles with the size of from 0.2 to 5 mkm which promote the qualitative change of the solution system organization. Such a spatial and temporal ordering of the system in the course of which the change in any of its parts (discrete factor) influences the interrelationship of other parts is assumed to be a system organization. For this reason the construction of a system organization model for the object being studied can be achieved only when all the possible interactions of separate structural units are taken into consideration.

In the course of electrodes polarization by means of alternating current a new mechanism of activating impact determined by the oscillating processes in the system appears. The origin of the process begins in a micro volume of water in an electrode sheath; then other parts of water volume are involved in an oscillating process as far as "pumping" of oscillating energy takes place. Since the process is an energetic one (the energy of oscillating process is proportional to a squared amplitude of oscillations), water activation doesn't occur according to a diffusing mechanism and, hence, it can be finished during a short period of time. At the same time, according to modern theories, the oscillations with low frequency are generated at the stage of a fine-crystalline structure formation. In this case the origin and destruction of dynamically developed kinetic compounds the quantitative accumulation of which leads to new qualitative changes in a cement system are assumed. If the frequency of external field coincides with that of characteristic oscillation of structural compounds in a cementing system, it can be a source of code information to the reproduction of "self-alike" compounds. The reasons for "self-alike" structures reproduction are not completely cleared, but its source is a rhythmicity of electromagnetic coding external impact. It means that imposition of external frequency field can change the conditions of structures "selection" and lead to a cement stone formation which possesses the properties that differ from those created in the absence of external impact.

2.2 Electromagnetic activation

Generators of swirling electromagnetic waves of different types and structures (cavitational converters, hydrocavitational generators, monovibrators, etc.) are used for electromagnetic activation of cement systems predominantly at the early stages of structure formation. The principle of action for such generators is based on the initiation of non-linear processes of phasic transformation in a liquid-phase medium by means of a high-gradient electric, magnetic or acoustic field. As a result the discontinuity of a liquid medium and, consequently, the formation of a new structural organization of water or water solution are observed.

2.3 The impact by means of magnetic field of constant magnets

The impact is fulfilled with the help of a confining liquid transmission through the magnetic field lines of constant magnets with their following injection into the volume of a cementing compound. The treatment of the liquid by means of magnetic field activates the following processes at the early stages of concreting: the size of cementing compound grain is decreased, the dissipation and hydration of a cement part is intensified, new structures which differ from those obtained in ordinary conditions are formed, and the excreting of smaller crystal phases is observed. It is determined that the hardness of compounds on the basis of cement is increased under compression, the porosity is decreased and the cold resisting property of hardening systems is increased while the duration of setting and flow ability of cement grout is changed. In this case the necessary conditions of forced response appearance are the following: laminar liquid flow regime, gradient structure of magnetic field lines, and optimal concentration of the solution and temperature of the liquid. Literary data don't have a distinct theoretical basis that can make their systematization quite difficult.

We can claim, however, that one of possible mechanisms of external impact consists of the destruction of system-structural organization of the liquid revealed in changing the ratio of polymer and monomer components of water structure. In this case the constant ratio of separate water molecule dipoles and their associates is maintained unless the system-structural organization of liquid "determined" by the external action is destructed by thermal energy. This implies that in spite of its small energetic component, the magnetic field is able to widen the spectrum of possible fluctuations of spatial distribution of liquid particles and to create the conditions which allow for some possible temporal configurations of water dipoles to be energetically more efficient and, hence, more resisting ones compared with the initial state. These statements are the results of comparison of interaction forces for structural components of water-salt solution and forces of magnetic field [3].

For example, if in a system there is a target flow of the substance with a constant direction and an average speed of $\bar{U}$, the particles with opposite charge must move in opposite directions under the influence of Lorentz force $\vec{F}_{\max} = ze\bar{U}B(\vec{B}\times\vec{U})$ that results in a short-term redistribution of ions leading to the formation of microregions of spatial charge along the direction of liquid flow (RSC). With appearance of RSC the force of electrical field starts acting for a short period of time: $F = zeE$, where E is the electric field intensity, ze- a particle charge, the direction of which coincides with that of magnetic force. E equals to (10^7 – 10^8) V/m. In this case the force of electric filed becomes comparable with the forces of ion - dipole (1) and dipole – dipole interactions (2).

$$F_{i-d} = d_m ze / 4\pi\varepsilon_0\varepsilon_1\varepsilon r^3 \approx 1{,}1\times10^{-13}H \quad (1)$$

$$F_{i-d} = d_m d_m / 4\pi\varepsilon_0\varepsilon_1\varepsilon r^4 \approx 4{,}10^{-15}H \quad (2)$$

whereE_{i-d}–is the force of ion-dipole interaction, d_m is a dipole moment of water molecule ($6{,}02\times10^{-30}$Cl×m), ε_0– permittivity of vacuum($8{,}85\times10^{-12}$ F/m);ε_1–permittivity of medium; r– radius of ion with ion atmosphere interaction region (in the region of long-distance hydration ≈ 10^{-9} m).

The calculation of particles speed values, which they obtain in the region of spatial charge due to the action of magnetic field, also shows its excess over the average speed of thermal molecule movement. This result indicates that the processes of system-structural destruction of liquid organization and the decrease of ion hydrated film which have been mentioned above can be possible. So, at the target stable direction of a flow we can observe the destabilization of the system from thermodynamic equilibrium on the back of thermal molecule movement, especially at high speeds of liquid flow and values of magnetic induction.

Analysis of liquid media activation methods studied in this paper allows confirming that the ideas about the destruction of water structural organization are the basis for the mechanism of the processes taking place in these media. The external impact results in dissymmetry that means some well-ordered deviation from symmetry of liquid initial state stabilized by thermodynamic conditions is detected. It is well-known that if in symmetrically situated points of the system there arises a difference of charges, currents, potentials, forces, flows, concentrations, and the conditions satisfy to their scalar or vector interaction, and the distance between interacting elements is quite short, the state of dissymmetry makes its work. Therefore, according to our views it is necessary to investigate structural peculiarities of water and water solutions used as a confining liquid in order to develop our ideas about mechanisms of activating external impact.

3. Methods and materials

Some results of the experimental study estimating structural processes in distilled water on the example of external impact of magnetic field compared with thermal treatment of liquid are shown below.

The experiments have been carried out at the device that allows measuring the conductivity and electric capacity of liquids at the frequency of current in the range of 1 to 300 kHz. The distilled water having the capacity of (1.3 – 3.0) mcC/sm has been used. The magnetic treatment of water has been done by means of its transmission through a funnel with two magnetic systems situated on both sides of its spout. Each system includes 6 magnets disposed gapless with the induction equals to 0.2 Tl, looking at each other with opposite poles. In order to compare the effect, water has been exposed to thermal impact without air because it is one of the factors which significantly changes structural organization of water. The measuring cell for conductivity estimation is a cylindrical glass vessel having a diameter of 7 sm, in which the investigated liquids of 200 ml have been placed, and two flat electrodes made of non-magnetic stainless steel (the square of each magnet equals to 22 sm^2) which are situated on the distance of 22 mm. The measuring cell for capacity estimation is a cylindrical glass vessel having a diameter of 2 sm and length of 200 mm on which two armature of condenser (with square of 22 sm^2) are situated with shifting relatively each other in parallel planes.

4. Results and discussion

In order to estimate possible mechanisms of water structure changes activated by magnetic field, electrical parameters have been measured after water heating from 20 °C to 60 °C and its following cooling until the initial temperature of 20 °C without air, after cooling until 10 °C and after heating up to 30 °C.

The results of experimental study are shown in Figure 1.

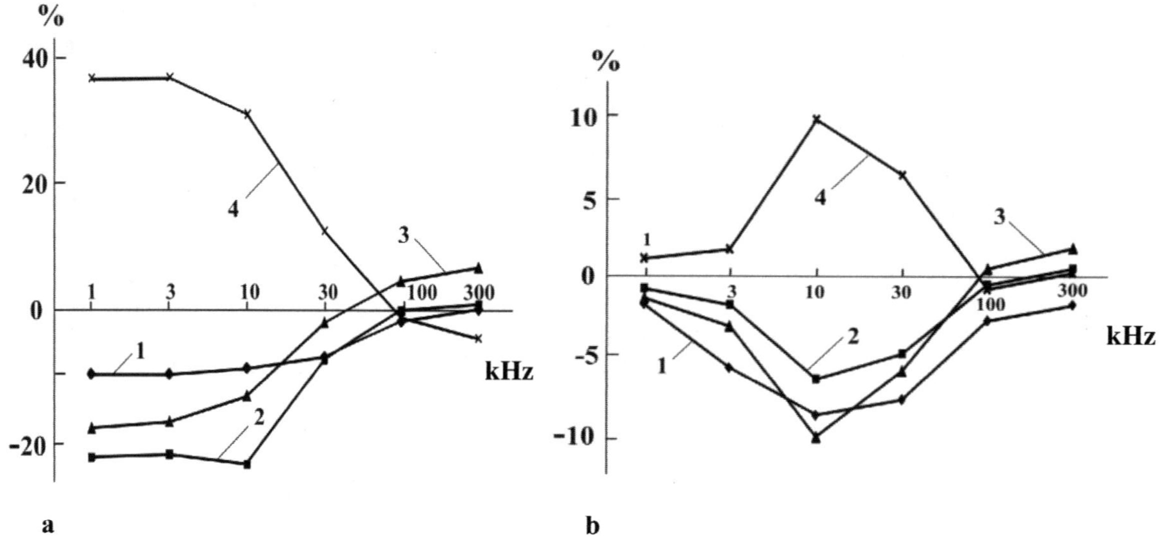

Figure 1. Relative changes of electric conductivity (a) and electric capacity (b) values of distilled water at different frequencies of reactive current in relation to the control sample: 1 – after treatment in magnetic field; 2 – after heating from 20 °Cto 60 °C and then cooling until the initial temperature of 20 °C; 3 – after cooling to 10°C; 4 – after heating up to 30 °C.

The results obtained reveal the decrease of electrical conductivity of distilled water after magnetic treatment of the liquid and they well coincide with the data of papers [5, 6]. In this case after heating of the liquid with its following cooling which is the same as in the experiments with activation made by magnetic field, the conductivity of distilled water is also decreased at the same frequencies of current. When the frequency of current is increased (100-300) kHz the effect is practically graded (Figure 1a). The dynamics of investigated parameters changes for distilled water at its heating up to 30 °C has an opposite direction. It means that electrical conductivity is increased at low frequencies and decreased at high ones. As far as the changes in electrical capacity of water at different external impacts concerns (Figure 1b), the presence of extreme points in the region of explorative current frequency of approximately 10 kHz should be mentioned that is probably determined by the existence of associates made of water molecules with high concentrations which are close to structural compounds according to their size [7]. It should be noticed the presence of correlating dependence of water electrical properties at different types of external impacts that can indirectly confirm a close mechanism of liquid activation.

Changes in conductivity and electrical capacity of distilled water after its activation by magnetic field testify about changes in its structure as a result of dissymmetry in a system. It is evident that after activation in distilled water such changes of structure at which the mobility of water dipoles is altered at the frequencies of explorative reactive current can occur.

Analysis of fulfilled experiments results shows that electrical properties of water before and after its activation by magnetic field or by means of heating are changed at different frequencies of reactive current in different ways. In some cases the increase of parameters values is observed, in other cases the decrease of them is detected. Apparently, external impacts can lead to both system activation due to the increase of energy store and to the reactivation, - the process of "mollification"due to the binding of structural units in harder or more organized compounds. We can assume that in the systems two processes are observed – formation of clusters and their destruction. The former process is accomplished by the decrease of free energy and reactivity of the system as a whole, the latter results in their growth.

The application of a confining liquid activation by means of magnetic field allows regulating kinetics of structure formation processes [8] and such technological properties as hardness under compression, duration of cement grout setting, water absorption and thermal resistance of the samples.

5. Conclusion

1. On the basis of comparative evaluation of the processes for the system activated by means of electromagnetic field it is shown that the ideas about the processes of liquid structural organization destruction as a result of dissymmetry in a system are the most important reasons for possible mechanisms of activating impact.
2. The linear correlation between changes in conductivity and electrical capacity of distilled water in conditions of its activation, cooling and heating is revealed.
3. It is assumed that activation of distilled water by means of magnetic field, heating with following cooling without air and also water cooling induces the similar changes of its structure which include, probably, the increase of water clusters size. In this case the mobility of water dipoles is decreased at the frequencies of explorative reactive current of (1 – 30) kHz, but it is increased at resonance frequencies of (30 – 300) kHz.

References

[1] Buhachenko A L 1999 *Chemistry Success* **68** 99
[2] Зятьков А И 1977 К вопросу о природе свойств магнитообработанной воды *Journal of Applied Chemistry* **50** 16
[3] Горленко Н П и Саркисов Ю С 2011 Низкоэнергетическая активация водно-дисперсных систем (Томск: Издательство Томского государственного университета)
[4] Пухаренко Ю В Аубакирова И У и Староверов В Д 2009. *Инженерно-технический журнал* **1** 40
[5] Akopyan S N and Airapetyan S N 2005*Biophysics* **50** 265
[6] Санкин Г Н и Тесленко В С 2000 Инерционность изменения электропроводности воды в слабых магнитных полях *Journal of Technical Physics* **76** 64
[7] Гончарук В В, Орехова Е А и Маляренко В В Влияние температуры на кластеры воды *Chemistry and Technology of Water* **30** 150
[8] Горленко Н П, Чернов Е Б и Саркисов Ю С 2010 Кинетический анализ процессов структурообразования в активированной системе «цемент-вода» *Вестник ТГАСУ* **2** 147

Water-repellent coatings for surface and 3D wood processing

N N Debelova[1], N P Gorlenko[1], G G Volokitin[1], Yu S Sarkisov[1], V P Dmitriyenko[1], E N Zavyalova[2] and P B Zavyalov[3]
[1]Tomsk State University of Architecture and Building, Tomsk, 634003, Russia
[2]National Research Tomsk State University, Tomsk, 634050, Russia
[3]National Research Tomsk Polytechnic University, Tomsk, 634050, Russia

E-mail: mackevichn72@mail.ru

Abstract. The paper presents the results of research in organic chemical compositions for hydrophobic protection of wood with the use of surface and three-dimensional coating techniques of impregnation and chemical compositions. Water absorption indicators, angles of contact on the surface of treated samples are detected herein. Kinetic equation of the moisture diffusion transition in capillary-porous structure of wood is suggested.

1. Introduction

Hydrophobic materials and coatings are substances of organic or inorganic nature. Their angle of contact exceeds 90°. These materials can be divided into two main groups: water-repellent coatings with three-dimensional impregnation capable of deep penetrating into products, and those with the surface impregnation which are applied in thin layers to the surface of hydrophilic materials thereby preventing moisture penetration. As a rule, protective properties of the latter are defined not only by the material characteristics as a whole, but by the properties and the structure of the surface layer. These coatings are applied to the surface by one or several molecular adsorptive and oriented layers. Unlike water-repellent coatings with 3D impregnation, the quality of such coatings is mainly defined by their adhesion to the product surface and resistance to the environmental factors. Water-repellent agents are practically used as low- and high- molecular organosilicon, fluor organic compounds, salts of fatty acids, surface-active agents and other compounds of organic and inorganic nature [1-3].

The aim of the present work is to study protective properties of water-repellent agents of organic nature in terms of the moisture penetration in cedar and assess the possibility of application of the suggested structures for restoration works.

2. Materials and methods

Cedar is chosen as an object of research. Cedar has density within 0.47 g/cm^3, a thin layer structure with rather narrow sapwood and high extraction of resinous substances. Cedar surface processing is carried out with a brush by application of water-repellent agent solution or fusion, while 3D cedar processing by dipping in water-repellent agent solution and their additional vacuum processing (10^{-2}atm) so as to accelerate the liquid diffusion in the volume of a capillary-porous body. The vacuum processing is carried out for several hours up to a complete gas bubble release from the sample volume. The choice of water-repellent agent for a superficial processing is conditioned by the application of compositions with chemical properties corresponding to that of the wood and thereby exerted minimum disturbance of physicochemical properties of natural wood. The work studies solutions of cellulose acetate and rosin of rather high light and weather resistance, and fusions of

IOP Conf. Series: Materials Science and Engineering **71** (2015) 012009 doi:10.1088/1757-899X/71/1/012009

modified amorphous polyethylene and atactic polypropylene. The latter are wastes from petrochemical production and characterised by high adhesive strength to a wood surface [4].

3. Results and discussion

Amorphous polyethylene and atactic polypropylene are modified by oxidation of polymer fusions at 180 °C temperature. As a result, products thermal-oxidative degradation achieve the enhanced polarity and, as consequence, higher values of adhesive strength. In Figure 1, the IR spectrum is presented for amorphous polyethylene modification.

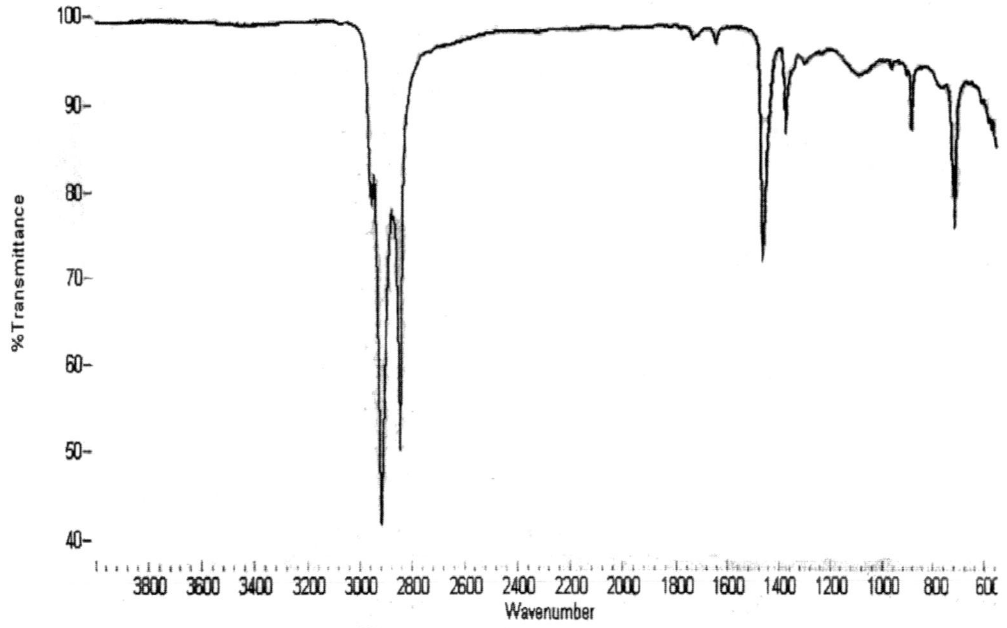

Figure1. IR-spectrums of absorption of modified amorphous polyethylene.

The analysis of curves of the IR spectrum shows that oxidizing processes run rather intensively. In comparison with the original samples, the greatest changes of absorption peaks are observed within 600–1300 cm^{-1} region. This spectrum region characterizes mainly the mixed type of chemical components of the environment. Thus, such groups as ketonic, aldehydic and ester are found in the oxidized amorphous polyethylene that interfaces with non-saturated communications.

Asymmetric ($V_{as}CH_2$) and symmetric stretch vibrations (V_sCH_2) are shown in the vicinity of 2926 and 2853 cm^{-1}, respectively. The methyl group includes two types of bending vibrations. The first type is symmetric bending vibrations at which deformation of C-H bond occurs in a sample phase. The second is asymmetric bending vibrations at which deformation of C-H bond occurs outward the phase. Symmetric bending vibrations (δ_sCH_3) occur near 1375 cm^{-1}, while the asymmetric ($\delta_{as}CH_3$) occur near 1450 cm^{-1}. Also, the intensive absorption is observed in the vicinity of 1200–1420 cm^{-1}, caused by O-H deformation and C-H stretch vibrations. These bands change their position which, however, can be determined by the high intensity of absorption. The existence of polar bond C-O causes the emergence of the intensive absorption band within the range of 1000–1200 cm^{-1} due to the contribution of this group to skeletal vibrations. Moreover, the intensive absorption bands in the region of 1250–1400 cm^{-1} are connected with planar bending vibrations of OH-group [5].

Thus, at the product modification, the quantitative and qualitative composition of polar groups is changed in the medium of oxidized amorphous polyethylene that leads to the increase of the adhesive strength and resistance to the environmental factors.

Table 1 presents experimental data on kinetics of cedar water absorption. These data show that kinetics of water absorption has the explicit induction period caused by the structural properties of cedar.

The calculation and analysis of wood kinetic curves show that the process of moisture transfer proceeds by the diffusive mechanism and can be obtained from:

$$C = a + b\times\exp(-k_1\times t) + c\times\exp(-k_2\times t), \tag{1}$$

where C is the total content of absorbed moisture; k_1, k_2 are constants of diffusion speeds, *a, b, c* are constant coefficients, (b+c) is the maximum moisture content absorbed by the sample, (b+c) is theoriginal moisture content in the sample, *t* is the time period of moisture absorption.

Table 1. Water absorption kinetics at surface cedar processing.

Water-repellentagent	Absorbed water ratio, min					
	5	10	15	20	30	3600
Cellulose acetate solution	0.01	0.01	0.01	0.02	0.04	0.35
Rosin solution	0.02	0.02	0.02	0.03	0.05	0.44
Modified atactic polypropylene	0.02	0.02	0.03	0.03	0.03	0.11
Modified amorphous polyethylene	0.03	0.05	0.05	0.05	0.06	0.21
Test sample	0.63	0.75	0.75	0.75	0.88	1.52

The kinetic equation parameters for the experiment conditions are given in Table 2.

Table 2. Kinetic equation parameters for cedar.

Kinetic parameter values				
a	b	c	k_1	k_2
34	-12	-15	-0,03	-0,05

As Table 2 and kinetic equation (1) show, the process of water absorption is characterized by two diffusion constants that means that water absorption is carried out in two ways. In authors' opinion, it is connected with the existence of capillaries and pores of different sizes and, as consequence, different speed of their filling with water.

It is known that the main indicator of the water-repellent agent efficiency is the value of the angle of contact which, in turn, depends on the surface structure of the samples. The values of the angle of contact of hydrophobic materials under study are given in Table 3.

Table 3. The contact angle of hydrophobic materials depending on the contact point between water drop and sample surface.

Water-repellent agent	Contact angle, min						
	0	5	10	15	20	30	60
Cellulose acetate solution	98	98	98	97	97	96	94
Rosin solution	96	96	96	96	95	94	92
Modified atactic polypropylene	119	119	119	119	118	117	115
Modified amorphous polyethylene	124	124	124	123	123	122	118
Test sample	34	2	0	0	0	0	0

It should be noted, that these results are obtained for a rough wood surface. It is known [6], that besides a chemical composition, the efficiency of water-repellency treatment depends on the surface texture, coating application technique, modes of homogeneous or heterogeneous wetting at which the liquid contacts with the entire surface or its part. To compare, the obtained compositions are applied to a glass substrate. As the experimental data show, the angle of contact increases by 5-10 % as compared to the rough wood surface. Since heterogeneous wetting is used for wood and homogeneous wetting for glass, the phenomena observed are considered as regular and can be obtained from [6]:

$$\cos\Theta = f \times r \times \cos\Theta_0 + f - 1, \quad (2)$$

where Θ andΘ_0 are angles of contact of homogeneous and heterogeneous wetting, respectively; f is the wetted surface area; $r = S_0/S$ is the roughness coefficient; S_0, $S$are true and apparent surface areas, respectively.

According to equation (2), in presence of the surface roughnesses and decrease of values f and r, the angle of contact also decreases. Therefore to achieve a highly hydrophobic state, it is necessary to create a specific surface texture. Different methods of creating such textures can be found in literature. One of them is a low-temperature high-speed plasma etching of a surface that leads to a formation of nanodisperse carbon having a high chemical attraction to organic wood composition.

As a result, not only a surface texture is formed, but also high operational and decorative properties, water permeability decreases, fungus diseases are eliminated, abrasion resistance increases. Moreover, the speed of surface processing is 1–2 s that reduces energy costs and improves cost efficience of the coating creating process [7]. The similar results are also observed when processing various wood surfaces using laser and radiation effects.

Figure 2 illustrates the surface topography of cedar before and after its processing with low-energy plasma.

Low-energy plasma treatment, as shown in Figure 2, eliminates the surface texture and forms closely packed dispersed carbon particles having a certain degree of crystallinity that provides the increase of hydrophobic properties of the wood surface. However, in compositions under study, the ange of contact (Θ) increases insignificantly on the pretreated surface.

Figure 3 presents hydrophobic properties of the surface covered by the cellulose acetate solution for pure wood and preprocessed low-energy plasma.

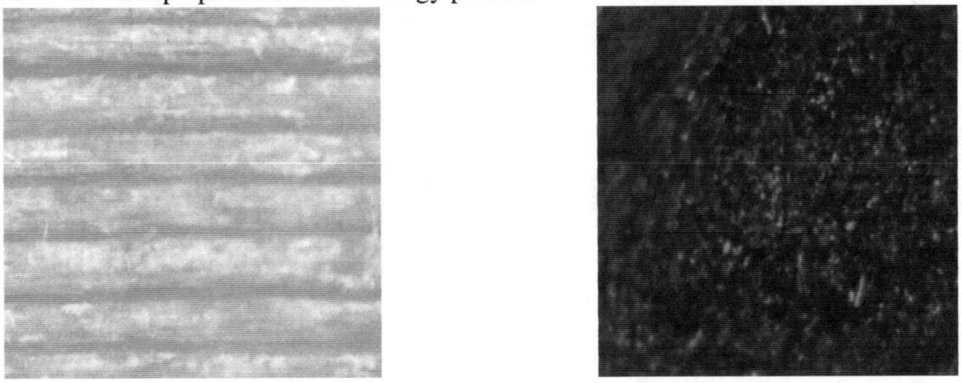

a b

Figure 2. Micrographs ofcedar pure surface (a) and afterlow-energy plasma treatment (b).

It can be explained first, by the fact that at closely packed dispersed carbon particles, irrespective of their size, it is impossible to create highly hydrophobic surfaces [6], and, second, that layers self-organized under a high-energy effect have a plurality of structural defects affecting water repellency. Optimization of the plasma treatment process will allow elimination of these limitations.

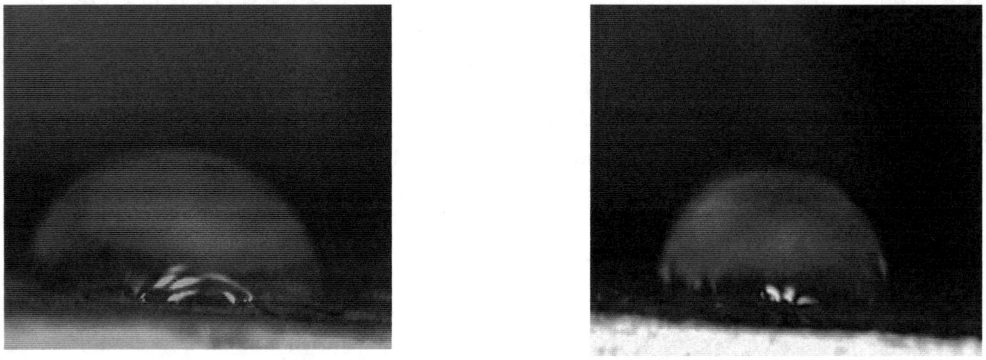

a b

Figure 3. Microphotos of water drops on pure wood surface after water-repellency treatment (a) and pretreated with low-energy plasma (b).

Despite the small value of its density showing the maturity of the pore structure, the three-dimensional wood impregnation with water-repellent agents is not practically observed under common conditions. The maximum moisture penetration comes to fractions of millimeters. In vacuum processing, the depth of penetration in water-repellent agents increases up to several millimeters. This can be explained by a thin-layer structure of cedar, the existence of hydrophobic functional groups in the chemical composition which prevent water-repellent agent molecules from diffusion into the volume of the solid body and due to some other reasons. The effective wood impregnation at great depth is observed when using an oxalaldehyde aqueous solution. However, water-repellent properties are practically absent. In further research it is supposed to use this substance as transport for other molecules possessing water-repellent properties. A scenario when the internal pore surface is modified by groups of the similar origin while the surface by another can be thus implemented. In authors' opinion, the lesser the size of pores the stronger this effect is.

One can assume that the adsorbed layer of the water-repellent agent will be formed by immobilized molecules which have no the possibility of moving as whole, and, in practice, their random distribution will occur within the capillary-porous body. At that, water-repellent agentmolecules can be adsorbed equally possible in any region of the surface, while the angle of contact will be determined just by the nature of end groups.

4. Conclusions

1. It was shown that at a thin-layer coating with water-repellent agent of the sample surface, the values of the angle of contact range between 96 and 98°. Taking into account the chemical compatibility of the suggested compositions with wood components, they can be used while restoration of wood products.
2. It was detected that compositions based on modified amorphous polyethylene and atactic polypropylene are the effective water-repellent agents. Wood surface coating with the suggested compositions allows decreasing water absorption by several times.
3. It was shown that wood processing with oxalaldehyde aqueous solution in vacuum installation allowed the three-dimensional sample impregnation.
4. It was shown that water adsorption by wood occurred by the diffusion mechanism and was obtained from $C = a + b\times\exp(-k_1\times t) + c\times\exp(-k_2\times t)$.

References

[1] Семенов В В 2009 Гидрофобизация древесно-стружечных и древесно-волокнистых плит кремний-органическими мономерами и жидкостями *Химия растительного сырья* **4** 177

[2] Verchinina G S 1989 Применение кремний-органических соединений в строительстве (Москва: Мир)

[3] Под ред. Кротикова 1991 Органосиликатные и кремний-органические материалы в практике строительных, противокоррозионных, защитно-декоративных, ремонтных и реставрационных работ *Материалы научно-практической конференции* 76 с

[4] Дебелова Н Н, Горленко Н П и Подшивалов И И 2009 Технология гидрофобной защиты капиллярно-пористых строительных материалов и изделий *Вестник ТГАСУ* **2** 121

[5] Smith A 1982 *Прикладная ИК -спектроскопия* (Москва: Мир)

[6] Boynovich L B and Emelyanenko A M 2008 Гидрофобные материалы и покрытия: принципы создания, свойства и применение *Achievements of chemistry* **77** 619

[7] Volokitin G G, Volokitin O G, Shekhovtsov VV, Maslov Ye A and Malyuga N V 2010 Plasma Process Modeling of Wood Surfaces *Vestnik of TSUAB* **3** 220

IOP Conf. Series: Materials Science and Engineering **71** (2015) 012010 doi:10.1088/1757-899X/71/1/012010

Application features of additives based on metakaolin in concrete

A A Kirsanova[1] and L Y Kramar[1]
[1]Federal State Budgetary Educational Institution of Higher Professional Education South Ural State University (National Research University), Chelyabinsk, 454080, Russia

E-mail: kramar-l@mail.ru

Abstract. The present paper is devoted to the influence of additives based on metakaolin (U-YF, UM-YF and YF-UMD) on speed concrete strength development in the early stages of concrete hardening, as well as the strength increase in 28 days. The authors have proved that metakaolin gauging in concrete should not exceed 3%. Introduction of 5% of metakaolin or more entails the fault in concrete strength in the later stages of concrete hardening and decreases its resistance to the influence of sulfate and frosty environments. The most effective of the developed additives are UM-YF and UMD-YF which provide high sulfate and frost resistance to the concrete (up to 800...1000 cycles). The above mentioned influence of additives on concrete properties is connected with an intended formation of structure of the cement matrix of concrete that is resistant to various aggressive environments.

1. Introduction

For highly efficient construction materials active mineral additives are widely used including granulated blast slag, silica fume, rice hull ash, and other additives; metakaolin is of particular interest lately.

Metakaolin (MTK) is a metastable reaction product of dehydroxylation of kaolin at the temperature of 650 – 850 ºC. The research on the influence of additives on the properties of MTK cement compositions was initiated in the middle of the twentieth century, but until now various issues on its application are still unclear. Disagreement among researchers relates to the formation of various compositions of hydrate neoplasms. Taking into account the defect structure and the weak link of the aluminum ions in the lattice, we can assume that in the presence of water MTK dissociates into aluminate and silicate parts, wherein there is a great probability of recovery of the lattice with the addition of different elements to form compositions characteristic of the "Al_2O_3–SiO_2–CaO–H_2O" system. Thus, F.L. Glekel in his work has noted that calcium hydroaluminates of various basicity and composition, gidrogelenit, calcium hydrosilicates, which are developed in dependence on the concentration of calcium ions in the liquid phase, may be the pozzolanic reaction products in the presence of MTK [1]. According to [2,3] and other researchers the $Al(OH)_3$ gel may occur in the system "Al_2O_3–CaO–H_2O" at an early stage of interaction which thereafter crystallizes in highly hydroaluminates of calcium, hydrogarnets, hydrogehlenits, and calcium hydrosilicates (depending on the acidity and other factors). It is not feasible to identify patterns in the formation of the phasic composition of cement matrix in the presence of addition of MTK because the "Al_2O_3-CaO-SiO_2-H_2O" system is quite complex and the formation of connections is determined by the conditions of the reaction, as well as the quality of ingredients and manufacturing technology. Besides the phasic

composition, there are disputable differences in MTK gauging. Great number of researchers indicate the MTK dosage as about 10 [4, 3], 20 [5], or even 30% [6] of the mass of cement which they believe will improve the instant and final strength of cement formations. In addition, S.A. Zakharov [4] noted that MTK gauging 1.5 ... 2% of the whole weight of cement is sufficient to improve the water resistance cement compositions. Moreover, statements of different researchers on the impact of addition of MTK on resistance to aggressive environments are ambiguous. Metakaolin is an aluminosilicate additive which includes Al_2O_3: SiO_2 = 1:2, while the state standards of Russia 55224-2012 and 22266-94 strictly regulate limit of C3A content in cement – not more than 7%, in sulphate-resistant cements – not more than 5%. Limitation of the amount of aluminate in cement is determined by the lower opportunity to form metastable hydroaluminates of calcium in the cement matrix [2, 7]. As a conclusion, the researchers note that MTK reduces the consumption of cement in concrete, accelerates the hydration and hardening of cement compositions resulting in a high, early, and final strength of concrete and reveals the increased water requirement, so that it should be used in conjunction with water-reducing admixtures.
The aim of the current study is to investigate the possibility of obtaining cement compositions with high strength and sulphate and frost resistance by modifying the complex additives based on MTK.

2. Materials and methods

Complex additives previously developed on the basis of MTK - U-YF (1.5-3.5% of MTK, 0.6% of SP-1), UM-YF (1.5-3.5% of MTK, 5% of SF, 1.2% of SP-1) and the UMD-YF (1.5-3.5% of MTK, 10% of SF, 1.2% of SP-1) were used. For the purpose of research the authors used MTK produced by 'Plast-Refey' in accordance with the specification TU 5729-095-51460677-2009, silica fume (SF) of Novokuznetsk factory (TU 5743-048-02495332-96) and superplasticizer SP-1 produced by "Polylayer" Novomoskovsk (TU 5870-005-58042865-2005). Moreover, the authors used the cement CEMI 42.5 produced by "Nevyanskiy cementnik"; quartz sand from Belonosovsky field by the state standard of Russia 8736-93 and gravel of 5-20 mm fraction in accordance with the state standard of Russia 8269.0-97. The authors carried out the testing of the strength characteristics of the cement and concrete in accordance with the state standard of Russia 310.4-81 and 10180-2012; specific surface area of cement matrix was evaluated by the BET method; the porosity was determined according to the state standard of Russia 12730.4-78; frost resistance was assessed by the third accelerated method by repeated freezing and defrosting in accordance with the state standard of Russia 10060.2-95. To study the water and frost resistance in accordance with the state standard of Russia 24211-2008 the authors developed heavy concrete samples. To study the structure of cement matrix the authors used the scanning electronic microscope JeolJSM-700 1F. The sulphate resistance was tested in accordance with the state standard. The solution concentration is determined at 1000 mg/l = 10 g/l. Water resistance is determined by the accelerated method according to the state standard of Russia 12730.5-84.

3. Results

Study of strength characteristics of cement matrix (Figure 1) revealed that the introduction of additives of U -YF leads to an increase in its strength for 2 days of hardening in 2 times compared to no additives composition. By the 28 days of hardening with the use of complex additive of U -YF strength cement matrix above 80% compared to non-additive composition. UM-YF and UMD-YF modifiers slightly underestimate the strength development in 7 days compared to the addition of U -YF, but by 28 days the strength of concrete with the abovementioned additives is increased by 10%. Application of 5% of MTK is accompanied with the loss of strength of cement in the later stages of hardening which is associated with recrystallization aluminate metastable phases in a stable one and is consistent with Kuznetsova's data [2]. Application of designed complexes reduces open porosity of cement matrix (Figure 2), increases the specific surface area of hydrated phases and creates a more dense packing. This is due to the filling of intergranular space of highly dispersed mineral supplements and decrease the W/C ratio, as well as due to the formation of hydrated phases in the form of a tightly welded plates, hexagonal and cubic hydroaluminates, hydroaluminosilicates and that is confirmed by electron microscopy analysis (Figure 3).

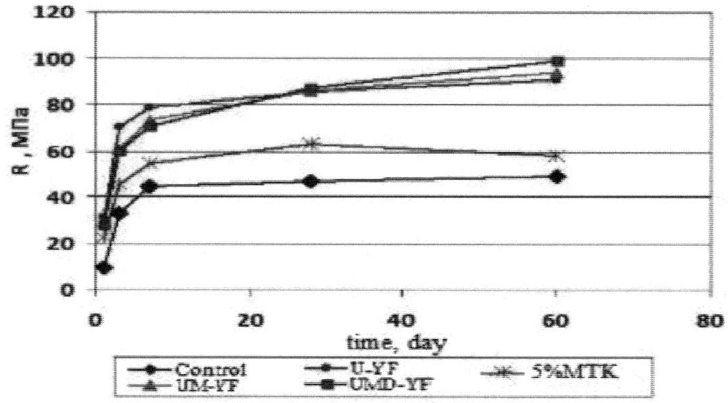

Figure 1. The compressive strength of cement matrix.

Study the effect of supplementation with MTK possible to establish the following. Application of MTK at 2 ... 3% by weight of cement does not lead to a decrease in sulfate-resistant cement and has level similar to one without additives.

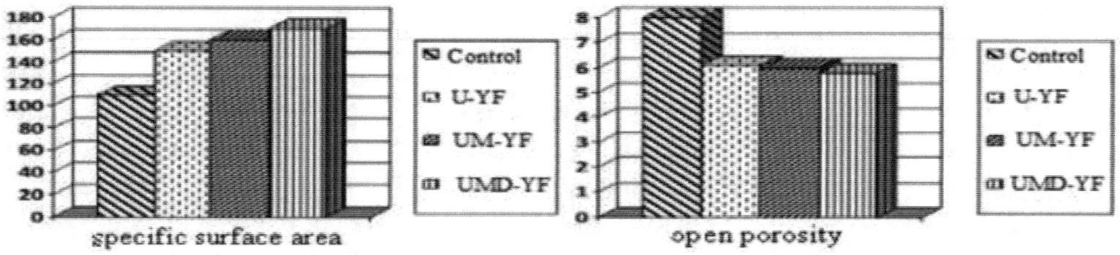

Figure 2. Specific surface area (m^2/g) and an open porosity (%) formed of cement matrix.

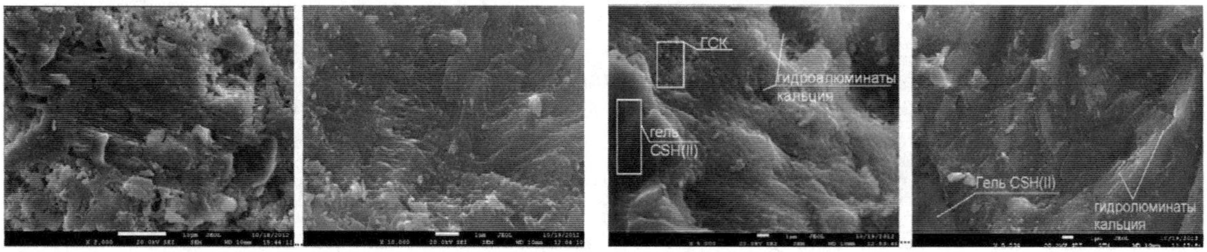

A) B) C) D)

Figure 3. Structure of cement matrix: a) Control; b) with the addition of U-YF; c) with the addition of UM-YF; d) supplemented with UMD-YF.

Introduction of additives 5% by weight of cement MTK (Figure 4) causes a decrease in cement matrix sulphate resistance Kh.s. <0.9 after 1 month of testing, sulphate corrosion due to the interaction of sulfate ions with aluminate phases and calcium hydroxide cement matrix to form ettringite and gypsum.

With the addition of 5% in the MTK structure of cement is constant recrystallization aluminate phases under the influence of sulfates, which leads to disruption of contacts in the cement matrix and strength to discharges [8]. Introduction of additives to complex "MTK + SP-1" promotes the formation of cement matrix and stable hydroaluminates hydrogarnets but this creates a more dense structure of cement matrix (Figure 3, d) due to active hydration, all this leads to a reduction in permeability and increased resistance to sulphate corrosion (Figure 4). The study revealed a significant waterproofing concrete complex additives influence on its seal, by accelerating the hydration of cement hydrate phases and increase in cement matrix. This effectively resulted in a decline of porosity and water resistant concretes provided when administered supplements U-YF brand W18, using complexes UM-YF and UMD-YF over W20, whereas no additives and composition with 5% MTK brand waterproofing had no more than W8.

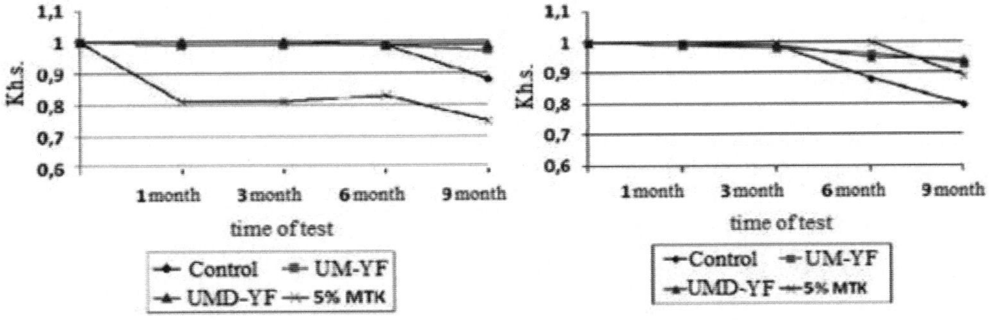

A) Changes in the flexural strength B) Change in compressive strength

Figure 4. Effect of complex additives on Kh.s. in sulfate environment conditions.

Application of MTK without 5% super plasticizer increases frost concrete brand (F 300) in comparison with non-additive composition (F 200), which is associated with accelerated cement hydration process and therefore a seal structure.

Developed complex additives - modifiers can significantly improve the frost resistance of concrete. Since the introduction of additives have U-YF hardiness increases by three times (F 600) for cement compared with non-additive composition, the use of additives UM-YF and UMD-YF improves frost resistance of concrete to F800-1000. All this testifies to the active influence of additives included in the complex formation of frost on concrete structure which is due to compaction and reduction of open porosity, and due to the formation of hydrated phases presented as hydro silicates calcium (GSK), resistant to cyclical changes in the acidity of the medium pore space reduction leaching of portlandite under cyclic temperature changes and facilitate stress relaxation during thermal cycling.

4. Conclusions

1. MTK dosage must be strictly limited to the production of concrete with high demands on performance hardiness and resistance to sulphate corrosion. MTK should be administered no more than 3%;
2. Developed additives are effective accelerators providing increased strength at 2nd day of curing over 50% compared to the strength at 28 days with no additives composition.
3. The most effective of the developed additives that provide high resistance to frost and sulfate resistance of concrete are UM-YF and UMD-YF, allowing to get the hardiness of F800…1000 which is connected with the peculiarities of formation structure of concrete.
4. U-YF, UM-YF and UMD-YF additives can be used for fast setting and high-strength concrete with ordinary cement at normal hardening in the field or in the factory without heat-moisture treatment or with low-temperature heat treatment to 50 °C.
5. Complex supplements of UM-YF, UMD-YF allow producing effective concretes with high strength and durability and can be used in the production of concrete for hydraulic structures, road construction, bridge construction and in the construction of other important structures.

References

[1] Глекель Ф Л 1975 *Физико-химические основы применения минеральных добавок* 198
[2] Kuznetsova T V and Talaber J 1988 *Alumina cement* 272
[3] Dworkin L I 2007 *Metakaolin in mortars and concretes* 215
[4] Захаров С А 2007 Высокоактивный метакаолин - современный минеральный модификатор цементных систем *Строительные материалы* 56-57

[5] Justice J M 2005 *Evaluation of metakaolins for use as supplementary cementitious materials* (Georgia Institute of Technology) 149
[6] Newman J and Choo B S 2003 *Elsevier Advanced Concrete Technology (Constituent Materials)* 280
[7] Kuznetsova T V 1986 *Aluminate and sulfoaluminate cements* 208
[8] Moskvin V M, Alekseev S, Ivanov F M and Guzeev E A 1980 *Corrosion of concrete and reinforced concrete, methods of protection* 536

System approach to peat raw materials for production of building materials

N Kopanitsa[1], A Kudyakov[1], M Kovaleva[1] and G Kopanitsa[2]
[1]Tomsk State University of Architecture and Building, Research Institute of Building Materials, Tomsk, 634003, Russia
[2]National Research Tomsk Polytechnic University, Institute of Cybernetics, Tomsk, 634050, Russia

E-mail: kopanitsa@mail.ru

Abstract. The paper presents systematic approach to the definition of requirements to peat raw materials to use them as a basis for building materials. The introduced approach to the classification of peat based on the composition and structure is suitable for research and development of building materials and technological processes. The approach allows using the same technique for peat as for traditional raw materials.

1. Introduction
The intention to increase energy efficiency of buildings and operation of houses requires research and promotion of building materials of higher quality at lower cost applicable in the building envelopes. Thus, production of building materials made of local environmentally friendly raw materials becomes demanded. Russian raw materials base is characterized by significant amount of peat which is a renewable raw material. In its natural state peat has low heat conductivity, porosity and antiseptic properties which makes it efficient use in construction.

Technology of peat based building materials production which is being developed by the authors takes into account the potential of peat due to its composition and structure and ability to regulate these properties. Peat's compound consists of organic and organic-mineral complexes and by modifying these complexes improvement of the binding properties of peat and increase the speed of formation material's composition is reached. Therefore peat can find wide application in production of wall structures, heat insulation building materials and multifunctional additives that provide strength and hydrophobic properties for cement-based building mixtures. Correspondingly, theoretical study, development and implementation of the techniques for modification of peat and peat based mixtures are important. These will improve the technologies of peat based building materials production. A systematic approach will provide good basis for determining quality characteristic for the walling building materials production on all stages of their life cycle.

2. Materials and methods
Analysis of the performed studies shows that the most worthwhile application of modified peat is as an active component (modifying additive, binding) in composite building materials.

Artificial structuring of peat can be treated as one of the ways of controlling the properties of natural dispersed materials as it affects the strength, hydro physical and structural properties of materials. Figures 1 and 2 present the approaches to the regulation of structure and properties of peat based composite (micro and macro level of peat-wooden composites) walling materials.

The formation of peat based composite materials with predetermined and predictable properties is possible if we know how the modification of single structural elements and their interaction affects the process of the structure's formation

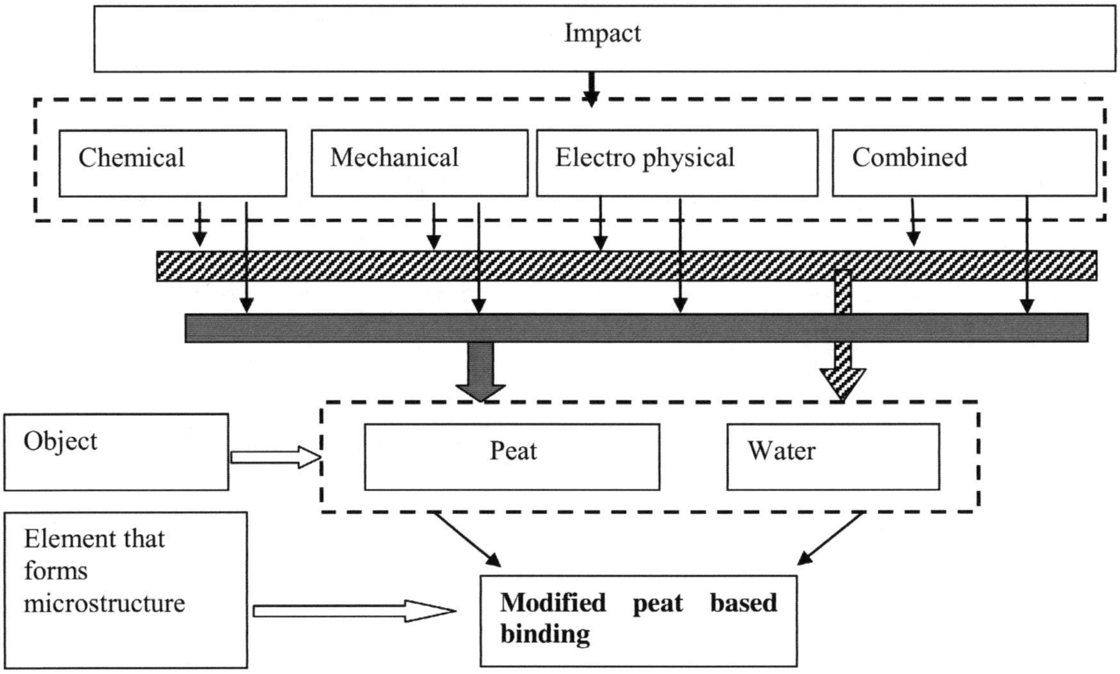

Figure 1. Types of modifying impact on the components of peat based walling materials to form a microstructure.

Thus control over the structures of peat compositions during the formation of energy efficient materials as well as control over the end-use requirements is associated with different ways of modification that can be performed on the different steps of the production cycle: raw material – molding sand – product.

The presented models are resulted from the analysis and generalization of multiple experimental studies. This paper presents and proves a dual-component (filler and peat based binding) model of a composite material. The filler provides a special core while peat based binding material glues the filler's grains and fills the core's spaces. The properties of the peat based binding can be regulated by modifying of peat.

Peat contains a large number of active functional groups of various compounds which allows controlling the structure and forming peat based composite materials with predictable and controllable properties. This hypothesis is based not only on the knowledge of the separate components of natural raw material, but on the regularities of interaction of functional groups with different arming and core forming components of the composition that are widely applied in building materials.

One of the main factors of the formation of rational hardening structures in composite building materials is compatibility of active (binding) and core components. The compatibility considers a material compound, structure and deformation properties.

The selection of filler was performed based on the following criteria: (i) low average density; (ii) high porosity; (iii) strength much higher than of the peat based binding; (iv) similar deformation properties; (v) similar chemical and group compound.

The analysis resulted in the selection of wooden filler as a core formatting material for the peat based binding. At low temperature wood behaves similar to pear regarding its material compound and temperature deformations [1-4].

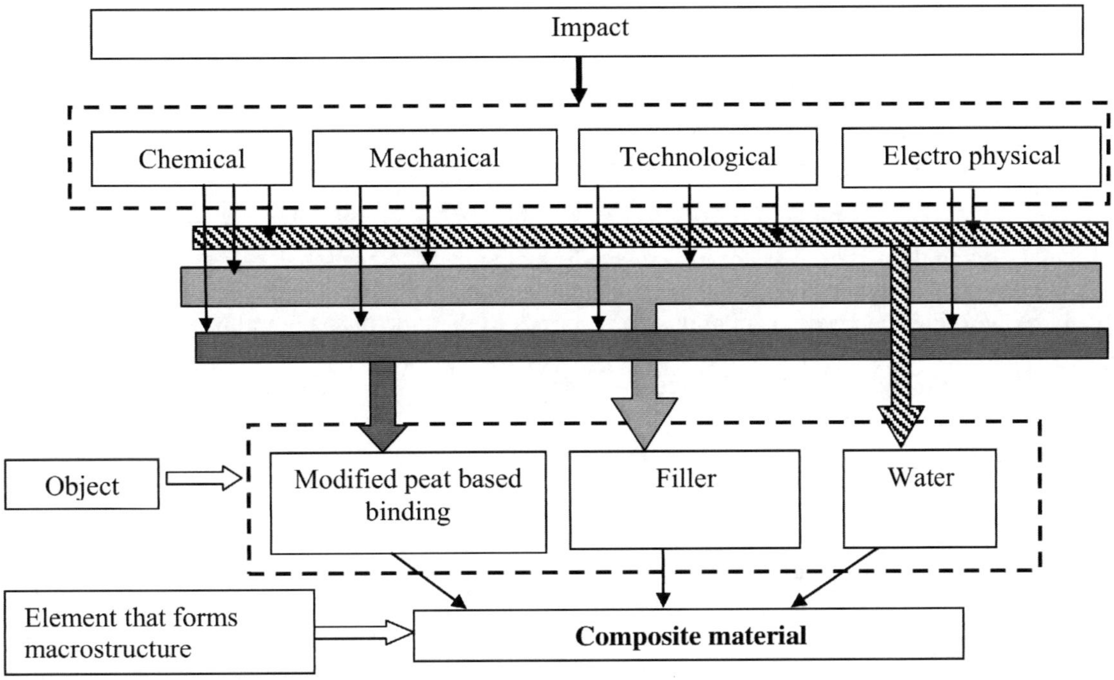

Figure 2. Types of modifying impact on the components of peat based walling materials to form a macrostructure.

3. Research model

It is necessary to mention that the proposed model considers that the binding and the filler contain lignin, humin substances, cellulose, sugar and mineral materials. The structures of the peat and wood are heteroporos, polydisperse, multilayered. The structures are related to the coagulation i.e. moving high elastic structures, where the interaction between single elements is driven by the hydrogen bonds of water molecules. These features allow physical and chemical interactions during the structure formation of the homogeneous composite materials.

The specific features of the peat-wooden composites in comparison to cement based concrete are: the larger phase separating surface; the number of physical and physical-chemical contacts in the volume unit of the material; higher porosity; geometrical and size characteristics of pores that change during preparation, formation and hardening.

The particles of the filler of different size are being enveloped and glued by the peat binding during the formation and drying processes. The peat mass obtains binding properties as a result of mechanical-chemical activation. This provides strong connections between structural components.

Smaller fractions of the filler and peat binding make a condition for the core to be glued. The proposed physical model pays much attention to the physical-chemical interaction of watered particles of the binding and filler. The similar model was used for cement based concretes in [5,6].

The above mentioned processes have not been studied for the cases when peat based binding is used as gluing element and wood is used as filler. The proposed model of peat based composite material allows producing a macro porous heat insulation material [7-11]. This material can be potentially inconsistent with the environment especially if humidity changes which can cause change in main specifications: strength, deformability, heat conductivity, frost resistance. Additional structural elements such as micro pores and arming fibers are being formed to increase the quality of the material (Figure 3).

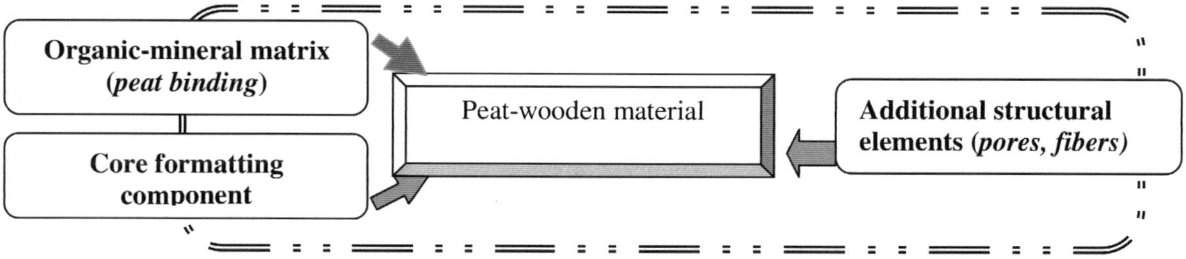

Figure 3. Physical model of a peat-wooden material.

The main controlling processes for structures' formation of composite peat-wooden materials are similar to cement systems and correspond to the general principles of material science [12-16]:

1. Matrix and filler components relations corresponds to the type of the binding. This provides the optimal ratio of the components according to the filler's strength criteria considering type of the filler, its dispersity, morphology of its particles and its compound.

2. Main structure formation processes for the peat-wooden composite materials develop in time. The following phenomena can be observed: (i) formation of a special core from the filler's grains; (ii) micro reinforcement of a peat matrix; (iii) formation of structural mechanical, physical, mechanical-chemical bindings in the chain "peat binding – filler - bulking agent" depending on the mineral-organic ration of peat.

3. Balance of structural bindings at formation of composite peat-wooden material was provided by a contact interaction by: (i) mechanical linkage; (ii) capillary contraction.

4. The following control factors are applied at the formation of the required structure of peat-wooden materials: (i) peat-binding-filler ratio; (ii) liquid body percentage in the mixture; (iii) conditions and options forsol, gel, and colloid states introduction, which causes formation of condense and crystal contacts in the system.

Structural model of peat-wooden material depends on the matrix and filler components as well as reinforcing elements as presented in Figure 4.

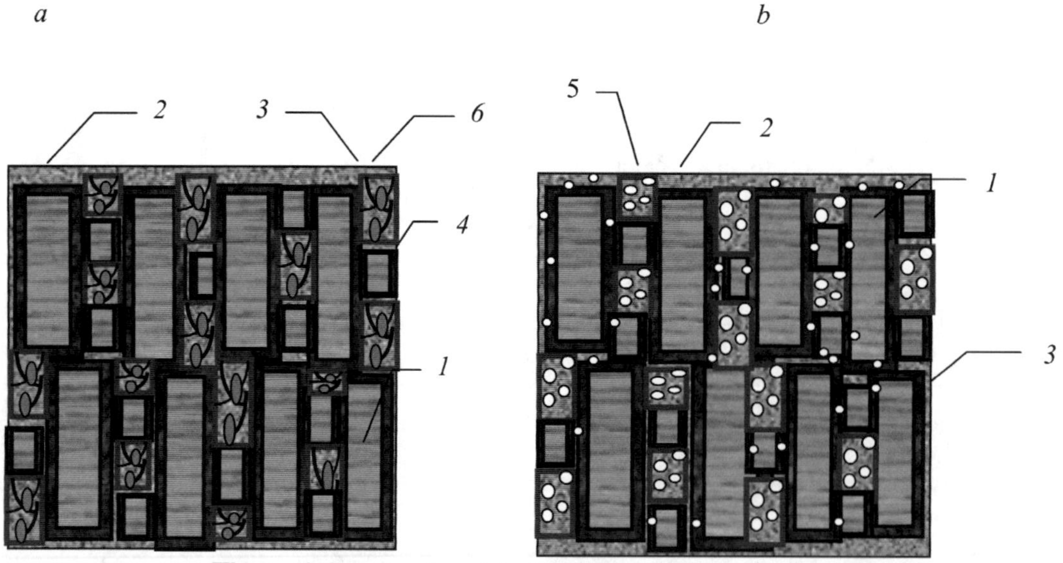

Figure 4. Structural model of a composite peat-wooden material:
a– construction-heat insolation; *b*– heat insolation; *1* – wooden filler; *2* – peat binding; *3* – contact zone; *4* – fibers; *5* – pores; *6* – micro filler

A basal binding system can be used for heat insolation peat-wooden materials. While porous system is applied for constructing heat insolation materials. Porous system requires that the extra reinforcing components such as micro fibers and fillers are added to matrix. Basalt binding is formed while peat binding porisation. The proposed structural model allowed proving a process that controls structure and properties of energy efficient peat-wooden walling materials. The control is possible on micro and macro level (Figure 4).

4. Conclusion

The paper proposed proved two component physical model of heat efficient insolation material. Wooden filler provides a spacious core. Peat binding glues grains of the filler to fill the spaces of the core. The properties of the peat binding can be controlled using different modification techniques. A selection of wood as a core formation material for peat based composite materials was scientifically proved. Peat and wood have chemical and structural affinity.

Structural model of peat as a raw material for building products was introduced. The model allows proving a selection of the way to modify peat and provide possibility to produce building materials with demanded properties. It was proved that the main peat raw selection criterion is a ratio of organic and mineral functional groups.

References

[1] Никитин В М 1978 Химия древесины и целлюлозы (Москва: Лесная промышленность)
[2] Fengel D and Wegener G.1984 Wood (Chemistry, Ultrastructure, Reaction) (Berlin) 512
[3] Yaman S Pirolisis of biomass to produce fuels and chemical feedstocks 2004 *Energy Conversion and Management* **45** 651–671
[4] Demirbas A Combustion characteristics of different biomass fuels 2004 *Progress in Energy and Combustion Science* **30** 219–230
[5] Ахвердов И Н 1981 Основы физики бетона (Москва: Стройиздат) 464
[6] Кудяков А И, Смирнов А Г и Петров Г Г 1987 Проектирование и использование заполнителей с оптимальной межзерновой пустотностью в *Известия вузов* **7** 135–138
[7] Копаница Н О, Кудяков А И и Ковалева М А 2009 Торфодревесные теплоизоляционные строительные материалы (Томск: ССТ) 183
[8] Копаница Н О, Саркисов Ю С и Рыжников А Б 2002 Эффективные строительные материалы на основе модифицированных торфов *Строительные материалы* **7** 12–14
[9] Копаница Н О и Калашникова М А 2007 Исследование вяжущих свойств нижних торфов при производстве теплоизоляционных материалов *Вестник ТГАСУ* Vestnik TGASU **1** 111–116
[10] Касицкая Л В, Саркисов Ю С и Горленко Н П 2007 Торфяные ресурсы Томской области и пути их использования в строительстве (Томск) 290
[11] Копаница Н О, Кудяков А И и Саркисов Ю С 2007 Рациональное использование торфа в строительных технологиях *Строительные материалы* **12** 32–34
[12] Кудяков А И и Копаница Н О 2005 Системный подход при разработке материалов для многослойных ограждающих конструкций *Строительные материалы* **12** 12-14
[13] Копаница Н О 2010 Структурное моделирование свойств торфа как сырья для производства строительных материалов Вестник ТГАСУ **2** 162–168
[14] Саркисов Ю С и Кузнецова Т В 2009 Синергетика и принципы неравновесного строительного материаловедения *Техника и технология силикатов* **4** 2–6
[15] Верещагин В И, Рихванов Л П и Саркисов Ю С 2009 Синергетические принципы создания строительных композиционных материалов полифункционального назначения *Известия ТПУ* **315** 12–15
[16] Чернышов Е М, Потамошнева Н Д и Кукина И В 2010 Строительные композиты с контактно-конденсационными нано- и микроструктурными матрицами из искусственного портландита **2** 308–321

IOP Conf. Series: Materials Science and Engineering **71** (2015) 012012 doi:10.1088/1757-899X/71/1/012012

Foam concrete of increased strength with the thermomodified peat additives

A I Kudyakov [1], N O Kopanitsa[1], Ju S Sarkisov [2], A V Kasatkina[1] and I A Prischepa[1]
[1]Tomsk State University of Architecture and Building, Department of Construction Engineering and Technology, Tomsk, 634003, Russia
[2]Tomsk State University of Architecture and Building, Department of Chemistry, Tomsk, 634003, Russia

E-mail: kopanitsa@mail.ru

Abstract. The paper presents the results of research of foam concrete with thermomodified peat additives. The aim of the research was to study the effect of modifying additives on cement stone and foam concrete properties. Peat additives are prepared by heat treatment of peat at 600 °C. Two approaches of obtaining additives are examined: in condition of open air access (TMT-600) and in condition of limited air access (TMT-600-k). Compressive strength of a cement stone with modifiers found to be increased by 28.9 – 65.2%. Introducing peat modifiers into foam concrete mix leads to increase of compressive strength by 44–57% at 28-day age and heat conductivity of foam concrete decreases by 0.089 W/(m•°C).

1. Introduction

Enhancing production of foam concrete with natural hardening and broadening of its application in construction has significant importance due to the demand for energy efficiency and necessity of energy intensity reduction while walling materials production [1]. Performance of foam concrete can be enhanced by ensuring the required strength while reducing thermal conductivity of material when selecting raw materials, composition design, preparation of raw mixture, products molding that can be achieved by maximum saturation of the material's volume with air bubbles. This can be reached by application of foam with the bubbles of size ensuring tight packaging or by saturation of sand-cement partitions with microporous disperse agents increasing the strength of cement stone [2]. Problem can be solved by introducing porous organic-mineral active modifiers into cement-sand mixture. Previous research resulted in development of method for production of modifiers based on peat and examined their influence on the properties of cement composite materials [3].

Specific feature of raw material base of many Russian regions is the presence of large reserves of peat. Composition of peat is various and represented by numerous functional groups, including both organic and mineral compounds, which provide different ways of its processing to obtain components of required composition and properties for the production of building materials of different purposes [4, 5].

The main purpose of current research is the development of formulations and scientific substantiation of technological methods for thermomodified peat additives and foam concrete of increased strength with the additives production.

IOP Conf. Series: Materials Science and Engineering **71** (2015) 012012 doi:10.1088/1757-899X/71/1/012012

2. Materials and methods

For the study of foam concrete following initial materials were used:

- Portland cement from "Topkinsky Cement" plant, ЦЕМ I 42,5H type - cement complies with requirements of All Union State Standard (GOST) 31108-2003;
- Silica sand from Kudrovskoe field, size modulus 1.44 - sand complies with requirements of Russian National State Standard (GOST) 8736-93;
- Foaming agent Benoteh PB-C;
- Additives based on peat: TMT-600 and TMT- 600-k.

Peat organic-mineral additives TMT-600 and TMT-600-k were obtained by heat treatment of peat of low type at 600 °C for 2 - 4 hours. After the heat treatment the resulting material was ground up in a planetary mill to the specific surface area 161 m^2/kg. Preparation of foam concrete mixture was carried out by one-step method in a laboratory foam concrete mixer with volume of 10 liters. A mixture of sand, cement and water (80 %) was stirred for 2 minute. Then the foaming agent was injected into the mixer with 20% of mixing water and the mixture was stirred for an additional 3 minute. Ready foam concrete mixture was poured into metal molds 10x10x10 cm. Formed foam concrete samples were kept in the standard wet conditions for testing. Testing and evaluation of the quality of foam concrete were carried out according to the requirements of All Union State Standard (GOST) 25485-89.

A hydraulic press was used to determine the compressive strength. IR of spectroscopic studies of cement stone samples with addition of peat modifiers were performed on FT-IR spectrometer Nicolet 5700, and electron microscopic studies were carried out on the microscope SEM SamScan-4. Electron microscopic studies of foam concrete samples were carried out on the microscope Philips SEM 515.

3. Results

Technology for the production of peat modifiers was earlier developed by the authors of current article. It is established that during peat thermal processing in open air conditions at 600 ° C (TMT-600) organic-mineral complexes in the form of fibers are formed. These complexes provide the effect of micro-reinforcement of cement stone structure. The maximum increase of compressive strength of cement stone (to 28.9%) is observed when the content of peat modifiers is 6 % by mass [6].

In order to improve the efficiency of organic-mineral additive method for production of peat modifier it was investigated on laboratory installation in conditions of limited air access. The pre-drying peat was heated in the sealed reactor to 600 °C for 2 - 4 hours. The resulting product was ground up in a planetary mill to the specific surface area 161 m^2/kg. The additives were introduced into the cement in an amount of 0.5% of cement weight. As a result of reactions of hydrolysis, hydration, coordinate restructuring of organic compounds, changes of correlation between the functional groups (amino-phenol, carboxylic et al.) both quantitative and qualitative nature take place, nanoscale structures of fulleroid type with high surface energy are formed. It determines activity of the synthesized additive TMT- 600 towards cement.

IOP Conf. Series: Materials Science and Engineering **71** (2015) 012012 doi:10.1088/1757-899X/71/1/012012

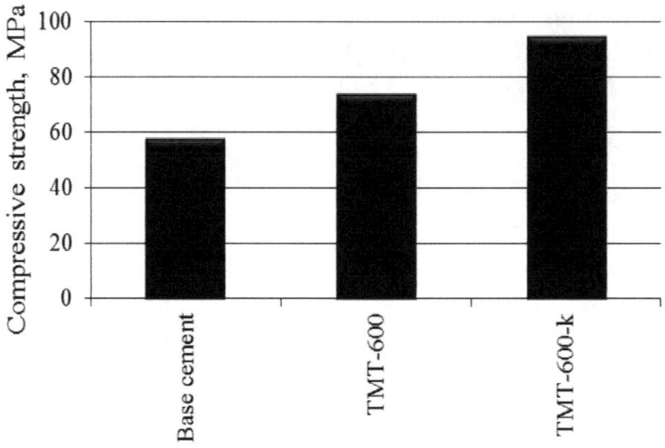

Figure 1. Strength characteristics of cement stone samples modified by additives TMT-600 and TMT-600-k.

Figure 1 presents the results of research of strength characteristics of cement stone modified by peat additives obtained in various ways. It is established that the effect of the compressive strength improvement of the cement stone by applying additives TMT-600-k is higher than by applying similar additives obtained in open air condition. Thus, at a content the additive TMT-600 in an amount of 6 % of the cement mass compressive strength increased by 28.9%, while at a content of the additive TMT-600-k of 0.5 % strength of cement stone increased by 65.2%.

To assess the impact of peat additives TMT-600 and TMT-600-k on the foam concrete strength in different periods of hardening, composition of foam concrete mix previously developed for industrial applications was used (per 1 m^3): 416 kg of cement; 265 kg of sand; 0.9 kg foaming agent, and 230 L of water. Peat modifiers are introduced in the amount of 25 kg. When introducing additives TMT-600 and TMT-600-k in foam concrete mixture compressive strength of foam concrete increased by 45-56 % in 3-days age, by 33-46% in 7-days age, and by 44-57 % in 28-days age (Figure 2).

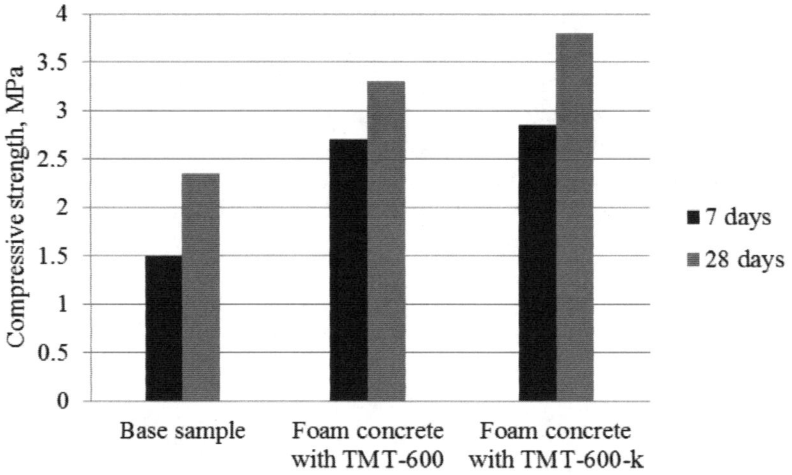

Figure 2. Changing of the strength of foam concrete modified by additives.

For the improvement of energy efficiency of the foam concrete applied to arrange thermal protective enclosing structures of buildings structural porosity and thermal conductivity are essential. The research results of the effect of additives TMT-600 and TMT-600-k on the foam concrete porosity are shown in Figure 3.

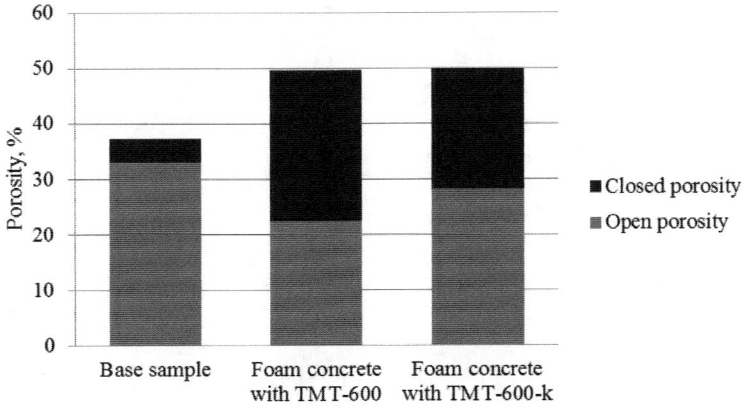

Figure 3. Influence of the additives TMT-600 and TMT-600-k on the foam concrete porosity.

The results of the data analysis showed that the total porosity of the foam concrete modified by peat additives TMT-600 and TMT-600-k increased up to 12 %, compared to basic samples without additives. At the same time the samples with peat additives are characterized by prevalence of closed porosity of 27%. Thermal conductivity of modified foam concrete with peat additives decreased by 14% with insignificant decrease of average density (Table 1).

Table 1. Influence of additives TMT-600 and TMT-600-k on the foam concrete properties.

Composition	Average density, kg/m^3	Thermal conductivity, W/mK
Basic sample	757	0,105
Foam concrete sample with TMT-600	733	0,092
Foam concrete sample with TMT-600-k	730	0,089

4. Discussion

The study results showed that peat modifiers significantly accelerate the process of structure formation of cement stone and porous cement-sand mortar of foam concrete. Thus, at a lower dose (0.5%) additive TMT-600-k provides increase of foam concrete strength at 28-days age almost two times higher than additive TMT-600 at a dosage of 6%. To explain the structure formation mechanism of cement stone with peat additives the IR spectroscopic studies of cement samples were conducted at 28-days of hardening (Figure 4.).

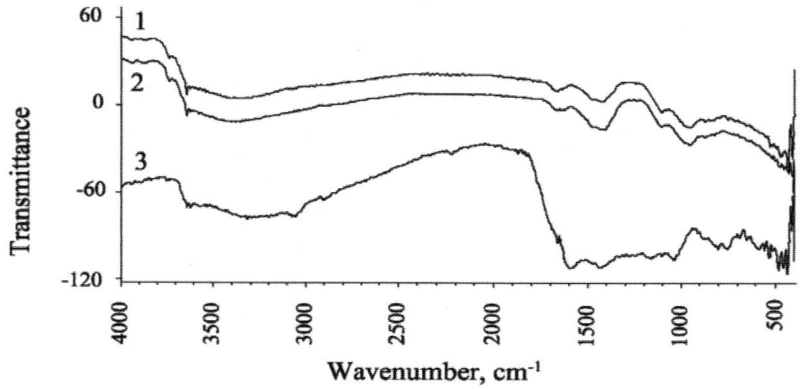

1 - Base cement

2 - Base cement + TMT-600

3 - TMT-600

Figure 4. IR spectrogram of cement stone and modifying additive.

On the IR spectrum following absorption bands were identified:

- bands in the range of (3500-3400) cm^{-1}, due to stretching vibrations of the free group OH in the adsorbed water;
- shift of the absorption bands of OH groups in the range of (3940 – 3800) cm^{-1}. Offset band of OH groups is sensitive to the organic compounds adsorption and the shift, broadening and an increase of the integrated intensity of the surface OH groups band is the spectral manifestation of the interaction of molecules with minerals active hydroxyl groups, which is carried out through the formation of new hydrogen bonds;
- bands at 3400 and 3258 cm^{-1}, should be attributed to the asymmetric valent and the first overtone of the bending vibration of the molecules adsorbed water.

According to the analysis, results of IR spectra of the cement stone with additives and of the base cement stone at the age of 28-days hardening are identical. The results of IR spectroscopic analysis indicated that the peat additives are actively involved in the process of cement stone structure formation only on early terms of hardening.

Figure 5. Electron micrographs of the cement stone with additive TMT-600-k.

In the electron micrographs of the cement stone with additives TMT-600-k (Figure 5) a significant amount of fibrous framework structures was identified, i.e. interwoven fibers of mineral or organic-

IOP Conf. Series: Materials Science and Engineering **71** (2015) 012012 doi:10.1088/1757-899X/71/1/012012

mineral composition thickness from 552 nm to 10 microns and a length of up to 100 microns. The fibers have randomly oriented structure and it is likely that their presence in the cement matrix provides a reinforcing effect, which leads to an increase in strength. Fibrous inclusions have a homogeneous dense border with cement stone. Thus, the introduction of additives TMT-600 and TMT-600-k into the cement matrix ensures reinforcing and structuring effect in cement stone.

According to the study results of the foam concrete porous structure with an average density of D800 both the base sample (Figure 6 a) and sample with the addition of TMT-600 (Figure 6 b) the influence of peat additives on the pore structure of foam concrete was established by means of electron microscopy method. Foam concrete with additives predominantly has closed pores unified in size (mainly 0.05 mm) and evenly distributed over the volume. General and closed porosity has increased by 27 %. Decrease of pore size contributes to reduction of foam concrete thermal conductivity from 0.105 W/ (m·°C) up to 0.09 W/(m·°C), i.e. by 14 %. The picture of the basic sample shows that the pores are through; they arranged randomly which reduces material thermal insulation properties. When introducing porous active mineral additives TMT-600 and TMT-600-k in cement matrix, effect of filling sand-cement walls with micropores is observed while increasing the cement stone strength.

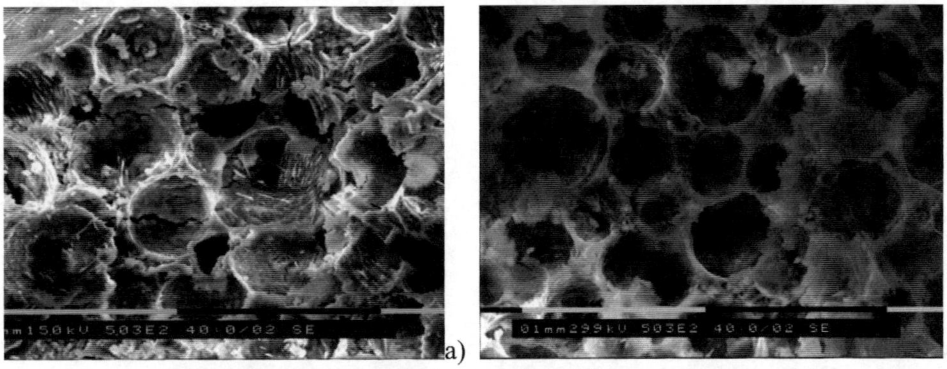
a) b)

Figure 6. Electron micrographs of foam concrete:
a) base sample; b) modified sample with additive TMT-600.

Thus, the introduction of peat additive into foam concrete mixture allows increasing foam concrete strength by 44 – 57 %, increasing porosity by 27 %, and decreasing thermal conductivity by 14 %.

5. Conclusion

By thermal treatment of peat at 600 °C in open air conditions or in conditions with limited air access effective additive for modifying structural cementitious mixtures was obtained.

Introduction of peat modifiers into foam concrete mix leads to improved quality of porous structure and speeds up the structure formation process that enhances the grade strength and reduces the thermal conductivity of foam concrete.

Acknowledgement

The research performed with the financial support of the Ministry of Education and Science of the Russian Federation within Governmental Decree No.218 of "Measures of state support for cooperation of Russian higher education institutions and organizations, for implementation of complex projects of high-tech production", integrated project No. 02.G25.31.0022, "Production technologies of energy- and resource-efficient housing of economy class based on the universal framed prefabricated architectural and construction system".

References

[1] Kobidze T E, Korovyakov V F, Listov S V and Samborskii S A 2006 *Building materials* **4** 40-41
[2] Morgun L V 2003 On the regularities of concrete structure formation with reinforcing fibers News of higher educational institutions. Construction 8 58-62
[3] Kudyakov A I, Kopanitsa N O, Prischepa I A and Shangin S N 2013 *Vestnik of TSUAB* **1** 172-177
[4] Kasitskaya L V, Sarkisov Yu S, Gorlenko N P, Kopanitsa N O and Kudyakov A I 2007 Peat resources of Tomsk area and the ways of their use in civil engineering (Tomsk: Scientific & Technical Translations)
[5] Мисников О С, Тимофеев А Е и Черткова Е Ю 2010 Труды Инсторфа **2** 15-33
[6] Kopanitsa N O, Sarkisov Yu S and Kasatkina A V 2013 *Vestnik of TSUAB* **4** 172-177

IOP Conf. Series: Materials Science and Engineering **71** (2015) 012013 doi:10.1088/1757-899X/71/1/012013

Effect of mixing water magnetic activation cycle on cement stone structure

S A Kugaevskaya[1], Yu A Abzaev[2], V N Safronov[1], Yu S Sarkisov[3], N P Gorlenko[3] and T A Ermilova[1]
[1]Tomsk State University of Architecture and Building, Building Materials Department, Tomsk, 634003, Russia
[2]Tomsk State University of Architecture and Building, Higher Mathematics Department, Tomsk, 634003, Russia
[3]Tomsk State University of Architecture and Building, Chemistry Department, Tomsk, 634003, Russia

E-mail: somana@mail.ru

Abstract. The paper presents results of investigations of hydration processes and structure formation of the cement paste matrix mixed with water activated by magneto static field using water treatment cycle technology. It is shown that crystallization of phases occurs in the cement-water system at different rates, and phase redistribution in the structure of the cement paste matrix is described before and after magnetic activation of mixing water. Also, modeling of the cement-water system and calculations of amorphous and crystalline phases using the Rietveld refinement method before and after magnetic activation show that strength properties of the cement paste matrix depend not only on quantitative but also qualitative relationship between phases.

1. Introduction

Today, consumer's market of cement-based concretes is one of the most attractive markets of building materials. Therefore development of functional cement-based materials is important to improve their functional performance and lowering material consumption.

Currently there is a broad range of techniques that allow purposefully controlling the structure and properties of cement systems. One of these techniques is activation of mixing water using external physical effects [1-5]. However, in spite of this perspective technique, it has not been widely developed within the concrete technology due to the absence of a safe theory that would predict the development of structurization processes in hardening cementitious systems. This work is the first to focus on fine interaction effects between water and cement including a dissolution consistency of existed and neoformed phases before and after activation; the ratio between amorphous and crystalline phases which form the structure of the cement paste matrix, etc.

This paper mainly focuses on quantification of the paste matrix phase content, detection of its structurization properties obtained from cement paste mixed with activated water using the treatment cycle technology.

2. Experimental procedure

Cement paste matrix used for this experiment was prepared from the type M 500 Portland cement-based paste mixed with water activated by a magnetostatic field with 40 mT magnetic density with the use of water treatment cycle technology. Magnetic activation included 15 and 25 cycles. Flat-surface specimens having 20×20×20 mm size were selected for this experiment.

XRD analysis was carried out by DRON-4-07 diffractometer which was modified for digital signal processing. Measurements were conducted using copper radiation (*K*) and Bragg-Brentano X-ray optical scheme. Specifications for the DRON-4-07 included 0.02^0 scanning step; 17 – 92^0 range for angles to be scanned.

The preliminary stage has the task to identify the structure of the cement paste matrix referring to the Crystallography Open Database (COD) [6]. Detected reference phases having high figures of merit (FOM) [4] are presented in Table 1 which also gives numbers, chemical formulas, lattice parameters, and space groups and types.

Table 1. Reference phases.

COD number	Chemical formula	FOM						Space group and type
		a, (nm)	b, (nm)	c, (nm)	α	β	γ	
100-1769	H_2O_2Ca	0.3589	0.3589	0.4911	90^0	90^0	120^0	P-3ml, trigonal
120-0010	$O_{20}Al_{7.04}Ca8Fe_{4.96}$	0.5584	1.4600	0.5374	90^0	90^0	90^0	Ima2, orthorhombic
900-0097	$C_6O_{18}Mg_6$	0.4633	0.4633	1.5016	90^0	90^0	120^0	R-3C, trigonal
900-0966	$C_6O_{18}Ca_6$	0.4988	0.4988	1.7061	90^0	90^0	120^0	R-3C, trigonal
900-3354	$O_{20}Al_{8.93}Ca_8Fe_{3.07}$	0.5299	1.4443	0.5509	90^0	90^0	90^0	Ima2, orthorhombic
900-8367	$O_{179.98}Si_{36}Ca_{108}$	3.3083	0.7027	1.8499	90^0	94.120^0	90^0	Cm, monoclinic
901-1411	$O_{64}Al_{28}Ca_{24}$	1.1970	1.1970	1.1970	90^0	90^0	90^0	I-43d, cubic
901-2682	$O_{24}Si_8Ca_9$	0.5074	1.1212	0.6753	90^0	90^0	90^0	Pmmm, orthorhombic
901-2792	$O_{16}Si_4Ca_6$	0.5507	0.6751	0.9305	90^0	94.597^0	90^0	P2/m, monoclinic

The quantitative phase analysis (QPA) of phase content by the Rietveld refinement method is used to determine intensities of each reference phase presented in Table 1 [7, 8]. As a result, the ratio between the theoretical and the experimental intensities of diffraction maxima is refined. The intensity reproducibility is detected by convergence criteria [7].

In these terms, the structural, profile, and experimental parameters involved in theoretical intensity are compared and refined using the non-linear least-square method. The QPA by the Rietveld method is carried out using the software Reflex [9]. In reference phases the following parameters are refined: spatial atom distribution in crystal lattices through redistribution of atomic relative coordinates; lattice;

reflections; zero shifts; background intensity, etc. Figures 1-3 contain plots of theoretical and experimental XRD histograms of phases and integrated intensity.

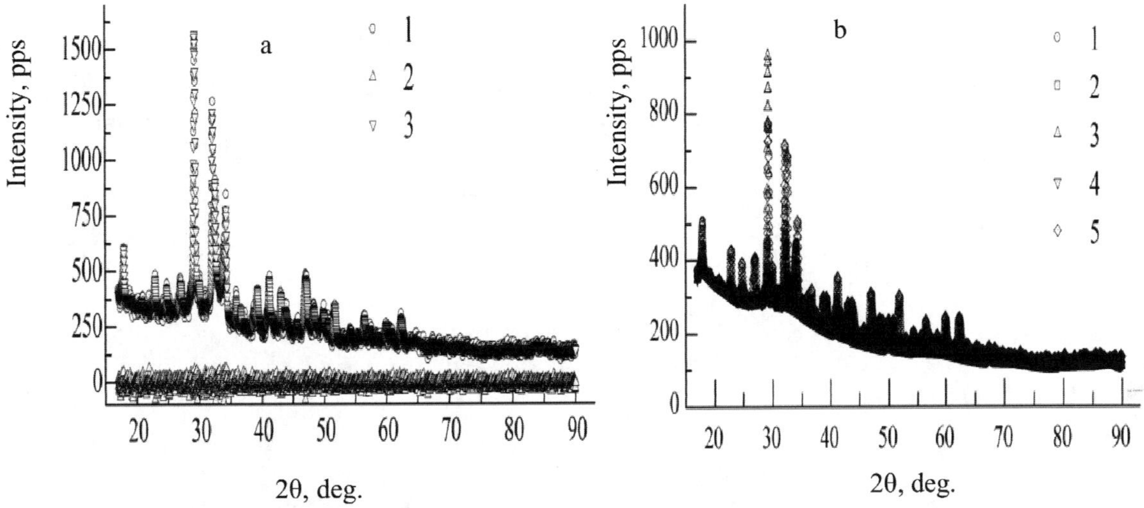

Figure 1. QPA of cement paste matrix in its original state: a) XRD histograms: *1* – theoretical, *2* – experimental, *3* – difference; b) theoretical XRD histograms of phases: *1*-6[$CaO \times CO_2$]; *2* - 36[$3CaO \times SiO_2$]; *3* - 6[$CaO \times 4SiO_{2.5}$]; *4* -$CaO \times H_2O$; *5* - 2[$4CaO \times Al_2O_3 \times Fe_2O_3$].

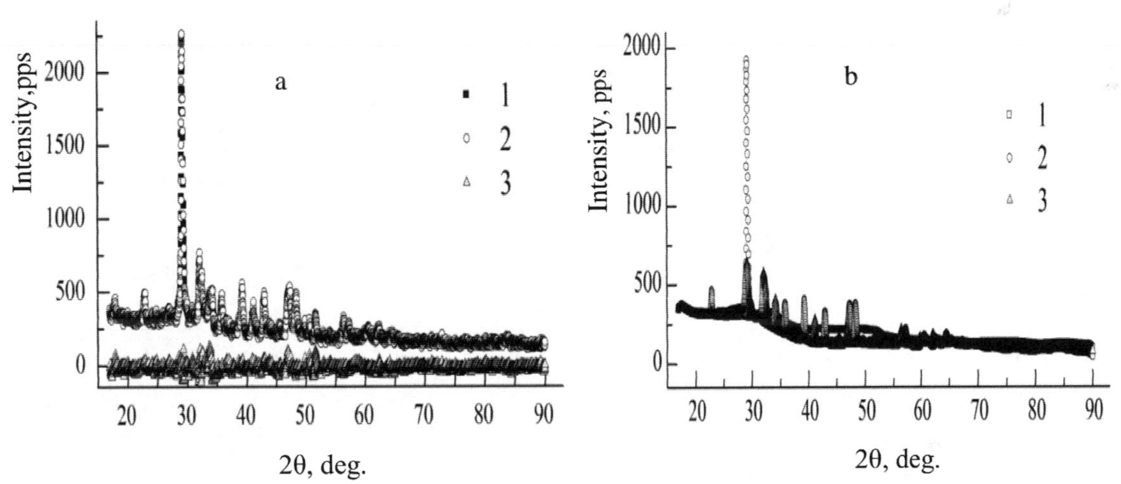

Figure 2. QPA of cement paste matrix at 15-cycle magnetic activation of mixing water: a) XRD histograms: *1* – theoretical, *2* – experimental, *3* – difference; b) theoretical XRD histograms of phases: *1* - 4[$6(CaO) \times 3(Al_2O_3) \times AlO$]; *2* - 6[$CaO \times CO_2$]; *3*- 36[$3CaO \times SiO_2$].

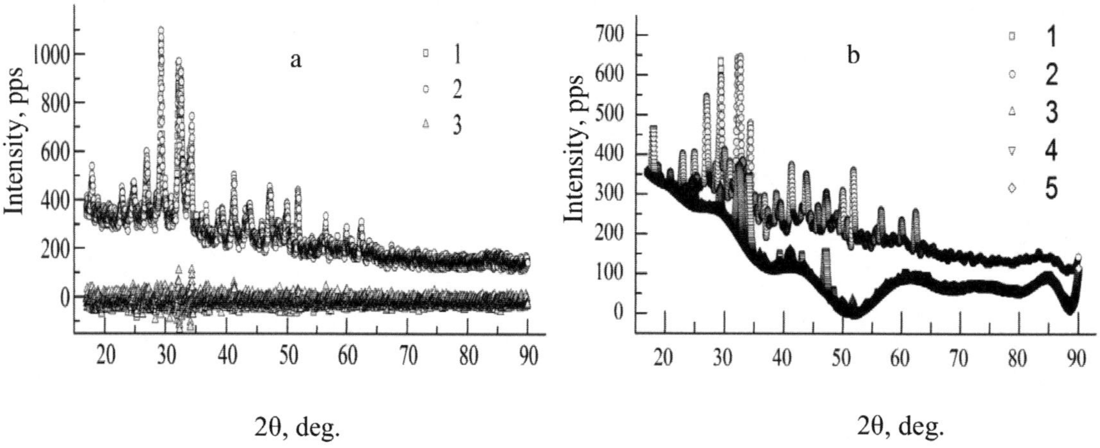

Figure 3. QPA of cement paste matrix at 25-cycle magnetic activation of mixing water: a) XRD histograms: *1* – theoretical, *2* – experimental, *3* – difference; b) theoretical XRD histograms of phases: *1* -6[$CaO \times CO_2$], *2* - 36[$3CaO \times SiO_2$]; *3* - 6[$CaO \times 4SiO_{2.5}$], *4* - $CaO \times H_2O$; *5* - 2[$4CaO \times Al_2O_3 \times Fe_2O_3$].

As shown in Figures 1-3, experimental XRD patterns are full of both intensive and weak overlapping peaks in crystalline phases. Also, the cement paste matrix contains a large amount of the amorphous phase. The quantitative phase analysis shows the substantial convergence between integrated theoretical and experimental intensities achieved for the cement paste matrix (Figures 1-3a). Two independent convergence criteria R_{wp} and R_p [7, 9] obtained by different ways are calculated for the original state of the paste matrix (6.46 and 4.92%); the paste matrix at 15-cycle magnetic activation (8.51 and 6.55%); and the paste matrix at 25-cycle magnetic activation of mixing water (6.92 and 5.41%) shown in Figures 1a, 2a, and 3a respectively. The difference between theoretical and experimental values of integrated intensities is insignificant. In this work, the QPAof the cement paste matrix by the Rietveld method allows determination of phases which contribute mush to the integrated intensity at refinement of the above stated parameters. The QPA results and a list of major phases in the cement paste matrix are presented in Tables 2-4.

Table 2. Properties of cement paste matrix in its original state.

COD number	Chemical formula	Intensity (%)	FOM			Space group and type
			a, (nm)	b, (nm)	c, (nm)	
900-8367	36[$3CaO \times SiO_2$]	49.64	3.3108	0.7041	1.8510	Cm, monoclinic
100-1769	$CaO \times H_2O$	17.02	0.3622	0.3622	0.4952	P-3m1, trigonal
901-2792	6[$CaO \times 4SiO_{2.5}$]	15.05	0.5489	0.67446	0.9226	P2/m, monoclinic
900-0966	6[$CaO \times CO_2$]	10.90	0.4991	0.4991	1.7162	R-3C, trigonal
120-0010	2[$4CaO \times Al_2O_3 \times Fe_2O_3$]	5.89	0.5562	1.4517	0.5371	Ima2,orthorhombic

Table 3. Properties of cement paste matrix after15-cycle magnetic activation of mixing water.

COD number	Chemical formula	Intensity (%)	FOM			Space group and type
			a, (nm)	b, (nm)	c, (nm)	
901-1411	4[$6CaO \times 3(Al_2O_3) \times AlO$]	51.03	1.1970	1.1970	1.1970	I-43d, cubic
900-0966	6[$CaO \times CO_2$]	25.10	0.4988	0.4988	1.7061	R-3c, trigonal
900-8367	36[$3CaO \times SiO_2$]	21.61	3.3083	0.7027	1.8499	Cm,monoclinic

Table 4. Properties of cement paste matrix after 25-cycle magnetic activation of mixing water.

COD number	Chemical formula	Intensity (%)	FOM a, (nm)	b, (nm)	c, (nm)	Space group and type
900-8367	36[$3CaO \times SiO_2$]	82.43	3.3083	0.7027	1.8499	Cm, monoclinic
901-2792	6[$CaO \times 4SiO_{2.5}$]	5.39	0.55075	0.67509	0.93055	P2/m, monoclinic
100-1769	$CaO \times H_2O$	4.76	0.3589	0.3589	0.4911	P-3m1, trigonal
120-0010	6[$CaO \times CO_2$]	6.53	0.5584	1.4600	0.5374	Ima2, orthorhombic

The QPA results show that the experimental XRD histogram of the original state of the paste matrix displays the superposition of contributions to the intensity from such phases as [$3CaO \times SiO_2$], $CaO \times H_2O$ (calcium hydroxide), [$CaO \times 4SiO_{2.5}$] (larnite), [$CaO \times CO_2$] (calcite), [$4CaO \times Al_2O_3 \times Fe_2O_3$] (tetracalciumaluminoferrite). As it follows from Table 1, the integrated intensity of these phases is close to unity and equals to ~ 0.985 that proves rather a high accuracy of phase quantification for the cement paste matrix. After 15 cycles of magnetic activation of mixing water (Table 3), the integrated intensity in the paste matrix comprises contributions from $6CaO \times 3(Al_2O_3) \times AlO$, $CaO \times CO_2$, and $3CaO \times SiO_2$ phases, while after 25 cycles (Table 4) it comprises contributions from $3CaO \times SiO_2$, $CaO \times 4SiO_{2.5}$, $CaO \times H_2O$, $CaO \times CO_2$ phases. The integrated intensity as shown in Table 4 is close to unity and equals to ~0.991. Thus, the QPA results show a high reproducibility and validity of theoretical and experimental data.

The QPA technique is used to detect parameters of the unit cell in calculating the theoretical intensity for each phase allowing for reference values. Refined phase parameters are given in Tables 2-4. Atomic relative coordinates in phase lattices are refined also, however, they are not given herein due to large data volumes.

The degree of crystallinity which coincides with the relative superposition of refined intensities in crystalline phases is described by experimental XRD histograms. It is stated that in the paste matrix the crystalline phase is 0.39 in the original state, 0.39 after 15-cycle activation, and 0.46 after 25-cycle activation of mixing water. Crystallinity grows with the increase of the number of cycles. The amorphous phase is characterized by a short-range order of atom arrangement, while translation symmetry is not observed. Diffraction patterns of amorphous phases demonstrate spread reflections of weak intensity which provide the main contribution to the background intensity.

The QPA results show that magnetic activation of mixing water has a strong effect on hardening of the paste matrix (see Tables 2-4). It is reflected, first of all, in the list of major phases of the paste matrix which are present in its original state, and secondly, in a change of their relative amount. In the original state, $3CaO \times SiO_2$ (alite) is the major phase for the paste matrix after 25 cycles of water treatment. At the same time, after 15 cycles of water treatment, $6CaO \times 3(Al_2O_3) \times AlO$ is the dominant intermediate phase. It is possible to assume that at 15-cycle magnetic activation the intensity of crystallization of $3CaO \times SiO_2$ phase significantly decreases, while at 25-cycle activation it increases due to a suppression of chemical reactions which produce $CaO \times H_2O$, $CaO \times 4SiO_{2.5}$, and $CaO \times CO_2$ whose amount decreases in the integrated intensity (see Tables 2-4). Also, one can suppose that activated water suppresses crystallization of $4CaO \times Al_2O_3 \times Fe_2O_3$ phase which is present only in the original state. The different activity of water has the different effect on crystallization of $CaO \times CO_2$ and $CaO \times 4SiO_{2.5}$ phases. The amount of the first phase increases after 15 cycles and decreases after 25 cycles of water activation. Crystallization of $CaO \times 4SiO_{2.5}$ phase is completely suppressed after 15 cycles and partially suppressed after 25 cycles of water activation.

A significant redistribution of the amount of crystallized phases as well as formation of new phases in the course of the hardening process results in a variation of mechanical properties of the cement paste matrix. Data obtained on the structure of unit cells of described phases are the basis for further investigations, in particular, theoretical calculations of cell energy, elastic constants, simulation of amorphous phases with a view to predict strength properties of the cement paste matrix and their experimental testing.

IOP Conf. Series: Materials Science and Engineering **71** (2015) 012013 doi:10.1088/1757-899X/71/1/012013

3. Conclusions

The quantitative phase analysis of the cement paste matrix by the Rietveld method showed that magnetic activation of water significantly affects the structurization processes in hardening cementitious systems. First of all, it is connected with the different intensity of crystallization of different phases and, as a consequence, a formation of phases differ from the original state, and secondly, with phase redistribution. It was stated that in the paste matrix the amount of amorphous phase is rather considerable. At the same time, this amount decreased with the increase of the number of cycles of magnetic activation.

References

[1] Бахир В М 1983 Физическая природа явлений активации веществ *Известия АН УзССР* **1** 60-64
[2] Blokhra R L 1999 *J. Chem., Sect. A: Inorg., Bio-inorg., Phys., Theor. Anal. Chem.* **38** 366-368
[3] Горленко Н П и Саркисов Ю С 2011 *Низкоэнергетическая активация дисперстных систем* (Томск: Издательство Томского государственного архитектурно-строительного университета) 264
[4] Сафронов В Н, Саркисов Ю С и Кугаевская С А 2009 Цикловая магнитная активация газонаполненных жидких сред затворения цементных систем *Вестник ТГАСУ* **4** 89-99
[5] O'Brien K N 1982 *J. Electrochem. Soc.* **129** 1266-1268
[6] Crystallography Open Data base Date Views 12.05.2014 online source [www.crystallography.net]
[7] Клопотов А А, Абзаев Ю А, Потекаев А И и Волокитин О Г 2012 *Основы рентгеноструктурного анализа в материаловедении* (Томск: Издательство Томского государственного архитектурно-строительного университета) 275
[8] Torre A, Cabeza A, Calventa A Bruque S and Aranda M AG 2001 *Anal. Chem.***73** 151-156
[9] Абзаев Ю А, Саркисов Ю С, Клопотов А А, Клопотов В Д и Афанасьев Д А 2012 Полнопрофильный рентгеноструктурный анализ клинкерного минерала C4AF *Вестник ТГАСУ* **4** 200-209

IOP Conf. Series: Materials Science and Engineering **71** (2015) 012014 doi:10.1088/1757-899X/71/1/012014

Thermal highly porous insulation materials made of mineral raw materials

A Mestnikov[1]
[1]Professor of the Department of Production of building materials, products and construction of the North-Eastern Federal University, Yakutsk, Russia

E-mail: mestnikovae@mail.ru

Abstract. The main objective of the study is to create insulating foam based on modified mineral binders with rapid hardening. The results of experimental studies of the composition and properties of insulating foam on the basis of rapidly hardening Portland cement (PC) and gypsum binder composite are presented in the article. The article proposes technological methods of production of insulating foamed concrete and its placement to the permanent shuttering wall enclosures in monolithic-frame construction and individual energy-efficient residential buildings, thus reducing foam shrinkage and improving crack-resistance.

1. Introduction

It is known that in modern conditions, the role of inorganic natural materials in the manufacture of building materials has priority. Particularly suitable for the cold climate in most regions of Russia are insulating materials which are produced using low-power technology [1] including cellular concrete and foamed concrete.

Research in the area of foamed concrete is aimed at creating new types of synthetic and protein foaming agents, new multicomponent compositions with various mineral and chemical additives, as well as creating dry mixtures for producing foamed concrete using mechanical activation of raw materials, etc. However, the results of such studies are not always acceptable for construction in the northern regions remote from industrial centers [2].

Therefore, relevant issue for each region is development of optimal compositions for insulating foamed concrete depending on the availability of source components and raw materials in accordance with their technical and economic efficiency.

2. Materials and methods

The basic component of the activated binding mixture used for production of insulating foamed concrete are burnt rocks, characterized by the following chemical composition,% wt.: SiO_2 – 65,05; TiO_2 – 0,70; Al_2O_3 – 16,62; Fe_2O_3 – 6,06; FeO – < 0,25; MnO – 0,07; MgO – 1,80; CaO – 2,45; K_2O – 2,52; Na_2O – 3,16; P_2O_5 – 0,18; SO_3 – 0,65; CO_2 – < 0,20; ignition losses – 0,28. Inhomogeneity of burned rocks properties is explained by the different degree of burning. Activity of burning rocks increases with increasing fineness of grinding. Powdered supplement of the burning breed completely passed through sieve № 008, its specific surface area (S_{sp}) is 350-400 m^2/kg. In the course of earlier studies [3] it was found that burnt rocks are typical representatives of the group of hydraulic substances of glinite kind. Hardening of binders based on burned rocks should be regarded as processes in polymineral system the $Al_2O_3 - CaO - SiO_2 - Fe_2O_3 - H_2O$.

In the selection of composition of foamed concrete the following materials were used: Portland cement (PC) of "Yakuttsement" company, characterized by the following chemical composition, wt%: SiO_2 – 21,16; Al_2O_3 – 5,45; Fe_2O_3 – 4,72; CaO – 64,85; MgO – 2,71; R_2O – 0,75; CaO_{free} – 0,15; other material was gypsum binder matter (GBM), made of gypsum stone of Olyokminskiy and Dapparayskiy deposits. Content in the gypsum rock $CaSO_4 \cdot 2H_2O$ is: Olekminsky deposit – 80.80% (III grade); Dapparayskiy deposit – 93.96% (I grade).

As silica component river sand was used, satisfying the requirements of GOST 8736-93 "Sand for construction works. TU" (Russian State Standard).

As the blowing agent synthetic foaming agent brand PB-2000 (Russia) was used for obtaining foam concrete by equipment "Stromros" and proteinaceous foaming agent "FOAMSEM" (Italy); and for obtaining foamed concrete by classical way the equipment "SOVBI" was used.

Standard methods were used to determine the physical and mechanical properties of binders and concrete with the use of modern techniques and equipment: XRD and DTA obtained materials carried on the X-SRS-3400 spectrometer and diffractometer D8 Discover with GADDS for phase analysis; study on thermal conductivity was carried out on HFM 436 Lambda on samples with dimensions 300x300x100 (300) mm.

In the selection of foam formulations mathematical planning of the experiment was used. Multivariate polynomial models allow us to solve the vast majority of engineering problems in materials science and technology, so these ES models should be regarded as basic to practice of engineering in the industry.

3. Insulating foam on the basis of rapidly hardening Portland cement

To solve the problem we have optimized composition of Portland cement, modified by the introduction of complex additives based on gypsum and burnt rocks (rapidly hardening Portland cement).

Acceleration of paste hardening is shown with decreasing content of burned rock in the complex additive, which corresponds to increased content of gypsum binder matter (GBM). It should also be noted that timing of setting binder is hardly affected by the change in flow of complex additive itself. Thus, start of setting can be controlled within 25...60 min, and the end of setting time - 40...140 min.

Table 1. Comparative characteristics of thermal insulation foamed concrete of D300 grade at various binders.

Heat resistant foamed concrete	Composition, wt. %			Duration, h		Average strength on compression, MPa
	Portland cement	gypsum	burned rocks	start of setting	end of setting	
On PC 400 - DO	100	–	–	≥0,75–1,75	4,5–6	0,51
On PC-B	90	5,5	4,5	0,5	1,25	0,56

Optimum compositions of PC-B have values of compressive strength after 28 days of hardening in normal conditions in the range 15...17 MPa and the linear expansion of 0,7%, containing 10% of integrated additive content in its composition as well as burning breed within 50...60% by weight.

Design of formulations of grades D300 and D500 foamed concrete was made on the basis of produced rapidly hardening cement. According to the results, the graphs of dependence of the tensile strength of foamed concrete on binder composition (Figure 1) were designed.

Analysis of the graphs showed that the highest compressive strength was achieved when using PC-B of 10/45 in both cases, i.e. at flow rate of 90% of Portland cement, 4,5% of burnt rocks and 5,5% of gypsum binder in the total mass of the composite binder. In this case setting and hardening time of foamed concrete mix is significantly reduced and, strength characteristics increase (up to 10%)

compared to foamed concrete on conventional Portland cement PC 400-D0 (Table 1). According to the test results on samples of air shrinkage in bench and field conditions, obtained foamed concrete relates to dimensionally stable cellular concrete.

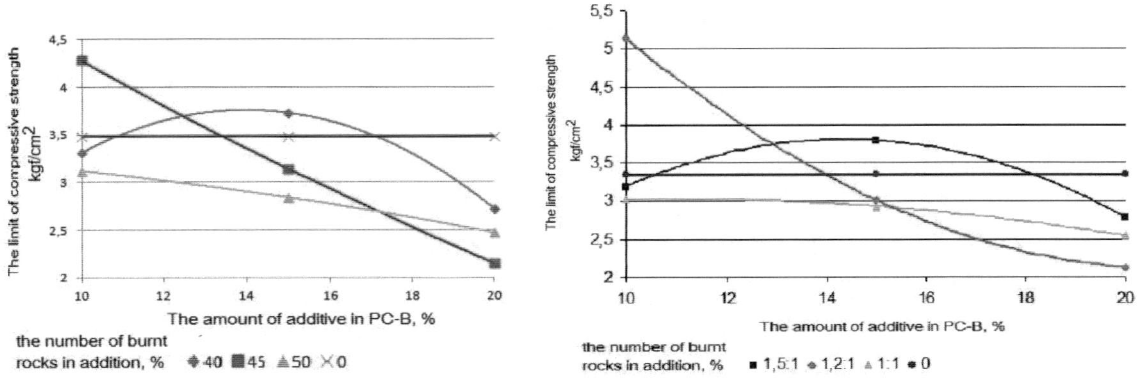

Figure 1. Dependence of the compressive strength of foamed concrete samples D 300 (28 days) and D500 (14 days) on the composition of rapidly hardening Portland cement PC-B.

4. Insulating foamed concrete based on composite gypsum binders

Optimum composition of composite gypsum binder with the strength of brand M200 using burnt rocks sifted through a sieve number 008 (with surface area of 350 m^2/kg), has starting and ending points of setting 7 and 7,5 min, respectively, compressive strength after 2 h. after mixing with water 4 MPa after 28 days – 20.5 MPa and 0,75 softening coefficient (Figure 3).

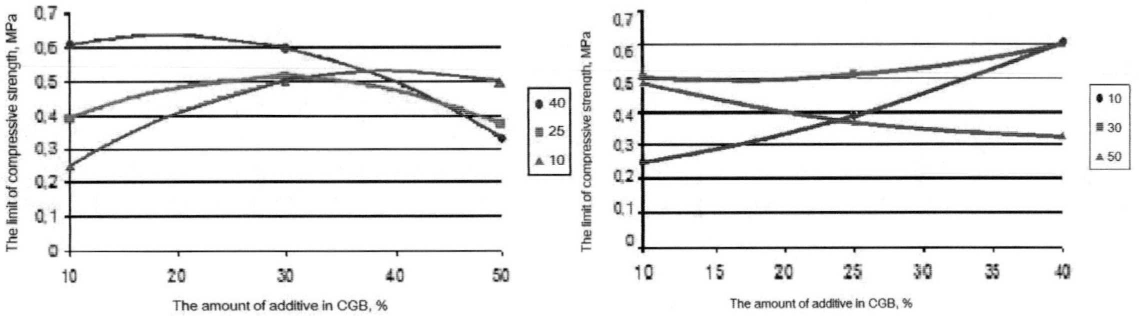

Figure 2. Dependence of the compressive strength of foamed gypsum concrete samples D400 aged 28 days on the flow of materials.

To confirm the phase composition of the foamed concrete samples based on PC-B and composite gypsum binder (CGB) X-ray analysis of the sample at the age of 3 years was carried out. Analyzing X-ray shown in Fig. 3, it must be concluded that the main cementing substance of the samples on the PC-B are low base calcium hydro silicates CSH (B) (d = 9,91; 7,59; 4,96; 3,79; 3,01; 2,67; 2,21; 2,07; 1,89; 1,63 $\cdot 10^{-10}$ m) and the dihydrate gypsum $CaSO_4 \cdot 2H_20$ (d = 7,59; 4,28; 3,06; 2,87 $\cdot 10^{-10}$ m).

For CGB the main cementing substance of the samples is gypsum dihydrate $CaSO_4 \cdot 2H_20$ (d = 7,59; 4,28; 3,79; 3,06; 2,87 $\cdot 10^{-10}$ m).

Radiographs of both binders samples contain lines of hydrosulfoaluminate of highly basic forms of calcium - ettringite $3CaO \cdot Al_2O_3 \cdot 3CaSO_4 32H_2O$ (d = 9,71; 5,60; 4,96; 4,69; 3,87; 2,78; 2,59; 2,22; 1,66 $\cdot 10^{-10}$ m), low basic calcium hydro silicates CSH (B) (d = 12,52; 9,91; 4,90; 3,07; 2,80; 1.83 $\times 10^{-10}$ m), calcium carbonate $CaCO_3$ (d = 3,00; 2,49; 2,28; 2,08; 1,91; 1,87 $\cdot 10^{-10}$ m), silica SiO_2 (d = 4,26; 3,34; 2,14; 1,99; 1,81; 1,53 $\cdot 10^{-10}$ m), as well as traces of hemihydrate gypsum $CaSO_4 \cdot 0.5 H_2O$ (d = 6,01; 4,35; 3,00; 2,80 $\cdot 10^{-10}$ m) and non-hydrated portland cement minerals (d = 2,3;

2,77 · 10^{-10} m). Electron micrographs show that increase of strength characteristics of foam is associated with decrease in porosity of interporous partitions.

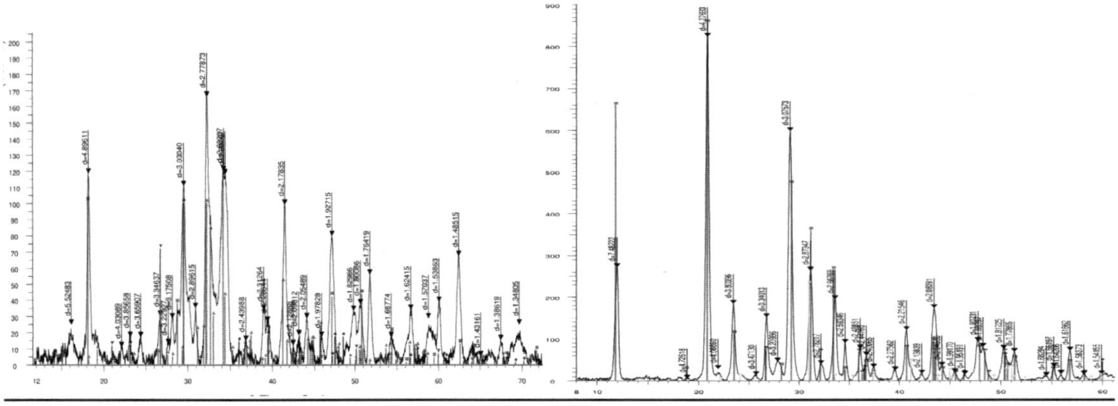

Figure 3. X-Ray pattern of the samples on the PC-B (aged 3 years) and CGB (aged 2 years).

5. Conclusion

For dimensionally stable insulating foam of D300 and D500 grades, maximum compressive strength of 0,43 MPa and 0,51, respectively (100% cement – 0,35 and 0,43 MPa, respectively) is achieved using the initial composition containing 90% cement, 5,5% and 4,5% gypsum burnt rocks in the total mass of dry mix, that can accelerate the process of foamed concrete mix hardening 2-3 times and provide high-quality thermal insulation layer of permanent formwork walling [4].

Introduction of additives consisting of Portland cement and fine ground burnt rock M400 with a specific surface of 350 m/kg^2 to the CGB allows to adjust the start and end time of paste setting (7,5 and 7 minutes, respectively), values of strength and water resistance (softening rate 0,75); compositions of CGB containing 10 to 50% additives, which include 10 to 40% of burnt rock have maximum strength and values of compressive strength 21,0 and 20,5 MPa respectively.

Technology of construction [4], as well as construction of wall fences [5] in the cast-frame construction and individual construction were developed and implemented to reduce costs and speed up construction time, improve the energy efficiency of buildings operated in cold climate of Yakutia with high quality of construction and installation works.

Acknowledgements

The author expresses gratitude to the staff of the department, to A. Egorova, Ph.D., Associate Professor, to B. Rozhin, Head of Laboratory and to A. Kardashevsky, graduate student for practical assistance in the implementation of laboratory and field studies.

References

[1] Burchell J D and Kelly A 1983 New inorganic materials *Scientific American* **7** 51-59

[2] Местников А Е, Егорова А Д, Кардашевский А Т и Корнилов Т А 2009 Технология возведения энергоэффективных стеновых конструкций жилых зданий на Севере *Строительные материалы* **4** 118-120

[3] Книгина Г И 1966 *Строительные материалы из горелых пород (*Москва: Стройиздат) с 207

[4] Егорова А Д и др. Патент Российской Федерации на изобретение №2361985 Способ теплоизоляции и облицовки поверхности стен плитками

[5] Егорова А Д и др. Патент Российской Федерации на полезную модель №84035 Строительный стеновой блок

Study on basalt fiber parameters affecting fiber-reinforced mortar

A A Orlov[1], T N Chernykh[1], A V Sashina[1] and D V Bogusevich[1]
[1]National Research South Ural State University, Chelyabinsk, 454080, Russia

E-mail: kosheen_s@mail.ru

Abstract. This article considers the effect of different dosages and diameters of basalt fibers on tensile strength increase during bending of fiberboard-reinforced mortar samples. The optimal dosages of fiber, providing maximum strength in bending are revealed. The durability of basalt fiber in an environment of cement, by means of microscopic analysis of samples of fibers and fiberboard-reinforced mortar long-term tests is examined. The article also compares the behavior of basalt fiber in the cement stone environment to a glass one and reveals that the basalt fiber is not subject to destruction.

1. Introduction

Fiber concrete, as well as regular concrete, is composite material including fiber rebar distributed in the volume. Dispersive fiber reinforcement to a great extent allows compensation of concrete shortcomings - low tensile strength and low strain at fracture. Fiber concrete has several times higher tensile and shearing strength, impact and fatigue strength, fracture and crack resistance, frost resistance, waterproofing capacity, heat stability and fire resistance [1, 2, 3, 4].

Current study is aimed at defining the most effective dosages and diameters of basalt fiber when reinforcing cement composites.

2. Methods and materials

BFC 300 (compatible satisfying requirements of Russian National Standard GOST 31108–2003 – ЦЕМ II/A-Ш 22,5 H) produced by "Magnitogorsk Cement and Refractory Works", JSCO was used in the study to define the most effective dosages and diameters of basalt fiber. The aggregate used was sand wash sized to 0, 16…2 satisfying requirements of Russian National Standard GOST 8736-93 "Sand for construction works. Technic specifications". Fiber reinforcement was done with basalt fiber produced by "Russian basalt", fiber was produced according to TOR-5952-002-91341008-2012, the diameters of fiber used were 10, 15, 20, 25 um, length 12 mm. To compare the behavior with basalt fiber when exploited in alkaline medium of cement matrix, glass fiber made of alkali-nonresistant glass specially produced by "Russian basalt" was also used.

Test methods used in the study are shown in Table 1.

Table 1. Test methods.

Test mode	Test method
Normal consistency, cement slurry setting time	GOST 310.3-76
Standard sand-cement mortar consistency, flexural strength	GOST 310.4-81
Longevity of fiber in cement matrix	Electronic focused beam microscope

3. Results

It is known that cement fibers reinforce only mortar phase of material and therefore they have lower reinforcement effect than cement slurry or mortar. Thus the experiment defining the most effective dosages and diameters of fiber was carried out on sand-cement mortars. To assess the influence of basalt fiber over the mortar strength two-factor experiment was planned. Its factors were:

- x-factor – fiber dosage, factor ranged on 3 values 0,7, 1,1 and 1,5 kg/m^3 of concrete mix;
- y-factor – fiber diameter, range values 15, 20 and 25 um.

Composite flexural strength gives initial response to volume micro reinforcement, thus considered to be the primary response. It has been controlled since the first day up to vintage age.

To conduct an experiment a composition of BFC, sand wash and mixing water in the ratio 1:3:0,56 (№ 10 in Table 1) considered to be controlled. Other compositions included basalt fiber according to the experiment plan (Table 2).

Experiment plan and the study results are listed in Table 2, the influence of basalt fiber dosages and diameters over flexural strength is graphically presented in Figures 1, 2.

Table 2. Experiment plan and the study results.

Item number	x-factor code	Fiber dosage, kg/m^3	y-factor code	Fiber diameter, um	Flexural strength, mPa		Flow, cm
					day 1	day 28	
1	-1	0,7	-1	15	0,92	6,85	108,5
2	0	1,1	-1	15	0,49	7,12	108,5
3	1	1,5	-1	15	0,56	7,06	109
4	-1	0,7	0	20	1,58	7,73	108,5
5	0	1,1	0	20	1,29	7,98	109
6	1	1,5	0	20	1,14	7,07	109,5
7	-1	0,7	1	25	1,75	7,8	108,5
8	0	1,1	1	25	1,85	7,61	109
9	1	1,5	1	25	1,75	7,18	110
10	-	0	-	-	0,8	6	108

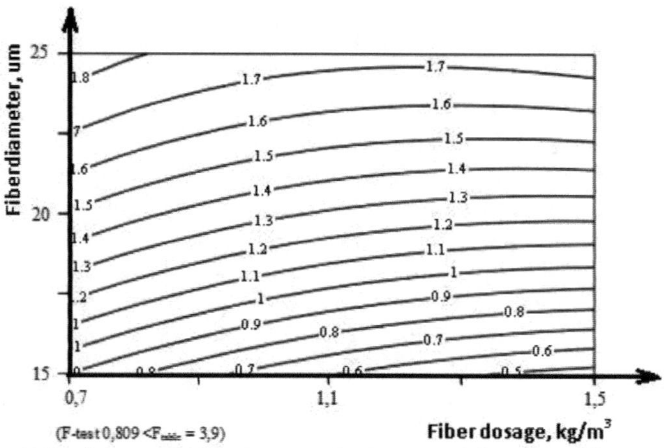

$$M(x,y) = 1,29 - 0,13 \cdot x + 0,56 \cdot y + 0,07 \cdot x^2 - 0,12 \cdot y^2 + 0,09 \cdot x \cdot y$$

Figure 1. Dependence of flexural strength in the 24-hour hardening on the fiber dosages and diameters.

The dependence obtained indicates that fiber diameter and dosage increasing in the 24-hour hardening leads to flexural strength increase, moreover it becomes 2 times stronger in comparison to control composition without fiber. In such event it is fiber diameter that largely affects sample strength. Strength increase with bigger diameter is probably connected with the fact that bigger diameter fibers are less likely to create beams (including vacuos) during formation.

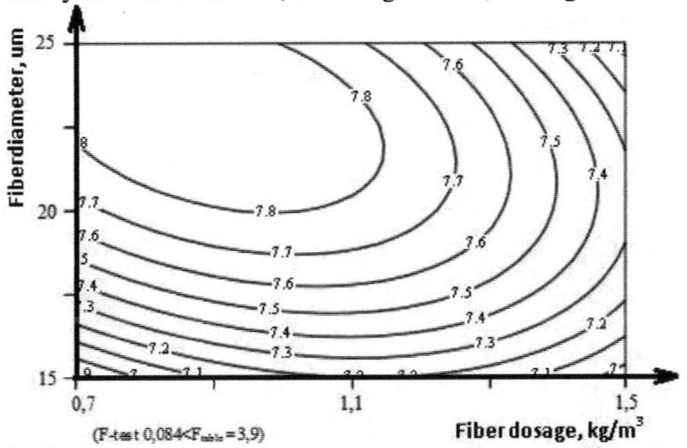

$$M(x,y) = 7,78 - 0,17 \cdot x + 0,26 \cdot y - 0,29 \cdot x^2 - 0,32 \cdot y^2 - 0,21 \cdot x \cdot y$$

Figure 2. Dependence of flexural strength in the 28th day hardening on the fiber dosages and diameters.

In the 28th day optimum region can be defined in the figure of dependence of flexural strength on the fiber dosages and diameters, the strength in which is at the maximum point for fiber with the diameter of more than 20 um in dosages from 0,7 to 1,0 kg/m^3.In condition of optimum parameters of mortar reinforcement more than three times strength increase is reached. The crucial factor is still fiber diameter, but when dosages used are more than 1 kg/m^{3}flexural strength get lower. This may be due to the fact that the excess of optimum dosages leads to an uneven distribution of the fibers in the material. Part of the fiber bunches into beams of 5…15grains, inside of which the fibers do not adhere to the cement matrix and when the material is being damaged they are easily pulled out, which ends up in lower flexural strength.

To estimate resistance of basalt fiber to alkaline medium of cement matrix a microscopic analysis of samples of cement matrix with fiberboard grains was performed. Fiberboard grains were in cement samples for a long time (4 months, under conditions of 100% humidity) (Figure 4).For comparison, Figure 3 shows photomicrographs of glass fiber of non-alkali glass.

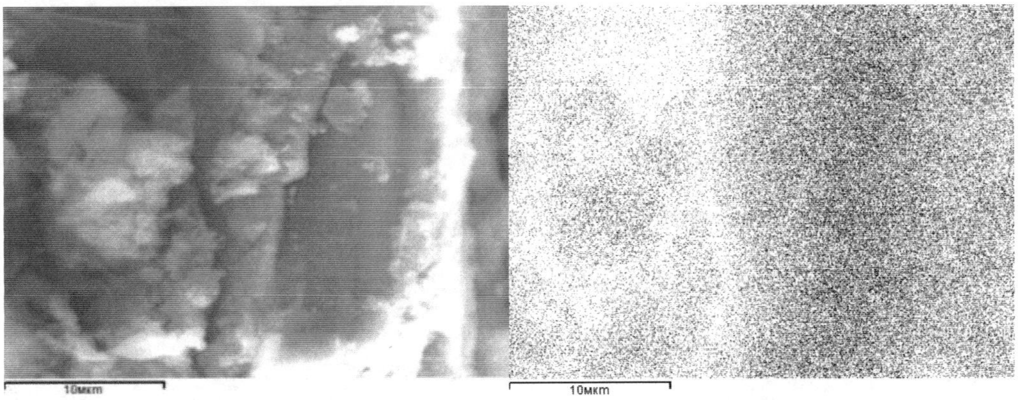

a) photomicrograph b) silicon distribution

Figure 3. Photomicrograph of glass fiber in the cement matrix.

As seen in photomicrograph (Figure 3a) hydrate layer was formed on glass fiber. And silicon distribution is uneven; hydrate layer differs from glass fiber and cement hydration product in chemistry indicating the chemical processes occurring on the surface of the fiber.

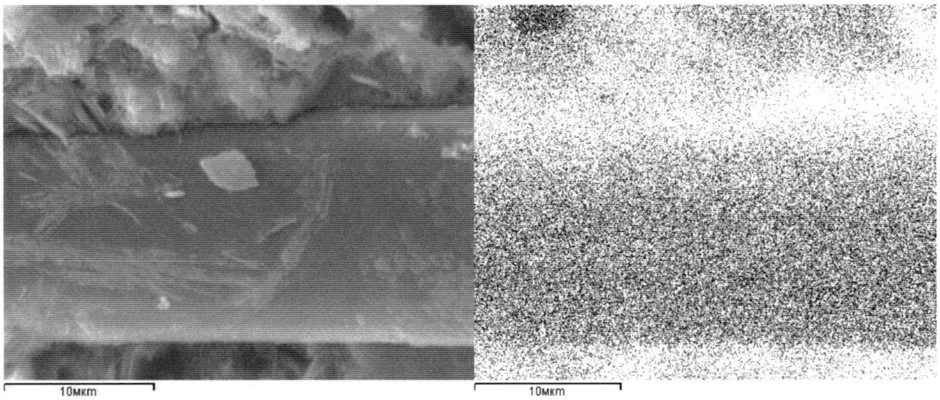

a) photomicrograph b) silicon distribution

Figure 4. Photomicrograph of basalt fiber in the cement matrix.

Basalt fiber micrograph has clear boundaries; there are no visible traces of destruction. This is confirmed by distribution pattern of silicon, there is clear boundary between the cement matrix and the fiber, i.e. there is no zone of variable composition, and fiber is well preserved in the alkaline medium of the cement matrix.

The results of the microscopic analysis of the samples with fiber are confirmed by bar (4x4x16) tests at different hardening ages. Test results of bar specimens are tabulated in Table 3.

Table 3. The results of determining the strength of the samples with fiber in various hardening time.

Composition number	Fiber type, diameter	Flexural strength, mPa			
		28 days	2 months	3 months	4 months
1	without fiber	6,0	6,5	7,1	7,9
2	glass fiber, 13 um	7,27	7,10	6,2	4,55
3	basalt fiber, 25 um	7,8	8,5	8,9	9,4

4. Conclusions

The study indicated that basalt fiber is suitable for application in cement-based materials and having optimal parameters of micro rein for cement (diameter 20-25 um in dosages ranging from 0.7 to 1.1 kg/m^3) is able to increase their flexural strength more than 1.5 times. It is also shown that the positive effects of fiber usage remain after contacting with cement matrix for a long time.

References

[1] Крылов Б А и Королев К М 1979 *Фибробетон и его применение в строительстве* (М.: НИИЖБ)

[2] Chaohua Jiang, Ke Fan, Fei Wu and Da Chen June 2014 Experimental study on the mechanical properties and microstructure of chopped basalt fibre reinforced concrete *Materials & Design* **58** 187

[3] Kabay Nihat 2014 Abrasion resistance and fracture energy of concretes with basalt fiber *Construction and Building Materials* **50** 95

[4] Dylmar Penteado Dias 2005 Fracture toughness of geopolymeric concretes reinforced with basalt fibers *Cement and Concrete Composites* **27** 49

IOP Conf. Series: Materials Science and Engineering **71** (2015) 012016 doi:10.1088/1757-899X/71/1/012016

Foam concrete with porous mineral and organic additives

A Kudiakov[1], I Prischepa[1] and M Tolchennickov[1]
[1]Tomsk State University of Architecture and Building, Department of Building Materials and Technologies, Tomsk, 634003, Russia

E-mail: ingaprishepa@mail.ru

Abstract. The article presents results of studies of structural heat insulating foam concrete with porous mineral and organic additives. By mixing additives with the concrete the speed of the initial structure formation increases. The additives of ash loss and thermal-modified peat TMT 600 provide a stable increase of strength by compression and bending of foam concrete. In the dried foam concrete with the addition of TMT and ash loss thermal conductivity decreases by 20% and 7% respectively. The regularities of changes in the thermal conductivity at various moisture of foam concrete have been investigated.

1. Introduction

In the main areas of economic and social development of Russia including construction industry a lot of attention is paid to the acceleration of scientific and technical progress. Energy saving and resource conservation through the development and implementation of energy efficient building materials for building envelopes are especially important [1].

In the construction sector significant part of energy consumption includes the energy cost of building materials production, their transportation, construction of objects, and further operation of buildings [2]. For heat-efficient house it is necessary to find materials and products of the new generation to apply in building envelopes and in bearing structures. The criteria for the effectiveness of such materials should be their high thermal properties, enhanced reliability and durability, ease of technology solutions, the low level of production costs in the manufacture of products. Therefore development and improvement of the production of heat insulating foam concrete of a higher level of quality and steady properties is crucial problem to be solved. The application of natural hardening foam in buildings will improve the energy efficiency of building due to its higher thermal resistance of the walls and by decreasing energy intensity of building envelopes [3].

In order to improve the competitive ability of natural hardening foam, that is, the improvement of its quality and the reduction of its cost, it is necessary to focus on reduction of thermal conductivity of concrete as much as possible at the stage of the selection of raw materials, the design of composition and preparation of raw material mixture, forming of the products by providing the required compressive and flexural strength. Preparation of foam concrete with low thermal conductivity and density is achieved by mixing the cement paste with porous foam or the introduction of fine-grained additives, as well as by the acceleration of setting time of cement which will fix the structure in the same condition in which it was formed in the process of mixing and molding. Solving the problem of stabilization of foam concrete mixture with porous mineral and organic additives will provide a new competitive and efficient heat insulating material based on cement binder.

The purpose of research performed was to develop concrete compositions and processing methods of manufacturing of structural heat insulating non-autoclave foam with low thermal conductivity and improved levels of quality through the use of porous mineral and organic additives.

The main goal of this paper is to investigate the influence of porous mineral and organic additives on the technological properties, structure of foam concrete mix, as well as the technical and operational characteristics of the hardened foam.

2. Materials and methods

For studies of foam the following raw materials were used: Portland cement from Topkinsky plant M500 D0 (GOST 30515-97), CEM I 4,5 B (GOST 31108-2003), quartz sand with Mach -1.44 (GOST 8736-93), ash loss CHP-5, Novosibirsk (GOST 25818-91), thermal-modified peat additive TMT600 developed by the staff of the Departments of Chemistry and Building Materials and Technologies of Tomsk State University of Architecture and Building [4]. Thermal-modified peat additive comprises thermally steady organic mineral complexes providing reinforcing and structure forming effect in the cement stone. Thermal-modified peat additive is chemically active and promotes the formation of new compounds in the structuring process of cement paste that leads to an increase in strength of cement stone. As the mixing water tap water was used that corresponds to the requirements of GOST 23732-2011. According to the results of preliminary studies [5] for the preparation of higher quality foam it can be recommended to use synthetic foaming agent Benoteh PB-C, satisfying the requirements of GOST 24211-2003 and technical condition 2481-010-58771162-2007. Foaming agent Benoteh PB-C has a high foaming ability, foam stability and steady properties when incorporated into cement slurry and increases the speed of structure formation.

Laboratory studies of foam were carried out in the laboratories of the Department Building Materials and Technologies in TSUAB. Preparation of foam concrete mix was carried out by one-step process in the laboratory using foam concrete mixer PBS with the volume of 10 liters. Foaming agent was injected into the foam concrete mixer with the mixing water. Plastic strength of foam concrete mix was determined by using conical Rebinder plastometer. Molded foam samples were kept up in wet conditions until the standard test. Testing and evaluation of the quality of foam was conducted in accordance with GOST 25485-89 d. The average foam density was determined in accordance with GOST 12730.1-78. Foam test in determining the compressive strength and flexural strength was carried out on a hydraulic press by loading to failure in accordance with GOST 10180-2012, and thermal conductivity - in accordance with GOST 7076-99.

3. Results and discussion

Experimental evaluation of the influence of porous mineral and organic additives on the intensity of interactions between structural elements that occur immediately after mixing of the components was carried out according to the magnitude of the plastic strength of the concrete mix. Change of the plastic strength at an early stage of structure formation is an important technological characteristic of the concrete mix, allowing evaluating the formability (remolding effort) at a certain time interval of the cycle of manufacturing of products [6]. According to this criterion, it is possible to set the optimum time for removal of form of the molded products (prefabrication) or structures (monolithic construction), that, in practice will help to coordinate the timing when carrying out concrete works.

The results of determination of the plastic strength of foam concrete mix with thermally modified peat additive and ash loss are presented in Figure 1.

The results show that the higher rates of plastic strength are formed in the foam with porous mineral and organic additives in the initial setting periods of hardening. So after 45 minutes of hardening of the concrete mix with additives plastic strength is increased up to 56 % in comparison with foam concrete mixture without additives, and after 3 hours of hardening up to 65 %. Thus, the results of tests concluded the possibility of using additives TMT600 and ash loss to accelerate the initial structuring of foam concrete mix that can accelerate manufacturing process of products and increase the smoothness of foam structure by means of demixing reduction.

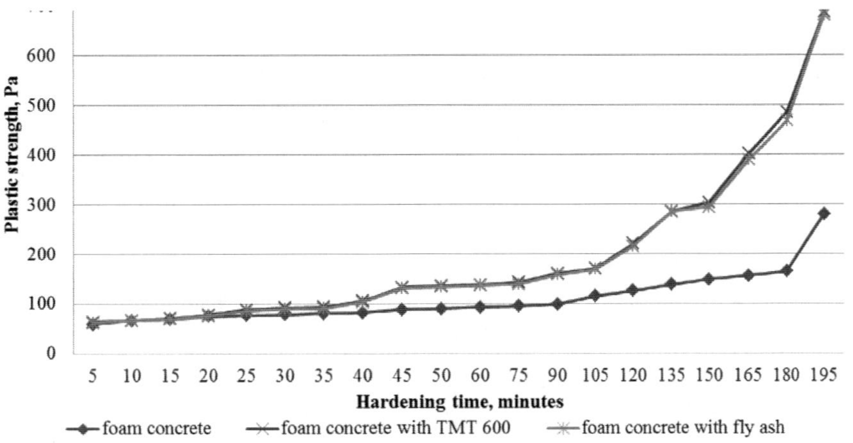

Figure 1. Plastic strength of concrete mixtures with additives.

Figures 2 and 3 show the results of a study of compressive strength and bending strength of branded foam at an average density of D700 with the addition of peat TMT600 and ash loss at the age of 14 and 28 days.

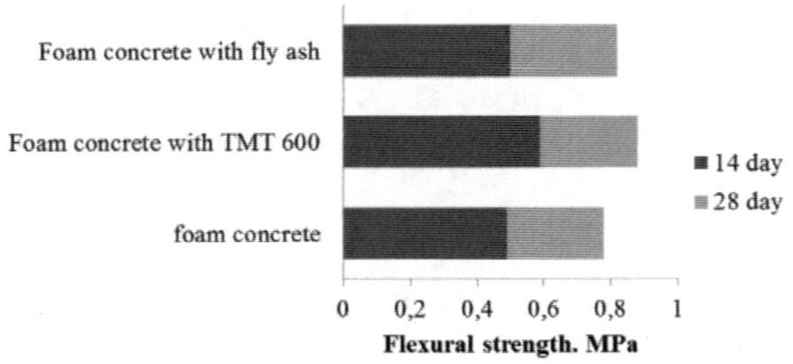

Figure 2. Kinetics of a set of bending strength of foam samples.

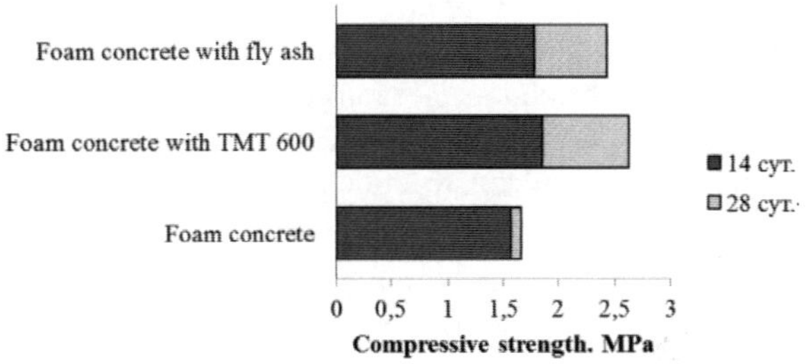

Figure 3. Kinetics of a set of compressive strength of samples of foam.

Analyzing the results (see Figures 2 and 3) one can see that when using ash loss in the age of 14 and 28 days the compressive and bending strength is raised up to 23 and 35 %, 10 and 12 %, respectively. In foam concrete with the addition of peat TMT 600 the compressive strength and bending strength is raised to 24 and 38 % and 15-20 % respectively.

For foam, used for the manufacture of building envelopes, thermal conductivity is of great importance. Products from foam are adapted to difficult climatic and economic conditions in Russia, namely, they have low thermal conductivity and high durability [7]. Thermal conductivity of porous materials, which are the foam concrete, is strongly influenced by the density of the material, type, size and location of pores and humidity [8]. Humidity has great influence on the value of the thermal conductivity. In Tomsk region, the average annual air humidity is in the range of 70-90 %.

Studies of the effect of moisture on the thermal conductivity of foam concrete have been conducted. To obtain foam with certain moisture, samples were kept in a desiccator over aqueous potassium hydroxide solution of various concentrations: 40 % potassium hydroxide solution (φ=70 %); potassium hydroxide solution 30 % (φ=80 %); potassium hydroxide solution 10 % (φ=93 %); water (φ=100 %) [9].

Thermal conductivity of foam concrete with additives TMT 600 and ash loss (Table 1) at 0 %, 20% and 7 % moisture is less respectively than the one without foam additives. This is due to the fact that the addition of TMT 600 and ash loss contribute to consolidation of cement stone structure in interporous partitions and provide increased micro porosity [10]. At 70 % moisture the thermal conductivity of foam with ash loss and 600 TMT is decreased to 30-32 % and at 90 % humidity to 25-

30 % compared with foamed concrete without additives.

Table 1. Thermal conductivity of foam concrete with the addition of peat and fly ash loss.

Thermal conductivity	Humidity, %			
	0 % W/mC	70 % W/mC	80 % W/mC	90 % W/mC
Foam concrete	0,14	0,2	0,23	0,24
Foam concrete with TMT 600	0,12	0,124	0,15	0,16
Foam concrete with fly ash	0,13	0,134	0,14	0,174

Using the obtained data the technological regulations of manufacturing of foam concrete mixes for prefabricated products have been developed.

4. Conclusion

Foam concrete with modifying mineral and organic additives have higher plastic strength by early structure formation, more homogeneous structure, compressive and bending rigidity and lower thermal conductivity in comparison with conventional foam concrete and can be recommended for use in the construction of thermal-efficient buildings.

Acknowledgement

The research performed with the financial support of the Ministry of Education and Science of the Russian Federation within Governmental Decree No.218 of "Measures of state support for cooperation of Russian higher education institutions and organizations, for implementation of complex projects of high-tech production", integrated project No. 02.G25.31.0022, "Production technologies of energy- and resource-efficient housing of economy class based on the universal framed prefabricated architectural and construction system".

References

[1] Матросов Ю А 2008 *Энергосбережение в зданиях. Проблема и пути ее решения* (Москва: НИИСФ) 496

[2] Zuhua Zhang, John L Provis, Andrew Reid and Hao Wang 2014 Geopolymer foam concrete: An emerging material for sustainable construction *Constr Build Mater* **56** 113-127

[3] Kudiakov A I, Kopanitsa N O, Prischepa I A and Shangin S N 2013 Structurally insulating foam with the addition of peat thermo modified *Herald Trace* 172-177

[4] Копаница Н О, Кудяков А И, Саркисов Ю С и Касаткина А В 2010 Влияние термомодифицированного торфа на свойства цементных систем *Сборник трудов «Научные исследования, наносистемы и ресурсосберегающие технологии в промышленности строительных материалов»* (Белгород, Россия) 166-170

[5] Прищепа И А, Шаньгин С Н и Курмангалиев Р М 2012 Влияние модифицирующих добавок на свойства пенообразователей цементных композиций *Сборник научных трудов IX Международной конференции студентов и молодых ученых «Перспективы развития фундаментальных наук»* (Томск, Россия) 776-778

[6] Morgun L V, Krylov O I and Shevaldykin P V 2003 Evaluation of the effect of surfactant concentration on the kinetics of plastic foam and strength fiber foam concrete *Proc. Conf. Building materials and technologies in the III millennium* (Rostov-on-Don, Russia) 52-56

[7] Udachkin I B 2002 Key issues in the development of the production of foam *Journal of Building Materials* **3** 8-10

[8] Ramamurthy K, Nambiar E.K.K. and Ranjani G.I.S. 2009 A classification of studies on properties of foam concrete *Cement Concrete Comp* **31** 388-396

[9] Лурье Ю Ю 1967 *Справочник по аналитической химии* (Москва, Россия)

[10] Прищепа И А, Толченников М В и Прищепа Б А 2013 Влияние добавок на структурно-механические свойства пенобетона (ПМТС-2013) *Материалы Первой Всероссийской научной конференции молодых ученых с международным участием "Перспективные материалы в технике и строительстве"* (Томск, Россия) 383-386

Aerated concrete with mineral dispersed reinforcing additives

G I Berdov[1], L V Ilina[1], I N Mukhina[2] and M A Rakov[1]
[1]Novosibirsk State University of Architecture and Civil Engineering, Department of Building Materials and Special Technologies, Novosibirsk, 630008, Russia
[2]Novosibirsk State University of Architecture and Civil Engineering, Department of Applied Mathematics, Novosibirsk, 630008, Russia

E-mail: nsklika@mail.ru, westcoast89@mail.ru

Abstract. To guarantee the production of aerated concrete with the lowest average density while ensuring the required strength it is necessary to use a silica component with a surface area of 250-300 m^2 / kg. The article presents experimental data on grinding the silica component together with clinker to the optimum dispersion. This allows increasing the strength of non-autoclaved aerated concrete up to 33%. Furthermore, the addition to aerated concrete the mixture of dispersed reinforcing agents (wollastonite, diopside) and electrolytes with multiply charged cations and anions (1% $Fe_2(SO_4)_3$; $Al_2(SO_4)_3$) provides the growth of aerated concrete strength at 30 - 75%. As a cohesive the clinker, crushed together with silica and mineral supplements should be used. This increases the strength of aerated concrete at 65% in comparing with Portland cement.

1. Introduction

Cellular concrete is an effective material for the frame structures and is used both in the form of panels and in the form of small building blocks, as well as in monolithic construction [1-4]. Despite the undoubted merits, cellular concrete is energy consuming building material. Cement is the most expensive concrete component and the silica component crushing and autoclaving are the most energy-intensive operations in cellular concrete technology.

The mass reduction after 2.0 -2.5 hours on 6.5-4.5% is a characteristic of the aerated concrete mixture based of natural sand with the specific surface 9.4 m^2/kg structure formation. The average density of aerated concrete based on the natural sand is 1140-1170 kg/m^3 and decreases with increasing the sand specific surface [5]. To ensure production of aerated concrete with average density 400-500 kg/m^3 the use of quartz sand with specific surface 250-300 m^2/kg (specific surface of no crushed sand is 3-19 m^2/kg) is required.

However, the clinker and the sand combining crushing have not been studied enough. To increase the hardness of materials based on Portland cement by introduction of mineral additives such as micro silica, fly ash, diopside etc. is widely used, helping realize the potential of Portland cement.

This work studied the effect of dispersed mineral additives and electrolytes on aerated concrete strength and density, and the effects of combing crushing the Portland cement clinker, sand and mineral additives to specific surface 280-300 m^2/kg.

2. Materials and methods

The cohesive has been the Portland clinker manufactured by "Iskitimcement", Ltd. company (Novosibirsk region), used in the production of Portland cement grade PC 400 d-20. It has the mineralogical composition,% mass: C_3S – 60-62, C_2S– 15-17, C_3A– 5-7, C_4AF– 14. Clinker was kept in the form of granules with size 5 - 20 mm for 7 days under standard conditions (temperature of 20 ± 2°C, humidity no more than 60 %) and 12 months in moist conditions (humidity 80-85 %).

As a mineral additives the next crushed rocks were used: wollastonite (Sinyukhinskoye deposit, mine "Veselyi", the Republic of Altai), diopside (Bugotakskoe deposit, Irkutsk region) that represent a production waste. Chemical compositions of wollastonite and diopside are shown in Table 1.

Wollastonite additive was admixed in 7%, diopside – 5% from the mass of clinker. Such content, as shown by previous experiments, is close to the optimal value [6]. The added electrolytes were salts with multiply-charged cations anions – $Fe_2(SO_4)_3$ or $Al_2(SO_4)_3$. The salts were mixed into the water in amount of 1% from clinker mass which corresponds to its optimal concentration to increase the cement stone strength [7].

Table 1. The chemical composition of the mineral additives.

Additive	The content of oxides, % of the mass								
	SiO_2	CaO	MgO	Al_2O_3	Fe_2O_3	Na_2O	K_2O	TiO_2	LOI
Wollastonite	53,4	34,7	0,3	3,1	2,4	-	-	-	6,1
Diopside	56,1	25,4	15,8	1,0	0,7	0,1	0,1	0,1	0,6

As a small filler the quartz sand from "Kamnerechenskij stone quarry",JSC,"NerudZapsib", CJSC (Novosibirsk region) was used. The mineral composition of the sand,% mass: quartz – 80-90, feldspar – 10-20. True density of sand grains was 2650 kg/m^3. The pour density of the sand was 1420 kg/m^3. Mass loss on roasting was 0.45%, contents of clay and silt particles – 0.5%.

Compressive strength and average density were determined on samples of size 100x100x100 mm, made of aerated concrete mixture, % mass: the Portland clinker– 27.23-28.36, silica component– 31.5, calx– 4.5, aluminum powder– 0.08, sulfanol– 0.001, water– 31.56, two-water gypsum stone– 2.27, diopside– 1.42-2.55, electrolyte $Fe_2(SO_4)_3$ or $Al_2(SO_4)_3$– 0.28. The samples hardened in conditions of heat and humidity processing with the regime: 3 hours the temperature rise up to 90°C, 8 hours isothermal aging at a given temperature and 3 hours reducing the temperature to 20°C.

3. Results and discussion

Consider the dependence of aerated concrete properties from a cohesive type. The strength and density of aerated concrete produced using Portland cement, crushed clinker, as well as clinker, crushed together with 50% sand have been defined (Table 2).

Table 2. Depending the strength and density aerated concrete on the cohesive type.

Cohesive type	Density, kg/m^3	Compressive strength, MPa
Cement	630	1.7
Clinker, crushed apart from sand	628	2.1
Clinker, crushed together with 50% sand	610	2.8

The clinker strength increased by 30% when crushed concrete was used comparing the Portland cement was applied. Even higher results were achieved when the clinker crushed together with 50% sand was used. While the aerated concrete density reduced from 630 to 610 kg/m^3, the compressive strength increased on 65%.

Later the joint crushing clinker with sand, mineral additives and carbonate component was applied. Water for dilution containing electrolytes was used to stir the dry aerated concrete mixture. The concrete test results are presented in Table 3.

Integration of 7% wollastonite or 5% diopside to aerated concrete mixture reduces the concrete density. The compressive strength increased by 11% with wollastonite and 18% with diopside additions. Strength increase may be explained by reinforcing effects of used additives.

A much greater increase in aerated concrete strength is seen while at the same time with wollastonite and diopside additives 1% electrolytes with multiply-charged cations and anions $Fe_2(SO_4)_3$ and $Al_2(SO_4)_3$ are used. Especially effective is the influence of the last salt. The growth in compressive strength is 57 % if one uses wollastonite and 75 % in the case of diopside.

Table 3. The strength and density of aerated concrete containing additives, made of Portland cement clinker.

The type and quantity of additives	Density, kg/m^3	Compressive strength, MPa	The strength increase, %
Clinker, stored 7 days under normal conditions			
Noadditives	610	2,8	-
Wollastonite7 %	600	3,1	11
Wollastonite7 % + 1 % $Fe_2(SO_4)_3$	615	3,9	25
Wollastonite7 % + 1 % $Al_2(SO_4)_3$	625	4,4	57
Diopside5 %	580	3,3	18
Diopside5 % + 1 % $Fe_2(SO_4)_3$	600	4,4	57
Diopside5 % + 1 % $Al_2(SO_4)_3$	610	4,9	75
Clinker, stored 12 months under damp conditions			
Noadditives	610	2,1	-
Wollastonite7 %	605	2,3	8,7
Wollastonite7 % + 1 % $Fe_2(SO_4)_3$	613	3,1	48,0
Wollastonite7 % + 1 % $Al_2(SO_4)_3$	620	3,4	62,0
Diopside5 %	590	2,4	14,3
Diopside5 % + 1 % $Fe_2(SO_4)_3$	603	3,4	62,0
Diopside5 % + 1 % $Al_2(SO_4)_3$	615	3,8	81,0

Effect of electrolytes containing ions of Fe^{3+} and Al^{3+} may be due to their intense exchange with clinker Ca^{2+} ion resulting in accelerated hydration alite and belite– silicate clinker minerals.

Sometimes, when the aerated concrete production is located in regions where there are no cement plants (Far North, Far East) the Portland clinker can be stored for a long time because the delivery is carried out during the short summer season. When using long-stored clinker aerated concrete strength significantly reduces (by 25 %). Adding dispersed minerals and electrolytes leads to increased strength as if clinker was stored 7 days under normal conditions (Table 3).

4. Conclusions

The strength of aerated concrete with additives of dispersed minerals and electrolytes that produced using long-stored clinker exceeds the strength of aerated concrete with no additives, produced using clinker stored 7 days under normal conditions. The hardness of aerated concrete manufactured from clinker crushed together with sand and mineral additives and stirred with salt solution increases in 2.9 times comparing with the hardness of aerated concrete produced from Portland cement without additives. In addition there is a decrease in the average density of aerated concrete with 630-615kg/m^3.

References

[1] Завадский В Ф 2005 Оптимизация параметров получения газобетона на новых видах дисперсных наполнителей *Известия вузов. Строительство* **4** 58-63

[2] Лесовик В С, Гладков Д И и Елистраткин М Ю 2002 Неавтоклавный газобетон повышенного конструктивного качества *Сборник трудов Международной научно-практической конференции «Рациональные энергосберегающие конструкции, здания и*

IOP Conf. Series: Materials Science and Engineering **71** (2015) 012017 doi:10.1088/1757-899X/71/1/012017

сооружения в строительстве и коммунальном хозяйстве» (Белгород, Россия) 135-137, 260
[3] Овчаренко Г И, Щукина Ю В, Борисов А А и Косенко К Г 2008 Сухие смеси для производства неавтоклавного газобетона *Сухие строительные смеси* **5** 8-10,76
[4] Sinica Marijonas, Laukaitis Antanas, Sezemanas Georgijus, Mikulskis Donatas, Keriene Jadvyga 2004 The influence of binding materials composition and fibrous addi-tives on the properties of porous concrete *J Civ Eng Manag* **10** 131-136
[5] Кудяков А И, Некрасов А В, Чубаков Н Г и Исаков Г Н 1996 Газобетон на песке естественной гранулометрии для строительства монолитных малоэтажных домов *Известия вузов. Строительство* **1** 46-48
[6] Бердов Г И и Ильина Л В 2010 Влияние количества и дисперсности минеральных добавок на свойства цементных материалов *Известия вузов. Строительство* **11-12** 11-16
[7] Бердов Г И и Ильина Л В 2010 Активация цементов действием добавок электролитов *Современные наукоемкие технологии* **9** 108-110

IOP Conf. Series: Materials Science and Engineering **71** (2015) 012018 doi:10.1088/1757-899X/71/1/012018

Cement clinker structure during plasma-chemical synthesis and its influence on cement properties

N Sazonova[1], N Skripnikova[2], A Lucenko[2] and L Novikova[1]
[1]Angarsk State Technical Academy, Department of Industrial and Civil Engineering, Angarsk, 665835, Russia
[2] Tomsk State University of Architecture and Building, Department of Applied Mechanics and Materials Science, Tomsk, 634003, Russia

E-mail: n.a.sazonova@mail.ru

Abstract. The aim of this study was to determine the degree of influence of cement clinker cooling modes, synthesized in a low-temperature plasma, its structure and physico-mechanical properties. The raw mixture consisting of marble, sand, ash from thermal power plants and pyrite cinders were used, which are characterized by saturation factor (1,045); silicate (2,35) and alumina (1,22) modules. It was found that the use of different cooling rates of fused cement clinker entails changes associated with the mineralogical composition (increase of alite of 8.7-19,2 %), morphology (variation of the mineral alite aspect ratio of 6,7–17,5), density of the structure (change in distance between the minerals from 1 to 7,5 microns), grindability, specific surface area (2600–3650 cm^2/g) and, in consequence, the activity of cement (56,9–73,2 MPa). Disorientation of alite mineral blocks against each other, a significant amount of microcracks, affect the increase in cement specific surface area of 14,3–21,6 %, which leads to activity growth of the system. Along with this, with the rapid cooling of the samples, alite $54CaO \cdot 16SiO_2 \cdot Al_2O_3 \cdot MgO$ is formed, with single units of the structure, more deformed relatively to C_3S, which has a positive effect on the hydraulic cement activity.

1. Introduction

Crystals of clinker minerals, formed during the non-equilibrium crystallization of a multicomponent melt in a low-temperature plasma, undergo considerable changes [1]. Significant impact on the morphology modification of clinker minerals is made by technological modes, where the main one is cement clinker cooling mode [2–4], whose correction allows not only to influence the shape, size of new growths, but also their structure and most importantly properties of the binders obtained on them. The present work is a separate part of a series of experiments in the study of cement clinker, synthesized in a highly concentrated heat fluxes. The aim of this study was to research the structural changes of clinker minerals when the raw material mixture is exposed to highly concentrated heat fluxes, with cooling modes being adjusted, and to study their impact on the activity of the synthesized binder.

2. Materials and research methods

The following raw materials were used: marble, sand, ash from thermal power plants, pyrite cinders, with the help of which the raw mixture was made. Modular characteristics of the charge are represented by the saturation factor (SF) equal to 1.045; silicate modulus (n) – 2,35; alumina module (p) – 1,22 [5]. The chemical composition of the mixture by weight, %: SiO_2 – 12,98; Al_2O_3 – 3,03; Fe_2O_3 – 2,49; CaO – 43,86; MgO – 3.2; LOI – 0,25.

The study of materials and cement clinker synthesized in a low-temperature plasma (LTP) was performed by using chemical analysis method in accordance with the requirements of GOST 5382-91 "Cements and materials in cement production. Methods of chemical analysis". Sampling was carried out with the method of cone quartering. The material was prepared by drying to constant weight and grinding to full pass through a sieve № 0063. X-ray analysis was carried out using the equipment Shimadzu XRD-7000 X-ray Diffractometer. The diffraction peaks were studied in the range of 2θ, equal to 3–80°, with a tube with radiation Cukα, 40 kV and 30 mA. Sampling step was 0,05; speed – 1 deg/min. Microstructure analysis was performed on petrographic microscope POLAM-R312 with the magnification of 450× and a scanning electron microscope Phenom Pro in the range 20–10000×, resolution up to 17 nm. We studied polished thin sections of the fused clinker etched for 2–3 sec. with a 1 % solution of nitric acid, and its chippings.

Heat treatment of the raw mixture was made using plasma-chemical reactor [6], where the temperature reached 3000 °C. Fractional composition of the raw mixture was less than 140 microns. Technological modes of thermal treatment of the samples: the exposure time of highly concentrated heat fluxes was determined by complete melting of the mixture and was 140 sec; with current intensity of 140–160 A; voltage 180–200 V. Under otherwise equal conditions cement clinker was cooled in different ways: under the air pressure (sample №1), in the open air (sample №2), in a closed kiln (sample №3). Subsequently the synthesized fused cement clinker was milled using a planetary ball mill PQ-N2.

3. Physical and chemical analysis of cement clinker

As a result of the chemical analysis of the synthesized cement clinkers, it was found that the difference between the estimated and actual mineral content, which was caused by the occurrence of non-equilibrium solidification process, is significant. The amount of alite grows by 8,7–19,2 % against the estimated content (Figure 1), while C_2S, C_3A, C_4AF decreases respectively by 4,5–16,3 %; 2,0–3,2 %; up to 1,7 %. The maximum number of alite is produced in the sample №1 and makes 79,1 %.

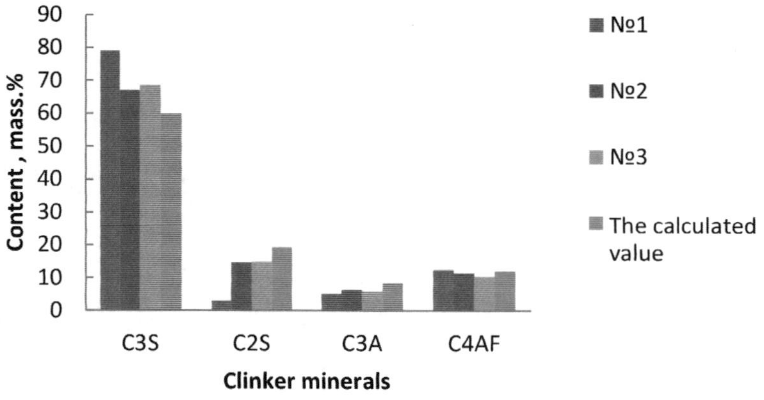

Figure 1. Estimated and actual content of the clinker minerals in samples No.1–3.

The microstructure of fused cement clinker, observed using an electron microscope and a petrography (Figures 2–4), varies considerably depending on technological regimes. The sample №1, quenched with cold air pressure after 140 sec exposure of highly concentrated heat fluxes contains mainly alite crystals and to a lesser extent belite. Alite has an irregular geometric shape of a platelet, which is characteristic of clinkers containing large amounts of mineral fluxes, as well as obtained by the process of fusion. In the thin sections alite newgrowths are evenly spaced in blocks that are disoriented relatively to one another at an angle of 30°. The sample has a very dense structure: the distance between the minerals is mainly of 1–2 microns.One can see inclusions in crystals of alite, whose number indicates the intensity of the melting process of the raw material mixture, when the fast-growing seed crystals "bypass" and include small particles of foreign matter. One can observe round or oval-shaped belite crystals that are enveloped by alite crystals. Their content is very slight. Alit has a size of (1–13)×(8–150) μm, belite is 1–7 μm (figure 2, a). Length/beam ratio of alite minerals (l/d) is 17,5 which indicates their high hydraulic activity [7]. Single cracks are present traversing both crystals and interphase, whose presence may be related to tensile stress which arise when the mixture is heated and compression strain with a significant thermal gradient during cooling of the melt. This effect may also indicate the transition from β-C_2S into γ-C_2S, because the latest modification has a lower density and increases in volume by 12 % while cooling [4]. However, X-ray analysis (figure 2, b) of the sample No.1doesn't show diffraction peaks characteristic of this modification, which confirms the presence of significant stress, reducing the strength of fused cement clinker and increasing their grindability.

a) b)

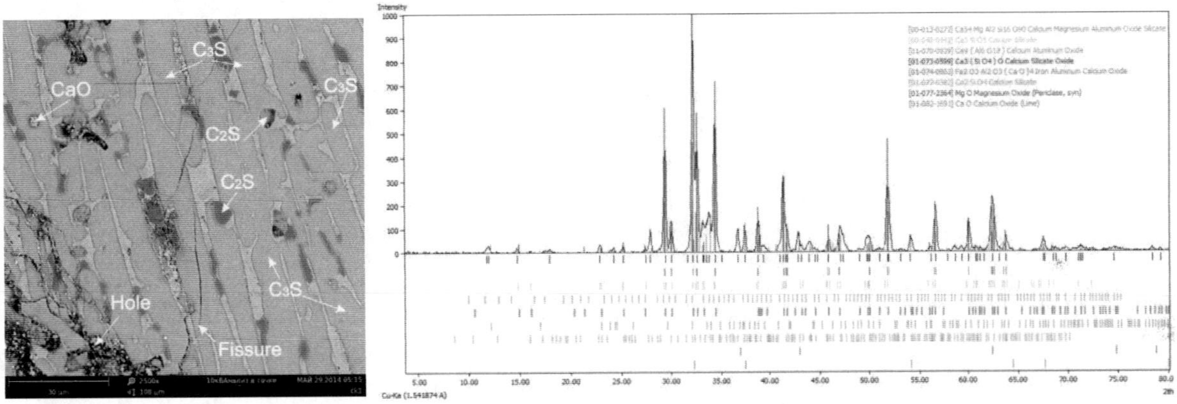

Figure 2. Fused cement clinker (sample No.1): a) structure, increase in 2500×; b) diffraction pattern.

As a result of X-ray phase analysis of cement clinker №1 it was found that the sample contains the diffraction maxima C_3S, β-C_2S, C_3A, C_4AF (Figure 2, b). The property that defines the cement clinker synthesized in a low-temperature plasma is the presence of alite having formula $54CaO \cdot 16SiO_2 \cdot Al_2O_3 \cdot MgO$, whose formation occurs during the replacement of two Si^{4+} ions with the two ions Al^{3+} and Mg^{2+} introduction into the interstitials of the lattice site. Individual elements of the structure of the compound are more deformed relatively to C_3S, which increases their hydraulic activity [8].

The morphology of sample No.2 (Figure 3), cooled at room temperature, is distinguished by less dense monadoblastic structure having fewer enveloped belite crystals. The structure is more ordered against sample №1 and has a greater number of disoriented alite blocks with the angle between them varying from 30 to 90°. The distance between the minerals varies from 1 to 3 microns. Alite and belite size is 1,5–2 times smaller than in the sample No.1 and is respectively (2–10)×(10–100) μm, 1–3 microns. However, there is a reduction ratio l/d, which is equal to 14. C_3S crystals have an irregular geometric oval and plate shape; C_2S are dense oval. On the thin section, some crystals of alite are perforated, which may indicate slow cooling. The sample is transpierced with a web of microcracks, whose number greatly exceeds the sample No.1 and should lower the micro-hardness of the samples more, which indicates an increase in the internal stresses during cooling of the melt. The X-ray pattern shows clinker minerals similar to sample №1. There is a slight decrease in the diffraction peaks belonging to $C_{54}S_{16}AMg$ showing that the synthesis of the represented compound is influenced not only by the highly concentrated heat fluxes, but also by the cooling mode.

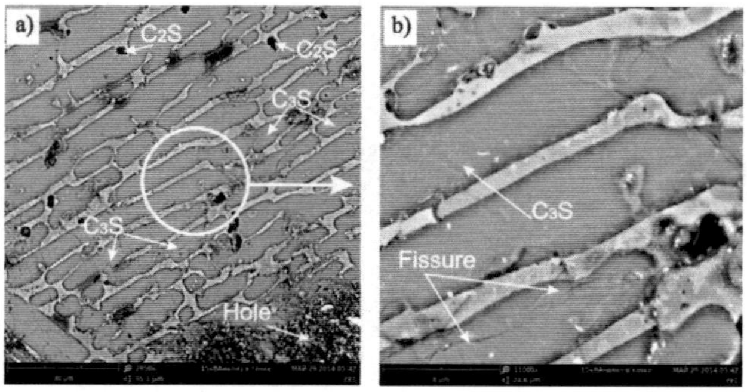

Figure 3. Structure of fused cement clinker (sample No.2), increase in: a) 2850×; b) 11,000×.

Increasing time for cooling samples leads to aggregation of alite crystals (sample No.3), having dimensions of (10–30)×(40–160) μm, the ratio l/d is 6,7 and irregular geometric shapes: plate and perforated multi-faceted (Figure 4, a). The number of additional inclusions on the surface of alite increases. The structure is less dense relatively to the previous samples: the distance between the minerals varies from 2 to 7,5 microns. Cracks in the minerals are nearly absent. In the interstitial material we can clearly see crystallized minerals tricalcium aluminate and tetracalcium alumoferrite. X-ray diffraction analysis of samples No.3 (Figure 4, b) revealed that there was a synthesis of the main clinker compounds. However, the diffraction peaks $C_{54}S_{16}AMg$ are difficult to identify because in most cases they are overlapped by lines C_3S, whose number is dominating.

a) b)

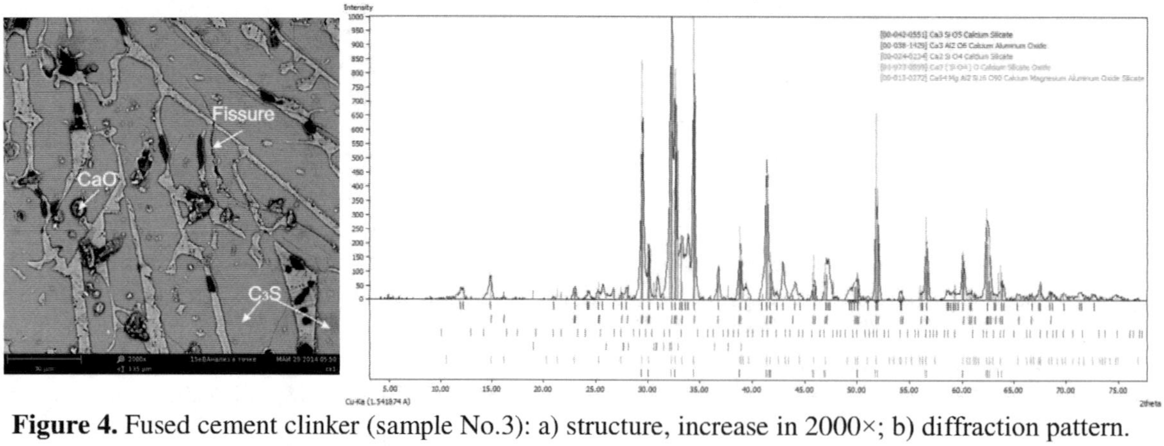

Figure 4. Fused cement clinker (sample No.3): a) structure, increase in 2000×; b) diffraction pattern.

4. Building and technical properties of cement

Dispersing capacity of cement is one of the key properties that influence the rate of hydration, kinetics of supersaturation of the liquid phase and heat release in the formation of hydrated compounds [9]. At this stage we were interested to follow the changes of the specific surface of the samples during the milling, depending on the method of cooling. With this objective in view, the fused clinker was previously crushed in the laboratory jaw crusher to a fraction of 4–10 mm. Equal test portions of clinkers synthesized under different conditions, were subjected to grinding in a laboratory planetary ball mill, under identical process conditions. Milling fineness of the cement was evaluated by the residue on the sieve number 008, and the specific surface area, which made 4,5–5,1 % (3000–3150 cm^2/g); 3,2–3,6 % (3600–3650 cm^2/g); 8,4–9,2 % (2600–2750 cm^2/g) for the samples number 1, 2, 3 respectively. Thus, the specific surface area of cement clinkers No.2 exceeds sample No.1 on 14,3–21,6 %; and sample No.3 on 32,7–38,5%. The results indicate a significant reduction in the time of cement clinker grinding, whose structure is dominated by cracks that run both through the minerals as through the interstitial material. Grindability is affected not only by the presence of pores, micro and macro-cracks, but also by the type of the packing of clinker minerals and their size. Structure solidification and increase of the crystal size increases the time of cement clinker milling. Therefore, the natural cooling of the melt in the air provides optimal conditions for the formation of significant internal stresses, which decrease the mechanical strength of the cement clinker and make the structure uniform and porous.

The time of cement setting, depending on the mineralogical composition, ambient temperature, additives of gypsum, milling fineness, etc., has been determined in accordance with GOST 310.3-76. The content of calcium aluminate has a special influence on this characteristic. However, the fused cements are less sensitive to the content of the aluminate phase, because of the nonequilibrium mineral crystallization and the presence of the glass phase, and can be easily adjusted with gypsum [7]. In this regard, the additive of gypsum in amount of 5% was added during the grinding of the fused cement clinker, which provided the normal setting time, which varied in the following ranges: start – 1 hour 30 minutes, the end – 5 h 50 min. Fundamental differences in the samples with identical specific surface area were not observed.

Despite the use of raw material mixture with a saturation ratio greater than 1 and the presence of increased amounts of MgO, the cements produced from a fused clinker withstand soundness test during solidification, steaming and autoclave treatment at a pressure of 2,1 MPa. Ensuring of the soundness during the solidification of the studied cements may be explained by the non-equilibrium process of crystallization of minerals in the conditions of the continuous process when magnesium oxide either does not crystallize in the form of periclase and remains in the glass phase, or is intruded additionally in the crystal lattice of the other components [1, 10]. Free calcium oxide is extremely rare in the structure of the samples and does not form clusters and has a size less than 5 microns, which has no negative influence on the soundness during solidification of the processed binder.

Water demand of the studied cements produced from samples No.1, 2, 3 is stable and essentially depends on their specific surface, whose increase raises the water demand. In this case, the normal density, characterizing the rheological properties of cement paste and used to determine the rational value of W/C, was determined in accordance with the requirements of GOST (Russian national state standard) and was 0,28–0,32.

Thus, the application of different rates of cooling of cement clinker entails a chain of changes associated with both mineralogical composition, morphology, grindability, specific surface area and, as a consequence, the cement activity. Homogeneity of the sample No.1, predominate formation of alite minerals $54CaO \cdot 16SiO_2 \cdot Al_2O_3 \cdot MgO$ with a ratio l/d equal to 17,5 and density of the cement clinker matrix affect favorably on the strength characteristics of the cement. For example, at the age of 3 days they are 17,2–18,0 MPa, at the age of 7 days – 50,4–52,1 MPa; at 28 days – 71,5–73,2 MPa. Thus, the introduction of foreign ions (Al^{3+} and Mg^{2+}) into the crystal lattice of alite is accompanied by a disturbance of its structure and increases the activity of its interaction with water. It is of interest that the strength characteristics of the samples obtained by using the cement clinker No.2, increase by 10–14 %. Microstructure of the sample is similarly represented by alite minerals, but with a smaller ratio l/d equal to 14, which should reduce the activity of the cement. However, due to the larger mineral blocks disorientation against each other, more microcracks increasing the specific surface of the cement by 14,3–21,6 %, the activity of the system increases. Therefore, these factors are essential. In this regard, the strength characteristics of the samples No.3 are significantly inferior to their predecessors and being 3 days old they are 10,3–11,8 MPa; 7 days – 39,9–40,5 MPa; 28 days – 56,9–59,3 MPa.

5. Conclusion

According to the obtained data, as a result of physical-chemical and physical-mechanical studies it may be noted that the cement clinker synthesized in a low-temperature plasma has increased the amount of alite minerals due to non-equilibrium solidification processes, raising defectiveness in the lattice of tricalcium silicate, which positively influence the cement quality. It has been established that one of the parameters for assessing the activity of a binder, based on the fused cement clinker, may be not only the morphology of minerals; length-beam ratio of alite mineral (l/d); sealing of crystals in thin section, but also the presence and nature of microcracks arising in the structure, which have a significant influence on the microhardness of the samples, the time of their milling and specific surface area. Air-cooling of a fused clinker leads to fine-grained matrix model composed mainly of alite minerals, with less amount of belite. Despite the larger ratio l/d of alite and less dense structure of the sample No.1, cooled under the air pressure, the activity of cement No.2, whose clinker is cooled naturally, is higher. This is due to a web of microcracks in the structure helping to increase the specific surface of binder (3600–3650 cm^2/g) and strength (78,6–83,4 MPa).

Reducing the rate of cement clinker cooling leads to considerable enlargement and clearer form of alite crystals, which is accompanied by a marked decrease in the activity of the system – 56,9–59,3 MPa. Herewith, the mineral, as a chemical compound, continues to maintain its active state, but due to the fact that its crystals are approaching the maximal sealing and saturation of bonds not neutralized previously by charge of intruded ions, it affects adversely the activity of its interaction with water.

References

[1] Skripnikova N and Sazonova N 2013 *Vestnik of ISTU* **8** 33
[2] Maki I, Ito S, Tanioka T, Ohno Y 1993 *Cement Concrete Res* **23** 1078
[3] Maki I, Tanioka T 1994 *Proc. of the 16th Int. Conf. on Cement Microscopy* 113
[4] Hewlett P 2006 *Lea's chemistry of cement and concrete* 1035
[5] Volokitin G, Skripnikova N, and Sazonova N 2011 *Vestnik of TSUAB* **4** 146
[6] Volokitin G, Skripnikova N, Pozdnjakova N 2008 *Vestnik of TSUAB* **4** 106
[7] Timashev V 1986 *Selected Works. Synthesis and hydration of cementitious materials* 424
[8] Yamaguchi G, Uchikawa H, Takagi Sh and Koike H 1962 *Influence of alkalis in clinker upon the hydration of tricalcium solid solution in Portland cement* (*Yogyo Kyokai shi*) **70** 147
[9] Sarkar Shondeep L *1990 Cement Concrete Res* **20** 398
[10] Skripnikova N, Sazonova N and Volokitin G 2013 *European Science and Technology* **5** 476

IOP Conf. Series: Materials Science and Engineering **71** (2015) 012019 doi:10.1088/1757-899X/71/1/012019

Shrinkage deformation of cement foam concrete

A I Kudyakov[1] and A B Steshenko[1]
[1]Tomsk State University of Architecture and Building, Department “Building Materials”, Tomsk, 634003, Russia

E-mail: steshenko.alexey@gmail.com

Abstract. The article presents the results of research of dispersion-reinforced cement foam concrete with chrysotile asbestos fibers. The goal was to study the patterns of influence of chrysotile asbestos fibers on drying shrinkage deformation of cement foam concrete of natural hardening. The chrysotile asbestos fiber contains cylindrical fiber shaped particles with a diameter of 0.55 micron to 8 microns, which are composed of nanostructures of the same form with diameters up to 55 nm and length up to 22 microns. Taking into account the wall thickness, effective reinforcement can be achieved only by microtube foam materials, the so-called carbon nanotubes, the dimensions of which are of power less that the wall pore diameter. The presence of not reinforced foam concrete pores with perforated walls causes a decrease in its strength, decreases the mechanical properties of the investigated material and increases its shrinkage. The microstructure investigation results have shown that introduction of chrysotile asbestos fibers in an amount of 2 % by weight of cement provides the finely porous foam concrete structure with more uniform size closed pores, which are uniformly distributed over the volume. This reduces the shrinkage deformation of foam concrete by 50%.

1. Introduction

Much attention is paid to research and development for energy conservation and its efficient use in the development program of Russia up to 2020 [1]. Currently, about 20% of energetic resources are spent for heating of buildings, about 45% of heat energy is lost through the building envelopes. Therefore, the development of structural heat-insulating materials, that are energy efficient both at the production stage and during the operation, is very important. The representative of energy-efficient wall materials is foam concrete. The research on improving production processes of foam concrete and its quality for precast and monolithic building envelope is relevant. Foam concrete is multifunctional building material that is made from inexpensive and readily available raw materials without using of imported components. Foam concrete has a relatively low average density, low thermal conductivity, high durability and it is environmentally friendly [2].

However, non-autoclaved foam concrete has significant shortcomings. For ensuring high level of quality in building structures which are made of foam concrete it is necessary to control a large quantity of technological parameters: the quality and quantity of raw materials, water solid ratio, rheology, temperature and pH environment. The process of structure formation of cellular concrete is difficult to control and regulate. The actual conditions of foam concrete structure formation often deviate from the optimum, which leads to the emergence of defects in its structure and reduction in the quality. Shrinkage strain refers to such defects. Moreover the smaller the density the greater the shrinkage deformation property is. Shrinkage strain of foam concrete with density of $300 \div 400$ kg/m^3 can be up to 7 mm/m. Such high values of shrinkage strain impose very severe restrictions on the scope of the foam concrete products of natural hardening and efficiency of their use in monolithic construction [3]. The problem of obtaining non-shrinking cellular concrete of natural hardening is very relevant, but is still not solved despite of several studies [4,5, 6] has been carried out on that subject.

The primary cause of shrinkage deformation of foam concrete of natural hardening are cracks occurring in concrete's interporous partition walls during the technological processing of raw materials and hardening of the cement stone. The propagation rate and size of cracks in concrete, according to P.G. Komohov [7], are determined by its structural properties that are formed by the structure and properties of raw materials and processing methods that ensure their interaction and continuity. The energy of any loaded volume is converted into crack and leads to the destruction of the material, therefore the concrete structure must include not only the durable and fragile base, but viscoelastic inclusions, capable of absorbing the growth energy of propagating crack. Fibers are visco-elastic components, i.e. a reserve of plasticity that increases the limit of the material destruction and also has a positive effect on the features of the structure during the preparation of fiber reinforced foam mixtures by one-stage technology, and its structure formation during natural hardening [8].

Many fibers are used in concrete technology such as steel, carbon, mineral and other fibers. Earlier, the type and amount of fibers to improve the efficiency of foam concrete was determined [9]. This paper presents the results of the research of shrinkage deformation of cement foam concrete with micro-reinforcing additive in the form of chrysotile asbestos fiber with an optimal content of 2% by weight of cement.

The research goal is to study the patterns of influence of chrysotile asbestos fibers on drying shrinkage deformation of cement foam concrete of natural hardening.

The research object is the hardened foam concrete with chrysotile asbestos fibers. The processes of the production of the modified foam concrete with fibers are under investigation.

2. The study of chrysotile asbestos fibers

Taking into account the wall thickness, effective reinforcement can be achieved only by microtube foam materials, the so-called carbon nanotubes, the dimensions of which are of power less than the wall pore diameter. One material exist in nature which comply to these requirements and can be used as dispersing reinforcement: chrysotile $Mg_3Si_2O_5\,(OH)_4$. This material in the fluff form of fibers has a tubular structure, and the fiber diameters are comparable in size to the thickness of the walls [10]. In this research chrysotile asbestos fiber grade A-6K-30 of Bazhenovsky field was used.

According to the data obtained from SEM Quanta 200 3D investigation it can be said that the chrysotile asbestos composed of fiber shaped agglomerations length up to 1.35 mm with a complex inner structure containing microstructures of a cylindrical shape. In Figure 1, a, scanning electron microphotograph is given of chrysotile asbestos fiber. This sample contains cylindrical fiber shaped particles with a diameter of 0.55 micron to 8 microns (Figure 1, a) which are composed of nanostructures of the same form with diameters up to 55 nm and length up to 22 microns (Figure 1, b).

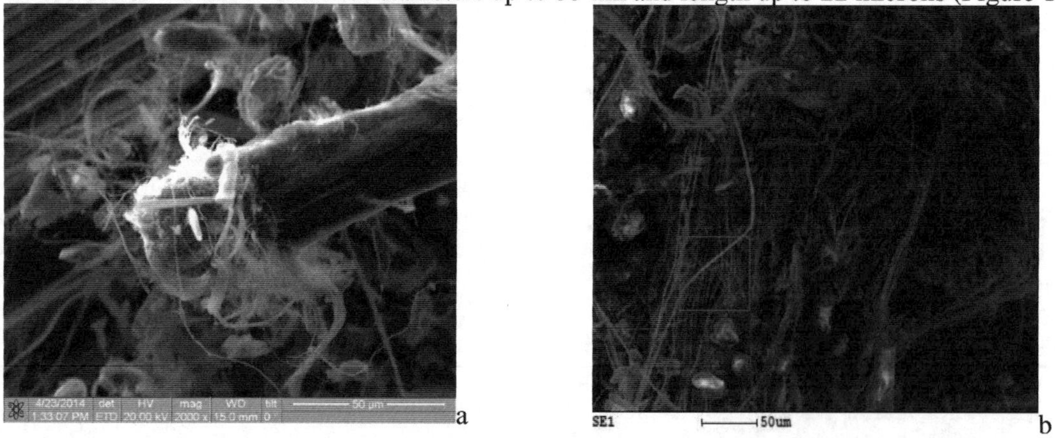

Figure 1. Microstructure of the chrysotile asbestos fiber: a – a fragment; b – general view.

With the additional consoles of SEM Quanta 200 3D, power-dispersion analyzer, it could be managed to get the elemental microanalysis of chrysotile asbestos fibers (Figure 2). The chrysotile asbestos fiber grade A-6 K-30 Bazhenovsky field is represented by oxides of silicon and magnesium (Figure 2). In addition, the chemical composition of the fiber includes a small amount of oxides of calcium, aluminum and iron.

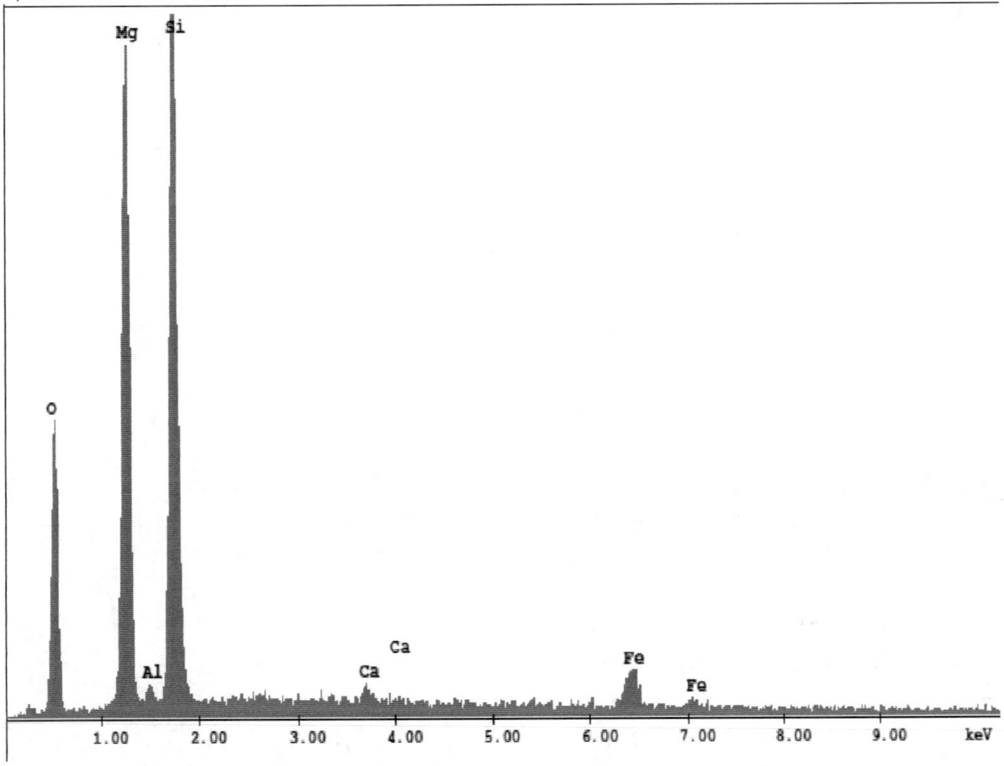

Figure 2. Elemental microanalysis of the chrysotile asbestos fibers.

3. Shrinkage of foam concrete reinforced chrysotile asbestos fibers

These fibers have been used as high-strength reinforcing element of the cement matrix to improve the physical and mechanical properties of the cement foam concrete of natural hardening. In the study was used a mixture of foam concrete, which includes: Portland cement of Topkinsky plant M500 (GOST 101178-85), quartz sand of Kudrovskoe field of Tomsk region with a fineness modulus of 1,8 (GOST 8736-93), foam agent PB-2000 (TC 2481-185-05744685-01), tap water (GOST 23732-79). The foam concrete mixture was produced by one-step technology in the laboratory mixer. For testing shrinkage strain finished concrete mix is placed in metal molds 4x4x16 cm. Molded foam concrete samples are incubated in closed desiccators over water. The testing and the evaluation of the quality of foam concrete were conducted in accordance with Russian State Standard GOST No 25485-89.

As follows from the results of the study adding chrysotile asbestos fibers in the foam concrete mixture significantly reduced shrinkage deformation of this material, up to 50% (Figure 3).

During the mixing, the fibers evenly distributed throughout the cement-sand mixture that provides spatial reinforcement throughout the foam concrete volume. Thus provides an optimization of the composition that prevents the probability of the formation and developing intrastructural defects in foam concrete [11]. Distributed chrysotile asbestos fibers in the cement foam concrete act as nucleation sites, which on the one hand lead to the development of fibrillar structure on the pore walls, which in turn ensures its continuity and uniformity (Figure 4. b), while on the other hand ensures an orderly structure orientated supermolecular sheath about the fiber. This reduces the shrinkage of foam concrete.

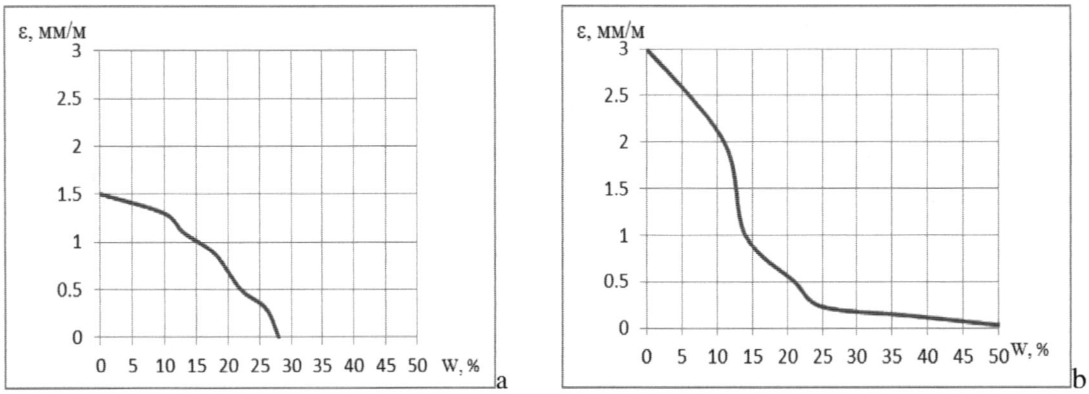

Figure 3. The curves of drying shrinkage of the cement foam concrete D500: a – fiber reinforced; b – without fibers.

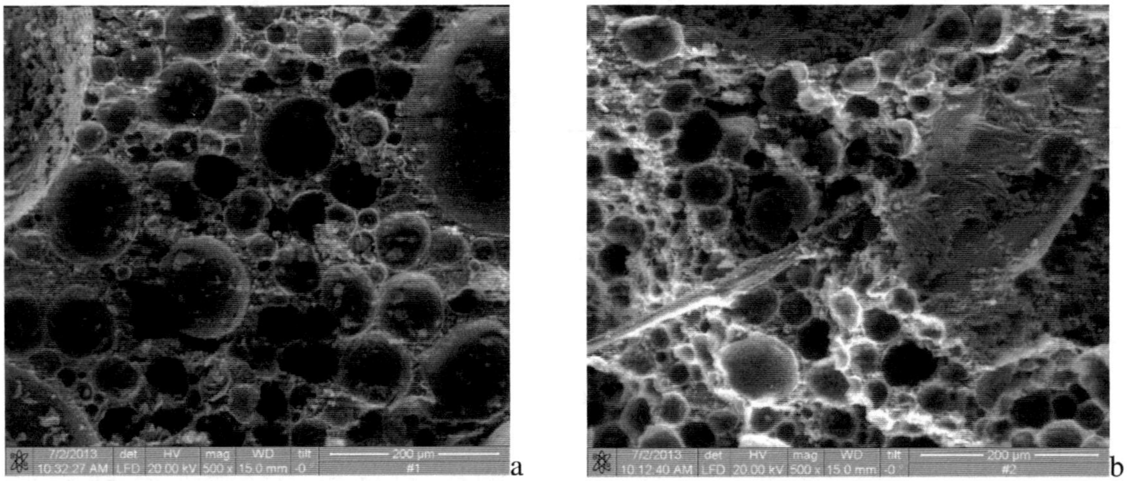

Figure 4. The structure of the cement foam concrete D 400: a – without fibers, b – fiber reinforced.

The microstructure investigation results have shown that foam concrete structure with chrysotile asbestos fibers are finely porous, with lots of uniform size and closed pores, which are uniformly distributed over the volume. The chrysotile asbestos fibers ensure the absence of pore wall percolation and stabilize its structure (Figure 4. b). In concretes without fibers, due to intensive percolation of wall pores (Figure 4. a) causes them to combine into larger ones, which in turn increases the shrinkage of the foam concrete and leads to deterioration of its operational properties. The presence of not reinforced foam concrete pores with perforated walls causes a decrease in its strength; it decreases the mechanical properties of the investigated material and increases its shrinkage. The stabilization of foam concrete structures with chrysotile asbestos fibers is mainly due to the reinforcement effect and the formation of the fibrillar structures, and due to the ordering effect of the a supermolecular structure in the pore walls.

A significant decrease in the values of shrinkage strain of the foam concrete during the natural hardening can predict a reduced level of stresses formation in the structure of foam concrete and, as a consequence, increased robustness of products made of this material. Regulations for manufacture of foam concrete with chrysotile asbestos fibers were developed.

4. Conclusions

The results of the investigation of the chrysotile asbestos fibers carried out by using electron microscopy showed that the fibers contain oxides of silicon and magnesium, they have the cylindric form, their diameter is of 0.55 micron to 8 microns. The fibers are agglomerated to fiber shaped agglomerates with a diameter up to 55 nm and length up to 22 microns. The introduction of the reinforcing additives (chrysotile asbestos fibers) is necessary to produce the foam concrete of natural hardening and its products with low shrinkage strain.

The findings suggest that the introduction of the chrysotile asbestos fibers to foam concrete mixture in amount of 2 % by weight of cement provides the finely porous structure with more uniform size closed pores, which are uniformly distributed over the volume. Experimental studies have established that this ensures a decrease in shrinkage deformation of cement foam concrete by 50%.

Acknowledgement

The research performed with the financial support of the Ministry of Education and Science of the Russian Federation within Governmental Decree No.218 of "Measures of state support for cooperation of Russian higher education institutions and organizations, for implementation of complex projects of high-tech production", integrated project No. 02.G25.31.0022, "Production technologies of energy- and resource-efficient housing of economy class based on the universal framed prefabricated architectural and construction system".

References

[1] Стратегия-2020: Новая модель роста – новая социальная политика [online resource] http://2020strategy.ru/

[2] Morgun V N 2003 On the development of strains in reinforced foam concrete based cements with expansive agents *Construction materials* **9** 10

[3] Steshenko A B and Kudyakov A I 2014 Air hardened foamed concrete *LAP LAMBERT Academic Publishing* 3

[4] Vasilovskaya N G, Endzhievskaya I G and Kalugin I G 2010 Management structure of cellular fiberconcrete *News of higher educational institutions. Construction* **11-12** 12-13

[5] Kahmer H 2001 Fibrous concrete successfully used in structural precast component production *Concrete Plant + Precast Technology* **8** 26 – 31

[6] Rakhimbaev S M, Degtev I A, Tarasenko V N and Anikanova T V 2007 On the question of reducing the shrinkage deformation of foam concrete products *News of higher educational institutions. Construction* **11-12** 41-44

[7] Komohov P.G. 1991 Physics and mechanics of destruction during formation of strength of cement paste *Cement and its Applications* **7-8** 4-10

[8] Bogatina A Y 2005 Structural reinforced foam concrete for buildings civil type *PhD thesis* 86

[9] Steshenko A B and Kudyakov A I 2014 Heat insulating reinforced air hardened foamed concrete *Vestnik of TSUAB* **2** 127

[10] Yakovlev G, Keriene J, Gailius A and Girniene I 2006 Cement Based Foam Concrete Reinforced by Carbon Nanotubes *Mater. Sci+* **12** 147

[11] Morgun L V 2003 On the regularities of concrete structure formation with reinforcing fibers *News of higher educational institutions. Construction* **8** 58-62

IOP Conf. Series: Materials Science and Engineering **71** (2015) 012020 doi:10.1088/1757-899X/71/1/012020

Primers role in plastering systems on concrete surfaces

H B Fischer[1]
[1]Bauhaus-Universität Weimar, F.A. Finger-Institut für Baustoffkunde, Weimar, 99421, Germany

E-mail: hans-bertram.fischer@uni-weimar.de

Abstract. A drastic reduction in time frames between the manufacturing process of concrete units and the rendering phase (including prior priming) does not allow the concrete to dry well. This fact is also underlined by changes in concrete technology (denser concrete and denser concrete surfaces). The tests showed that the reduction of drying time (storage time) had a significant influence on the bonding properties of gypsum plaster on concrete surfaces. In such cases it is absolutely necessary to use an appropriate primer no matter what the processing temperature (2 °C to 20 °C) might be. In this publication the varying primer quality is shown.

1. Introduction

Indoor plasters have the primary task of covering substrates (walls and ceilings) for thermal, hygienic and aesthetic purposes. Gypsum plasters are remarkable for a variety of distinct advantages. Their most significant advantage is that they are able to create a pleasant room atmosphere. Their physical properties allow for the regulation of room climate, and their low thermal conductivity leads to a sense of cosiness. They are quick and easy to process; they dry easily. Apart from these factors, gypsum plasters also contribute to the improvement of fire protection qualities. Gypsum plasters can be applied on almost all kinds of substrates (background surfaces):

- Concrete
- Walls made of brick, lime-sandstone or aerated concrete
- Light excelsior boards (wood wool)
- Hard foam boards

Since the 1960s, their application has generally been based on dry mortars, and is carried out with the help of machines.

The adhesive mechanism of plasters on surfaces can be described as follows (cf. KHODA [23] and OHNEMÜLLER [11]):

- Adhesion of fresh plaster based on the negative pressure effect during roughcasting of plaster, air is dispelled and a zone of low pressure that enhances adhesion is created;
- Adhesion of fresh and hardened plaster based on interlocking surface coarseness of the substrate leads to interlocking with the plaster the crystals emerging from the hydration process in the interface layer also continue to partly develop in the pores of the substrate, thus resulting in additional mechanical interlocking;
- Adhesion of hardened plaster based on chemical bonding crystal in-growth between plaster and substrate requires similar or chemically related plastering surfaces; this, however, is a seldom occurrence.

Despite diverse and long-standing know-how [1-23], fractures between gypsum plasters and concrete surfaces sometimes occurred. Possible reasons cited were as follows: ([03, 04, 05, 07, 08, 09, 10, 12, 13, 16, 20, 21, 23]): Influences from within the plaster; concentration of anhydrite in interface layer; changes of structural gypsum crystal (i.e. change in crystal habits); plaster thickness; pH value; type of additives; longitudinal changes through total moisturisation; problematic plastering surface; too high moisture levels (unsatisfactory drying of plaster); too low capillarity; too low coarseness; concentration of particular chemical combinations on the surface (for concrete) e.g. portlandite, vaterite, alkalis; surface alkalinity (for concrete); deformations; cement quality (for concrete): cement composition, hydration level; separating agent for furring (for concrete); loose impurities (dust); influences of surrounding conditions; thermal load; air expansion in pores; failure to bond due to moisture influences.

A study of pertinent scientific literature shows that highly differentiated individual factors influence gypsum bonding on concrete surfaces. Contradictory data cannot be explained otherwise.

The scope of this study has been to examine some of the factors described in the cited literature and to clarify their influences on bonding characteristics in present conditions. The following points were examined within the framework of this study:

- Given similar conditions, do different primers significantly influence adhesion characteristics of gypsum plasters on smooth concrete surfaces;
- To what extent the adhesion is dependent on the application temperature of the primer;
- The storage period of the concrete specimens, and thus the moisture conditions of the substrate, were varied for this purpose. Apart from that, the influence of additional moisturisation (on the back) of the concrete specimens and surface coarseness were also determined.

2. Preparation of specimens

Factory-made concrete plates (from WITTMER + KLEE, Waghäusel) with dimensions of 1000 mm x 700 mm were used for the study of the influence of various primers. These specimens were freed of surface residues and stored horizontally with the smooth surface facing upwards. Plaster application was carried out immediately on 6 plates (5 days after manufacturing), and 35 days later on the rest of the 6 plates (intended partial carbonation, superficial drying).

The client put eleven easily available primers (synthetic emulsions based on styrene acrylate and vinyl acetate polymer) at our disposal. These were prepared according to the instructions provided (partly diluted with water, and homogenised through stirring). They were then applied to the smooth side of the concrete plates (area of application 500 mm x 500 mm, respectively) with rollers. A comparative surface (reference test) was left untreated.

One day after the primers had been applied gypsum mortar made from specially supplied manual plaster of low quality with a water/plaster ratio of approximately 0.52 was mixed (2 measures of ready-made gypsum plaster, 1 measure of stucco, in order to determine the influence of the primers exactly). This mortar was applied manually at a thickness of approximately 8 mm on the horizontal specimens. Processing time was around 20 minutes.

3. Test methods

Tests for *adhesive strength* were carried out in accordance with the German Industrial Norm 18555 (DIN 18555), "Test of mortars with mineral bonding agents; hard mortar; determination of adhesive strength".

4. Results and discussion

Generally speaking, almost all the primers fulfilled expectations that were attributed to them in the gypsum plaster/concrete system. It can be assumed that a gypsum plaster possesses sufficient bonding characteristics with the concrete when adhesive strength is measured at 0.2 N/mm^2. However, a limiting value for guidelines and norms does not exist. It also became clear that a primer must definitely be used in the conditions chosen by us.

In a majority of the specimens, fracturing occurred in the interface between gypsum plaster and primer (adhesion fracture). In the case of some primers with high adhesive strength (primers 2 and 7), the fracture occurred partly in the interface between primer and mortar, and partly in the mortar itself (adhesion-cohesion fracture).

Plaster adhesion of non-stored concrete units was very high with some primers (2, 7 and 11). A lack of adhesion during the application of gypsum plaster without primer, as also was the case with primer 9, could be determined. Primers 1 and 8 showed low adhesive values.

It can be assumed that five-week storage leads to superficial drying and partial carbonation of the surface. Due to these reasons, adhesion characteristics of the plasters changed. All specimens that originally showed low or average adhesive qualities increased their adhesive strength significantly. Only the reference test (specimen without primer) showed absolutely no adhesive characteristics even in this case.

The test results could prove the differences in quality of the 10 primers used. We therefore recommend that users get appropriate information from manufacturers as a form of reassurance. Sufficient adhesive strength (0.2 N/mm^2) should be achieved by primers in the concrete/primer/gypsum plaster interface. In order to determine the efficiency of a primer it is not only necessary to determine adhesive strength - fracture patterns are equally important.

Table 1. Adhesive strength of gypsum plasters on concrete depending on type of primer.

Primer no.	Application of primer without storage	Application of primer after 35-day storage
	Adhesive strength, N/mm^2	Adhesive strength, N/mm^2
0	≈ 0	≈ 0
1	0,19	0,10
2	1,37	1,06
3	0,58	0,77
4	0,54	0,70
5	0,36	0,80
6	0,41	0,77
7	1,35	1,10
8	0,19	0,66
9	≈ 0	0,16
10	0,34	0,30
11	1,23	1,19

IOP Conf. Series: Materials Science and Engineering **71** (2015) 012020 doi:10.1088/1757-899X/71/1/012020

Figure 2. Depiction (ESEM) of a section of gypsum plaster with primer (substrate removed!). Top: Transition area containing primer became denser. Gypsum/primer 10 to 20 µm (primer 1 Betokontakt). Bottom: Clear denseness of transition area containing primer. Gypsum/primer 15 to 60 µm (primer 2 Quarzgrund)

5. Conclusions

On the one hand, dihydrate crystals of the plaster must grow into the substrate in order to achieve sufficient adhesive properties. This requires certain capillarity of the substrate, and a sufficient moisture gradient between plaster and concrete. On the other hand, detachment and changes in crystal habits caused by subsequent drying must be kept at a minimum so that the gypsum/concrete bonding is not weakened. For this reason the substrate must possess sufficient capillarity; it should not, however, be too moist.

Damages at a later stage cannot be ruled out when time frames between the manufacturing process of concrete walls and the rendering phase are extremely short.

All tests on this topic were commissioned by the Federal Association of Gypsum and Gypsum Board Industry e.V., Germany (Bundesverband der Gips- und Gipsbauplattenindustrie e.V.). The authors are grateful for the interesting assignment and fruitful discussions. We would also like to take this opportunity to thank Mr. Forg (of Heidelberger Maxit) and Professor Dr. Hummel (of Gebrüder Knauf Westdeutsche Gipswerke, Iphofen) for their support.

References

[1] Ludwig U, Kuhlmann J 1974 *Tonindustrie-Zeitung* **98** 1
[2] Lucas G 1974 *Tonindustrie-Zeitung* **98** 5
[3] Poch W 1974 *Tonindustrie-Zeitung* **98** 7
[4] Haerter M 1974 *Tonindustrie-Zeitung* **98** 13
[5] Lange J 1974 *Tonindustrie-Zeitung* **98** 15
[6] Kollmann H, Strübel G and Trost F 1977 *Zement-Kalk-Gips* **30** 224
[7] Hurbanic M, Schicht R and Scholze H 1977 *Tonindustrie-Zeitung* **98** 232
[8] Engelke H 1979 *Zement-Kalk-Gips* **32** 560
[9] Grunau E B 1980 *Stuckgewerbe* 17
[10] Ostrowski C. 1980 *baustoffindustrie* 8
[11] Scholze H, Hurbanic M and Engelke H 1981 *Zement-Kalk-Gips* **34** 560

[12] Manns W, Wisotzky T and Zimbelmann R 1983 *Schriftenreihe des Otto-Graf-Institutes* **77** 1
[13] Grün I 1983 *Bau* 367
[14] Moisset J 1984 *Tonindustrie-Zeitung* **108** 289
[15] Aeppli E 1984 *Tonindustrie-Zeitung* **108** 383
[16] Khoda F 1985 *Tonindustrie-Zeitung* **109** 694
[17] Moldan D 1986 Ursachen für mangelhaftes Haften von Maschinenputzgips auf Beton (RWTH Aachen, Dissertation)
[18] Khoda F 1986 *WTA Seminar 'Schadensfreier Putz'* 59
[19] Koslowski T and Ludwig U 1986 *WTA Seminar 'Schadensfreier Putz'* 64
[20] Böttcher R 1988 *Der Stuckateur* 6
[21] Schütz P 1992 *BMI, Universität Innsbruck* 37
[22] Khoda F 1992 *BMI, Universität Innsbruck* 45
[23] Khoda F 1992 *Stuck · Putz · Trockenbau* 10

IOP Conf. Series: Materials Science and Engineering **71** (2015) 012021 doi:10.1088/1757-899X/71/1/012021

Influence of the processed sunflower oil on the cement properties

A U Fleysher[1], V V Tokarchuk[1] and V A Sviderskiy[1]
[1]National Technical University of Ukraine "Kiev Polytechnic Institute", Department of Chemical Technology of Composite Materials, Kiev, 03056, Ukraine

E-mail: watrushkoo@mail.ru

Abstract. Used oils (vegetable oil, animal oil, engine oil, etc.), which are essentially industrial wastes, have found application as secondary raw materials in some braches of industry. In particular, the only well-known and commonly-used way of utilizing wastes of vegetable oils is to apply them as raw materials in the production of biodiesel. The goal of the present study is to develop a conceptually new way of vegetable oil wastes utilization in the building industry. The test admixture D-148 was obtained from the processing of wastes of sunflower oil and it mainly consists of fatty acid diethanolamide. The test admixture was added to the cement system for the purpose of studying its influence on water demand, flowability, setting times, compressive strength and moisture adsorption. The test admixture D-148 at the optimal content 0.2 weight % causes 10% decrease in water demand, 1.7 time increase in flowability (namely spread diameter), 23% increase in grade strength and 34% decrease in moisture adsorption. The results of the present investigation make it possible to consider the final product of the waste sunflower oil processing as multifunctional plasticizing-waterproofing admixture.

1. Introduction

Food waste is food products that partially or completely have lost the initial customer appeal during production, processing, consumption or storage. They are divided into two categories: production waste (from food industry) and consumption waste (from households and different food services, e.g. cafes, restaurants, schools and hospitals).

Today there exist different methods of utilization of food industry waste in the building industry. Such waste is used as chemical admixtures, mineral additives and raw materials for alternative fuels.

It has been proposed to use eggshell, sugar cane bagasse and rice husk ash as mineral additives. The method of eggshell utilization was proposed in [1]. It considers reusing eggshell waste as an alternative raw material in soil-cement bricks. The results show that eggshell waste can be used in soil-cement bricks as a partial replacement for Portland cement in the range of up to 30 wt%.

The author [2] has proposed the use of sugar-cane bagasse ash in concrete. Sugar-cane bagasse is a fibrous waste-product of the sugar refining industry. Bagasse ash mainly contains aluminum ion and silica, so it is possible to use sugar-cane bagasse ash as cement replacement material (a pozzolanic additive) to improve the quality and reduce the cost of construction materials, such as mortar, concrete pavers, concrete roof tiles and soil cement interlocking block. The results of the fore-mentioned investigation show that the cement could be advantageously replaced with sugar-cane bagasse ash up to maximum limit of 10%.

Rice husk is an agricultural residue which accounts for 20% of the 649.7 million tons of rice produced annually worldwide. There have already been made a number of attempts to produce active

IOP Conf. Series: Materials Science and Engineering **71** (2015) 012021 doi:10.1088/1757-899X/71/1/012021

mineral additives by burning and grinding of the rice husk [3]. The results of the investigations have proved the possibility of replacement of up to 20 weight % of Portland cement with rice husk ash without a decrease in the strength, and the possibility of replacement of up to 10 weight % with the strength significantly increased.

It was proposed to use milk whey, molasses, sugar industry wastes and protein-contained products (for example, hydrolyzed blood) as chemical admixtures.

The most well-known chemical admixture which is widely used in the building industry (at least in the post-Soviet areas) is sulfite-yeast brew, being the yeast industry waste obtained during the processing of sulfite-cellulose liquor into nutritional yeasts. Such an admixture was firstly used as a plasticizer in the early 1930-ies [4, 5].

The members of the Chemistry department of Penza Institute of Engineering and Building have conducted the research and development work aimed at finding the possible ways of utilizing milk whey as chemical admixtures in concretes and mortars. In the building industry the milk whey has found application as a typical hydrophilic surfactant. The chemical composition of the milk whey allows it to be used as a chemical admixture for mortars and concretes, intended for modifying the rheological properties, retarding the setting and hardening times, wetting the hydrophobic aluminium powder used in cellular concrete manufacture, accelerating the lime hydration, etc. [6].

Molasses is a by-product material which is obtained from paper and sugar industries. The composition of molasses differs depending on the source it has been obtained, such as from reed or beet. Beet molasses consists of sucrose and other sugars (ash and nitrous materials) and water. About 4-8% of beet remains in molasses after the process. Molasses can be used as a water reducer or a retarding admixture, which satisfy the ASTM standards [7]. Also, molasses is used as a plasticizing admixture [8]. For example, for a long time concentrated molasses yeast brew has been used as a plasticizing admixture [9]. Such an admixture is the waste product of the nutritional yeast manufacturing and consists of the mix of humic substances and mineral salts.

Casein is the principle protein in bovine milk, which is mainly responsible for the functionality of milk proteins. As a chemical admixture, the application of casein can be tracked back to ancient Rome [10]. At that time, builders added various organic additives to mortars to adjust certain mortar properties. For example, the addition of sugar, fruit syrup and blood extended the workability time of mortars. The addition of malt, beer and urine could improve the durability and frost resistance, etc. Casein as well as some casein contained products were generally used as plasticizer or stiffening agent, which could improve the workability or adjust the consistency of cementitious mortars. Today in the construction industry casein is by far the most widely used dispersant in self-levelling underlayments.

Foam forming substances obtained from the protein-contained products can be used as an effective admixture during foam concrete production [11]. Such protein-contained products include blood, skin, bones, horns and hoofs, feathers, fish scale, oil cake of oil crops and milk products and after processing (hydrolysis, evaporation, chemical modification) they give extremely stable foam as a foam admixture. The example of such admixtures is the protein foam admixture CreenFroth-P.

The purpose of the present article is to extend the range of raw materials resources for the building industry, especially enlarging the number of waste products in general and food waste products in particular that could be used in cement and concrete industries. One of the possible approaches is the utilization of waste vegetable oil as a chemical admixture for hydraulic binding materials. Waste vegetables oils are already widely used in biodiesel production [12-14]. The largest producers of such types of food wastes are food service areas – cafes and restaurants. So, the one of the objectives of the present study is the development of a conceptually new way of waste vegetable oil utilization, namely in the building industry.

IOP Conf. Series: Materials Science and Engineering **71** (2015) 012021 doi:10.1088/1757-899X/71/1/012021

2. Experimental program

2.1. Materials

Ordinary Portland cement type I/500 and fractioned river sand (0,315-0,900) were used in the investigation.

Test admixture (D-148) is a product of the chemical reaction between the waste sunflower oil components and diethanolamine in the presence of the alkali promoter (0.5 weight %) and glycerine. The reaction was conducted in the presence of the excess amount of diethanolamine. The final product (D-148) consists of the following components: sun oil fatty acid diethanolamide, small amount of sun oil fatty acid salts, small amount of diethanolamide and glycerol.

2.2. Test methods

The influence of the admixture on the cement system was estimated according to the following characteristics: compressive strength on 1, 3 and 28 days of hardening, water demand, setting times, spread diameter according to mini-slump test, and moisture adsorption.

For compressive strength, water demand, setting times and moisture adsorption tests cement samples have been prepared in the following way. The test admixture dissolved in water was added to the cement and mixed for 15 min in the lab ball mill. For mini-slump test the already prepared cement samples were mixed with sand for 5 min in the lab ball mill.

Compressive strength of cement after 1, 3 and 28 days of hardening was tested on cube samples with dimensions of 20*20*20 mm. Water demand and setting times were measured with mini-Vicat apparatus.

Mini-slump test is carried out by using a mini cone to evaluate flowability of cement-sand suspension. The truncated cone is placed on a smooth and non-absorbing metallic plate filled with cement-sand suspension and lifted. The resulting final diameter of the fresh cement-sand mix sample is the mean value of two measurements made in two perpendicular directions. The measured spread diameter is related to plasticizing effect of the admixture. The higher the spread diameter, the higher the plasticizing effect of the admixture.

For moisture adsorption test the method of determining hydrophobic property was used. According to this method powder cement samples were put in the exsicattor with 100 % relative humidity for 15 weeks. The mass gain of the cement was measured. The lower the mass gain, the higher the waterproofing properties of the cement samples.

3. Results and discussions

Dependences between water demand, setting times, compressive strength, spread diameter of the cement-sand mortar and the test admixture content are presented in Table 1.

Table 1. Comparison of the physical-mechanical characteristics of the cements with different amounts of the test admixture.

D-148 content, weight %	Spread diameter, mm	Water demand, %	Setting times, h-min		Comprassive strength, MPa / day		
			initial	final	1	3	28
0.00	120	25.0	0-45	3-00	17.8	30.5	39.5
0.04	179	24.0	0-45	3-20	20.8	32.4	35.4
0.08	198	24.0	0-55	3-30	16.4	23.6	33.3
0.20	201	22.5	1-00	3-40	16.5	26.3	48.5
0.60	195	18.0	1-30	4-00	12.1	21.4	24.0

Dependence between the cement moisture adsorption and admixture content is presented in Figure 1.

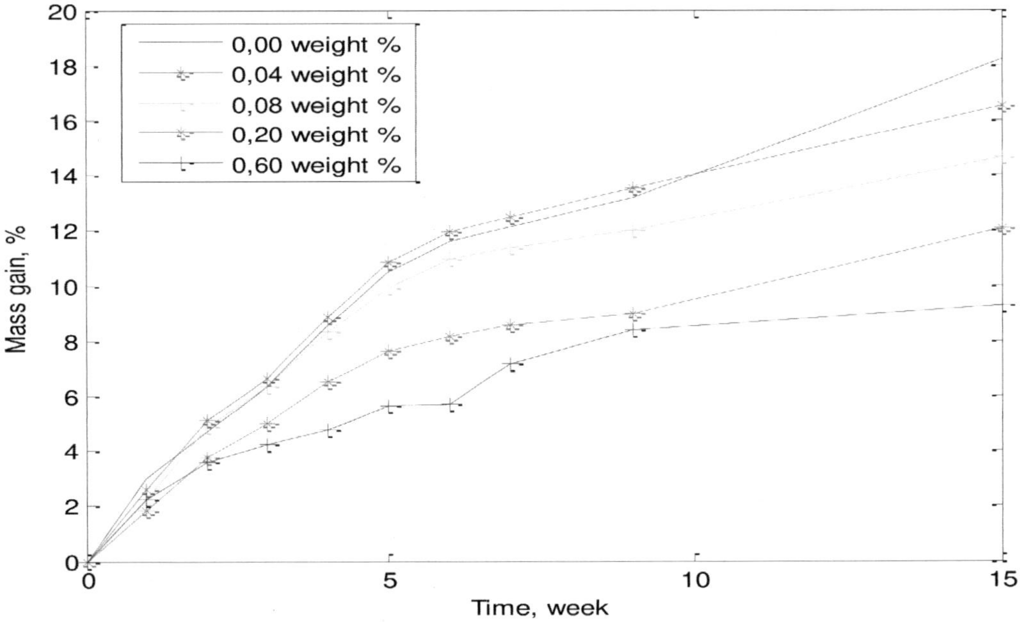

Figure 1. Dependence between moisture adsorption and the test admixture content.

According to the table data, namely the changes of water demand of cement pastes and spread diameter of the cement-sand mortars, the test admixture can be considered as a plasticizing/water-reducing admixture. A plasticizing admixture is a chemical compound added to concrete/mortar before or after it is mixed in order to produce concrete/mortar using less water while maintaining the material's workability, or to produce concrete/mortar with higher workability while maintaining the constant water demand.

As already known, plasticizing admixtures are broadly classified into two categories: normal and high range [15]. The normal plasticizing admixtures are also called "plasticizers" or water reducers, while the high range plasticizing admixtures are called "superplasticizers". While the normal plasticizing admixtures can reduce the water demand by 5 – 10%, the high range plasticizing admixtures can cause a reduction of 15 – 40%.

The test admixture causes the maximum water demand to decrease up to 28 % (at admixture content 0.6 weight %). This result makes it possible to consider the test admixture as superplasticizer (at admixture content 0.6 weight %). The samples with 0.2 weight % admixture content show optimal plasticizing results (the largest spread diameter and the water demand decreased up to 10 %) that allows considering the test admixture as plasticizer or water reducer.

The main feature of plasticizing admixtures is the plasticizing effect, namely the ratio of changes in concrete/mortar workability after adding the plasticizing admixture to its composition. But plasticizing admixtures cause the decrease in the early strength along with increasing workability. That is why in the studies of admixtures one must take into account the term "rational plasticizing effect", meaning the plasticizing effect that is not accompanied with strength decrease [16].

The main physical-mechanical characteristics (water demand, spread diameter and compressive strength) being considered, one can observe that optimal content of the test admixture (D-148) falls within the limit of 0.2 weight %. At such admixture content the samples have the largest spread diameter and lower (by 10 %) water demand. Early compressive strength is decreased negligibly (by 14 % after 3 days of hardening) but grade strength is higher by 23 % or almost by one grade.

The test admixture not only decreases early strength but also retards setting times, which is more obvious at higher admixture contents. But at optimal admixture content (0.2 weight %) early strength decrease and setting times retardation are negligible.

The test admixture gives the cement hydrophobic properties, this being especially apparent at 0.2-0.6 weight %. The samples with 0.6 weight % of the test admixture have the best hydrophobic properties, in other words, the lowest mass gain (49 % lower in comparison with the control sample).

The optimal admixture content is 0.2 weight %. At this content cement samples have the largest spread diameter, lower water demand (by 10 %), higher grade strength and additionally hydrophobic properties. After 15 weeks moisture adsorption of the cement samples with 0.2 weight % of the test admixture was 34 % lower in comparison with the control ones.

4. Conclusions

The results of the investigation enable to arrive at the following conclusions:

1. The final product of the waste sunflower oil processing can be used as a chemical admixture in cement, mortar and concrete.
2. The test admixture D-148 is normal range plasticizing admixture, which at the optimal content causes 10% decrease in water demand, 1.7 times increase in spread diameter and 23 % or almost one grade increase in grade strength.
3. The test admixture gives the cement hydrophobic properties and causes moisture adsorption to decrease by 34 % (at optimal content).
4. The test admixture obtained during the processing of food industry wastes can be utilized in the building industry as a multifunctional plasticizing - waterproofing admixture.

Acknowledgements

The authors of the work wish to thank O.I. Vasylkevich (Department of Organic Chemistry and Technology of Organic substances, National technical university of Ukraine "Kiev Polytechnic Institute", Kiev, Ukraine) for consulting and providing the investigation with raw materials.

References

[1] Amaral M C, Siqueira F B, Destefani A Z and Holanda J N F 2013 Soil–cement bricks incorporated with eggshell waste *Proceedings of the ICE - Waste and Resource Management* **166** 137-141

[2] Shrinivasan R 2010 Experimental Study on Bagasse Ash in Concrete *IJSLE* **5** 60-66

[3] Habeeb G A and Mahmud H B 2010 Study on properties of rice husk ash and its use as cement replacement material *Materials Research* **13**

[4] Kalashnikov V I, Tarakanov O V, Kuznetsov U S, Volodin V M and Belyakova E A 2012 Concretes of the new generation based on the dry very fine-grained powder mixes *Magazine of Civil Engineering* 8

[5] Bolobova A V and Kondrashchenko V I 2000 Use of yeast fermentation wastes as a biomodifier of concrete *Applied Biochemistry and Micribiology* **36** 205-14

[6] Utilization of milk whey in building [online resource] http://brazilwaxing.org/molochnaya-syvorotka/3468-ispolzovanie-molochnoy-syvorotki-v-stroitelstve.html

[7] Yildirim H and Altun B 2012 Usage of molasses in concrete as a water reducing and retarding admixture *IJEMS* **19** 421-26

[8] Jumadurdiyev A, Hulusi Ozkul M, Saglam A R and Parlak N 2005 The utilization of beet molasses as a retarding and water-reducing admixture for concrete *Cem. Conc. Res.* **35** 874-882

[9] Aphanasiev N F and Tseluyko M K 1989 Admixtures for concretes and mortars *Bydivelnyk* (Kiev, Ukraine) 128

[10] Bian H and Plank J 2013 Fractionated and Recombined Casein Superplasticizer in Self-Leveling Underlayments *Adv. Mat. Res.* **687** 443-448

[11] Foam admixtures obtained from natural organic compounds [online resource] http://www.ibeton.ru/a23.php
[12] ElSoih N E M 2011 The manufacture of biodiesel from the used vegetable oil *A thesis submitted to the Faculty of Engineering at Kassel and Cairo Universities for the degree of Master of Science*
[13] Zhang Y, Dube M A, McLean D D, Kates M 2003 Biodiesel production from waste cooking oil *Bioresource Technology* **89** 1-16
[14] Sales A 2011 Production of biodiesel from sunflower oil and ethanol by base catalyzed transesterification [online resource] http://www.diva-portal.org/smash/get/diva2:443295/FULLTEXT01.pdf
[15] Cement – superplasticizer compatibility [online resource] http://www.theconcreteportal.com/compat.html
[16] Kalashnikov V I, Tarakanov O V, Kuznetsov U S, Volodin V M and Belyakova E A 2012 Concretes of the new generation based on the dry very fine-grained powder mixes *Magazine of Civil Engineering* **8**

IOP Conf. Series: Materials Science and Engineering **71** (2015) 012022 doi:10.1088/1757-899X/71/1/012022

Physical chemical studies of dispersed aluminosilicate wastes for obtaining the burned building materials

I Y Iuriev[1], N K Skripnikova[1], G G Volokitin[1], O G Volokitin[1], A V Lutsenko[1] and P V Kosmachev[1]
[1]Tomsk State University of Architecture and Building, Department of Applied Mechanics and Material Sciences, Tomsk, 634003, Russia

E-mail: yiywork@mail.ru

Abstract. This paper presents results of the studies that determined that grinding can be one of the ways to modify aluminosilicate wastes. The optimal grinding modes were defined in laboratory conditions. Physical and chemical studies of modified ashes were carried out by means of X-ray phase analysis, differential thermal analysis and microscopy. The results have shown that modified ashes of thermal power stations when being applied in production of ceramic brick influence positively the processing properties of raw materials and the ready products.

1. Introduction

High development rates of industrial and civil engineering both at domestic and international markets are followed by the increase in production output of building materials as well as products of different kinds and different application. Ceramic industry takes one of the leading positions at the market of building materials. From the one hand, the field of ceramic building materials faces a problem of untimely delivery of high-quality raw materials. From the other hand, the increase in energy consumption leads to increase in the amount of wastes from coal burning – bottom ashes. Therefore one of the objectives of modern science is studying the possibility of their recycling into different industries. The problem arises in different processing characteristics of ashes (physical, chemical and mineral properties). The widest application of thermal power station wastes are in production of different kinds of building materials: cement, concrete, wall materials, slag wool [1-3]. The most relevant is the application of ashes in production of ceramic brick [4].

2. Materials and methods

It is known that chemical and granulometric composition of ashes obtained from burning of black coal sufficiently differs from each other, even within the limits of one ash disposal area [5]. Moreover, applications of ashes without the preliminary treatment for instance in production of ceramic brick increases the risk of obtaining products with unstable properties [6]. This is one of the key reasons of rare cases of application of thermal power station wastes in production of building materials, especially at the Russian market.

In this regard a relevant issue is studying the modification methods of aluminosilicate ashes wastes to improve their processing characteristics that enable obtaining ceramic building materials with good and stable properties.

The goal of the presented research was to study the structure and phase composition of the preliminary-treated aluminosilicate wastes and their influence on brick properties.

2.1. Physical-chemical studies of aluminosilicate and clay brick
The raw materials under study are ashes from burning the black coals at the electric station in the city of Tomsk (Russia) and clay of Verkhovoe field in Tomsk region. Chemical composition of raw materials is given in Table 1.

Table 1. Chemical composition of raw materials.

Item	Oxides content, m %							
	SiO_2	Al_2O_3	Fe_2O_3	CaO	MgO	R_2O	TiO_2	Loi
Clay	64,80	12,10	4,53	5,90	3,08	2,97	0,70	5,92
Ash	61,80	22,40	3,95	3,00	1,81	1,34	1,25	4,45

Table 1 shows that content of aluminum oxide 12,1 % in clay makes it acid (Al_2O_3 – < 14 %). CaO and MgO in clay in the amount of 5,9 % and 3,0 % are most likely found as carbonate compounds. At the indicated quantitative combination of aluminum oxide characterizing the raw materials with its fire resistance and fluxing agents - magnesium oxide, calcium oxide and ferrous oxide the raw materials can be classified as low-melting.

High content of Al_2O_3 (22,4 %) and SiO_2 (61,8 %) in ash can be the reason of mullite-like compounds crystallization. Loss of less than 5% while calcination shows low amount of residual fuel thus minimizing the shrinkage processes while burning.

The earlier obtained results [7] showed that one of the ways to improve aluminosilicate ashes is dispersion.

Plastic deformation while grinding leads to the change in crystal structure of quartz. At the same time its reaction capacity grows sharply and the temperature of thermal dissociation decreases. That can lead to the reduction of temperature of formation the principal minerals specific to ceramic products.

Use of mills where mechanical impulse from grinding media is brought by the simultaneous combination of impact, pressure and shear is the most effective [8]. Planetary mill refers to such kinds of mills. The optimum grinding modes were defined in the laboratory conditions (Table 2).

Table 2. Grain size composition of ashes depending on the time of grinding.

No.	Time of grinding, hours	Content of particles, %, fractions, um					
		<3	3-5	5-7	7-10	10-14	>14
1	1	84,1	6,9	3,0	2,2	1,5	2,3
2	2	86,3	7,2	2,5	1,9	1,0	1,1
3	4	92,1	3,4	1,9	1,5	0,8	0,3
4	6	93,1	2,7	2,0	1,3	0,7	0,2
5	8	93,2	2,6	2,0	1,3	0,7	0,2
6	10	93,4	2,5	1,9	1,3	0,7	0,2

It should be noted that the optimum grinding time is 4 hours as further grinding shows almost no difference in grain size composition. This can be explained by the inverse process of conglomeration – aggregation of particles under action of adhesive forces and also high strength of mono-crystal particles of small sizes [9].

The obtained ash powder was studied by scanning electrone microscope TM3000 at the zooming of x200 and x1000. Ashes microstructure is shown in Figure 1.

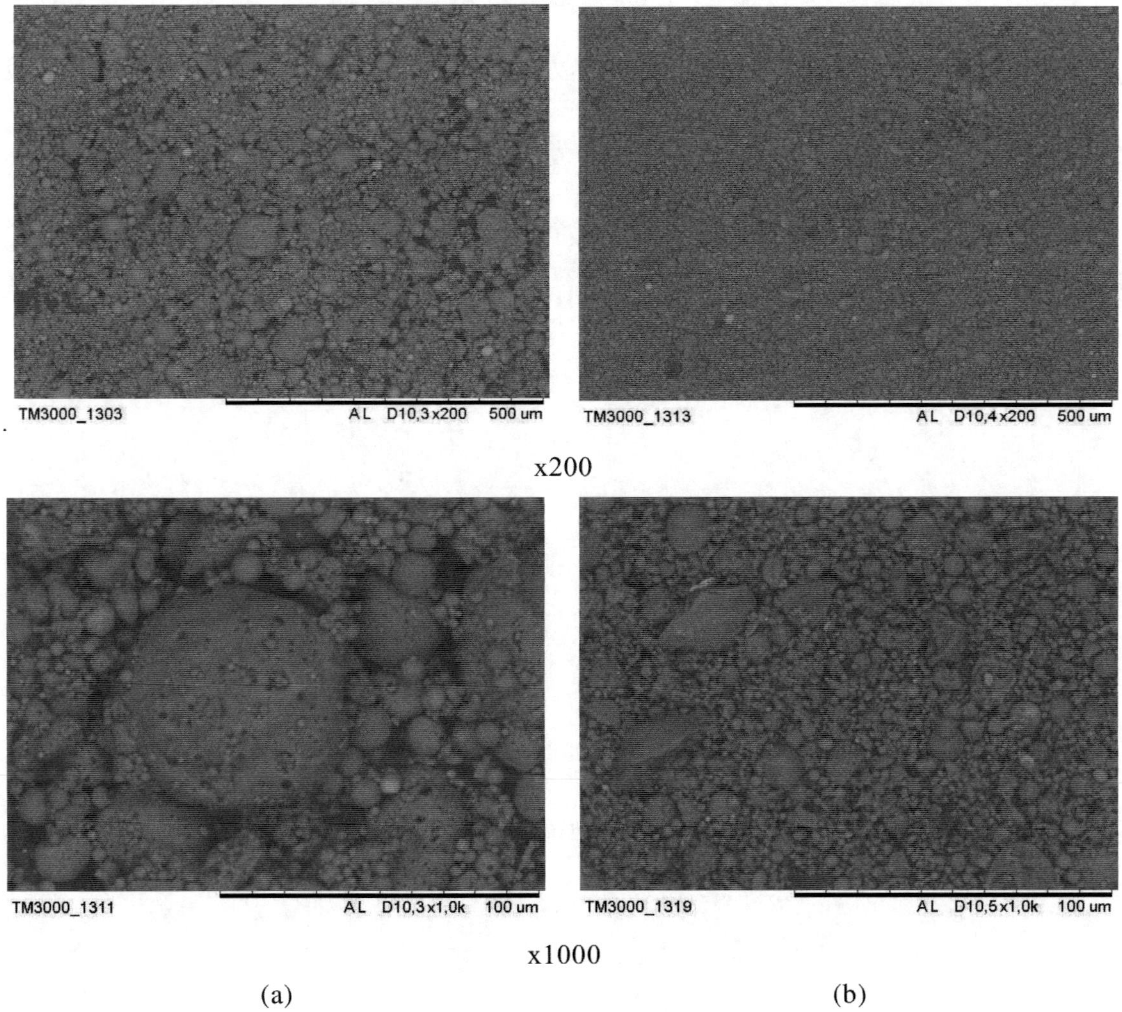

Figure 1. Microphotographs of ashes; (a) ashes before grinding, (b) ashes after grinding.

Microphotographs as shown in Figure 1 demonstrate the structure and the nature of particles of aluminosilicate raw materials before and after the grinding. Ashes in the city of Tomsk are peculiar with domination of spherical particles of the regular shape with close and open porosity of less than 10 um in size and also small content of large porous inclusions of irregular shape. After grinding (Figure 1b) mainly occurs the fragmental destruction of large inclusions with the formation of flour particles. Spherical particles of the regular shape that are quartz-containing compounds do not change their size due to high hardness. The structure becomes more homogenous and dense thus contributing to the increased reaction capacity while ceramics sintering.

Thermal studies of raw materials were held using thermal analyzer SDT Q600 at the speed of temperature increase of 20 degrees per minute, with interval 20–1100 °C. The results of thermal analysis are presented graphically in Figure 2.

Figure 2 shows that clay of Verkhovoe field is characterized by 4 endo-effects and 2 exo-effects. Endo-effect at the temperature interval 30–150 °C is connected with the removal of adsorbed moisture and interpacket water from the structure of clay minerals. Effect 500°C – dehydration of hydrous micas. At the temperature 575 °C polymorphic transformation of quartz occurs. The interval 620–750 °C is characterized with loss of hydroxyl groups by crystal lattice of montmorillonite and its destruction and the decay of carbonate inclusions. Exothermic effect at the temperature of 355 °C is connected with oxidation of organic substances. 910 °C – recrystallization of amorphous decomposition products and formation of new minerals.

At thermal treatment of ashes in interval 40–200 °C the loss of water adsorbed by the developed surface of particles occurs. Carbonate dissolution is observed at the temperature of 697 °C. Burning out of the residual fuel – 500–700 °C. The nature of intensiveness and weight loss indicates the amount of non-burned residues representing the coal particles as well as carbon residues and half-carbon residues. The relatively small exothermic effect with maximum 933,9 °C reflects crystallisation of mullite-like compounds in aluminosilicate phase.

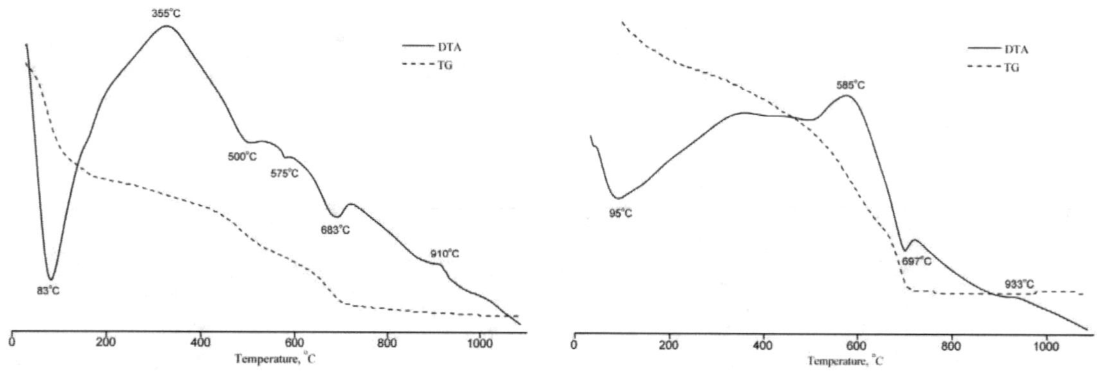

(a) (b)

Figure 2. DTA curves and TG of raw materials; (a) clay, (b) thermal power station ashes.

X-ray structure analysis of ashes after grinding and clay was held using X-ray diffraction meter Rigaku MiniFlex. The results of X-ray phase analysis is given in Figure 3.

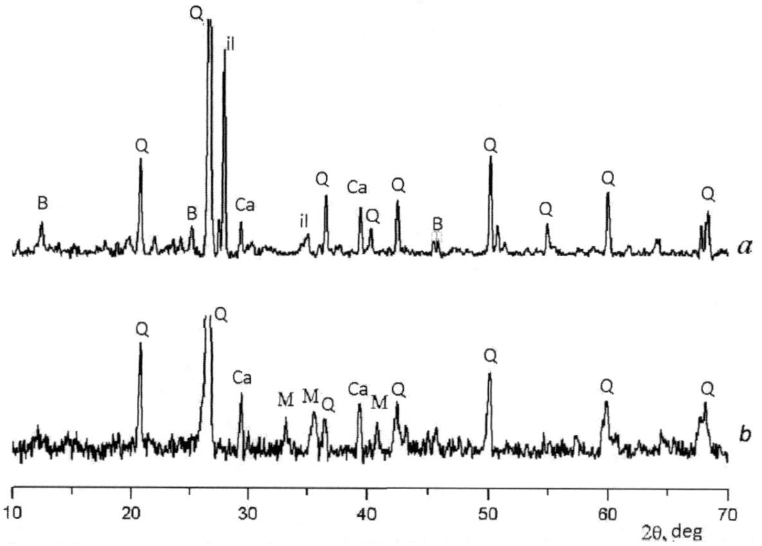

Figure 3. X-ray patterns of raw materials; (a) clay, (b) thermal power station ashes.
Q – quartz, B – biotite, Ca – calcite, il – illite, M – mullite.

X-ray phase analysis showed that clay from the city of Tomsk is presented by quartz, illite, biotite and calcite phases. This raw material can be referred to the group of hydromicaceous-montmorillonite clays. The ash except for quartz is defined by the clearly observed mullite phase. It can further serve while clinkering of ceramic fragment as the center of initiating crystallization of mullite-like compounds providing strength properties to the ready products.

2.2. Obtaining laboratory samples of ceramic brick

To define the influence of dispersed ash on the properties of clay brick and ceramic products a number of experiments have been held. Laboratory samples of ceramic brick were prepared by semi-dry compaction method adding ashes into furnace charge in the amount of 50%. The samples were moulded and burned in the muffle furnace at the temperature of 950–1000 °C. The experiment has shown that when using minced ashes the strength of clay brick increases by 30 % in comparison with the samples from unmilled raw materials. Air shrinkage decreases by 30 %, fire shrinkage - by 50 %. Water absorption of ready ceramic samples reduces by 20 %. As the burning temperature increases up to 1000 °C the properties of products undergo almost no changes thus allowing pointing out the completeness of the basic phase formation processes at the temperature of 950 °C.

2.3. Physical-chemical studies

The carried out research enables to conclude that the main phase changes while burning out the ceramic brick samples made from reasonable compositions with dispersed ash in furnace charge are as follows: the formation of initial melt from clay matter and several low-melting ashes grains; occurring mullite- and anortite-like compounds being the strengthening phases.

The main strengthening structural changes are: healing of pores with the formed glass phase and contraction of non-melted raw materials particles at the same time forming the solid product.

x1000 x5000

Figure 4. Microphotographs of ceramic samples structure of the reasonable composition (ashes content 50 %) with the use of dispersed ashes of thermal power station.

Generally while sintering of compositions when mixing thermal power stations ashes and clay matter the strengthening phases occur (mullite and anortite) that can be explained by existence of these phases in the initial ashes. When burning out the compositions with Tomsk ashes solid-phase sintering prevail while when obtaining the ceramic products with ashes from the city of Seversk – liquid-phase sintering prevails. Presence of clay in furnace charge and the amount of clay matter that can provide the defined raw materials is crucial when forming the ready product while burning process as it is the envelop for non-plastic inactive particles of ashes.

3. Conclusions

The carried out analysis of bottom ash wastes of thermal power stations has shown that one of the methods to enhance the quality of raw materials is dispersion using grinding in mills which are peculiar with the combined action of impact forces, pressure and shear forces influencing the grinding material.

Resulting from the physical-chemical analysis of ashes in the city of Tomsk it can be concluded that modified ash powder can contribute to the intensive formation of mullite-like compounds while combined burning with clay. It should be noted that the temperature of mullite crystallization according to differential thermal analysis is 933 °C, i.e. to obtain the regular ceramic brick the burning temperature no more than 950–960 °C is enough and that was proved experimentally.

Low content of combustible residues in ash and large amount of micro-dispersed particles less than 3 um contributes to obtaining of products with decreased porosity and that in its turn influence positively on water permeability and frost resistance of ready products.

Introduction of modified ashes into furnace charge, those ashes serving as leaner, leads to the reduction of shrinking deformation of ceramic product and broadens the sintering range.

Therefore dispersion of bottom ash wastes of thermal power stations by means of grinding is an advanced way of enhancing the quality of raw materials of poor quality for its further application in production of ceramic brick. Its introduction in different ratio enables to adjust processing properties both of raw materials and of ready ceramic product as well. Mass use of thermal power stations ashes can reduce the environmental impact and reduce total costs for ashes disposal.

References

[1] Volokitin O G, Volokitin G G and Skripnikova N K 2014 *Glass Ceram+* **70** 340
[2] Bondareva N S, Volokitin O G, Morozova O O and Sheremet M A 2013 *Thermophysics and aeromechanics* **20** 621
[3] Potekaev A I, Klopotov A A, Ivanov Y F and Volokitin O G *Russ Phys J* **56** 914
[4] Скрипникова Н К и Юрьев И Ю 2013 *Вестник ТГАСУ* **2** 245
[5] Дворкин Л И и Дворкин О Л 2007 *Строительные материалы из отходов промышленности* (Москва: Феникс) 368
[6] Путилин Е И и Цветков В С 2003 *Обзорная информация отечественного и зарубежного опыта применения отходов от сжигания твердого топлива на ТЭС* (Москва: Союздорнии) 60
[7] Скрипникова Н К и Юрьев И Ю *Актуальные проблемы современности* **70** 84
[8] Prokofiev V Y and Gordina N E 2012 *Glass and ceramics* **69** 65
[9] Bakunov V S and Lukin E S 2008 *Glass and ceramics* **65** 117

IOP Conf. Series: Materials Science and Engineering **71** (2015) 012023 doi:10.1088/1757-899X/71/1/012023

Load effects calculation according to EN 1991-2 Load Model 1

V Slavchev[1]
[1]University of Structural Engineering and Architecture VSU "Lyuben Karavelov", Sofia, 1373, Bulgaria

E-mail: slavchev@vsu.bg

Abstract. In the present paper, fast calculation method for determination the effects of EN 1991-2 LM1 is presented. The influence lines method was used for calculation. Both bending moments and shear forces at typical section can be easy calculated by using one table.

1. Introduction

Road traffic actions – vertical loads (except fatigue), defined in EN 1991-2 [1]:

- Load Model 1 (LM1) is represented as concentrated and uniformly distributed loads, which cover most of the effects of the traffic of lorries and cars. This model should be used for general and local verifications and is the main load model for bridges with spans more than 7 m;
- Load Model 2 (LM2) can be predominant in the range of loaded lengths up to 3 m to 7 m and normally is used for load effects calculation for short structural members;
- Load Model 3 (LM3) is representing special vehicles (e.g. for industrial transport) which can travel on routes permitted for abnormal loads;
- Load Model 4 (LM4): A crowd loading should be used only for some transient design situations.

Load models defined in EN 1991-2 should be used for the design of road bridges with loaded lengths less than 200 m. The length of 200 m corresponds to the maximum length taken into account for the calibration of LM1.

Loads due to the road traffic, consisting of cars, lorries and special vehicles (e.g. for industrial transport), give rise to vertical and horizontal, static and dynamic forces. The defined load models do not describe actual loads. They have been selected and calibrated so that their effects (with dynamic amplification included where indicated) represent the effects of the actual traffic in European countries.

Because of LM1 is the main load model for bridge design (except fatigue load models), and mention in the paragraph above, gives the author a reason to search a fast calculation method for TS that will give the same effects on simply supported bridges. The presented method below allows designers to use a table for their calculations.

Load Model 1 consists of two partial systems, as follows:

1.1. Double-axle concentrated loads (tandem system: TS), each axle having the following weight:

$$\alpha_Q Q_k \quad (1)$$

1.2. Uniformly distributed loads (UDL system), having the following weight per square metre of notional lane:

$$\alpha_q q_k \quad (2)$$

Load Model 1 should be applied on each notional lane and on the remaining areas.
On notional lane Number *i*, the load magnitudes are referred to:

$$\alpha_{Qi}\, Q_{ik} \tag{3}$$

$$\alpha_{qi}\, q_{ik} \tag{4}$$

On the remaining areas, the load magnitude is referred to:

$$\alpha_{qr}\, q_{rk} \tag{5}$$

Where: α_{Qi}, α_{qi}, and α_{qr} are adjustment factors that should be selected depending on the expected traffic and possibly on different classes of routes. The values should be accepted in the national annexes.

The values of Q_{ik}, q_{ik} and q_{rk} are shown in Table 1 [1].

Table 1. Load Model 1: characteristic values.

Location	**Tandem system *TS*** ***Axle loads Q_{ik} (kN)***	***UDL* system** ***q_{ik} or q_{rk} (kN/m^2)***
Lane Number 1	300	9
Lane Number 2	200	2,5
Lane Number 3	100	2,5
Other lanes	0	2,5
Remaining area (q_{rk})	0	2,5

The details of Load Model 1 are illustrated in Figure 1.

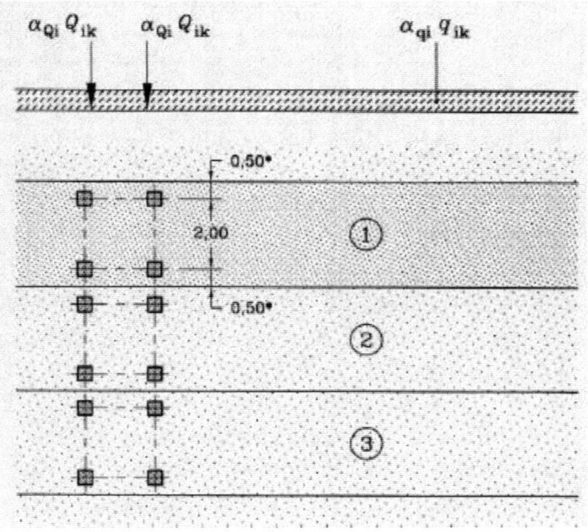

Figure 1. Load Model 1.

2. Proposed method

Bending moment (M_y) and shear force (V_z) calculation, at section "s" at a distance "x" from the support of a simply supported beam loaded with two point loads on a distance 1,2 m (the same like an *LM1 TS* in a longitudinal direction), based on the influence line method is shown on Figure 2.

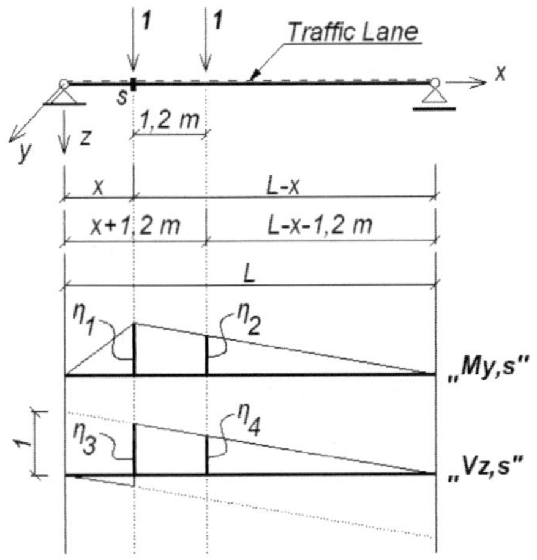

Figure 2. Load effects calculation.

The influence lines are drawn with their positive values above the reference line.
Load effects from that model are as follows:

$$M_{y,s} = \eta_1 + \eta_2 \tag{6}$$

$$V_{z,s} = \eta_3 + \eta_4 \tag{7}$$

Where, $\square_1$ to $\square_4$ are the corresponding influence line ordinates under the point loads.

$$\eta_1 = x\,(L - x)\,/L \tag{8}$$

$$\eta_2 = x\,(L - x - 1{,}2)\,/L \tag{9}$$

$$\eta_3 = (L - x)\,/\,L \tag{10}$$

$$\eta_4 = (L - x - 1{,}2)\,/ \tag{11}$$

The maximum bending moment from such point loads is not at the middle of the span. Location of the section with maximum bending moment can be found from the equation:

$$dM\,/\,dx = (2\,L - 4\,x - 1{,}2)\,/\,L = 0 \tag{12}$$

The section with maximum bending moment from *LM1 TS* is at distance from the support:

$$x = L\,/\,2 - 0{,}3\text{ m} \tag{13}$$

A method that represents tandem system *TS* with equivalent uniform loads is presented in [2]. The method allows designers to use only uniform loads for bending moment and shear force calculation for simply supported beams. It also gives a possibility for easy preliminary calculations of prestressing force needed to balance the equivalent uniform load. Some calculations and result comparison of load effects in bridge superstructure between common and proposed [2] methods were reported and commented in [4].

3. Load effects calculation from any value of *TS*

To simplify load effects calculation author propose a table which can be used by designers. Bending moments ($M_{y,s}$) and shear forces ($V_{z,s}$) can be calculated with formulas, as follows:

$$M_{y,s} = k_{My} \sum \alpha_{Qi}\, Q_{ik} \tag{14}$$

$$V_{z,s} = k_{Vz} \sum \alpha_{Qi}\, Q_{ik} \tag{15}$$

Values for k_{My} and k_{Vz} at typical sections for simply supported beams can be found in Table 2.

When there is a girder bridge superstructure system, the designer should calculate the load redistribution between beams first. There are several hand calculation methods and some details about such calculations, according to the transversal stiffness of the bridge superstructure with comments can be found in [3] and [5].

Comments for the influence of different load distribution in load effects calculation can be found in [4].

IOP Conf. Series: Materials Science and Engineering **71** (2015) 012023 doi:10.1088/1757-899X/71/1/012023

Table 2. Values for k_i at typical sections for simply supported beams.

Span, [m]	x = 0	x = L/8		x = L/4		x = 3/8 L		x = L/2		x = L/2 – 0,3 m
	k_{Vz}	k_{My}	k_{Vz}	k_{My}	k_{Vz}	k_{My}	k_{Vz}	k_{My}	k_{Vz}	k_{My}
6	18,0000	11,6250	15,5000	19,5000	13,0000	23,6250	10,5000	24,00	8,0000	24,5455
7	18,2857	13,8125	15,7857	23,2500	13,2857	28,3125	10,7857	29,00	8,2857	29,4737
8	18,5000	16,0000	16,0000	27,0000	13,5000	33,0000	11,0000	34,00	8,5000	34,4186
9	18,6667	18,1875	16,1667	30,7500	13,6667	37,6875	11,1667	39,00	8,6667	39,3750
10	18,8000	20,3750	16,3000	34,5000	13,8000	42,3750	11,3000	44,00	8,8000	44,3396
11	18,9091	22,5625	16,4091	38,2500	13,9091	47,0625	11,4091	49,00	8,9091	49,3103
12	19,0000	24,7500	16,5000	42,0000	14,0000	51,7500	11,5000	54,00	9,0000	54,2857
13	19,0769	26,9375	16,5769	45,7500	14,0769	56,4375	11,5769	59,00	9,0769	59,2647
14	19,1429	29,1250	16,6429	49,5000	14,1429	61,1250	11,6429	64,00	9,1429	64,2466
15	19,2000	31,3125	16,7000	53,2500	14,2000	65,8125	11,7000	69,00	9,2000	69,2308
16	19,2500	33,5000	16,7500	57,0000	14,2500	70,5000	11,7500	74,00	9,2500	74,2169
17	19,2941	35,6875	16,7941	60,7500	14,2941	75,1875	11,7941	79,00	9,2941	79,2045
18	19,3333	37,8750	16,8333	64,5000	14,3333	79,8750	11,8333	84,00	9,3333	84,1935
19	19,3684	40,0625	16,8684	68,2500	14,3684	84,5625	11,8684	89,00	9,3684	89,1837
20	19,4000	42,2500	16,9000	72,0000	14,4000	89,2500	11,9000	94,00	9,4000	94,1748
21	19,4286	44,4375	16,9286	75,7500	14,4286	93,9375	11,9286	99,00	9,4286	99,1667
22	19,4545	46,6250	16,9545	79,5000	14,4545	98,6250	11,9545	104,00	9,4545	104,1593
23	19,4783	48,8125	16,9783	83,2500	14,4783	103,3125	11,9783	109,00	9,4783	109,1525
24	19,5000	51,0000	17,0000	87,0000	14,5000	108,0000	12,0000	114,00	9,5000	114,1463
25	19,5200	53,1875	17,0200	90,7500	14,5200	112,6875	12,0200	119,00	9,5200	119,1406
26	19,5385	55,3750	17,0385	94,5000	14,5385	117,3750	12,0385	124,00	9,5385	124,1353
27	19,5556	57,5625	17,0556	98,2500	14,5556	122,0625	12,0556	129,00	9,5556	129,1304
28	19,5714	59,7500	17,0714	102,0000	14,5714	126,7500	12,0714	134,00	9,5714	134,1259
29	19,5862	61,9375	17,0862	105,7500	14,5862	131,4375	12,0862	139,00	9,5862	139,1216
30	19,6000	64,1250	17,1000	109,5000	14,6000	136,1250	12,1000	144,00	9,6000	144,1176
40	19,7000	86,0000	17,2000	147,0000	14,7000	183,0000	12,2000	194,00	9,7000	194,0887
50	19,7600	107,8750	17,2600	184,5000	14,7600	229,8750	12,2600	244,00	9,7600	244,0711
60	19,8000	129,7500	17,3000	222,0000	14,8000	276,7500	12,3000	294,00	9,8000	294,0594
70	19,8286	151,6250	17,3286	259,5000	14,8286	323,6250	12,3286	344,00	9,8286	344,0510

Span, [m]	**x = 0**	**x = L/8**		**x = L/4**		**x = 3/8 L**		**x = L/2**		**x = L/2 – 0,3 m**
	k_{Vz}	k_{My}	k_{Vz}	k_{My}	k_{Vz}	k_{My}	k_{Vz}	k_{My}	k_{Vz}	k_{My}
80	19,8500	173,5000	17,3500	297,0000	14,8500	370,5000	12,3500	394,00	9,8500	394,0447
90	19,8667	195,3750	17,3667	334,5000	14,8667	417,3750	12,3667	444,00	9,8667	444,0397
100	19,8800	217,2500	17,3800	372,0000	14,8800	464,2500	12,3800	494,00	9,8800	494,0358

Note: Values in Table 2 are 10 times scaled.

4. Conclusion
Presented method for fast calculation of load effects from tandem system *TS* which is a part of the main load model LM1 simplify many hard calculations for the designers. It can be used not only for a preliminary calculations and obtaining balancing loads and also in many calculation tasks in the engineering practice in connection with the bridge design. This method also minimizes the risk of calculation mistakes in the load effects calculation and gives a possibility to easily calculate the maximum bending moment, which is not at the middle of the beam.

References
[1] EN 1991-2 Eurocode 1: Actions on structures - Part 2: Traffic loads on bridges
[2] Slavchev V 2012 Fast calculation model for EN 1991-2 Load Model 1 using equivalent uniform loads *1st Virtual Int. Conf. on Advanced Research in Scientific Areas* (Slovakia: ARSA-2012)
[3] Safronov V and Slavchev V 2004 Analysis about the size of bending moments in beam superstructure bridges with different spans and transversal stiffness from loads according standards published in different periods of time in Bulgaria *Int. Scientific Conf. VSU'2004* (Sofia, Bulgaria)
[4] Vasilev L 2014 Calculation of load effects in a bridge superstructure under different load models *14th Int. Scientific Conf. VSU'2014* (Sofia, Bulgaria)
[5] Славчев В 2014 *Проектиране на мостови пътни плочи – ръководство за проектиране* ISBN: 978-954-331-051-7

Mathematical modeling of steel fiber concrete under dynamic impact

N N Belov[1], N T Yugov[1], D G Kopanitsa[2], G D Kopanitsa[3], A A Yugov[3] and V V Shashkov[3]
[1]Tomsk State University of Architecture and Building, Department of Higher Mathematics, Tomsk, 634003, Russia
[2] Tomsk State University of Architecture and Building, Department of Metal and Wooden Structures, Tomsk, 634003, Russia
[3]National Research Tomsk Polytechnic University, Cybernatic Center, Tomsk, 634050, Russia

E-mail: kopanitsa@mail.ru

Abstract. This paper introduces a continuum mechanics mathematical model that describes the processes of deformation and destruction of steel-fiber-concrete under a shock wave impact. A computer modeling method was applied to study the processes of shock wave impact of a steel cylindrical rod and concrete and steel fiber concrete plates. The impact speeds were within 100-500 m/s.

1. Introduction

This paper presents a study that investigates the behavior of steel fiber concrete (SFC) under a shock wave impact. Physical mechanical properties of fiber concrete under static and dynamic loading are studied by a number of scientists [1, 2]. Except for results obtained for regular fiber concretes [3] also was considered operation of high-performance powder concrete reinforced with steel fiber under static and dynamic loading. In the present work SFC is modeled as a homogeneous two-phase steel-concrete mixture with an initial density of $\rho_0 = \nu_1\rho_{0c} + \nu_2\rho_{0\delta}$. Where $\nu_1, \nu_2, \rho_{0c}, \rho_{0\delta}$ are initial volume concentrations and densities of steel and concrete. A dynamic destruction of SFC is considered as a process of growth and fusion of micro defects under a stress that emerges during the impact process.

A special volume of a porous media υ is considered a sum of a special volume of material matrix υ_m, a special volume of the pores υ_p and a special volume υ_t that emerges due to the opening of cracks: $\upsilon = \upsilon_m + \upsilon_p + \upsilon_t$. The porosity of material can be characterized as relative volume of hollows $\xi = \xi_p + \xi_t$, or a parameter $\alpha = \upsilon/\upsilon_m$, that are dependent as $\alpha = 1/(1 - \xi)$. Where $\xi_p = \upsilon_p/\upsilon$, $\xi_t = \upsilon_t/\upsilon$ – relative volumes of pores and cracks.

2. Mathematical modeling of a collision between a cylindrical rod and a steel-fiber-concrete plate

The equation system that describes the move of a porous elastic-plastic medium is presented below [4-6]:

$$\frac{d}{dt}\int_V \rho dV = 0; \quad \frac{d}{dt}\int_V \rho u dV = \int_S n\cdot\sigma dS; \quad \frac{d}{dt}\int_V \rho E dV = \int_S n\cdot\sigma\cdot u dS; \quad e = \frac{s^i}{2\mu} + \lambda s;$$

$$s:s = \frac{2}{3}\sigma_T^2; \qquad p = \frac{1}{\alpha}\left[\frac{c_0^2\rho_0(1-\gamma_0\eta/2)\eta}{(1-s_0\eta)^2} + \rho_0\gamma_0\varepsilon\right], \tag{1}$$

where t – time; V – integration volume; S – surface; $\mathbf{n}$ – outward normal; ρ – density; $\boldsymbol{\sigma} = -p\mathbf{g} + \mathbf{s}$ – stress tenser; $\mathbf{s}$ – its deviator; p – pressure; $\mathbf{g}$ – metric tenser; $\mathbf{u}$ – velocity vector; $E = \varepsilon + \mathbf{u}\cdot\mathbf{u}/2$ – specific total energy; ε – specific internal energy; $\mathbf{e} = \mathbf{d} - (\mathbf{d} : \mathbf{g})\mathbf{g}/3$ – deformation strain deviator; $\mathbf{d} = (\nabla\mathbf{u} + \nabla\mathbf{u}^T)/2$ – strain-rate tenser; $\mathbf{s}^J = \dot{\mathbf{s}} + \mathbf{s}\cdot\omega - \omega\cdot\mathbf{s}$ – derivative of deviatoric stress (Jaumann-Noll); $\mu = \mu_0(1-\xi)[1-(6\rho_0 c_0^2 + 12\mu_0)\xi/(9\rho_0 c_0^2 + 8\mu_0)]$, $\sigma_T = Y_0/\alpha$ – effective shear modulus and yield strength; $\omega = (\nabla\mathbf{u}^T - \nabla\mathbf{u})/2$ – vortex tenser; ρ_0, c_0, μ_0, Y_0, s_0 – material matrix constants; $\eta = 1 - \rho_0\upsilon/\alpha$. The parameter λ is excluded by yield condition.

The matrix coefficients c_0 and s_0 of a linear dependency of a shock wave velocity D form a mass velocity u ($D = c_0 + s_0 u$) are defined by the shock adiabats of the mixture's components $D_i = c_{0i} + s_{0i}\cdot u$ $(i=1,2)$. A shock adiabat for the variable υ_m, p_m is presented below.

$$\upsilon_m(p_m) = \sum_{i=1}^{2} m_i \left\{ \upsilon_{0i} - \frac{1}{p_m}\left[\frac{c_{0i}}{s_{0i}}\left(\sqrt{\frac{s_{0i}p_m}{\rho_{0i}c_{0i}^2} + \frac{1}{4}} - \frac{1}{2}\right)\right]^2 \right\} \quad (2)$$

If the following relationship to the shock wave is applied:

$$D = \upsilon_0\sqrt{\frac{p_m}{\upsilon_0 - \upsilon_m(p_m)}};\quad u = \sqrt{p_m(\upsilon_0 - \upsilon_m(p_m))},, \quad (3)$$

one can build a dependency of shock wave velocity from the mass velocity to derive the coefficients c_0 and s_0.

The Gruneisen coefficient γ_0 for the mixture is defined as its components' Gruneisen coefficients γ_{0i}:

$$\frac{\upsilon_0}{\gamma_0} = \sum_{i=1}^{2} m_i \frac{\upsilon_{0i}}{\gamma_{0i}} \quad (4)$$

The shear modulus μ_0 and the yield strength Y_0 are defined as:

$$\mu_0 = 1/\left(\frac{\nu_1}{\mu_{01}} + \frac{\nu_2}{\mu_{02}}\right);\quad Y_0 = m_1 Y_1 + m_2 Y_2, \quad (5)$$

where $m_i = \dfrac{\nu_i \rho_{0i}}{\rho_0}$ - mass concentrations of steel (i=1) and concrete (i=2) in SFC; μ_{0i}, Y_i (i = 1,2) – shear modules and yield strengths for the mixture components.

Equations that describe the change of the parameter α under tension and compression are required to lock the system (1). The equation that defines the parameter α under an elastic deformation of SFC is presented below:

$$\rho_0\gamma_0\varepsilon + \frac{c_0^2\rho_0(1-\gamma_0\eta/2)\eta}{(1-s_0\eta)^2} + \frac{3\mu_0(\alpha-\alpha_0)}{8(1-\nu)N_0 R^3\alpha\alpha_0} = 0 \quad (6)$$

In deriving the equation (6) it was considered that no new cracks appeared during the loading process. The deformation was caused only by the growth of initial crack with a characteristic dimension R:

$$\dot{R}/R = F_1 + F_2, \quad (7)$$

where: $F_1 = (\alpha s_i - s_*)/\eta_1$ at $\alpha s_i > s_*$ and $F_1 = 0$ at $\alpha s_i \le s_*$;

$F_2 = (|\alpha p| - p_*)/\eta_2$ at p<0, $|\alpha p| > p_*$ and $F_2 = 0$ and $p \ge 0$, $|\alpha p| \le p_*$;

$p_* = p_0(1-R/R_*)$; $s_i = \sqrt{\frac{3}{2}\mathbf{s}:\mathbf{s}}$; $s_* = s_{01}(1 - R/R_*)$; $R_* = \beta / \sqrt[3]{N_0}$; $s_{01}, p_0, \eta_1, \eta_2, \beta$ – material constants; N_o – number of crack is the volume unite, ν – Poisson coefficient; α_0 - initial porosity of SFC.

It is considered that a fusion of micro cracks in SFC starts when the specific size reaches the critical value $R_* = \beta / \sqrt[3]{N_0}$ at a constant concentration N_o.

The process of fragmentation of the damaged by cracks material is described within a porous elastic-plastic medium model. The system (1) is locked by the equations that relate pressure p and porosity α under compression.

$$\left(p \ge \frac{2}{3}\sigma_T \ln\left(\frac{\alpha}{\alpha - 1}\right)\right)$$

$$\rho_0\gamma_0\varepsilon + \frac{c_0^2\rho_0(1-\gamma_0\eta/2)\eta}{(1-s_0\eta)^2} - \frac{2}{3}\sigma_T \ln\left(\frac{\alpha}{\alpha-1}\right) = 0 \tag{8}$$

and at decompression ($p \le -a_s \ln(\frac{\alpha}{\alpha - 1})$)

$$\rho_0\gamma_0\varepsilon + \frac{c_0^2\rho_0(1-\gamma_0\eta/2)\eta}{(1-s_0\eta)^2} + a_s \ln\left(\frac{\alpha}{\alpha-1}\right) = 0. \tag{9}$$

The fragmentation of the damaged by cracks material under tension load happens the relative volume of the hollows reaches the critical value $\xi_* = \frac{\alpha_* - 1}{\alpha_*}$. If the damaged material is being loaded by a compression impact then the fragmentation criterion will be the intensity value of the plastic deformations e_u^*:

$$e^*_u = \frac{\sqrt{2}}{3}\sqrt{3T_2 - T_1^2}, \tag{10}$$

where a_s – model parameter; T_1 and T_2 – the first and the second invariants of the deformation tenser.

The destructed material is modeled as a granulated medium that is tolerant to compression and not tolerant to a stretching load. In [3] design-experiment method was used to investigate the process of the impact interaction of steel cylindrical hitters (diameter d_0=7.65 mm, length = 23 mm) with a concrete plate (width = 200 mm) at 130÷700 m/s. The concrete consisted of one part of cement 400, 2 parts of fine send. The production time = 30 days. The diameter (D) of the facial break-away in the concrete plate and the depth of the crate (L) were measured during the experiment. The experimental results have matched the results of mathematical modelling. The presented study focuses on the modeling of the impact interaction of the above mentioned hitter and a SFC plate. The results of the mathematical modeling of the impact interaction are shown in Table 1 and Figure 1.

Table 1. Calculated penetration depth.

V, м/с	L_1/d_0	L_2/d_0	R, %	V, м/с	L_1/d_0	L_2/d_0	R, %
100	0,15	-	-	150	0,89	0,16	-82
200	1,65	0,67	-59	244	2,25	1,31	-42
300	2,76	2,06	-25	350	3,2	2,54	-21

Where V is the velocity of the impact; L_1/d_0, L_2/d_0 – relative crater depth in concrete and SFC plates; R – difference between craters' depth. Figure 1 presents the impact interaction of the cylindrical hitter with concrete and SFC plates at 350 m/s.

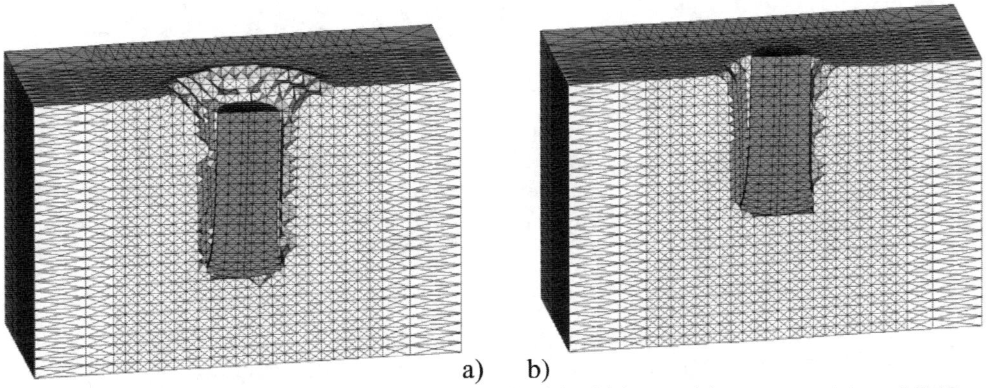

a) b)

Figure 1. Impact interaction of the cylindrical hitter with concrete (a) and SFC (b) plates at 350 m/s.

As shown in Table 1 the penetration depth of the steel cylinder in the concrete plate is deeper than in the SFC plate. The process of the crater origination in the concrete plate starts at about 90 m/s. Meanwhile the same process in the SFC plate starts at 135 m/s. The crater's depth in the concrete plate is 0.15 d_0 at 100 m/s. Meanwhile the impact at the same velocity the SFC plate causes 0,16 d_0 crater. The difference between two craters is 82%. The difference decreases with an impact's velocity growth. At 350 m/s the difference is 21%. In [6] a design-experiment method was used to investigate the process of the impact interaction of the above mentioned hitter with concrete and ferroconcrete plates at 300÷750 m/s. A hitter perforates 24 mm concrete plate at 495 m/s. Facial and rear breaks-away originate during the experiment. Their diameters are $d_f=3{,}0\cdot d_0$; $d_r=5{,}2\cdot d_0$. The hitter's velocity is $U_k=327$ m/s after it has left the plate.

Figure 2 presents the impact perforation of concrete and SFC plates by a cylindrical hitter at 495 m/s. A facial break-away $d_f/d_0=3{,}06$ and a rear break-away $d_r/d_0=2{,}53$ have originated in the plate. The velocity of the hitter after it has left the plate is 310 m/s. The difference between the modeling and experimental velocity measurements is 5%.One can observe the difference between rear break-away (30%). A relatively large difference can be explained by the limitations of the model where the maximum value of the barrier is 5,8 d_0 meanwhile it could reach up to 7,8 d_0 during the experiment. A facial break-away ($d_f=1{,}8\cdot d_0$) originates when a SFC plate is being perforated. A rear break-away diameter is $d_r=1{,}47\cdot d_0$. The velocity of the hitter after it has left a SFC plate is 285 m/s. This is 8 % lower in comparison to a concrete plate.

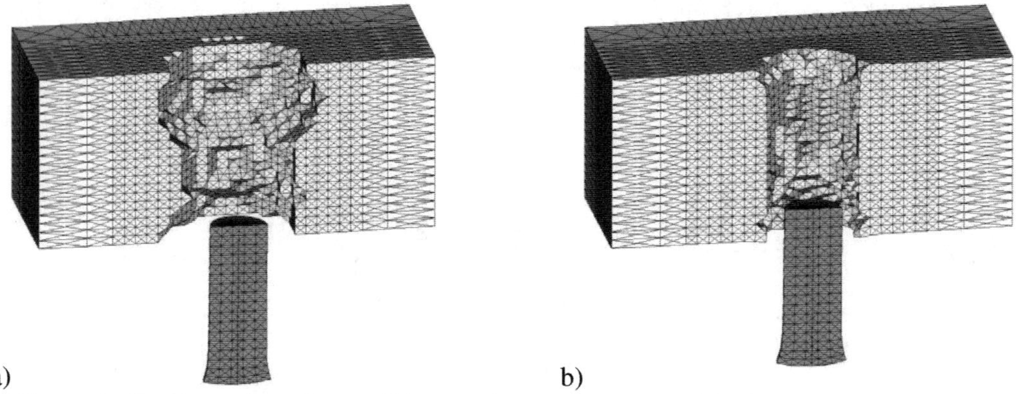

a) b)

Figure 2. Configuration of a steel hitter and a concrete (a) and SFC (b) plates at 495 m/s in t= 130 micro seconds.

Conclusions

The results of the study demonstrate the capabilities of the mathematical model to calculate stress-strain state and destruction of SFC structures under shock wave load. Comparison of the results for concrete and SFC showed that the application of steel fiber allows increasing a dynamic strength of a plate up to 500 m/c (10%).

References

[1] Талантова К В и Михеев Н М 2011 Исследование влияния свойств стальных фибр на эксплуатационные характеристики сталефибробетонных конструкций *Ползуновский Вестник* **1** 194-199

[2] Johnston Colin D 1982 Steel fibre-reinforced concrete-present and future in engineering construction *Composites* **13** 113-121

[3] Саргсян А Е 2013 Динамика и сейсмостойкость сооружений атомных станций (Саров, Россия: РФЯЦ-ВНННЭФ) 550

[4] Белов Н Н, Югов Н Т и Копаница Д Г 2005 Модель динамического разрушения мелкозернистого бетона *Вестник ТГАСУ* **1** 14-22

[5] Белов Н Н, Дзюба П В и Кабанцев О В 2008 Математическое моделирование процессов динамического разрушения бетона *Механика твердого тела* **2** 124-133

[6] Белов Н Н, Кабанцев О В и Копаница Д Г 2008 *Расчетно-экспериментальный метод анализа динамической прочности железобетонных элементов конструкции* (Томск: СТТ) 292

Behavior of reinforced concrete structures with metal and non-metal reinforcement at impulse loadings

P A Radchenko[1], S P Batuev[2], M E Goncharov[3], V S Plevkov[3] and A V Radchenko[4]
[1]Tomsk State University of Architecture and Building, Department of Applied Mathematics, Tomsk, 634003, Russia
[2]Tomsk State University of Architecture and Building, Department of Geoinformatics and Cadastre, Tomsk, 634003, Russia
[3]Tomsk State University of Architecture and Building, Department of Reinforced Concrete and Masonry Structures, Tomsk, 634003, Russia
[4]Tomsk State University of Architecture and Building, Institute of Cadastre, Economics and Engineering Systems in Construction, Tomsk, 634003, Russia

E-mail: radchenko@live.ru

Abstract. Results of the study on the operation of the reinforced concrete columns joints without reinforcement and reinforced by metal casing under short-term dynamic loading are presented in the paper. Program of experimental and numerical studies was developed. As a result of experimental studies schemes of deformation, cracking and fracture, as well as values of fracture loads of experimental samples were obtained. Present results are compared with data of the numerical calculations performed using the computational program based on finite elements method. Results of experimental and numerical calculations showed good convergence and reliability of columns joints strengthening by metal casing.

1. Introduction

The study of reinforced concrete columns joints operation is of a significant demand nowadays, as while constructing reinforced concrete frames of multistoried residential and civil buildings defects and failures occur reducing bearing capacity of the columns and their joints. This may lead to fracture of separate structures of the building or building as a whole [1]. Besides, dynamic impacts can be experienced by structures, which were not earlier calculated for dynamic loadings, as seismic activity is now increased in many regions of Russia and, especially in cities of Western Siberia [2, 3].

2. The results of experimental studies

The program of experimental and numerical studies which included three series of samples was developed for observation of operation of the concrete columns joints strengthened by metal casing, for identification of maximum load rate and nature of fracture under short-term dynamic loadings. Each series had five samples in scale 1:4 to real columns with different variations of joints and reinforcement in the joint area and external strengthening by means of metal elements. The following labels were used for experimental samples: K_d – a sample without joint; SK_d – a sample with a joint, with grids in joint area; SKD_d – a sample with a joint without grids in joint area; SKM_d – a sample with a joint with grids in joint area and reinforced by a metal casing; $SKMD_d$ – a sample with a joint without grids in the joint area and reinforced by a metal casing. All experimental samples were reinforced by spatial knitted frameworks. Principle reinforcement was made of 8 rods with 8 mm diameter of the class A-III (A-400). Confinement reinforcement (collars and grids) was made of a wire of Bp-I (B500) with a diameter of 3 mm. For prevention of local fracture in the adjacent areas confinement reinforcement is placed in the form of 5 grids with a gap of 20 mm. In case of reinforced concrete columns joints without defect (SK_d, SKD_d) five grids with a gap of 20 mm were placed in the joint area. In case of reinforced concrete columns joints with defect (SKD_d, $SKMD_d$) grids in a joint area were absent. All prototypes were made of concrete which corresponds to the class B20.

In order to obtain information of the behavior of the elements, patented pile driver with the measuring system devices of the Department of Reinforced Concrete and Masonry Structures of Tomsk State University of Architecture and Building was used [4] (Figure 1). It allows testing concrete element under short-term dynamic loading of the type presented in Figure 1a.

a)

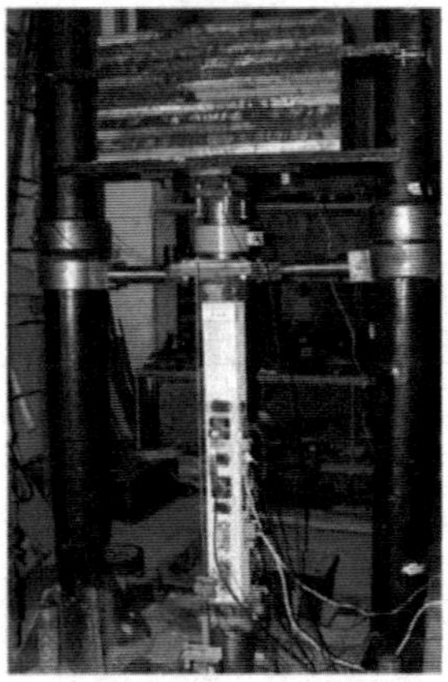

b)

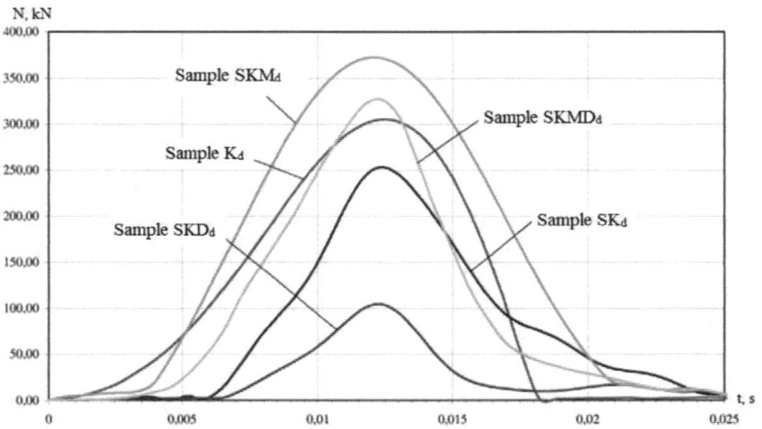

Figure 1. General view of the stand for testing of experimental samples under short-term dynamic load (a) and the nature of the load is applied to the experimental samples and its dependence on the duration (b).

The conducted experimental studies allowed to reveal nature of deformation and fracture of reinforced concrete columns and their joints (Figure 2) at dynamic loadings, and also to determine values of ultimate loads (Figure 1b) for experimental samples: for a sample K_d – 305.0 kN; sample SK_d – 253.0 kN; sample SKD_d – 104.4 kN; sample SKM_d – 371.6 kN; sample $SKMD_d$ – 326.5 kN.

3. Numerical simulation

Numerical calculations of the dynamic deformation under loadings of the type illustrated on Figure 1b were carried out in addition to experimental studies of reinforced concrete columns joints. Samples of numerical simulation fully conformed to the ones of experimental studies. In these studies method of finite elements applying finite-difference scheme of Johnson was used [5]. This approach allows us to study the fracture mechanism of the samples under dynamic loading.

In the numerical finite element model on the free surfaces boundary conditions with absence of shear stresses at the contact surfaces – the sliding conditions without friction were applied. The initial velocity of the load was 4.43 m/s and it fully coincided with experimental studies. The total number of finite elements in the sample was $1 \cdot 10^6$.

For the description of metal load behavior elastoplastic model was used and for the description of experimental concrete sample behavior brittle-plastic model was used. The medium was represented as quasi-homogeneous.

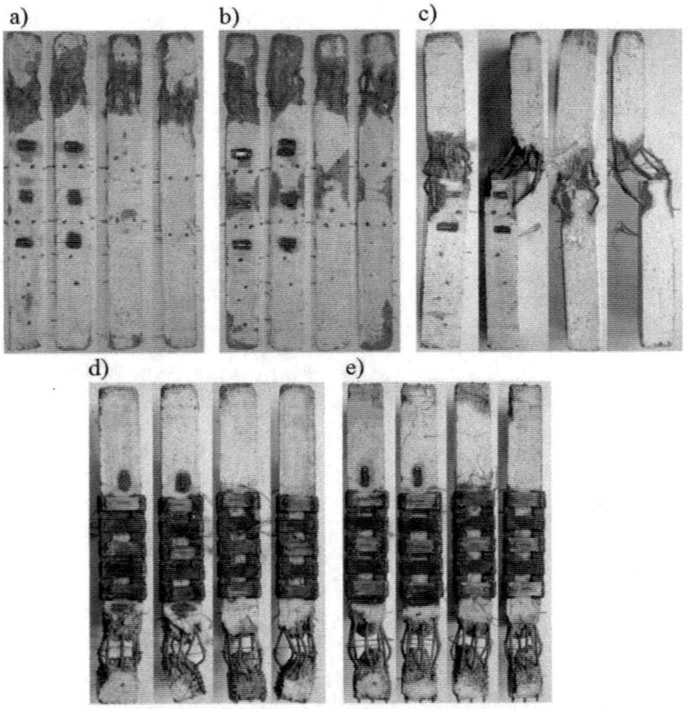

Figure 2. Nature of fracture of experimental samples: a) sample K_d; b) sample SK_d; c) sample SKD_d; d) sample SKM_d; e) sample $SKMD_d$.

The complex of equations describing non-stationary adiabatic movements of the compressible media in the Cartesian coordinate system of XYZ includes the following equations [5, 6]:

– continuity equation

$$\frac{\partial \rho}{\partial t} + div\rho\vec{\upsilon} = 0, \quad (1)$$

– equation of motion

$$\rho a^k = \nabla_i \sigma^{ki} + F^k, \quad (2)$$

where

$$a^k = \frac{\partial \upsilon^k}{\partial t} + \upsilon^i \nabla_i \upsilon^k ; \nabla_i \sigma^{ki} = \sigma^{ki}_{,i} + \sigma^{mi}\Gamma^k_{mi} + \sigma^{ki}\Gamma^m_{mi},$$

– energy equation

$$\frac{dE}{dt} = \frac{1}{\rho}\sigma^{ij} e_{ij}. \quad (3)$$

Here ρ – density of the media; $\vec{\upsilon}$ – velocity vector; a^k – acceleration; F^k – components of mass forces vector; Γ^k_{ij} - Christoffel symbols; σ^{ij} – symmetric components of the stress tensor; E – specific internal energy; e_{ij} – components of the symmetric strain rate tensor: superimposed dots mean a derivative on time; separation of symbol by comma – a derivative on the corresponding coordinate.

Components of a stress tensor in the sample before fracture correlate with coefficients of generalized Hooke's law which was expressed through strain rates:

$$\dot{\sigma}_{ij} = C_{ijkl} e_{kl}. \quad (4)$$

For failure criterion of the sample Hoffman's criterion [7] is used. This criterion allows using

IOP Conf. Series: Materials Science and Engineering **71** (2015) 012025 doi:10.1088/1757-899X/71/1/012025

various strength characteristics of material under compression and tension. The component of a stress tensor expressed through scalar functions, looks as follows:

$$C_1(\sigma_2-\sigma_3)^2+C_2(\sigma_3-\sigma_1)^2+C_3(\sigma_1-\sigma_2)^2+ \\ C_4\sigma_1+C_5\sigma_2+C_6\sigma_3+C_7\sigma_4{}^2+C_8\sigma_5{}^2+C_9\sigma_6{}^2\geq 1 , \quad (5)$$

where C_i – material constants.

It is supposed, that fracture of anisotropic materials in the conditions of intensive dynamic loadings happens as follows: in areas where the strength criterion is violated under compression ($e_{kk}\leq 0$), it is considered that material loses properties of anisotropy, and its behavior is described by hydrodynamic model, and the material keeps strength only under compression; stress tensor becomes in this case spherical ($\sigma_{ij}=-P$); in areas, where the strength criterion is violated at tension ($e_{kk}>0$), the material is considered completely fractured, and components of a stress tensor are set equal to zero ($\sigma_{ij}=0$).

Pressure in the material of samples fractured under compression is calculated by means of the equation of a state [9]:

$$P=\left[\exp\left(4\beta\frac{V_0-V}{V_0}\right)-1\right]\frac{\rho_0\alpha^2}{4\beta} . \quad (6)$$

Here ρ_0 – the initial density; V_0, V – initial and current specific volumes. Coefficients of this equation are calculated from the shock adiabatic curve: $D=\alpha+\beta u_m$, where $\alpha=1400 m/s$, and U_m – the mass velocity.

For the numerical studies calculation software developed by Radchenko A.V. and Radchenko P. A. was used.

Values of relative fracture volume V_c/V_t are shown on Figure 3 in gray gradations. Values V_c and V_t are given in the grid points: V_c – number of elements in node which satisfy fracture criteria; V_t – the total amount of the elements forming this node. Value $V_c/V_t=1$ corresponds to final fracture of a material in the node of the computational grid. Nodes, in which the material is completely fractured, are excluded from calculation for clearness.

4. Conclusion

Numerical calculations allowed revealing fracture schemes of experimental samples in different time periods (Figure 3) with good convergence with experimental studies.

The analysis of results of experimental and numerical studies of samples under dynamic loading showed reliability of strengthening of reinforced concrete columns and their joints. Increased load bearing capacity due to the metal casing was 40% in the test sample SKM_d compared with sample SKD_d and more, than three times at test of sample $SKMD_d$ in comparison with SKD_d sample. The strengthening by a metal casing allowed increasing the bearing ability of a sample with joint $SKMD_d$, having a defect, in comparison with SK_d sample by 29%.

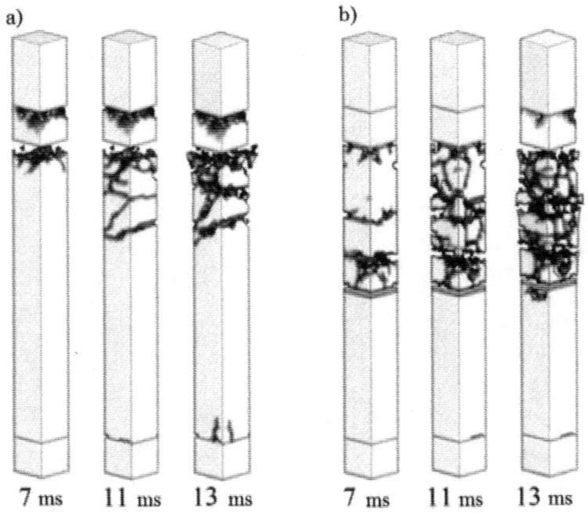

Figure 3. General view of destruction of an experimental sample in time at numerical experiment: a) without joint; b) with a joint of columns without reinforcing.

Acknowledgement
The research performed with the financial support of the Ministry of Education and Science of the Russian Federation within Governmental Decree No.218 of "Measures of state support for cooperation of Russian higher education institutions and organizations, for implementation of complex projects of high-tech production", integrated project No. 02.G25.31.0022, "Production technologies of energy- and resource-efficient housing of economy class based on the universal framed prefabricated architectural and construction system".

References

[1] Плевков В С, Балдин И В, Гончаров М Е, Максимов В Б и Ботьева И А 2010 Методика и результаты экспериментальных исследований стыков железобетонных колонн, усиленных металлическими элементами *ВИНИТИ РАН* **25-В2010** 43
[2] Плевков В С, Балдин И В, Фурсов В В и Балюра М В 2009 Некоторые особенности технического состояния железобетонных конструкций, оснований и фундаментов здания после длительного перерыва в строительстве *Вестник Томского государственного архитектурно-строительного университета* **3** 82-91
[3] Кумпяк О Г, Плевков В С, Копаница Д Г и Балдин И В 2000 Некоторые вопросы динамики железобетона *Вестник Томского государственного архитектурно-строительного университета* **1** 124-136
[4] Плевков В С, Однокопылов Г И, Балдин И В, Саркисов Д Ю, Гончаров М Е, Дзюба П В 2009 Патент РФ на изобретение № 2401424 МПК G01N 3/30 (2006/01) Стенд для испытания железобетонных элементов на кратковременное динамическое сжатие
[5] Johnson G R 1977 High velocity impact calculations in three dimension *J. Appl. Mech.* **44** 95-100
[6] Радченко А В 2003 Модель поведения хрупких анизотропных материалов при динамических нагрузках и ее приложения *Вестник Томского государственного архитектурно-строительного университета* **2** 179-193
[7] Radchenko A V and Radchenko P A 2011 Numerical modeling of development of fracture in anisotropic composite materials at low-velocity loading *J Mater Sci* **46** 2720-2725
[8] Ву Э М 1978 Феноменологические критерии разрушения анизотропных сред *Механика композиционных материалов* под ред. А.А. Ильюшина и Б.Е. Победри (Москва: Мир) 401-491
[9] Канель Г И и Щербань В В 1980 *Пластическая деформация и откольное разрушение железа "Армко" в ударной волне ФГВ* **16** 93-103

IOP Conf. Series: Materials Science and Engineering **71** (2015) 012026 doi:10.1088/1757-899X/71/1/012026

Complex research of molybdenum ore tailings

G Volokitin[1], N Skripnikova[1], O Volokitin[1], I Iuriev[1] and V Shekhovtsov[1]
[1]Tomsk State University of Architecture and Building, Applied Mechanics and Material Sciences Department, Tomsk, 634003, Russia

E-mail: volokitin_oleg@mail.ru

Abstract. The paper considers the processes of plasma chemical synthesis of silicate melts produced from quartz-feldspar raw materials with a view to obtain new construction materials having the advanced functional performance. Presented results illustrate physicochemical research findings (X-ray diffraction analysis, infrared spectroscopy, thermal analysis) related to quartz-feldspar raw materials and melts produced there from.

1. Introduction

Plasma technologies used in production of high-temperature silicate melts provide consistently high temperatures converting initial silicate materials to a liquefied state. Presently, plasma treatment is widely used for high-temperature melts obtained from bottom ash waste generated by thermal power plants. A silica modulus of these wastes is notably higher than that of raw materials (basalt) which are traditionally used to obtain silicate melts in producing mineral fibers. A subsequent research carried out in this field implies using raw materials obtained from beneficiation of different kinds of ores such as feldspar tailings, a chemical composition of which allows converting them into silicate melt using low-temperature plasma (LTP).

A content of quartz-feldspar raw materials obtained from molybdenum ore beneficiation ranges from 90 to 99%. A great deal of raw materials is accumulated in waste dumps of ore mining and processing enterprises [1-3]. A chemical composition of the raw material under study contains 62% of SiO_2 that corresponds to that one of industrial glass and can be used to synthesize silicate melts including those obtained in production of mineral fiber [4-5]. In order to produce high-quality mineral fiber, the melt should possess a uniform chemical composition. Traditional high-temperature melt technologies result in increased energy consumption due to a long-term thermal effect the raw materials are exposed to.

2. Experimental procedure

In existing electric-plasma installations fine particle top feeding is used to supply raw materials to plasma jet projected in a melting furnace [6]. However, according to experiments, in this case many fine particles do not participate in the process of melting. To avoid this problem, an electro-plasma apparatus with side feeding was invented which includes a worm feeder for raw material to be supplied to a zone of silicate melt formation.

Figure 1 presents color photography and a schematic view of this experimental electro-plasma apparatus intended for mineral fiber production using refractory silicate-based materials. This apparatus comprises such principal components as plasma torch, melting furnace, worm feeder, and forming hood.

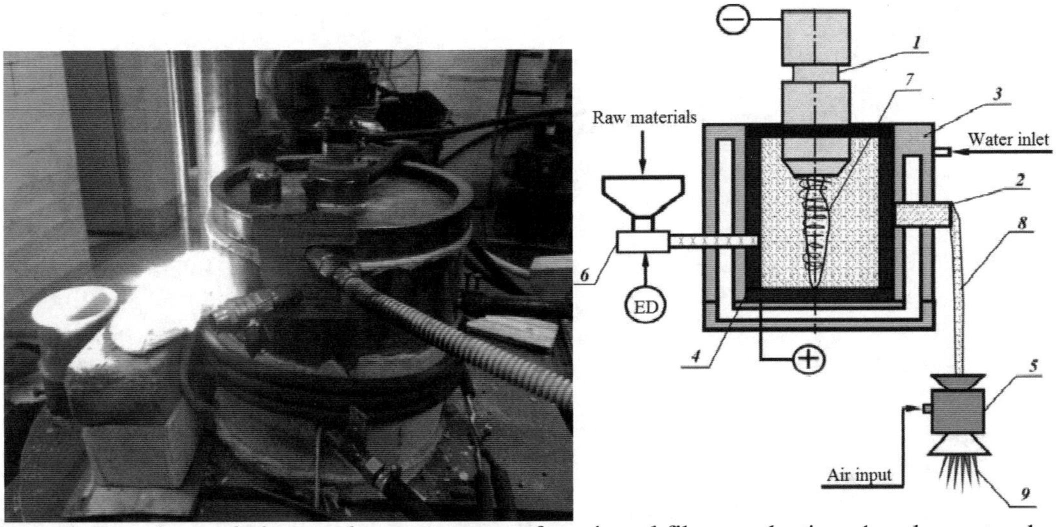

Figure 1. Experimental electro-plasma apparatus for mineral fiber production: 1 – plasma torch; 2 – pouring hole; 3 – water-cooled melting furnace; 4 – graphite electrode; 5 – forming hood; 6 – worm feeder; 7 – plasma jet; 8 – melt; 9 – mineral fiber.

Operation of the electric-plasma apparatus is based on the interaction between highly concentrated plasma flows and a raw material. Plasma torch is placed in melting furnace equipped with pouring hole. Graphite electrode is installed in the bottom of melting furnace and connected with the positive pole of DC power supply (anode), while the plasma torch is connected with its negative pole (cathode).

Once silicate melt reaches the level of the pouring hole, it will be discharged through it into forming hood. By means of batcher mounted into worm feeder driven by electric drive, powder raw material is fed to the melting furnace. Plasma jet is projecting between plasma torch and graphite electrode. A raw material starts to melt and homogenize under a strong thermal effect (3000-5000 °C). The raw material particles are then mixed with a high-temperature silicate melt and melted uniformly. The melt formed in the forming hood is converted into mineral fibers. Furthermore, this method of feeding of raw materials makes possible to avoid a loss of fine particles blowing by a LTP jet. This method provides a uniformity and viscosity of low melt within the whole volume of the melting furnace [7].

3. Results and discussion

In order to study processes which occur at high-temperature melt synthesis, physicochemical research included X-ray diffraction (XRD) analysis and infrared spectroscopy.

XRD analysis was carried out by DRON-4-07 diffractometer which was modified for digital signal processing. Measurements were conducted using copper radiation (K_α) and Bragg-Brentano X-ray optical scheme. Specifications for the DRON-4-07 included 0.02° scanning step; 20 – 92° range for angles to be scanned; 30 kV voltage; and 25 mA current. Powder tailing wastes were studied in the initial and amorphous states. Amorphous state was achieved as a result of the LTP flows.

Figures 2 and 3 show theoretical and experimental diffraction patterns for different states of ore tailings. As shown in these Figures, a good agreement between theoretical and experimental diffraction patterns is observed.

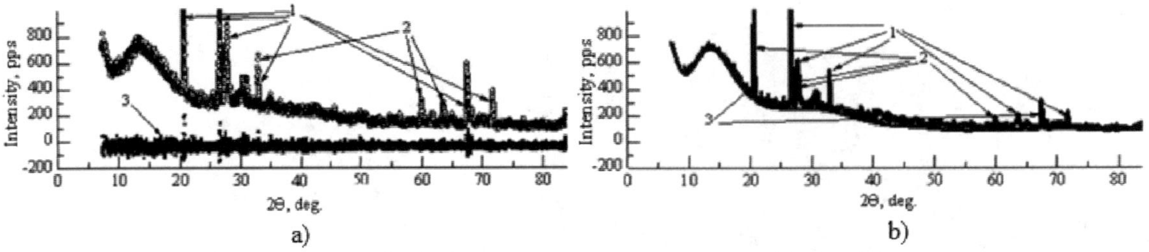

Figure 2. Numerical phase analysis of tailing wastes in initial state:
a) experimental (1); theoretical XRD pattern (2); difference between theoretical and experimental XRD patterns (3);
b) theoretical XRD patterns for phases: O_2Si (1); $O_{22}Al_{20}$ (2); O_6Si_6 (3).

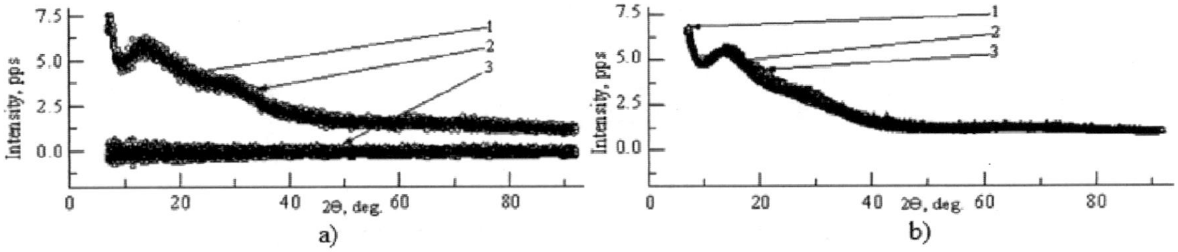

Figure 3. Numerical phase analysis of tailing wastes in amorphous state:
a) theoretical (1); experimental XRD pattern (2);
difference between theoretical and experimental XRD patterns (3);
b) phases: O_2Si (1); $O_{22}Al_{20}$ (2); $O_{192}Si_{96}$ (3).

As shown in Figure 2a, the main phases for molybdenum ore tailings are silica and feldspar. XRD analysis has shown that the melt product has no spikes typical for Figure 3a that indicates the absence of a crystalline structure, i.e. the melt becomes X-ray amorphous upon its cooling.

Infrared spectroscopy of ore mining wastes was conducted before and after plasma chemical synthesis. Fourier transform infrared (FTIR) spectra were detected with Nicolet 6700 infrared spectroscopy manufactured by Thermo Nicolet Corporation (USA). Results of spectroscopy are presented in Figure 4 for silica alumina raw materials and aluminosilicate glass.

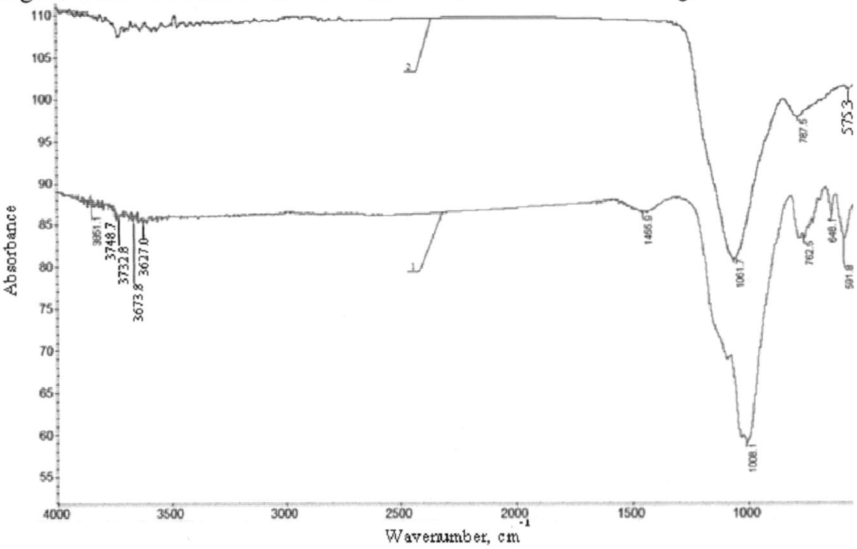

Figure 4. FTIR spectra: 1 – molybdenum ore tailings; 2 – melt product.

The main base unit of feldspar is a tetrahedral alumina-silicon-oxygen framework comprising $[SiO_4]^{4-}$ and $[AlO_4]^{5-}$. These tetrahedrons are transformed into chains which then form polyhedrons whose cavities are filled with metal cations (K, Na, Ca, etc.) and H_2O molecules, and neutralize a negative charge of the framework occurring at substitution of Si^{+4} by Al^{+3}. Infrared spectrum typical for feldspars (Figure 4, curve 1) is sharply changed after plasma treatment. The structure-sensitive band maximum displacement is observed towards ordering of the silicon-oxygen framework. The maximum 1008,1 cm^{-1} of the antisymmetric stretching absorption band shows a shift to the red resulting in 1061,7 cm^{-1} maximum. At the same time, the maximum 762,5 cm^{-1} of the symmetric stretching absorption band shows a shift to the blue resulting in 787,5 cm^{-1} maximum. A clear deformation band is then observed with 456,9 cm^{-1} maximum. Discrete stretching absorption bands lying between 400 and 700 cm^{-1} almost completely vanish due to the intensive vaporization of low melting components and aluminosilicates. As a result, O_2Si concentration in the melt increases that leads, in its turn, to a structural ordering of the silicon-oxygen framework (Figure 4, curve 2).

The thermal gravimetric analysis/differential thermal analysis (TGA/DTA) was carried out by Simultaneous Thermal Analyzer Netzsch STA 449 C Jupiter (Germany) at heating rate of 10 °C/min in air atmosphere.

Being a natural material, feldspar contains components comprising hydroxyl and carbonate groups. Thermal analysis has shown the presence of these thermally unstable compounds. TGA/DTA measurements (Figure 5) show that the initial feldspar sample has a gradual weight loss within the temperature range of 25–1000 °C.

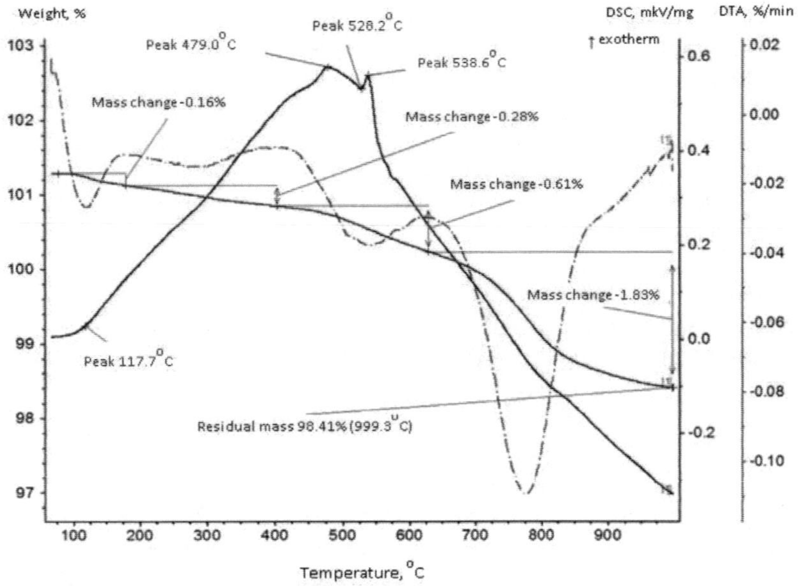

Figure 5. TGA curves of the initial feldspar sample.

DTA measurement as shown in Figure 1, gives grounds to suppose that the initial feldspar sample has up to 0,16 wt.% crystallization water with dehydration temperature of 117,7 °C.

Thermal analysis of plasma-treated feldspar shows its stability as shown in Figure 6. The insignificant weight growth is observed (0,35 wt.%) within the indicated temperature range. The feldspar sample is characterized by a thermodynamic stability represented by the differential scanning calorimetry (DSC) curve and a chemical inertness in air atmosphere within the temperature range of 25–1000 °C. Plasma treatment results in decay of thermally unstable impurities in the feldspar sample.

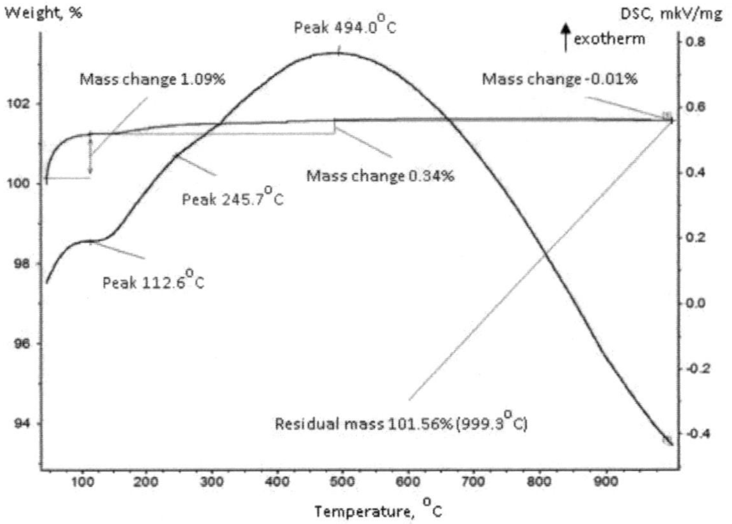

Figure 6. TGA curves of the plasma-treated feldspar sample.

4. Conclusions

Thus, physicochemical research findings have shown that plasma chemical synthesis of silicate melts produced from quartz-feldspar raw materials using low-temperature plasma allows obtaining the ordered system of aluminosilicate glass. Melt products are characterized by homogeneity and ordering, i.e. the process of melting has provided a complete melt homogeneization. Therefore, it is possible to produce glass-based construction materials by plasma chemical technique employed for melting aluminasilicates, including mineral fibers possessing the improved service properties (chemical, thermal, and water resistant), mechanical-and-physical and thermal properties, and durability. This is connected with the amount of aluminosilicate glasses SiO_2 (62%) and Al_2O_3 (15%) when the index of mineral fiber acidity has range between 9 ~ 12.

5. Acknowledgements

The authors acknowledge the contribution of Prof. Yu. Abzaev of Tomsk State University of Architecture and Building for great assistance and helpful comments. Thanks also go to M. Vorob'eva, the translator of TSUAB Publishing House.

References

[1] Skripnikova N K, Otmakov V I and Volokitin O G 2010 Processes occurring upon plasma chemical synthesis of refractory silicate materials *Glass and Ceram.* 19-21

[2] Vereshchagin V I, Buruchenko A E and Kashchuk I V 2000 Possibilities of using secondary raw materials for structural and glass ceramics production *Construction Materials* 20–22

[3] Kaz'mina O V, Vereshchagin V I, Abiyaka A N, Semukhin B S 2009 Low-temperature synthesis of granular glass from mixes based on silica-alumina-containing components for obtaining foam materials *Glass and ceramics* 341–344

[4] Sendrov E E 1990 *Processes of ordering aluminosilicate frameworks* (Moscow: Nauka)

[5] Volokitin O G, Maslov E A and Shekhovtsov V V 2014 Plasma treatment technology for silicate melt used in mineral fiber production *Advanced Materials Research* 233-236

[6] Volokitin O G, Timonov E V, Volokitin G G, Nikiforov A A and Chibirkov V K Plasma apparatus for synthesis of refractory silicate melt *Pat. Rus. Fed. N 2503628*

[7] Volokitin G G, Skripnikova N K, Volokitin O G and Zayats A A 2014 Physicochemical processes at plasma-chemical synthesis of silicate melts *Russ. Phys. J.* 111-114

IOP Conf. Series: Materials Science and Engineering **71** (2015) 012027 doi:10.1088/1757-899X/71/1/012027

Impact of zeolite-based nanomodified additive on the structure and strength of the cement stone

A D Egorova[1] and K E Filippova[1]
[1]North-Eastern Federal University, Yakutsk, 677000, Russia

E-mail: eg_anastasy2004@mail.ru

Abstract. Portland cement is the main binder in the building materials industry; its properties strongly influence properties of mortars and concretes. Some regions experience difficulties with delivery and storage of Portland cement, raising the need to develop an effective additive from the available raw materials. Such materials for the Republic of Sakha (Yakutia) are zeolite-containing rocks. Studies have shown that introducing of dibutylphthalate to the composition of modified additive during mechanochemical activation leads to achievement of up to 11% of total amount particles with the size of 3-30 nm. After introducing 0.5% of the obtained additives, the compressive strength of cement-sand slurry samples increases up to 28%. Positive effect of additives introduction is also observed at high flow rate of water (W / C = 0.7). Gaining strength reaches 23%, allowing the efficient use of additive for movable mixtures with enhanced strength properties. In general, the proposed supplement allows reducing the water flow in the solution without decreasing its mobility, and increasing strength properties, which makes it possible to obtain a whole class of solutions of modified cement binder. The market value of the developed additives is 18 rubles per 1 kg, making sound competition in the market of modifying additives.

1. Introduction

Materials and products based on Portland cement are an integral part of the entire building complex. Among them there are solutions as well as heavy concrete. On the vast territory of the Republic of Sakha (Yakutia), with poor infrastructure, the only cement plant cannot perform high-quality supply of raw materials to the remote districts. This problem hardly provides an opportunity to accomplish monolithic building, while the demand for it keeps on increasing every year (the construction of bridges, schools, kindergartens, cottage low-rise construction, etc.).

In the manufacture of precast concrete products and structures, an urgent task is to reduce consumption of expensive Portland cement without losing strength. For this purpose, the existing plants producing concrete products use classical additives such as superplasticizer-3, sodium formate, the PFM-NLK. Modern additives such as "Superplast" are not in high demand, as their use may considerably increase the cost of production.

One way to improve the performance properties of concrete is the introduction of active mineral additives (AMD) to the mixture. AMD-additives provide hydrolicity of the binder and the existing structure of resistance over time. Mechanism of action of a hydraulically active additive is mainly caused by their chemical interaction with the lime, formed by hydrolysis of C_3S during cement hydration. At the same time type C-S-H(B) low-basic calcium hydrosilicates are basically formed, calcium hydroaluminates and hydroferrites, which increase the gel component of the cement paste, improve strength and deformation properties of concrete. As minerals siliceous components can be used - natural AMD of sedimentary or volcanic origin, artificial additives (glinite, opus signinum, pulverized waste generated during burning of expanded clay), burnt rock, ash and slag, as well as inert additives (coal dust, sand, glass, etc.) [1].

In the production of building materials zeolite is used as an active mineral supplement since 1960s. This is due to the fact that zeolite is an effective material for reducing the flow of cement (15-30 %) for the production of light and heavy concrete, and as for preparation of solutions it can completely replace lime. This increases water-holding capacity of mortar and crack resistance. Portland cement additive zeolite in amount of more than 15 % provides excellent corrosion resistance of cement compositions relative to the chloride and sulphate brines. [2]

Extensive exploration showed that zeolites are widespread in nature, and some sedimentary rocks contain almost monomineralic layers, in which the original volcanic rocks 95% turned to a particular zeolite mineral. Some zeolites deposits can easily be explored as overlie the shallow or surface. Natural zeolites may not replace synthetic in those areas where raw material is of high purity, however, they are of interest to the industry in those cases where there are large amounts and, therefore, low cost of natural minerals are more important than the feed purity. [3]

Great possibilities of using natural zeolites in Yakutia, associated with the development of the field Honguruu, which were first discovered in Yakutia by K.E. Kolodeznikov, the employee of Yakut branch of Russian Academy of Sciences in 1978 in Honguruu settlement of Republic of Sakha (Yakutia). Zeolite materials of this deposit according to the data, are different from other known in CIS and Russian deposits of this type and characterized by high content of mineral clinoptilolite of Haylandite series (70-98%), with predominance of the chemical composition of the aluminosilicate sodium-calcium form.

The presence of local raw zeolite rocks in the Republic of Sakha (Yakutia) determines the feasibility of establishing effective integrated additives obtained by AMD mechanochemical activation - from fine zeolite rocks and plasticizing component capable of increasing the physical, mechanical and performance properties of cement-sand mortar and heavy concrete for severe climatic conditions.

2. Materials and methods

Complex additive based on zeolite is obtained by modifying natural zeolites and plasticizing additives. Due to its chemical and mineralogical composition, zeolite is able to be dispersed to the value of the specific surface of 2,500 kg/m^2 without plasticizers. Such indicators of specific surface were obtained by mechanical activation of zeolites in a planetary mill "Activator-25", developed at the Institute of Solid State Chemistry and Mechanochemistry, Novosibirsk Branch of the Russian Academy of Sciences.

As seen from the graph (see Figure 1), during grinding over 5 minutes a decrease in specific surface is observed due to the increased fineness of the particles and their self-aggregation. To prevent this phenomenon, plasticizer was introduced.

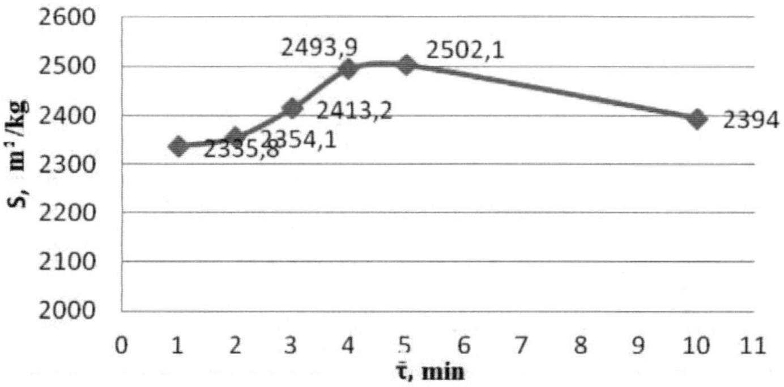

Figure 1. Graph showing dependence of specific surface value of grinding time period.

As a plasticizer, common liquid plasticizer was used - dibutyl phthalate (DBP), GOST 8728 (Russian State Standard). Owing to this, the activity of zeolite does not fall in time and joint mechanochemical activation allows you to receive nanofraction. When activated, the particle size of zeolites reach an average of 1.3 microns with 11 % of the total amount of the particles having sizes of 3-30 nm, i.e. with the first approximation it can be considered as an additive nanomodificator that is supported by studies of the sample with a laser particle size analyzer of HORIBA brand (Figure 2).

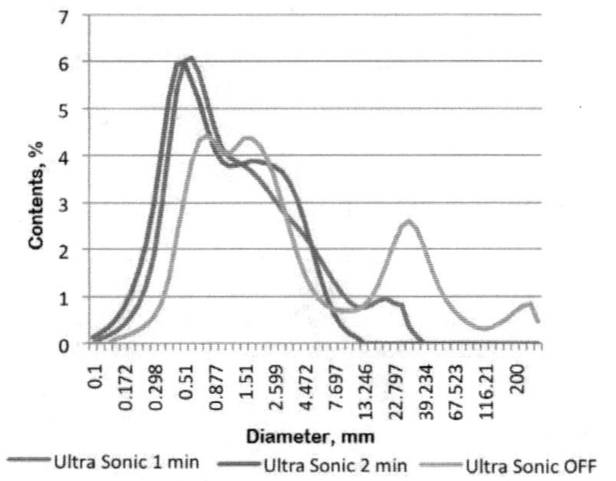

Figure 2. Results of the zeolite sample study: dibutyl phthalate (1:1) after 15 minutes of grinding on the planetary mill "RETSCH PM400".

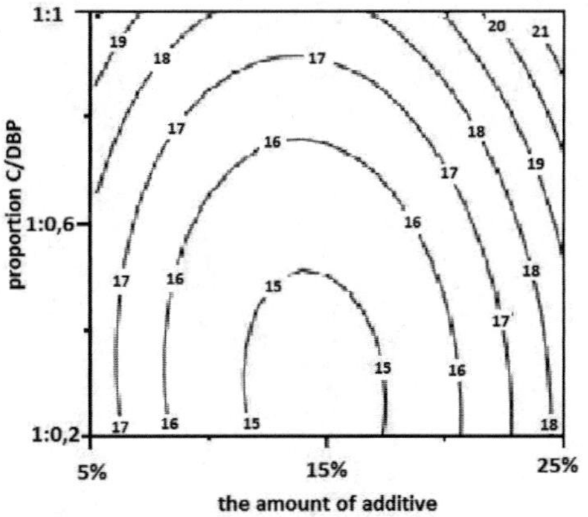

Figure 3. Nomogram of dependence of compressive strength of cement-sand mortar with nanomodified additive based on zeolite at the age of 28 days.

Further, the complex additive was prepared in different proportions of zeolite and dibutyl phthalate (DBP): 1:0.2; 1:0.6; 1:1, respectively.

3. Experimental

The prepared additive was mixed in different quantities of the cement-sand mortar. The additive prepared in a ratio of 1:0.2 is light green-tinted powder. Under laboratory conditions three standard samples of cement-sand mortar preparation time is 3 minutes. An additive prepared in a ratio of 1:0.6 has a high viscosity and dark green color. For obtaining a homogeneous suspension in water additive requires more time as compared to other ratios. Under laboratory conditions for the preparation of three standard test-beams of cement-sand mortar, slurry preparation time is 5 minutes at water temperature of 15°C. At higher water temperature during preparation, time of the aqueous slurry is significantly reduced.

Additives prepared in a ratio of 1:1, have average viscosity and they are more fluid compared with the above-described compositions of additives. Color of additives is green. Total suspension preparation time for obtaining three standard cement-sand mortar test-beams in the lab conditions is 1.5 minutes.

The density of cement-sand mortar does not depend on the amount of introduced complex additive; there is no definite relationship. The densest packing of the sample is provided at a ratio of zeolite and DBP 1:0.2, respectively, and the amount of injected additive is 5%. A lowest value is reached at a ratio of 1:0.6 and in the quantity of 15%.

The maximum value (22 MPa) of compressive strength of cement-sand mortar samples is achieved with an integrated flow additive 25% and the ratio of zeolite and DBP 1:1. This is due to the fact that the composition increases the amount of DBP and consequently decreases the amount of zeolite. Also, increasing the tensile strength is observed even at 5 % of complex additive administration that shows structural changes in the cement-sand mortar, which results in improved strength properties of the materials (see Figure 3).

The results obtained are confirmed by studies of the microstructure of cement paste solution with the scanning probe microscope "NtegraPrima" NTMDT (Figure 4).

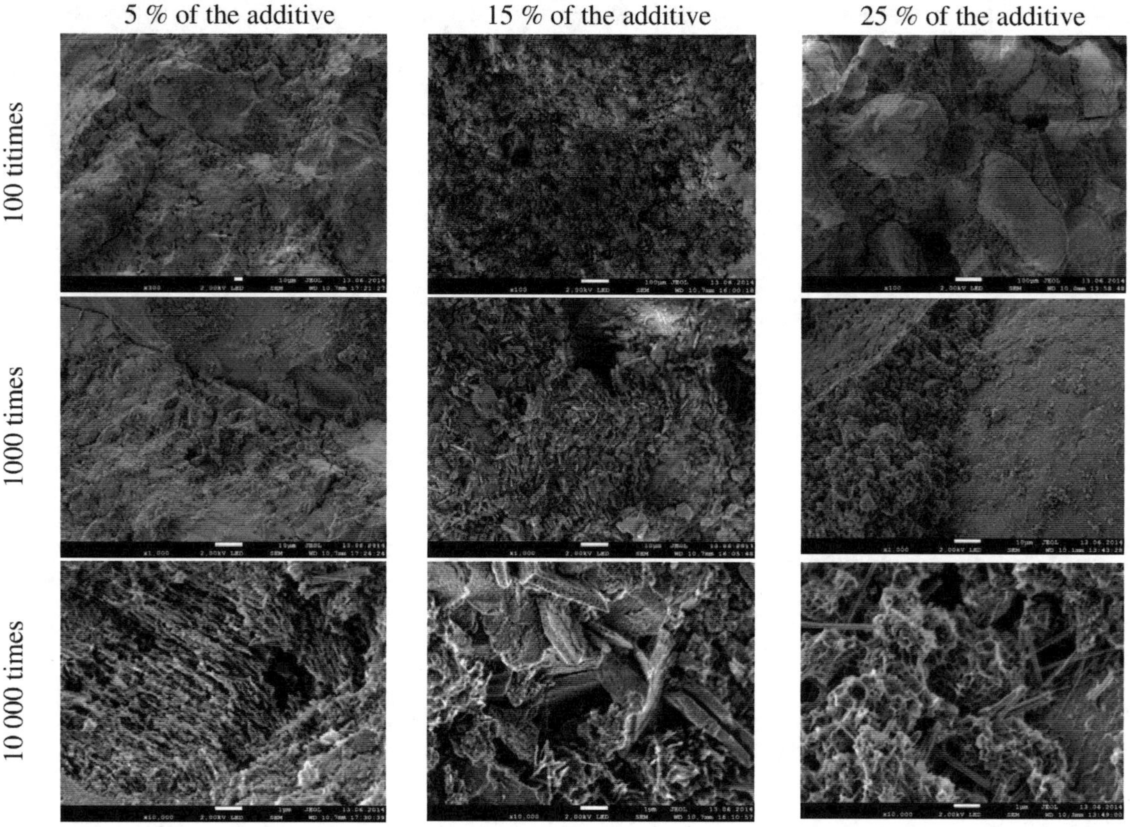

Figure 4. Micrograph structure of cement-sand mortar with nanomodified additive based on zeolite in the ratio of 1:1.

Analysis of the cement stone micrographs reveals that increase of its compressive strength while introducing additives in the amount of 5 to 25% is due to thickening of the structure. In the former case, the direction of new formations orientation, and in the second - the formation of the zeolite structure, which completely fills all of the intercrystalline space.

Even with introduction of the additive in the amount of 0.5%, increase in the compressive strength of the samples of cement-sand mortar is 28%. There is a positive effect of the introduction of additives at high flow rate of water (W/C = 0.7). Gaining strength in this case reaches 23%, which allows the possibility of efficient use of the additive for movable mixtures with enhanced strength properties.

Increasing the amount of additives and decrease in the flow of water in the composition of the cement-sand mortar causes increase of flexural strength. This is due to the fact that increasing the amount of additive increases the amount of plasticizing component which imparts organic features to the material; they are characterized by high tensile bending loads.

4. Conclusions

In general, the proposed supplement allows to reduce water flow of the solution without decreasing its mobility, as well as to increase the strength properties, which makes it possible to obtain a whole class of solutions of modified cement binder.

The market value of the developed additives is 18 rubles per 1 kg, which can make sound competition in the market of builders.

References

[1] Дементьев С Н 1989 *Новые подходы к изучению физико-химических свойств цеолита* (Новосибирск: ИГИГ) с 135

[2] Jeffrey D Rimer, Manjesh Kumar, Rui Li, Alexandra I Lupulescu and Matthew D Oleksiak 2014 Tailoring the Physicochemical Properties of Zeolite Catalysts *Catalysis Sci Technol* 51-57

[3] García R A, Serrano D P, Vicente G, Otero D and Linares M 2008 Catalytic activity of the beta zeolite with enhanced textural properties in the Friedel-Crafts acylation of aromatic compounds *Proc 4th Int FEZA Conf Zeolites and Related Materials: Trends, Targets and Challenges* (Paris, France) 1-5

[4] Nik Ahmad Nizam Nik Malek and Alias Mohd Yusof 2007 Removal of Cr(III) from Aqueos Solutions using Zeolite NaY Prepared from Rice Husk Ash *The Malaysian Journal of Analytical Sciences* **11** 76-83

[5] Под редакцией Сафронов А Ф, Колодезников К Е и Уаров В Ф 2004 Полезные искапаемые Сунтарского района и перспективы их промышленного освоения (Якутск: ЯФГУ Издательство СО РАН) 73-88

[6] Колодезников К Е 2003 *Цеолитоносные провинции востока Сибирской платформы* (Якутск: ЯФ Изд-ва СО РАН) 224

IOP Conf. Series: Materials Science and Engineering **71** (2015) 012028 doi:10.1088/1757-899X/71/1/012028

High-strength lightweight concrete mixtures based on hollow microspheres: technological features and industrial experience of preparation

A S Inozemtcev[1]
[1]Moscow State University of Civil Engineering, 129337, Moscow, Russia

E-mail: InozemcevAS@mgsu.ru

Abstract. The research results concerning dependencies between technological parameters and physical properties of structural lightweight concrete are presented in the article. High-strength lightweight concrete has unique performance characteristics: low average density (less than 1500 kg/m^3) and high compressive strength (more than 70 MPa). Hollow alumina-silicate microspheres with nanoscale modifier are used for obtaining these properties. It is shown in the article that the preparation of high-strength lightweight concrete in industrial conditions must be implemented using a turbine mixer having six paddles and engine power more than 39.2 kW. Oscillation frequency of more than 3000 rpm, vibro-compacting time less than 15 seconds, heat-humid treatment temperature approximately 60-65 ^{o}C and heat-humid treatment time 6-7 hours are optimal for production. The results of industrial mixing-test are presented.

1. Introduction

The recognized world research trend in direction of modern building materials is development of new composites having complex properties that increases application area and provides technical and economical efficiency of new technology [1-4]. One of such materials is high-strength lightweight concrete (HSLWC). This type of concrete developed by researchers from Moscow State University of Civil Engineering has low average density (less than 1500 kg/m^3) and high compressive strength (more than 70 MPa). The structure of HSLWC [5] can be described as closed-packed hollow microscale spherical particles of filler evenly integrated to multifractional cement-mineral matrix. Obviously, technological factors (time and rate of mixing, parameters of vibro-compacting, mode of heat-humid treatment (HHT) and other) have influence on quality of concrete. Therefore, research the influence of technological modes on physical and mechanical properties of high-strength lightweight concrete based on hollow microspheres is relevant task for improving of construction technologies.

2. Materials, methods and equipment

Portland-cement CEM-I 42.5, hollow glass and alumina-silicate microspheres (diameter is 0-100 mkm), quartz sand fraction of 0.16-0.63 mm, stone powder with a specific surface area of 700-800 m^2/kg, microsilica and polycarboxylate plasticizer were used for preparation of concrete mixture. Composition of HSLWC was chosen according to [6]. The mixing of concrete was carried out using the “Automix” mortar mixer. The “Advantest 9” servo-hydraulic press was used for determination of strength in 1 day after heat-humid treatment according to ASTM C39/C39M-14. The structure of HSLWC was examined on «Eclipse MA200» microscope.

3. Results and discussion

There are different equipment parameters during preparation of the concretes at the laboratory and plants. Therefore it is required to determine the character of changes in the physical and mechanical properties of the high-strength lightweight concrete based on hollow microspheres for different technical characteristics of the manufacturing equipment. Three modes of the mixing with different rate and time were selected (Table 1).

Table 1. Mixing modes of HSLWC with hollow microspheres.

No	Type of mixing	Mixing time, sec		
		Mode M1	Mode M2	Mode M3
1	Slow mixing	120	120	30
2	Fast mixing	–	–	90
3	Pause	10	10	10
4	Slow mixing	90	–	–
5	Fast mixing	30	120	120

Notes: slow mixing is stirrer speed about 140±5 rpm, fast mixing is about 285±10 rpm.

The inhomogeneous structure of the concrete is formed if preparation of mixture is carried out according to M1 mode. The data of the Table 1 and Figure 2 show that M1 mixing mode promotes the formation of agglomerates of cement particles which are adsorbing water only on the surface. The average density of concrete reaches 1390±5 kg/m^3 and compressive strength is up to 34.0 MPa (specific strength is R_{sp}=24.5 MPa, R_{sp} – specific strength of material: compressive strength divided by relative density). Figure 2 demonstrates the structure of high-strength lightweight concrete which was prepared in accordance with M1 mode. There is loose structure with a lot of pores of different sizes and inhomogeneous zones of the cement and mineral part (region of the fuzzy image on the microphoto). The best quality of structure of the HSLWC is obtained if the preparation of the mixture is carried out with increased time of the fast mixing.

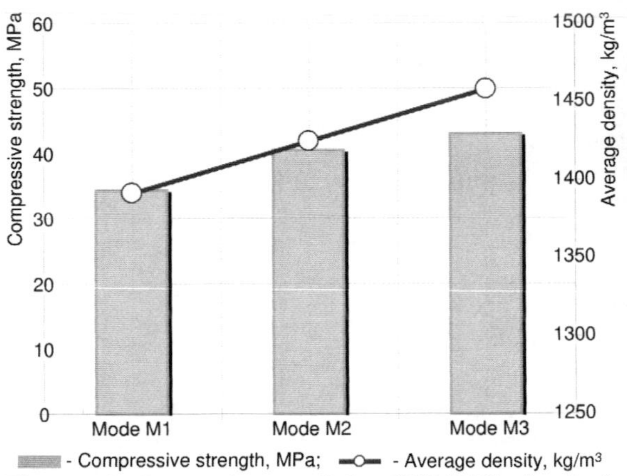

Figure 1. Dependence of the HSLWC physical and mechanical properties from mixing mode.

Figure 2. Microphotography of the HSLWC structure which was prepared by Mode M1.

For the M2 mode, the prepared samples are characterized by almost the same density (the excess is about 2-3 %) and compressive strength (excess is 5-7 %). The structure of such concrete presented on figure 3 is more ordered system of filler/cement-mineral part. There is denser packing but still has pores of big sizes. The preparation of HSLWC compositions using a mode M3 (201 seconds of the total time of the fast mixing with stirrer speed about 285±10 rpm) allows to obtain the homogeneous and uniform structure with small numbers of small pores (Figure 4). In this case the high-strength lightweight concrete has average density 1455±5 kg/m^3 and compressive strength of more than 42.0 MPa (specific strength more than 28.9 MPa) in 1 day after HHT.

Figure 3. Microphotography of the HSLWC structure which was prepared by Mode M2.

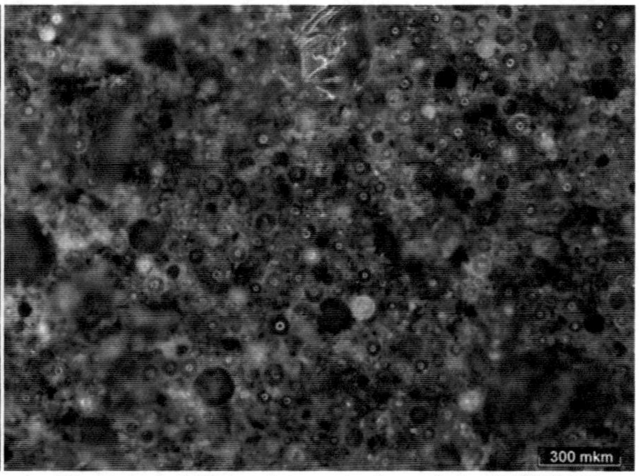

Figure 4. Microphotography of the HSLWC structure which was prepared by Mode M3.

The power calculation for production mixer is carried by formula [5]:

$$N = K_N \rho n^3 d_m^{\ 5}, \quad (1)$$

where ρ – density of the mixing medium, kg/m^3; n – rotating speed of stirrer, sec^{-1}; d_m – diameter of mixer, m.

For the turbulent mode there is no significant change of the parameter K_N; for the laminar mode this parameter depends on the Reynolds number:

$$R_e = nd^2 \vartheta^{1/2}, \quad (2)$$

where ϑ – kinematic viscosity of the medium, m^2/sec.

It is obvious that technical parameters of laboratory and manufacturing equipment are different. Thus, for the degree of homogeneity of the concrete mixture which provides required physical and mechanical properties of HSLWC have to be:

$$N_1 = K_{N1} \rho n_1^{\ 3} d_{m1}^{\ 5}. \quad (3)$$

The power of mixer allowing getting a concrete mix with similar quality:

$$N_2 = K_{N2} \rho n_2^{\ 3} d_{m2}^{\ 5}. \quad (4)$$

The N_2 can be found from the equation:

$$N_2 = K_{N2} n_2^{\ 3} d_{m2}^{\ 5} N_1 \left(K_{N1} n_1^{\ 3} d_{m1}^{\ 5}\right)^{-1}. \quad (5)$$

The laboratory equipment has the following characteristics:

n_1 = 4.75 sec^{-1}, d_{m1} = 0.2 m,
K_{N1} = 210, K_{N2} = 1, N_1 = 500 W,
n_2/d_{m2} = 0.2, d_{m2} = 2.7 m.

For these values the N_2 = 39.2 kW.

Because quality of high-strength lightweight concrete with hollow microspheres depends on packing density and uniform distribution of cement and mineral mixture, the application of vibro-compaction is expedient for moderately mobile concrete mixtures. The optimal compaction time for concrete mixture is defined by the relation of the best strength to minimal energy expense to lay such composition. The vibro-compaction of concrete mixture for formation was carried out in layers according to the modes (Table 2) for modeling the real manufacturing processes.

Table 2. Compacting modes of HSLWC mixtures with hollow microspheres.

No	Layer	Compacting time, sec			
		Mode C1	Mode C2	Mode C3	Mode C4
1	First	0	5	10	15
2	Second	0	10	20	30
3	Total	0	15	30	45

The results of experimental research show that vibro-compaction naturally leads to an increase of density coefficient of HSLWC mixture (table 3). It enabless to increase the strength in age of one day after heat-humid treatment. The application of compaction for 15 seconds increases the concrete strength more than by 15 % but modes C3 and C4 do not improve the technical characteristics of HSLWC. Moreover, the vibro-compaction for 30 and 45 seconds promotes additional entrained air, increases open porosity of products and worsens the quality of their surface.

Table 3. Influence of compacting time on physical and mechanical properties of HSLWC mixtures with hollow microspheres.

No	Layer	Compacting mode			
		Mode C1	Mode C2	Mode C3	Mode C4
1	Compacting time, sec	0	15	30	45
2	Average density, kg/m^3	1457.4	1463.5	1462.8	1463.8
3	Density coefficient	0.971	0.976	0.975	0.976
4	Compressive strength (age is 1 day after HHT), MPa	42.8	49.7	48.5	48.8

Note: density coefficient is a ratio of the actual density to project density

Thus, vibro-compacting within no more than 15 seconds is sufficient for manufacturing of HSLWC products based on the mixture with hollow microspheres.

The heat-humid treatment is a final stage of technological processes during production the reinforced concrete products. In this research work experimental and statistical modeling was involved for study of heat treatment modes of lightweight concrete with hollow microspheres. For the approximation of the data obtained during experiment with two-factor full factorial design the linear regression model $y=b_0+b_1x_1+b_2x_2$ was used. The experiment was carried out for two factors: x_1 (T) – temperature of the isothermal exposure, °C, x_2 (t) – time of the isothermal exposure, hours. The main level for first factor is 70 °C, and for second factor is 7 hours (intervals of variation are 10 °C and 1 hour, respectively). The model represents changes the strength of HSLWC based on hollow microspheres as a function of temperature and time of the heat-humid treatment. The parameters of regression model were obtained from experiment:

$$y = 44{,}3 - 0{,}64 \cdot x_1 + 1{,}57 \cdot x_2 \text{ or } R_{\text{com}} = 38{,}28 - 0{,}064 \cdot T + 1{,}57 \cdot t, \quad (5)$$

where R_{com} is compressive strength (age is 1 day).

The plot of equation (5) is on Figure 5.

The equation (5) and Figure 5 show that x_1 factor has a negative impact on strength of the investigated concrete. That is increase the temperature of the isothermal exposure up to 80°C leads to reduction of compressive strength in age of one day after HHT. It can be explained by the accumulation of the internal thermal stresses because of cement hydration processes and extra temperature gradient (the thermal conductivity of HSLWC is only 0.6 W/(m · K)). The visual examination of the specimens after heat-humid treatment (temperature is 80 °C) shows the local changing of the volume (Figure 6).

IOP Conf. Series: Materials Science and Engineering **71** (2015) 012028 doi:10.1088/1757-899X/71/1/012028

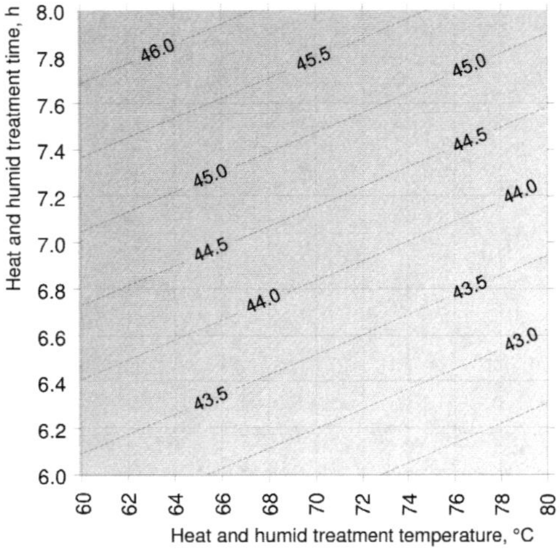

Figure 5. Influence of the HHT temperature and duration on compressive strength of the HSLWC.

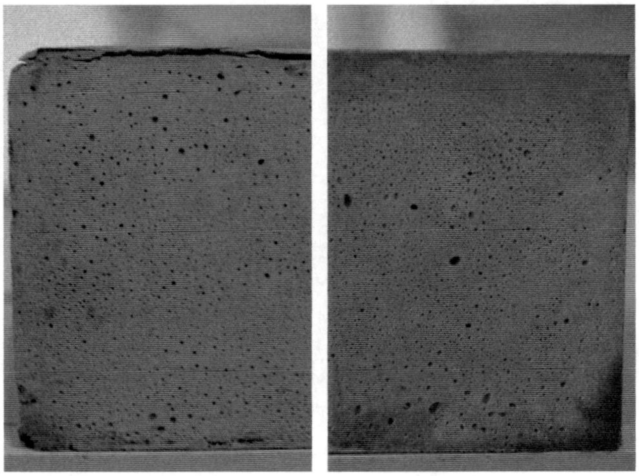

Figure 6. Photographs of the HSLWC samples after HHT T=80 °C (left) and T=60 °C (right).

The influence of the temperature on the processes of the deformations can be decreased, in particular, by means of optimization of the HHT mode: the holding time can be varied and temperature of the products (before heat treatment stage) can be stabilized.

As it follows from the experimental results presented on Figure 5, increasing of isothermal holding time leads to small changes of the strength of the high-strength lightweight concrete. This is due to completer hydration of the cement. It is important that there are no local changes of volume of specimens of the high-strength lightweight concrete after heat-humid treatment at temperature 60 °C. The stable structure without visible cracks, bundles and other defects is observed after HHT and the samples have high coefficient of quality k_{HHT}, $k_{HHT}=R_{com}/T_p$, where $T_p=1$ – concrete hasn't visual cracks or $T_p=100$ – concrete has visual cracks (Figure 7).

Thus, for manufacturing of the HSLWC products with hollow microspheres it is appropriate to use the following modes of the heat-humid treatment: isothermal holding time during 6-7 hours and temperatures are 60-65 °C.

a) b)

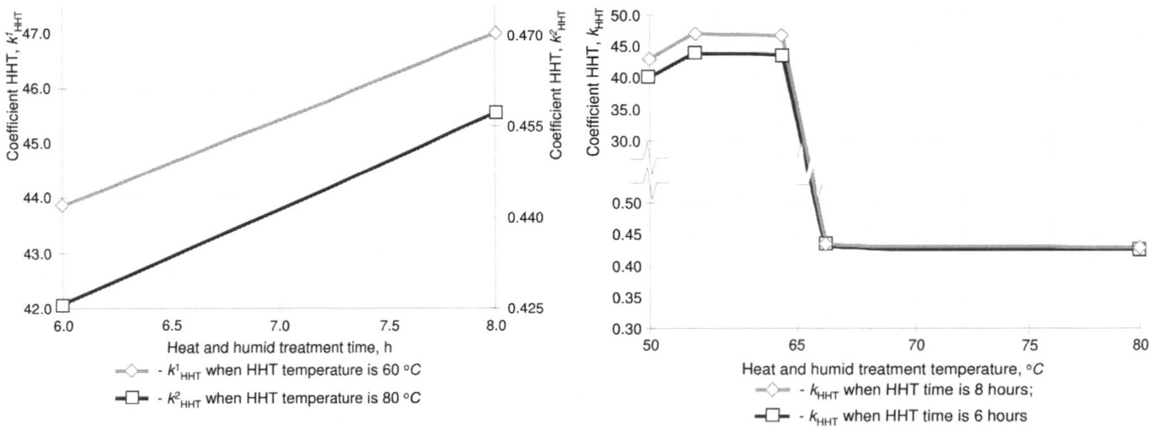

Figure 7. Dependence of coefficient k_{HHT} from a) HHT duration; b) HHT temperature.

As a result of the laboratory test the compositions of HSLWC based on hollow microspheres with design density equal to 1500 kg/m^3 was selected during calculation of mixture and production of the crossbars (of 1980×370×200 mm) in the industrial conditions at the plant for manufacturing reinforced-concrete products. The preparation of the high-strength lightweight concrete was carried using by the "SB-138B" Russian mixer of a power about 37 kW and shaft speed less than 23 rpm. The laying of concrete mixture into plywood forms was made using by a concrete-stacker (Figure 8).

Figure 8. The laying of HSLWC mixture.

Figure 9. Finished HSLWC product.

Formed products were subject to heat-humid treatment according to manufacturing mode with isothermal temperature of 80 °C. The compressive strength of 100 mm cubes formed from mixture of general molding was 40-42 MPa (in the age of one day after HHT). The average density of HSLWC (in the age of 28 days) was 1518 kg/m^3 and compressive strength was 55±2 MPa (these values correspond to B40 grade). The photo of finished product is presented on Figure 9. Technical and economic efficiency of such materials can be evaluated according to [7].

4. Conclusions

The results of research allow making the following conclusions:

- the preparation of HSLWC compositions requires the using of a turbine mixer having six paddles and engine power more than 39.2 kW;
- the oscillation frequency have to be more than 3000 rpm, vibro-compacting time have to be less than 15 seconds, heat-humid treatment temperature have to be approximately 60-65 °C and heat-humid treatment time have to be 6-7 hours; these values correspond to optimal manufacturing technological mode (and, therefore, to the best operational properties) of the products of mixtures based on hollow microspheres.

The industrial approbation of the high-strength lightweight concrete shows that developed technology can be adapted in existing production facilities using already available industrial equipments.

References

[1] Korolev E V 2013 Principle of realization of nanotechnology in building materials science *Construction Materials* **6** 60–64

[2] Payam S, Hafez G, Hilmi Bin M and Mohd Zamin J 2014 A comparison study of the mechanical properties and drying shrinkage of oil palm shell and expanded clay lightweight aggregate concretes *Materials and Design* **60** 320-327

[3] Korolev E V 2013 Model of integrated nanosized modifier for asphalt *Regional architecture and engineering* **3** 5–21

[4] Garkina I A and Danilov A M 2008 Materials with special characteristics quality control *Management problems* **6** 67-73

[5] Inozemtcev A S and Korolev E V 2014 Structuring and properties of the structural high-strength lightweight concretes with nanomodifier BisNanoActivus *Construction Materials* **1-2** 33–37

[6] Korolev E V and Inozemtcev A S 2012 Pat. RUS 2515450 *High-strength lightweight concrete* 3

[7] Bazhenov Yu M and Korolev E V 2009 Estimation of technical and economic efficiency of nanotechnologies in building materiology *Construction Materials* **6** 66–67

Surfaces of refractory products modified by low-temperature plasma in glass manufacturing

V A Vlasov[1], P V Kosmachev[1], N K Skripnikova[1] and G G Volokitin[1]
[1]Tomsk State University of Architecture and Building, Applied Mechanics and Material Sciences Department, Tomsk, 634003, Russia

E-mail: pvkosm@gmail.com

Abstract. The paper describes the production of strong and thermally resistant refractory materials modified by low-temperature plasma. Mechanical and physical tests are presented in this paper for determination of thermal resistance, density, and ultimate compressive strength. X-ray phase analysis and scanning electron microscopy are conducted to determine a phase composition of specimens before and after bake-out. It is shown that plasma treatment results in a formation of a non-porous vitreous layer which is X-ray amorphous and saturated with silicon dioxide and alumina oxide. Moreover, refractory materials exhibit 4-9% strengthening.

1. Introduction

Refractory materials and the equipment used for glass manufacturing allow considerably to increase service life of basic production facilities and reduce both energy and manufacturing costs. At the same time, the improved quality of refractories meets the world standards making them more available and attractive to customers. Applications of the advanced nanodispersed materials of high quality and innovative technologies provide the unique superior performance characteristics of manufactured products [1].

The requirements for glass manufacturing are being constantly improved including cost reduction, environmental standards (energy saving, NOx emission, etc.), the quality of glass containers and flat glass. Along with the use of contemporary glass manufacturing technologies, a choice of refractories intended for the construction of glass tank furnaces is another possibility of their well positioning because they considerably define the quality of products and extend the life of a furnace.

In glass tank furnaces, electrocast refractory materials are mainly used within zones contacting with glass mass. Zirconia-containing materials such as alumina/zirconia/silica (AZS) are usually suitable for electric furnaces because corrosion resistance of these refractories increases with the increase of zirconium oxide (ZrO_2). Therefore, in contact zones, for example, furnace throats, it is reasonable to use such materials as Refel 1240 or Monofrax CS-5 with the highest ZrO_2 content (41% and 39.5% respectively). The intensive forced convection within this zone is caused by a severe erosion of refractory. This fact should be thoroughly studied because the change of the furnace throat area disturbs the smooth operation of the furnace process resulting in glass defect formation. Areas adjacent to the furnace throat are also considered to be contact zones because the refractory components while entering the glass mass, say in a refining zone, do not melt and generate glass defects too. In not numerous contact zones, the use of such refractories as Refel 1532 or Monofrax CS-3 with ZrO_2 content of 32-34% seems to be sufficient. A well-chosen bake-out schedule for production of electrocast refractory materials is also a crucial aspect for contact zones in which they must exhibit superior corrosion resistance to glass melting. Baking parameters have a significant effect on the structure of material, and in particular on a crystalline bond between ZrO_2 and Al_2O_3 and phase distribution in the structure of refractories. A solid crystalline bond is a prerequisite for superior corrosion resistance of refractories.

Besides AZS 41, such products as Refel 1240 FVMo and Monofrax have shown good results in protecting refractory linings against chemical attack. Refel 1240 FVMo is a product, composed of fused cast 41% ZrO_2 material reinforced with a molybdenum insert. It has performed extremely well in refractory masonry of more than 20 glass tank furnaces. The similar behavior has been demonstrated by the refractory based on Cr_2O_3 obtained by isostatic molding technique [2].

Metpump AZS and Met-Silcast refractories find use, for example, as refractory blocks for those parts of furnaces which are subjected to severe high temperature operational conditions. However, exudation of the glassy phase from the surface of the refractory can occur in using these products.

This paper mainly focuses on the production of hot-strength, thermal shock and creep resistant refractory materials similar to Metpump AZS and Met-Silcast products designed by Magneco/Metrel, Inc., USA, with the formation of a fused layer on the material's surface using the low-temperature plasma (LTP).

2. Experimental procedure and results

The process of preparation of test specimens includes the following major stages: preparation of a raw mixture, molding, drying, and baking.

To obtain refractories of the types Met-Silcast and Metpump AZS 20, 14 wt.% and 10 wt.% mass ratios are respectively selected for these types of specimens. A binder is then added to the refractory powder in desired proportions and mixed until a uniform mixture is obtained. Compositions of the raw mixtures are given in the Table 1.

Table 1. Raw mixture compositions.

Composition	Refractory type	Binder (wt.%)
1	Met-Silcast	14
2	Metpump AZS 20	10

At the next stage, test specimens having 50x50x30 mm size are prepared using the vibratory compaction molding technique and vibratory compaction. Specimens are then dried and fused at temperature not higher than 950 °C.

Mechanical and physical properties of specimens such as thermal shock resistance, density, and ultimate compressive strength obtained are then tested.

For the thermal shock test, the surface of specimens is treated with LTP jet generated by a plasmatron at the following operating parameters: T ~ 3000-5000° C, V = 190 V, I = 340 A, plasma propagation velocity υ = 0,07 m/s) [3-6]. The quality of plasma-treated surfaces was detected by the scanning electron microscope (SEM) with 800x magnification. SEM images of the surfaces before and after LTP treatment are presented in Figures 1 and 2.

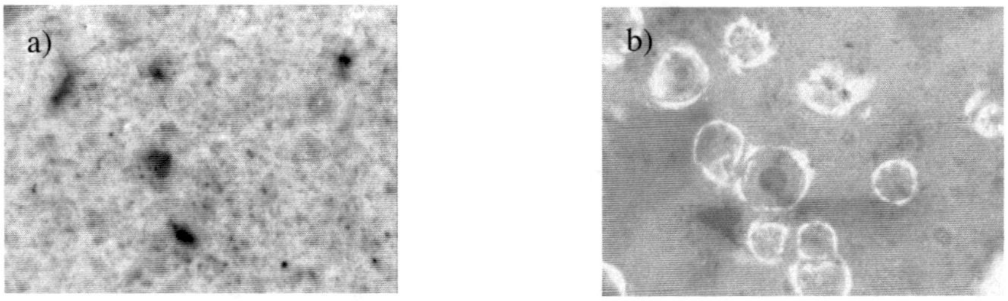

Figure 1. SEM of the type Met-Silcast specimen: before (a) and after (b) plasma-treatment.

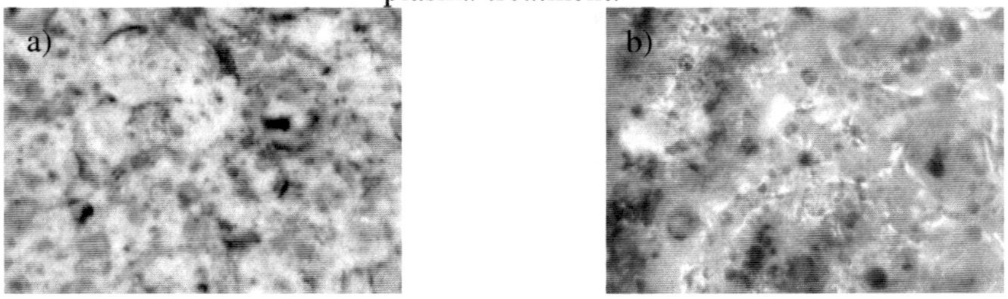

Figure 2. SEM of the type Metpump AZS 20 specimen: before (a) and after (b) plasma-treatment.

SEM analysis shows no crack formation after the plasma treatment. As illustrated in Figures 1 and 2, the porosity of the surface in both types of specimens decreases up to a complete pore elimination, and some of them are closed by fusion. The both types of specimens exhibit thermal shock resistance, however, the surface of the type Met-Silcast specimen is smoother and more uniform. As for the type Metpump AZS 20 specimens, their surface is uneven. This unevenness is the glassy phase, and it is possible to gain more even fusion choosing the appropriate mode of plasma treatment either by increasing its velocity or moving the specimen away from the plasma jet.

The density of refractory specimens is detected by the mass-volume ratio, and equals to 1770 kg/m^3 and 2960 kg/m^3 for the types Met-Silcast and Metpump AZS 20 specimens, respectively.

The test machine INSTRON-3382 equipped with the load transducer was used in this experiment to measure the ultimate compressive strength and the load applied. Measurements were conducted at 24ºC at 2 mm/min. The stress-strain curves of compression states are presented in Figures 3 and 4 for both types of specimens.

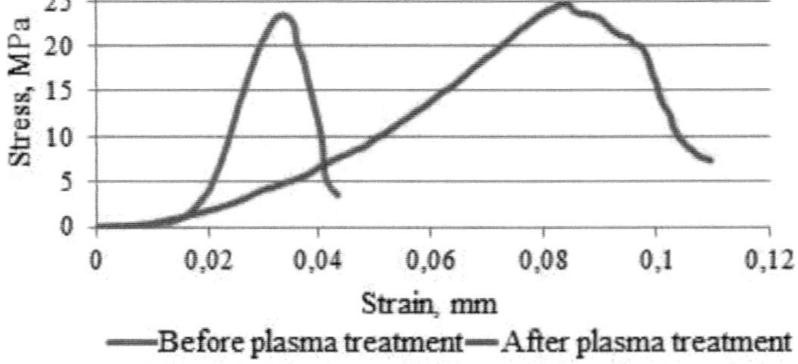

Figure 3. Stress-strain curves of compression for the type Met-Silcast specimen.

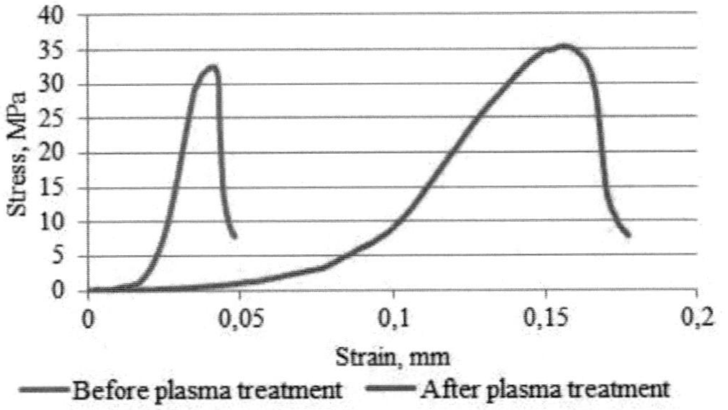

Figure 4. Stress-strain curves of compression for the type Metpump AZS 20 specimen.

The ultimate compressive strength for the type Met-Silcast specimen is 23,4 MPa. After the LTP it reaches 24,5 MPa, i.e. increases by 4%. The ultimate compressive strength for the type Metpump AZS 20 specimen is 32,5 MPa. After the LTP it reaches 35,4 MPa, i.e. increases by 9%.

Phase composition before and after the bake-out process was determined by the X-ray diffraction method. The resulting microstructure in the modified specimen surfaces was examined using scanning electron microscope Quanta 200 3D.

Figure 5 shows the X-ray diffraction diagram of the type Met-Silcast specimen before and after surface modification.

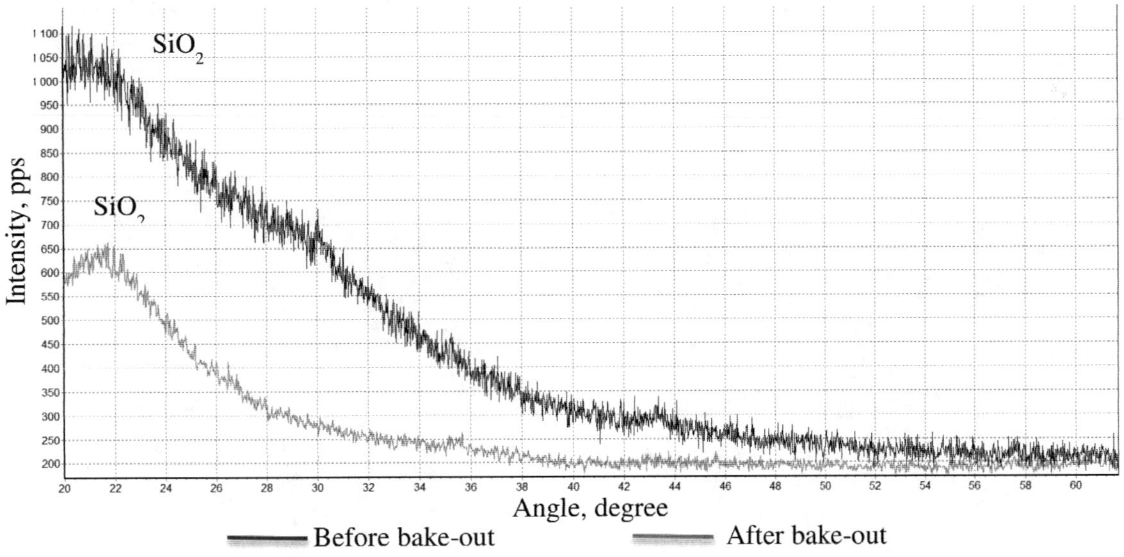

Figure 5. XRD patterns for the type Met-Silcast specimens.

XRD patterns presented in Figure 5, show that the amorphous behavior of the type Met-Silcast specimen microstructure is not changed before and after bake-out. These specimens are mainly characterized by a glassy phase saturated with silicon oxide.

Results of the qualitative X-ray diffraction analysis of the type Metpump AZS 20 investigated specimens are presented in Figure 6.

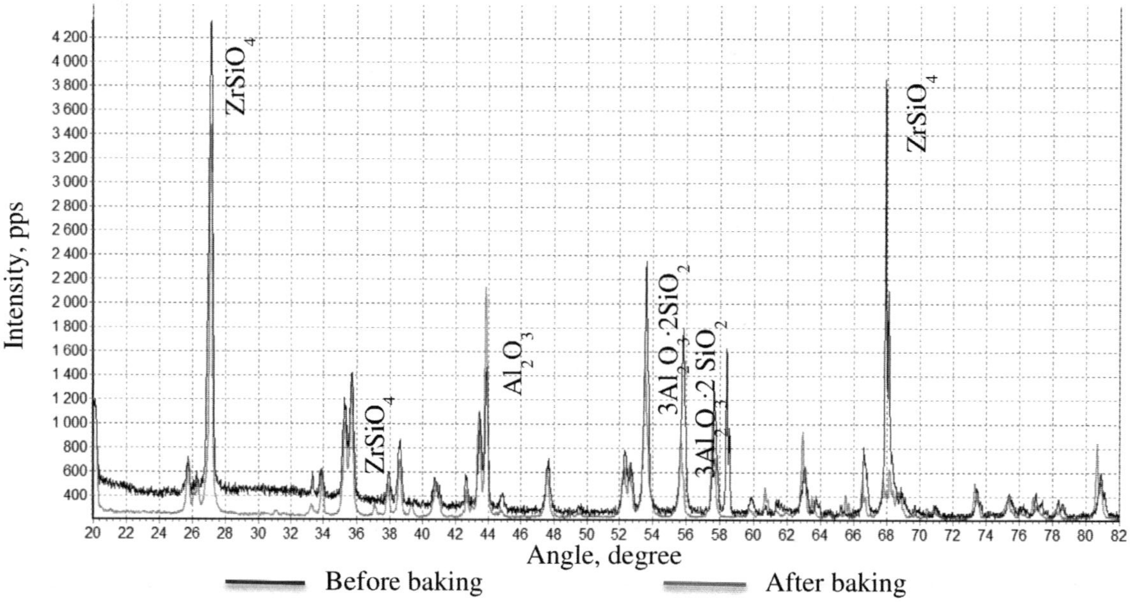

Figure 6. XRD patterns for the type Metpump AZS 20 specimens.

The original specimen of the type Metpump AZS 20 is represented by phases: zircon ($ZrSiO_4$, d = 0,328; 0,251; 0,138 nm), alumina (Al_2O_3, d = 0,208 nm), and mullite ($3Al_2O_3{\cdot}2SiO_2$, d = 0,171; 0,164 nm). The XRD analysis showed that after bake-out some of crystalline phases in this type of specimens were partially changed. This is because the recrystallization of zirconia-containing compounds and formation of mullite-like phases.

Figure 7 shows scanning electron micrograph surfaces of specimens modified by bake-out carried out at Tomsk Core Facility Centre for Materials Science.

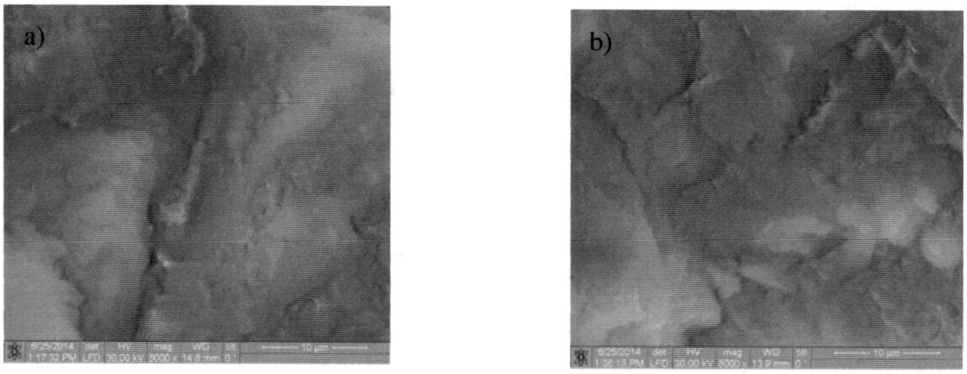

Figure 7. SEM of the types Met-Silcast (a) and Metpump AZS 20 (b) specimens: before (a) after bake-out.

SEM analysis proves results of X-ray investigations, i.e. the type Met-Silcast specimens comprise mainly a uniform homogeneous microstructure having small cracks in some places. As for the type Metpump AZS 20 specimens, they have a grain microstructure.

3. Conclusion

The plasma treatment of specimens of the type Met-Silcast resulted in the formation of a smooth non-porous vitreous layer. Strength properties of this specimen type are increased by 4%. Modification of the type Metpump AZS 20 specimens with plasma treatment resulted in a non-porous uneven vitreous layer. Its ultimate compressive strength increased up to 9%. The non-porous vitreous layer was X-ray amorphous and saturated with silicon dioxide (the type Met-Silcast specimens) and alumina oxide (the type Metpump AZS 20 specimens).

Acknowledgement

The authors acknowledge the Ministry of Education and Science of the Russian Federation for the financial support (Project No. 11.351.2014/K).

References

[1] Гузман И Я 2012 *Химическая технология керамики* (Москва: ООО РИФ «Стройматериалы»)

[2] Канюка И В 2008 *Огнеупоры для стекольной промышленности* (Черкассы: Черкасский НИИТЭХИМ)

[3] Lutsenko A V, Skripnikova N K, Volokitin G G and Turashev A S 2012 Production of glass – crystalline materials from silicate melts with the use of low-temperature *Vestnik of TSUAB* **3** 142–144

[4] Volokitin O G, Skripnikova N K, Volokitin G G, Shekhovtsov V V, Vereshchagin V I and Khaisundinov A I 2013 Mineral Fiber Produced in Aggregates of Low Temperature Plasma from Products of Burning of Coal and Combustible Shales *Constr. Mat.* **11** 44-47

[5] Volokitin G G, Skripnikova N K, Abzaev Yu A, Volokitin O G and Shekhovtsov V V 2013 Study of plasma-chemical synthesis of high-temperature silicate melts (part 1). Tailings f molybdenum ores beneficiation process *Vestnik of TSUAB* **4** 197–202.

[6] Volokitin G G, Skripnikova N K, Abzaev Yu A, Volokitin O G and Shekhovtsov V V 2014 Study of plasma-chemical synthesis of high-temperature silicate melts (part 2). Smelting of tailings of molybdenum ores beneficiation *Vestnik of TSUAB* **1** 80–84

IOP Conf. Series: Materials Science and Engineering **71** (2015) 012030 doi:10.1088/1757-899X/71/1/012030

Strength and deformability of concrete beams reinforced by non-metallic fiber and composite rebar

K L Kudyakov[1], V S Plevkov[1] and A V Nevskii[1]
[1]Tomsk State University of Architecture and Building, Department of Reinforced Concrete and Masonry Structures, Tomsk, 634003, Russia

E-mail: konst_k@mail.ru

Abstract. Production of durable and high-strength concrete structures with unique properties has always been crucial. Therefore special attention has been paid to non-metallic composite and fiber reinforcement. This article describes the experimental research of strength and deformability of concrete beams with dispersed and core fiber-based reinforcement. As composite reinforcement fiberglass reinforced plastic rods with diameters 6 mm and 10 mm are used. Carbon and basalt fibers are used as dispersed reinforcement. The developed experimental program includes designing and production of flexural structures with different parameters of dispersed fiber and composite rebar reinforcement. The preliminary testing of mechanical properties of these materials has shown their effectiveness. Structures underwent bending testing on a special bench by applying flexural static load up to complete destruction. During the tests vertical displacements were recorded, as well as value of actual load, slippage of rebars in concrete, crack formation. As a result of research were obtained structural failure and crack formation graphs, value of fracture load and maximum displacements of the beams at midspan. Analysis of experimental data showed the effectiveness of using dispersed reinforcement of concrete and the need for prestressing of fiberglass composite rebar.

1. Introduction

Current stage of scientific and technological development raised new requirements for building and construction materials: in addition to strength and deformation properties of the material, particular attention is now paid to the corrosion, electrical, magnetic and radio resistance. These requirements are applied in construction of buildings and structures of the energy, aerospace, medical and road-building industries. In this case, the main structural material is ordinary reinforced concrete which does not meet the above mentioned requirements [1].

One of the ways to solve this problem is to use non-metallic composite rebar as reinforcement of concrete structures. Fiberglass reinforced plastic (FGRP) bars have low specific weight, high strength, corrosion resistance, dielectric properties, magnetic inertness and radio transparency. However, these materials have some disadvantages which severely restrict fields of their application: a relatively high deformability, low fire resistance, the lack of sustainable technology of rebar prestressing. So, the aspects of designing, calculation and application of concrete structures reinforced with non-metallic composite materials are highly relevant [1 - 5].

2. Experimental procedure

Experimental studies have been held to determine flexural strength and deformability of concrete elements reinforced with FGRP bars, as well as reinforced with basalt and carbon fiber. The design of experiment is shown in Figure 1.

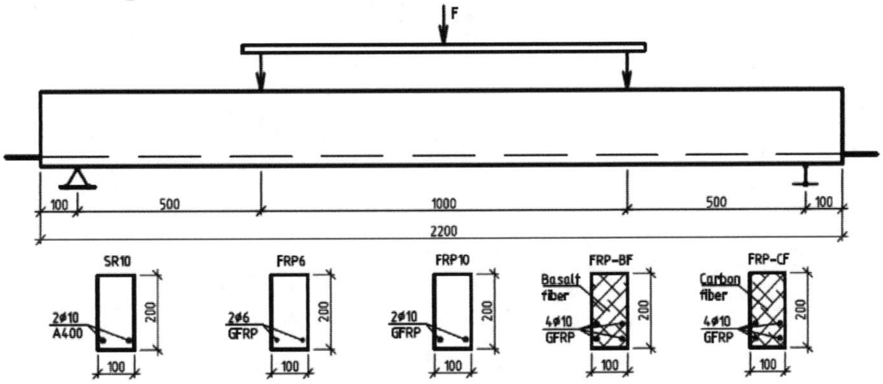

Figure 1. The design of experiment.

The initial data of experimental samples is given in Table 1, where μ_r - percentage of rebar reinforcement in cross section of concrete element; μ_f - the percentage of disperse reinforcement depending on the weight of the cement.

Table 1. The initial data of experimental samples.

Sample identifier	Number of samples	Specification	μ_r, %	Weight of sample, kg
SR10	1	Steel bars of mark A400, 2Ø10 mm	0,897	114,7
FRP6	1	FGRP, 2Ø6 mm	0,302	112,7
FRP10	1	FGRP, 2Ø10 mm	0,827	112,6
FRP-BF	1	FGRP, 4Ø10 mm; Basalt fiber, $\mu_f = 0{,}5\%$	1,654	112,3
FRP-CF	1	FGRP, 4Ø10 mm; Carbon fiber, $\mu_f = 0{,}2\%$	1,654	112,3

2.1. The design of experimental beams

For the purpose of studies flexural elements were designed and made, dimensions and design are given in Figure 2.

2.2. Properties of materials

For samples production fine grained concrete composition with a mobility class P2 and compressive strength with class B30 were used. Portland cement of M500 class was used as a hydraulic binder. Water-cement ratio equals to 0.65 [6].

As the longitudinal rebar reinforcement composite FGRP brand of "MONSTEROD" (HC "Composite", Moscow, Russia) with an external diameter of 6 mm and 10 mm was used. Fiberglass rods have ribbed surface for bonding with concrete. There were also used reinforcing steel rods of class A400. As the transverse reinforcement reinforcing wire of Bp500 class was used [6].

Preliminary experimental testing of tensile strength and deformability of fiberglass rods showed that the reinforcement has a high tensile strength (up to 1200 MPa), the longitudinal deformation of the rods are in the range from 2 to 3%, and modulus of elasticity is 35 - 45 GPa. However, these properties depend on the producer company. Specific weight of reinforcement is 1.95 g/cm^3, that allows reduction of the total weight of the structures up to 1.9% when replacing the steel reinforcement with composite rebar [7].

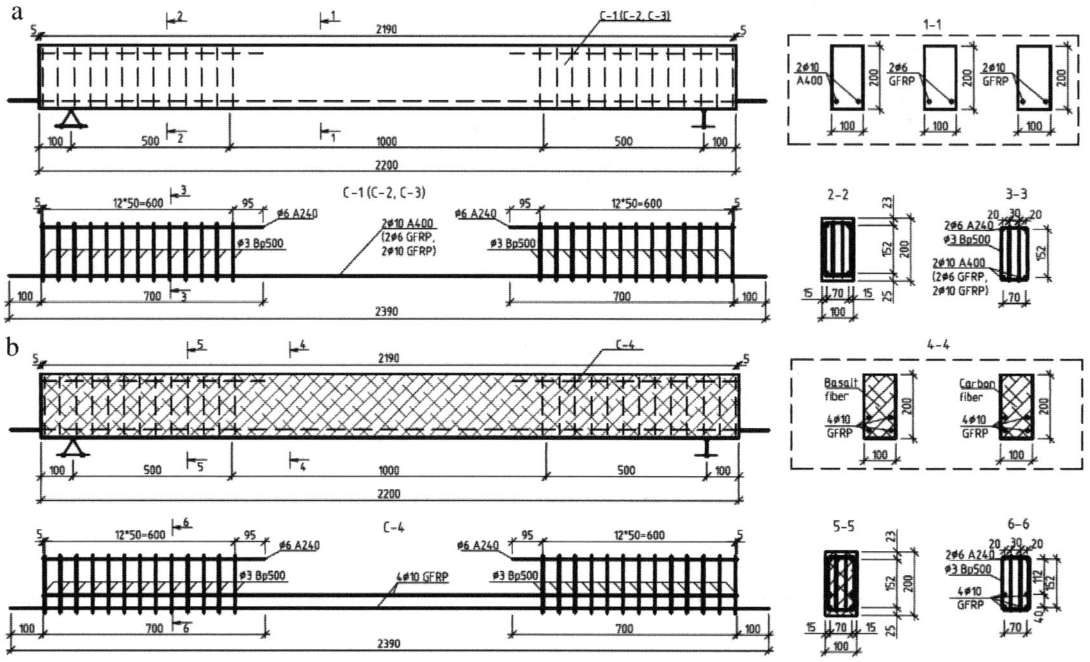

Figure 2. The design of the experimental beams:
(a) samples SR10, FRP6, FRP10; (b) samples FRP-BF, FRP-CF.

Basalt and carbon fiber were used as the disperse reinforcement. At first, before production of experimental flexural elements, the study was held to investigate strength and deformability of fiber-reinforced concrete based on basalt and carbon fibers. Further 48 fiber-concrete samples were fabricated and tested. The amount of fiber, chemical additives and production technologies were different.

The studies resulted in the developed production technology of fiber-reinforced concrete mixture that ensures good dispersion of fibers in concrete matrix. Testing of fiber-concrete samples showed that introduction of estimated amount of basalt fibers into concrete mixture enables to increase compressive strength of concrete up to 51.2% and tensile strength up to 28.8%. The introduction of the carbon fibers into concrete allows increasing compressive strength up to 43% and tensile strength up to 18% [8, 9].

2.3. Testing of samples under static load

Figure 3 shows the scheme of installation of devices, the applied load was measured by dynamometer (D), vertical deformations were defined by inductive displacement sensors (S1 - S5). Slippage of rebar into concrete was recorded by dial indicators installed on rebar (I1 - I4).

Experimental investigations of strength and deformability of flexural elements were performed on special bench (Figure. 4). Tests were conducted by static stepwise load according to the scheme of single-span beam jointed with the hinge support, and up to complete destruction of the samples.

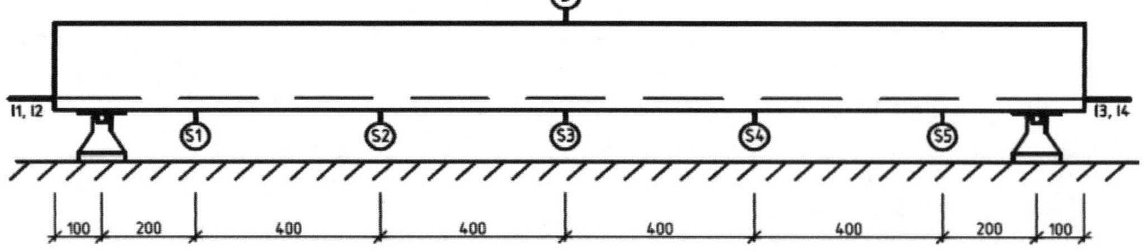

Figure 3. The scheme of installation of measuring devices.

Figure 4. General view of the test bench.

3. Results

The results of the tests enabled to obtain common crack formation patterns and failure of beams reinforced with steel (Figure 5, b), fiberglass rods Ø6 mm (Figure 5, c) and fiberglass rods Ø10 mm (Figure 5, g). Analysis of cracking and failure schemes showed that disruption of elements went by the cross section, normal to the longitudinal axis with formation of main crack within the area of pure bending. During the tests, the slip in the concrete of steel and fiberglass rods has not been identified.

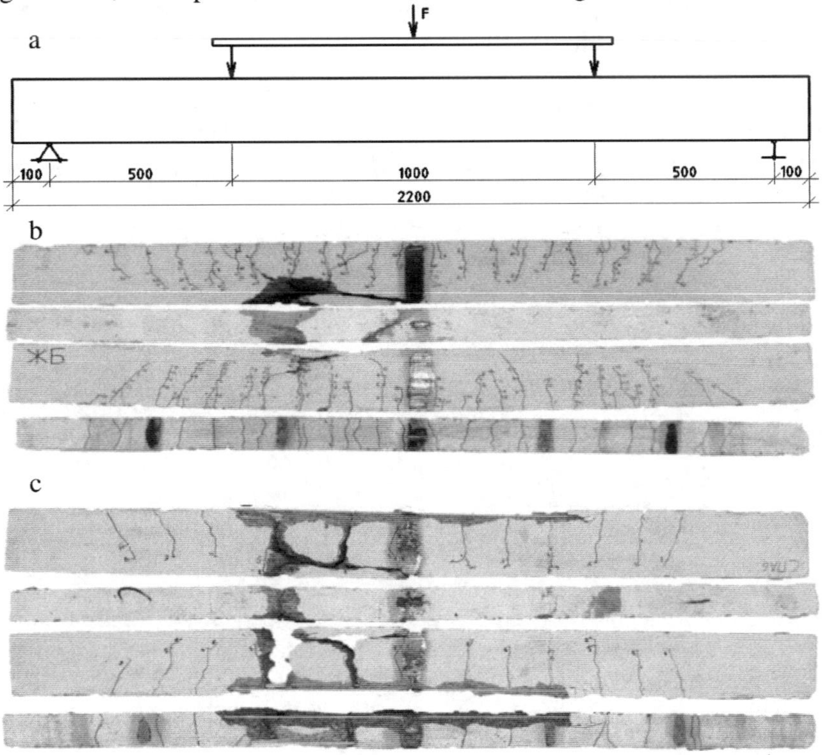

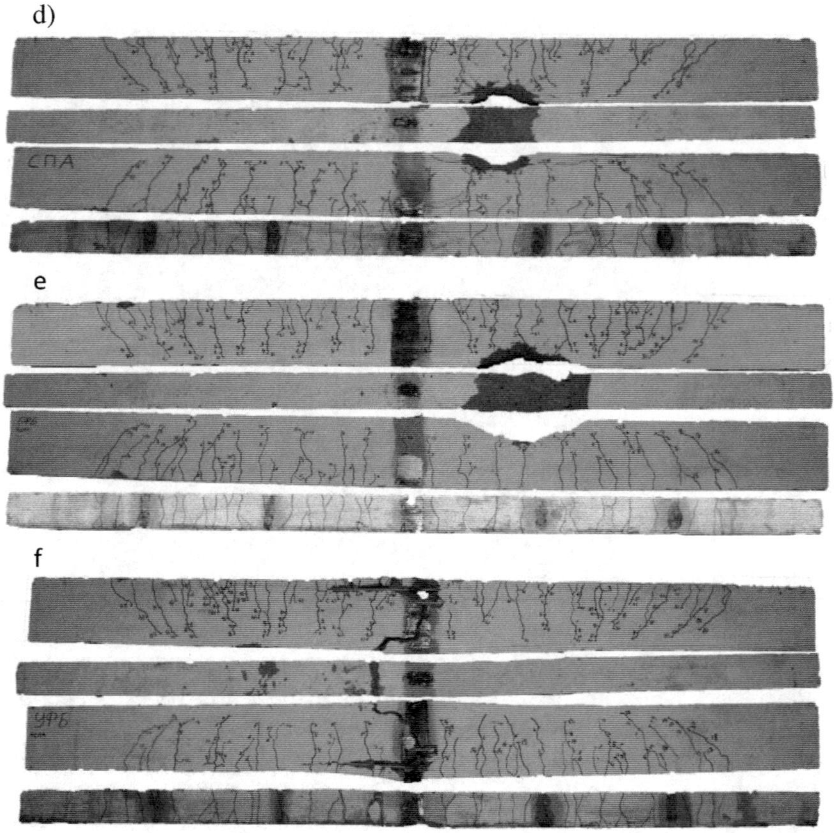

Figure 5. Beams testing by static load: the test circuit (a); characteristic schemes of cracking and failure constructions SR10 (b), FRP6 (c), FRP10 (d), FRP-BF (e), FRP-CF (f).

The main experimental results are given in Table 2, where: F_i - acting force; F_{ult} - calculated critical force; F_{max} - actual critical force; M_{max} - the ultimate bending moment; $a_{crc,i}$ - the width of the cracks under the current force; f_i - deflection of element in the current force; f_{max} - maximum deflection of the element in the middle of the span [7].

Table 2. The experimental results.

Sample identifier	$F_i = F_{ult} = 41{,}3$ kN		F_{max}, kN	M_{max}, kN*m	f_{max}, mm
	$a_{crc,i}$, mm	f_i, mm			
SR10	0,1	7,7	61,2	15,3	12,8
FRP6	1,8	37,1	64,4	16,1	64,5
FRP10	1,15	31,1	67,2	16,8	52,1
FRP-BF	0,25	15,4	85,2	21,3	40,4
FRP-CF	0,45	14,4	91,2	22,8	39,2

4. Conclusions

It was found out that the replacement of steel rods with FGRP leads to increase in the bearing capacity of concrete beams under static load up to 5 - 10%. At the same time, the deformability of the element increases by 4 - 5 times as compared with a steel reinforced concrete element, but when the load ceases, the elements reinforced with fiberglass rods get the initial geometry of elements. High deformability of flexural elements reinforced with composite rods may increase bearing capacity and durability of flexural elements subjected to dynamic loads, further research is needed in this area.

Under load $F_i = F_{ult}$ in beams reinforced with fiberglass rods the deflections and crack width unacceptable by construction codes took place. This indicates the necessity of the prestressing FGRP rebar if it is used in flexural structures.

Application of fiber-reinforced concrete based on basalt and carbon fibers can increase the strength of the compressed zone in a flexible element, thereby increasing the bearing capacity of the construction and increasing stiffness, as well as there is an increase in crack resistance.

Thus, the use of FGRP rebar in flexural structures, including in conjunction with dispersed fiber reinforcement makes it possible to increase bearing capacity of the element and deformability. Probably, glassfiber rebar prestressing will reduce the unacceptable deformation and cracking in flexural elements, but this issue requires further research.

Acknowledgement
The research performed with the financial support of the Ministry of Education and Science of the Russian Federation within Governmental Decree No.218 of “Measures of state support for cooperation of Russian higher education institutions and organizations, for implementation of complex projects of high-tech production”, integrated project No. 02.G25.31.0022, “Production technologies of energy- and resource-efficient housing of economy class based on the universal framed prefabricated architectural and construction system”.

References

[1] Степанова В Ф, Степанов А Ю и Жирков Е П 2013 *Арматура композитная полимерная* (Москва: АСВ) 200
[2] Varma M B and Pujari R S 2011 Flexural strength of concrete beam reinforced with GFRP rebar *International Journal of Earth Sciences and Engineering* **4** 994-996
[3] Alkhrdaji T, Wideman M , Belarbi A and Nanni A 2001 Shear strength of GFRP RC beams and slabs *Composites in Construction* 409-414
[4] Chia Hwan Y and JianBo H 2014 *The mechanical behavior of fiber reinforced PP ECC beams under reverse cyclic loading advances in materials science and engineering* (Hindawi Publishing Corporation) ID 159790
[5] Sungwoo S, Daewon S and Byumseok H 2009 Performance of concrete beams reinforced with GFRP bars *Journal of Asian Architecture and Building Engineering* **8** 197-204
[6] *СП 63.13330.2012 Бетонные и железобетонные конструкции. Основные положения. Актуализированная редакция СНиП 52-01* (Москва: НИИЖБ)
[7] Кудяков К Л, Невский А В и Кудяков В А 2014 Влияние воздействия щелочной среды на физико-механические свойства композитной арматуры на основе стеклянных и углеродных волокон *Труды XI Международной конференции студентов и молодых учёных Перспективы развития фундаментальных наук* (Томск: ТПУ) 788-790
[8] Кудяков К Л, Невский А В и Ушакова А С 2013 Влияние дисперсного армирования базальтовым волокном на прочностные свойства бетона *Труды X Международной конференции студентов и молодых учёных Перспективы развития фундаментальных наук* (Томск: Томский политехнический университет) 708-710
[9] Кудяков К Л, Невский А В и Ушакова А С 2014 Влияние дисперсного армирования углеродными волокнами на прочностные свойства бетона *Труды X Международной конференции студентов и молодых учёных Перспективы развития фундаментальных наук* (Томск: Томский политехнический университет) 799-802

Relation of dynamic parameters of brick masonry fragment at fracture under static and dynamic loading

D G Kopanitsa[1] and E S Useinov[1]
[1]Tomsk State University of Architecture and Building, Department of Metal and Wooden Structures, Tomsk, 634003, Russia

E-mail: useinov_em@mail.ru

Abstract. The article shows the results of experimental analysis of brick masonry fragments under static and dynamic loading. The measurements of fractures and natural vibration frequencies of samples have been carried out in the course of tests. It has been shown that at appearance of inelastic deformations and cracks, there is a change in natural vibration frequencies, received from the analysis of the corresponding spectra. Comparison of results of experiments of brick masonry fragments, received under action of static and dynamic loads has been carried out.

1. Introduction

Evaluation of analytic models of structures, experiencing the action of seismic load is very complex issue and it is necessary to analyze the results of numerous experiments to find the solution. Study of brick buildings is complicated because of the complexity of building behavior modeling under seismic load. Along with tests using dynamic machines, experimental studies on static loads impacts created by powerful jacks shifting the building footing according to the set scheme are conducted. The dynamic effect is achieved by the load loss. The seed of loading of building structure is relatively low. Measurements show that earthquake and overloads arising in building structures do not exceed 2g [6]. Low loading speeds slightly influence material properties; therefore the results of static experimentation are quite acceptable for justification of analytic models.

Detailed study of the brick building behavior at destruction can be simplified by considering the operation of its models. Relatively full scope of information on the operation of brick structures can be obtained by studying fragments of the building or its models. It has been established by M. Dzhabarov that decrease in natural vibration frequency of a wall fragment to 62% indicates its fracture [1]. In order to study change of dynamic parameters of brick masonry in the process of fracture, experimental investigation of the brick masonry fragments, described below, has been conducted.

2. Experimental investigation

For carrying out experiments the scheme with cross loading has been chosen. Such loading corresponds to the third stage of deformation of wall partition under action of horizontal component from seismic load causing masonry fracture in its plane. Sample fracture accrues under action of the main tensile stress.

Samples were obtained in laboratory conditions with chain system of bonding by the bricklayer with secondary qualification from the solid ordinary clay brick of M100 type and masonry mortar of M75 type.

4 samples were tested. Two of them were tested for static loading, two - for impact. Comparison of the received results followed the tests.

When conducting experiments on static force the load was applied from top to down along diagonal of the sample, put in load frame section of hydraulic press. Overall observation of tests is shown in Figure 1a. Fractures were measured by dial indicators and linear position sensors. Test of each sample had been carried out for 10 stages. Measurements of natural vibration frequency of sample were carried out while the increase of load at each stage of tests. These measurements allowed showing the influence of damage rate of tested samples on the change of their dynamic parameters.

Dynamic tests of studied samples were carried out on pile driver. Load action was accepted according to the scheme of static tests along a diagonal of sample (Figure 1b). The load on sample was performed by energy of falling weight. The impact accrued to the dynamometer, which was on distributing strongback. The mass of dropped weight was 500 kg, and drop height was ranging from 0,5 to 1,5 m. Series of impacts carried the samples of brick masonry to a limit state. Ultimate failure of samples or crack formation and crack opening more than 5 mm in brick masonry was accepted to a limit state. Measurements of accelerations, speeds and movements were performed during the experiment.

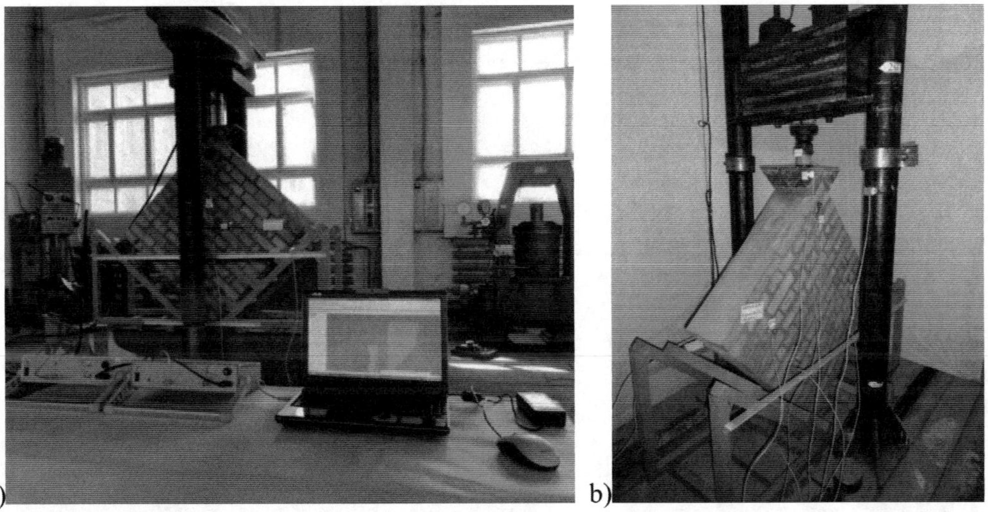

a) b)

Figure 1. Overall view of testing bed: a) tests for action of static load; b) tests for action of dynamic load.

Tests showed that all samples had identical sequence of crack formation and fracture with the appearance of the first crack in the sample center along the pressed diagonal, with further crack opening up to ultimate failure. Similar nature of fracture of masonry and vibrated brickwork panels is established in works [2, 3, 4].

In order to receive information on the behavior of the sample under loading, accelerometer fixing dynamic parameters in two planes on surfaces of brick masonry in direction of sample diagonals was installed. By online measuring results, the chart of accelerations has been plotted (Figure 2). Referring to the chart, it is seen, that brick masonry fracture is accompanied by release of stored energy in the form of accelerations of various intensity [5].

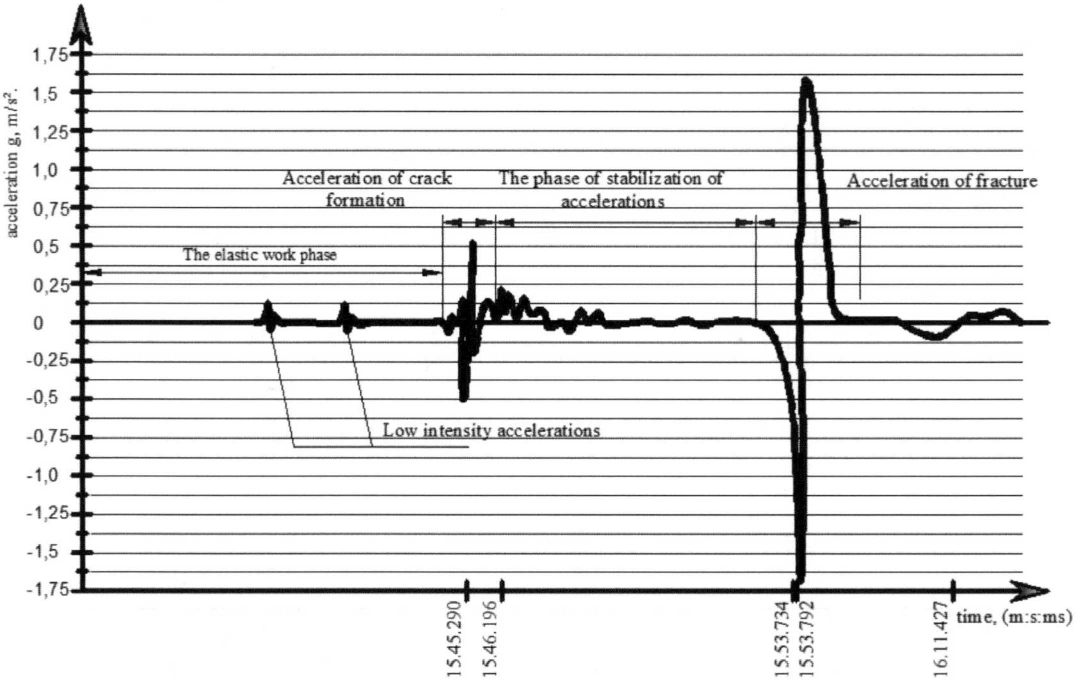

Figure 2. The chart of accelerations at masonry fracture under static load.

With load increase to 0,6 Pfr (Pfr - fracture load), chart shows low intensity accelerations, caused by embrittlement of material and sampling of interwelded pinhole. It didn't cause decrease in natural vibration frequencies of the fragment. The load increase to 0,95 Rfr caused the appearance of inelastic deformations and vertical crack to 2,4 mm with the corresponding rush in accelerations to 0,5 g. The phase of stabilization of accelerations in the process of which there was a redistribution of internal forces in the body of masonry is characterized by the insignificant growth of fractures. The further load increase (to Pfr = 280,0 kN) led to the fast growth of fractures and loss of bearing capacity of the sample. Opening of existing cracks and formation of new ones, at the time of sample fracture, (Figure 3) caused the appearance of accelerations to 1,6 g, with the corresponding decrease in natural frequencies.

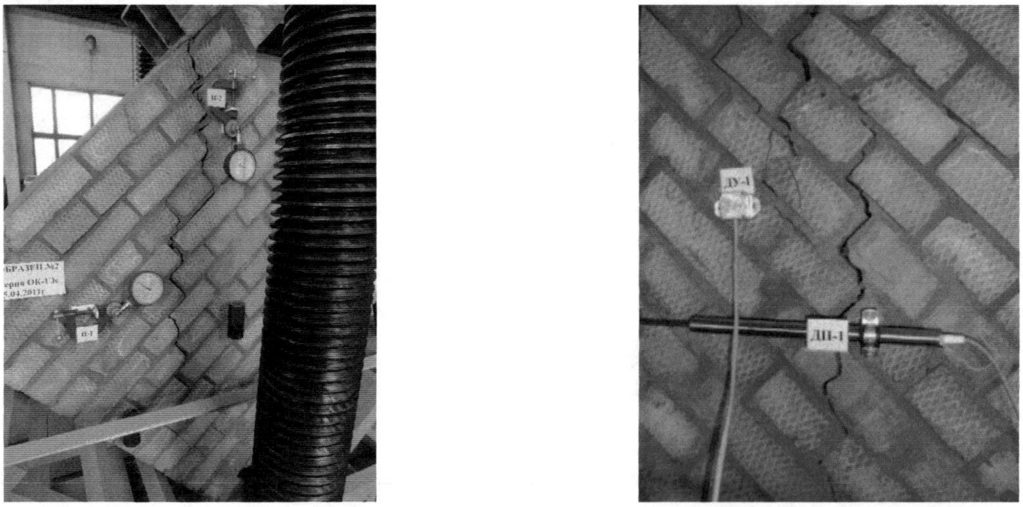

Figure 3. Fracture of samples under action of static force.

The measurement of natural vibration frequencies was carried out in the process of load increase at each stage of tests. Oscillations were excited by impact load and fixed by computing and measuring complex by means of accelerometer, installed on a lateral area of sample. The general spectrogram of natural vibration frequencies power of masonry fragment (Figure 4) has been plotted by results of measurements. The first measurement was performed while preparing of test sample. Oscillations were excited by impact load and fixed by computing and measuring complex. The "zero" load step is characterized by the corresponding range of frequencies.

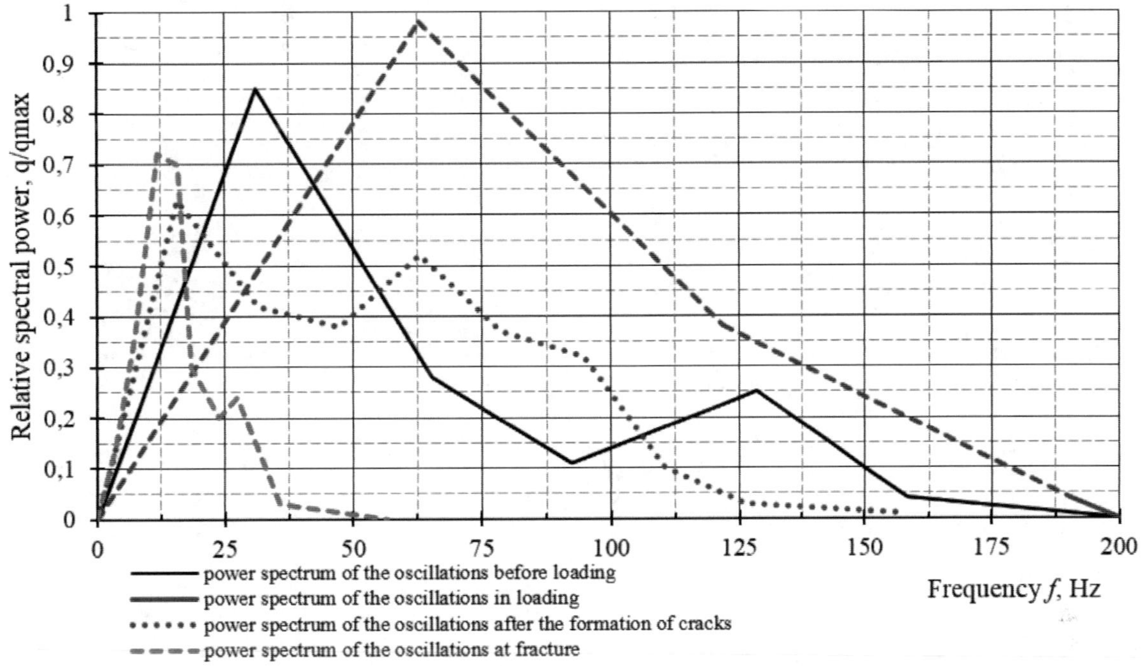

Figure 4. Ranges of vibration power of fragment at test for action of static force.

Natural vibration frequency on the eigentone of not loaded sample is equal to 31,1 Hz. The static load of 32,0 kN applied at the first stage caused double increase in frequency due to inclusion of lower and upper support in work (Figure 4). At load increase to 208,0 kN there were higher frequencies on the accelerogram at almost not changing ranges of natural frequencies.

The load increase to 272,0 kN caused the appearance of inelastic deformations and vertical crack to 2,4 mm with the corresponding range expansion and its shift to the area of low frequencies (Figure 4). At the same time the lower value of frequency decreased from 60 Hz to 15 Hz.

The load increase to 280,8 kN led to sample fracture with the corresponding natural frequency decrease of the sample to 11,1 Hz (Figure 4).

Experimental investigation showed that fracture of masonry fragment under the action of static force can conditionally be divided into 2 stages. The longitudinal crack along the pressed diagonal with 4-x fold decrease in natural vibration frequency on the eigentone was formed at the first stage. At the second stage with an insignificant load growth, there was development of cracks and sample fracture.

Dynamic tests were carried out in two stages for action of two consistently enclosed shock loadings. At the first stage, the impact was created by dropped weight of 500 kg from height of 0,75 m. Duration of load action was of 15 ms with the maximum value of 360,14 kN.

The first impact caused the crack appearance along the pressed diagonal the crack development was recorded by motion sensors installed in center portion of construction (Figure 6a). Crack opening width was of 1,7 mm.

The second impact was carried out on the sample with crack. Damping behavior of masonry promoted the change of the load chart. The load gradient decreased in relation to the first impact by 30%, respectively, exposure duration increased by 30% from 15 to 19 ms.

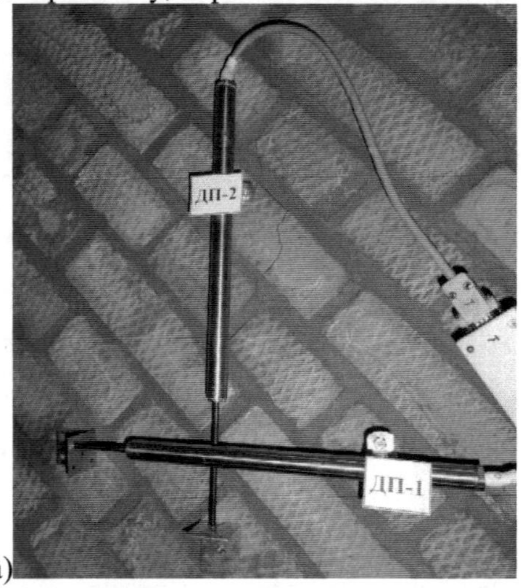

a)

б)

Figure 6. a) crack along the pressed diagonal after the first impact, b) masonry fragment fracture after the second impact

Response of the sample to real load is reflected in charts of accelerations of AS-2 and AS-3 sensors, installed on the main diagonal in direction of real load (Figure 7). The specified sections of charts reflect sample variations at the action of impact load (section I) and after weight rebound (section II).

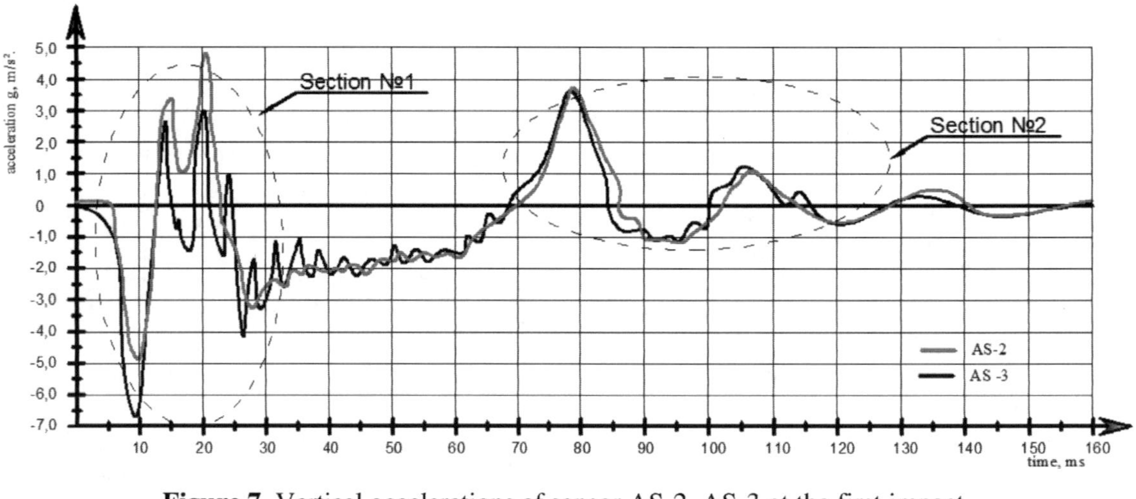

Figure 7. Vertical accelerations of sensor AS-2, AS-3 at the first impact.

Acceleration sensor-2 (AS-2) located closer to strongback showed smaller values of accelerations in comparison with the acceleration sensor-3 (AS-3) located in the zone of structural bearing. Amplitude of accelerations of the above-located sensor AS-2 is of 48,41 m/s^2. Impact action caused the acceleration of structural bearing to 67,4 m/s^2 (AS-3). The corresponding power spectrum of vibrations is shown in Figure 8.

The frequency of 62,34 Hz corresponds to the eigentone of vibrations. Amplitude of vibration spectrum of sensor AS-2 is 1,5 times lower in comparison with amplitude of vibration spectrum of the sensor AS-3. It corresponds to more intensive process of deformation of sample bearing.

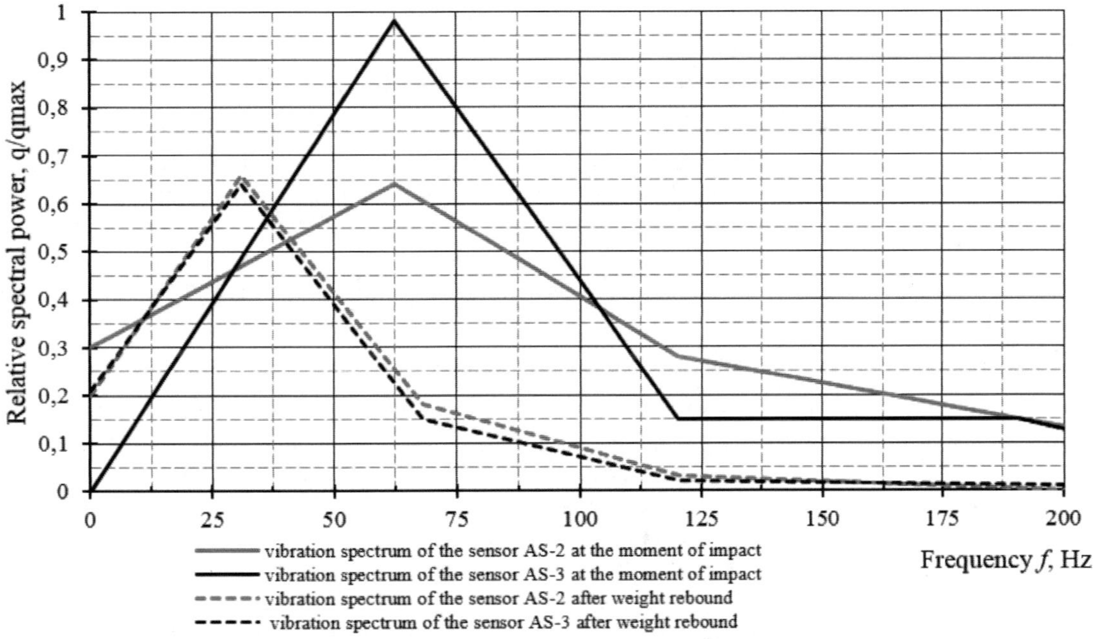

Figure 8. Power spectra of vibrations of sensors AS-2, AS-3 at the first impact.

There were vibrations of load-free construction after weight rebound the second section of the accelerations chart in Figure 7 corresponds to this process. According to power spectrum of vibrations in Figure 8, the frequency of 31,3 Hz corresponds to the eigentone. Decrease in vibration frequencies to 31,3 Hz reflects the cracking in construction body with aligning of power spectra of vibrations of sensors AS-2, AS-3.

Drop height of weight was reduced to 0,5 m at the second stage of tests, the mass of weight of 500 kg was not changed. Load time was of 19 ms with the maximum value of 210,8 kN. Load action on the second impact destroyed the sample in two approximately equal parts (Figure 6). Maximum value of accelerations of 49,3 m/s^2 was recorded in the first vibrations cycle. Action of the second impact was on the right part of fragment separated by crack. According to the power spectrum of vibrations, vibrations of the separated part generally happened on the first and third natural vibration frequency. Base frequency was of 31,11 Hz, and frequency in the third form of vibrations was of 92,7 Hz.

3. Conclusions

Experiments showed that the crack formation in brick masonry both under static and dynamic loading is accompanied by the change of dynamic parameters. Appearance of decompactions and cracks in masonry at load increase is characterized by shift of power spectrum of vibrations to the area of low frequencies.

Dynamic load action at speed up to 5 m/s has no essential influence on the nature of formation of cracks in comparison with crack formation under action of static load.

Sample fracture of brick masonry under action of static force is characterized by three stages. There were microcracks with weak impulses of accelerations to 0,125 m/s^2 at the first stage. The crack of limited length characterized by surge in accelerations to 0,5 m/s^2 appeared at the second stage. Final sample fracture happened when the main crack appeared, it divided the sample into two parts with the corresponding impulse of accelerations with the amplitude of 1,6 m/s^2.

The scheme of sample fracture of brick masonry under action of static force qualitatively coincided with the scheme of sample fracture under action of dynamic load.

Comparison of experimental results showed, that fracture of test samples both under and dynamic load began at the same natural vibration frequency with the bottom tone of 62,84 Hz.

Formation of cracks in brick masonry caused the change of dynamic parameters of samples and was characterized by the shift of power spectrum of vibrations to the area of low frequencies with corresponding decrease in vibration frequency.

Acknowledgement

The research performed with the financial support of the Ministry of Education and Science of the Russian Federation within Governmental Decree No.218 of "Measures of state support for cooperation of Russian higher education institutions and organizations, for implementation of complex projects of high-tech production", integrated project No. 02.G25.31.0022, "Production technologies of energy- and resource-efficient housing of economy class based on the universal framed prefabricated architectural and construction system".

References

[1] Джабаров М 1986 Методы усиления кирпичных зданий пневмобетоном и штукатурными слоями в сейсмических районах: *Диссертация на соискание ученой степени кандидата технических наук* (Душанбе: Академия наук Таджикской ССР Институт сейсмостойкого строительства и сейсмологии)

[2] Кожаринов С В 1980 Исследование деформаций кирпичной кладки при действии горизонтальных нагрузок *Динамика и сейсмостойкость зданий и сооружений* 127-134

[3] Коноводченко В И 1967 Усиление стен кирпичных зданий для повышения их сейсмостойкости *Сейсмостойкость крупнопанельных и каменных зданий* (Москва) 180-186

[4] Тонких Г П, Кабанцев О В и Кошаев В В 2007 Результаты статических испытаний каменной кладки, усиленной железобетонной аппликацией *Вопросы безопасности военной деятельности, создания и функционирования объектов военной инфраструктуры Сб. научных трудов под ред. Латушкина С. Н., Малофеева Ю. В.* (Москва: ЦНИИ) 105-116

[5] Усеинов Э С 24–27 апреля 2012 г. Образование трещин в кирпичной кладке при сжатии// *Труды IX Международной конференции студентов и молодых учёных Перспективы развития фундаментальных наук* (Россия, Томск: Национальный Исследовательский Томский политехнический университет*)* 800-802

[6] Фахриддинов У 2004 Сейсмозащита многоэтажных кирпичных зданий в районах высокой сейсмической опасности: *Диссертация на соискание ученой степени доктора технических наук* (Москва: Московский институт коммунального хозяйства и строительства)

Reinforcement of asphalt concrete pavement by segments of exhausted fiber used for sorption of oil spill

V N Lukashevich[1] and I N Efanov[1]
[1]Tomsk State University of Architecture and Building, Tomsk, 634003, Russia

E-mail: vnluc@yandex.ru

Abstract. The paper is aimed at construction of the experimental road pavement made of dispersed reinforced asphalt concrete. Electronic paramagnetic resonance, infrared spectroscopy and fluorescent bitumen studies were used to prove that disperse reinforcement of asphalt concrete mixtures with fibers of exhausted sorbents reduce the selective filtration of low polymeric fractions of petroleum bitumen and improve its properties in the adsorption layer. Sesquioxides are neutralized as catalysts aging asphalt binder. This leads to improvement in the elasticity of bitumen films at low temperatures and provide better crack resistance of coatings to reduce the intensity of the aging of asphalt binder, and, therefore, to increase the durability of road pavements. The experimental road pavement made of dispersed reinforced asphalt concrete operated during 4 years and demonstrated better transport-performance properties in comparison with the analogue pavements.

1. Introduction

The main disadvantage of asphalt concrete as a material for pavements is considerable influence of air temperature on its strength properties (low resistance to shear forces at high positive temperatures and lack of fracture toughness at low temperatures). The mentioned disadvantage results in pavements deformations such as cracking, rutting, buckling and shifting.

Dispersion-reinforcement of asphalt concrete with discrete segments of chemical fibers is able to improve its rheological, physical and mechanical properties, thereby reducing probability of asphalt concrete pavements deformation occurrence and it has been confirmed by the research held in Russia, France, Switzerland, Netherlands, Poland and United States. There have been proved that introduction of dispersion-reinforcement into asphalt concrete mixture increases both its shear strength at temperature of 50° C by 25-30% and its tensile strength at negative temperatures by 40-80%. In this case deformability is improved up to 90-200% at temperature below zero and it mostly influences fatigue strength (improvement reaches by 200-500%) [1-3].

Disposal of fibrous sorbents wastes used for containment and oil skimming is an important of today. Existing methods of used sorbent disposal have disadvantages related to contamination of environment. Sliced fibers of waste sorbents used as dispersion – reinforcement in asphalt concrete are suggested being more environmentally friendly.

2. The theoretical studies of cross-linking processes of asphalt concrete

It is well known that asphalt concrete aging is caused by physico-chemical processes in mastic asphalt while asphalt mixtures preparing and pavements operation. The main processes affecting mastic asphalt aging intensity are selective filtration of oil bitumen components in pores and capillaries of mineral materials, change of oil bitumen fractional composition under influence of climatic factors and sesquioxides as catalyzators of aging [4].

Oils being the least viscous components of bitumen penetrate in pores of mineral materials in the process of selective filtration. Mastic gums fill small pores and asphaltenes are adsorbed on the mineral materials surface. Changing fractional composition of adsorbed layers of bitumen under influence of climatic factors and sesquioxides is accompanied by transition of oils to mastic gums and mastic gums to asphaltenes. Adsorption layers of oil bitumen on the mineral materials surface are impovered with low molecular weight fractions becoming more viscous, and consequently more brittle under temperatures below zero. This effect happens due to selective filtering components of bitumen, fractional composition changes of its films and negative influence of sesquioxides, accelerating processes of negative fractionation. Crack resistance of asphalt concrete pavements is reduced and the terms of their operation are shortened.

These negative processes can be eliminated using dispersion - reinforced asphalt concrete mixtures with fiber slices of their sorbents. They contain hydrocarbon feedstock collected at skimming that occur in case of a tanker sinking, accidents at oil pipelines and drilling rigs, spilling sludges and other wastes. Using this technique it's possible to process mineral materials with organic binders in two steps [5]. Fibrous sorbents containing the controlled oil collected are added to mineral material (the first stage) then oil bitumen is added (the second one). Mineral material contacting with fibrous sorbents is processed with their hydrocarbon feedstock (mostly oil, fuel oil, mastic gums of different origin, shale and carbolic sludges) at the first stage. Perfect adhesion of the organic binder to the mineral material's surface is provided due to this feedstock containing large amount of surface-active substances (phenols, ketones, carboxylic acids). These components interacting with the mineral material's surface ensure availability of chemisorption bonds and waterproof-soluble compounds formation on this surface. At the same time they neutralize sesquioxides that are catalyzators of aging. The obtained organo-mineral mixture is treated with oil bitumen at the second stage.

There have not been any selective filtration of the oil bitumen components in pores and capillaries of mineral material since these pores and capillaries have already been filled with components of organic binder at the first stage. Consequently, adsorption layers of oil bitumen on the mineral materials surface won't be impoverished with low molecular weight fractions and this fact positively impacts their elasticity under negative temperatures.

Furthermore due to large amount of these fractions in the adsorption layer of oil bitumen transition of oils to mastic gums and mastic gums to asphaltenes will be less intense. As a result, crack resistance of asphalt concrete and pavements durability are increased.

3. Experimental procedure

According to the theory of professor F.G. Unger free radicals of mineral materials surface can be the centers in which the asphaltenes are deposited. Then their combination and further increase of amount happen. Since asphaltenes are almost 100% concentrate paramagnetic materials, intensity indicator of the aging process of the oil disperse system can be paramagnetic centers concentration in it, indicating concentration of asphaltenes [6].

The investigations of mastic asphalt aging by electronic paramagnetic resonance method (EPR) have been conducted using the radiospectrometer "RADIOPAN" SE / X-25-44, frequency of 9 GHz and wave length of 3.2 cm. The spectra of EPR mixtures have been obtained immediately after mixing components, as well as mixtures subjected to aging at 160 °C in a thermo-stabilized cell for 6 hours. Concentration of paramagnetic centers in mixtures has been defined. The investigation results are presented in Table 1.

Table 1. Impact of technology of bituminous mixture preparing on concentration of paramagnetic centers

No	Composition mixture	Concentration of paramagnetic centers in the mixture,г$^{-1}$	Relative concentration of paramagnetic centers in the mixture
	before aging		
1.	Granite + Bitumen	0,24x10^{17}	
2.	Granite + Mastic gum + Bitumen	0,3x10^{17}	
3.	Limestone + Bitumen	4,5x10^{17}	100 %
4.	Limestone + Mastic gum + Bitumen	0,5x10^{17}	11 %
	after aging		
5.	Granite + Bitumen	0,7x10^{17}	100 %
6.	Granite + Mastic gum + Bitumen	0,3x10^{17}	42 %
7.	Limestone + Bitumen	6,9x1017	100 %
8.	Limestone + Mastic gum + Bitumen	5,0x1017	72 %

The data of the Table 1 demonstrate that modification of the granite surface with coal tar pitch doesn't considerably effect on the concentration of paramagnetic centers. One can see that the concentrations in mixture No. 1 (prepared according to the traditional technology) and No. 2 (prepared using the proposed technology) are approximately the same. Apparently it happens due to the fact while granite surface processing with oil bitumen there is no active interaction between them.

It's obviously that concentration of paramagnetic centers is changed after aging very significantly. Granite treated with bitumen contains 0,7x10^{17} paramagnetic centers. The same granite treated with coal tar pitch first, and then with oil bitumen contains only 0,3x10^{17} paramagnetic centers, which is 42%. It testifies that less asphaltenes (by 42%) have been appeared in the mixture and the intensity of the granite mixture aging with organic binder using the proposed technology is significantly lower.

Effect of the proposed technology on intensity of limestone mixtures and organic binders interaction is more discernable. Mixture of limestone and bitumen which hasn't been subjected to aging is paramagnetic centers concentration within 4,5x10^{17}g^{-1} which testifies active interaction of bitumen and limestone surface. Coal tar pitch has been treated with limestone first and then with bitumen. Limestone free radicals interact with coal tar pitch components and lose their activity. After addition of oil bitumen to this mixture the number of centers where asphaltenes are associated and combined decreases. As a result concentration of paramagnetic centers (testifying asphaltene concentration) in the mixture prepared by two-stage technology (mixture No. 4) is reduced to 0,5x10^{17} g^{-1} which is only 11%.

Study of mixtures of limestone with organic binders subjected to aging testifies concentration of paramagnetic centers decreasing if two-stage technology of addition of organic binders is used. Concentration of paramagnetic centers in mixture No. 8 (prepared according to the proposed technology) is 28% lower than in mixture No. 7 (prepared using the traditional technology).

The method of luminescence analysis [7] has been used to study selective filtering of organic binder components in case of its interaction with mineral materials surfaces. This method is based on oil bitumen components' ability to luminesce under ultraviolet (UV). Each component luminesces with its color using certain characteristic features. Selective filtering of organic binders components those treat fine pore rocks has been studied during the experiment. Some portions of crushed stone have been taken as samples. The first one has been dried to a constant weight and treated with organic binder. The second portion has been treated with oil bitumen and another portion has been treated with oil. The third portion has been treated with oil first and then – with oil bitumen. Each portion has been cut and examined selective filtration of organic binders components after cooling.

The selective filtration process has been observed using the universal biological microscope MBI-15 with ultraviolet source. The microscope is equipped with a digital camera and electronic devices which automatically determine exposure.

The conducted experimental studies of the crushed stone section of fine-porous limestone that had been treated with oil bitumen before have revealed fractionation of oil bitumen components.

Oils penetrate deeper into mineral material and mastic gums are closer to the surface. Asphaltenes are adsorbed on the surface. Thus, as has been expected, oil bitumen films providing link between mineral particles of asphalt bitumen mixture are essentially impoverished with low molecular weight fractions. It causes their brittle increasing and consequently acceleration of asphalt concrete aging. Luminescence of crushed stone of fine-porous limestone treated with oil has shown that oil penetrates into mineral material deeper than oil bitumen. Radiation of its components is expressed with less bright and less saturated with colors. Luminescence of crushed stone of fine-porous limestone treated with oil first and then with oil bitumen testifies absence of oil bitumen components filtration in mineral material, as pores and capillaries have been already filled with oil components.

Investigations of the sesquioxides interaction processes with active ingredients of products absorbed by sorbents as well as the processes of mastic asphalt aging have been carried out using IR spectroscopy. The results of these studies are presented in Figures 1 and 2.

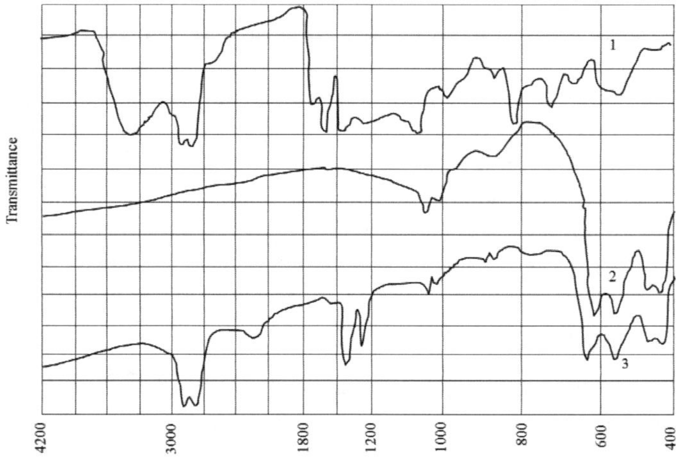

Figure 1. IR- spectra: 1-shale sludges; 2 - ferric oxide; 3-ferric oxide modified with shale sludges.

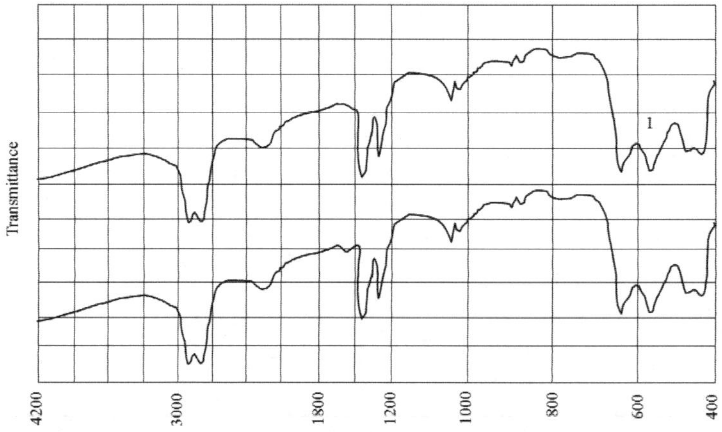

Figure 2. IR-spectra of mastic asphalt after aging: 1- ferric oxide + oil bitumen;
2 – ferric oxide + shale sludges + oil bitumen.

Figure 1 shows that there are broad absorption bands in the spectrum of shale mastic gum. They are extending from 3000 to 3799 cm^{-1}, showing oscillations of free or associating OH groups and confirming the presence of phenolic compounds. There are absorption bands of 1050 and 1100 cm^{-1} in the spectrum of shale mastic gum associated with the stretching vibrations of C- OH [8]. Disappearance of these absorption bands at transition from mastic gum to its mixture with ferric oxide can be explained by carboxylic acids active interaction with oxide resulting in neutralization of the catalyzator of aging as organic binder.

Effect evaluation of the interaction processes of shale sludges and ferric oxide on binder aging intensity has been made later (Figure 2). The depth of carbonyl absorption at 1600 cm^{-1} indicating the presence of aromatic compounds in the mixture has been adopted as a criterion for evaluation aging intensity. Analyzing this figure it should be noted that mixtures of ferric oxide and oil bitumen are characterized of less intense carbonyl absorption band rather than mixture of oil bitumen and oxide modified with shale sludge. This testifies higher concentration of aromatic compounds in the mixture of oil bitumen and ferric oxide modified with shale sludge. It can be explained by processes slow up of asphaltene formation from low molecular weight fractions of oil bitumen due to neutralization of sesquioxides as catalyzators of aging during their interaction with phenols.

The suggested technology was tested in a pilot operating construction (site of the national department of highways "Gorno -Altayavtodor"). The experimental section of the road using asphalt concrete pavement dispersion reinforced with fibers obtained from waste fibrous sorbents subjected to centrifugation after skimming in 2010.

The construction process is shown in Figure 3. Operation of the pilot area for four years has proved that its performance is significantly better than similar ones of the standard zone.

Figure 3. A section of the road using asphalt concrete pavement dispersion reinforced with segments of chemical fibers obtained from waste fibrous sorbents.

4. Conclusion

There has been conducted experimental research using the methods of electron paramagnetic resonance and fluorescent bitumen studies. They testify that at dispersion reinforcement of asphalt concrete mixtures with fibers of sorbents used for containment and oil skimming it's advisable to enter first sorbents containing recovered oil products into the mineral material, mix them and then oil bitumen can be added to this mixture.

Advanced processing of mineral materials with oil containing in the fibrous sorbents provides modification of their surfaces. Oil products penetrate into pores and capillaries filling them. In case of the further mixture processing with oil bitumen selective filtration of its components in pores and capillaries of mineral materials doesn't happen because they have been already filled with oil contained in the fibrous sorbents. As a result adsorption layers of bitumen on the surface of mineral materials are not impoverished with low molecular weight fractions that improve their elasticity at low temperatures, providing increased cold check resistance in winter and consequently durability of pavements.

References

[1] Акулич А 1987 *Структура и свойства дисперсно-армированных асфальтобетонов: Автореферат диссертации на соискание научной степени кандидата технических наук* (Минск: Белдорнии) 17
[2] Гамеляк И 1992 *Разработка методики конструирования дорожных одежд со слоями из дисперсно армированных асфальтобетонов. Автореферат диссертации на соискание научной степени кандидата технических наук* (Москва: Госдорнии) 23
[3] Мерзликин А и Гамеляк И 1990 *Конструирование, расчет и испытание дорожных одежд* (Москва: Труды Союздорнии) 17-25
[4] Железко Е и Печеный Б 1970 *Труды Союздорнии* **46** 137-142
[5] Лукашевич В 2003 Строительные материалы **1** 24-25
[6] Унгер Ф и Андреева Л 1995 *Фундаментальные аспекты химии нефти. Природа смол и асфальтенов* (Новосибирск: Институт химии нефти Сибирского отделения РАН) 192
[7] Флоровская В и Авчинникова Л 1970 *Люминесцентная микроскопия битуминозных веществ.* (Москва: Издательство Московского университета) 80
[8] Белами Л 1971 *Новые данные по ИК – спектрам сложных молекул: Пер. с англ.* (Москва: Мир) 319

IOP Conf. Series: Materials Science and Engineering **71** (2015) 012033 doi:10.1088/1757-899X/71/1/012033

Applicability of the model of particles drift for the study of the disperse phase transfer in a gas flow

O V Matvienko [1,2] and E V Evtyushkin[1]

[1]Tomsk State University of Architecture and Building, Department of Theoretical Mechanics, Tomsk, 634003, Russia
[2] Tomsk State University, Department of Physical and Computational Mechanics, Tomsk, 634050, Russia

E-mail: matvolegv@mail.rul

Abstract. Two different approaches are currently available for the analysis of the behavior of the solid particles in flows. These are termed Eulerian and Langrangian. In the Lagrangian method on the trajectories of the individual size fractions are evaluated by solving time dependent ordinary differential equations. In the Eulerian approach, partial differential equations for the conservation of mass and momentum are written for each of the particles fractions, which are solved together with the equation of the liquid flow. The particles drift model is based on the algebraic slip velocity approach. In this approach the relative velocities between particles and liquid are evaluated by consideration of the forces acting on the particles. The analysis of results enable to conclude that if $Kn_p < 1$, $Re < 1$ using model of particles drift gives the error that is not exceeding 1%.

1. Introduction

Multiphase flows may be encountered in various forms in industrial practice, as for example, transient flows with a transition from pure liquid to a vapour flow as a result of external heating, separated flows (i.e. stratified flows, slug flows, or film flows), and dispersed two-phase flows where one phase is present in the form of particles, droplets or bubbles in a continuous carrier phase (i.e. gas or liquid). Such dispersed two-phase flows are encountered in numerous technical and industrial processes, as for example in particle technology, chemical engineering, and biotechnology.

For the description of properties of polyphase flows two methods based on the way of Lagrange and Euler [1] are used. Within approach of Lagrange equations of motion of the separate particles, considered as the material points, in the form of the second Newton's laws in which right members there are forces operating on a particle in a stream are written out. Despite apparent simplicity of the description of driving of particles within approach of Lagrange, this method possesses, at least, two essential shortcomings. The first of them is bound to computing difficulties, bound to need to solve huge number of equations of motion for an assembly of particles. So for the description of space movement N of particles it is required to solve 6N equations. The problem becomes even more composite if there is a need of model operation of driving of particles taking into account their interaction. The second problem is bound to difficulty of the accounting of stochastic nature of particles driving in a stream with turbulence. The approaches applied now based on application of a

Monte-Carlo method [2] demand carrying out the whole series of calculations, so that the result of their averaging had the objective character.

The equation of motion for small particles in a viscous quiescent fluid (i.e. for small particle Reynolds-numbers, which is also referred to as Stokes flow) goes back to the pioneering work of Basset [3], Boussinesq [4] and Oseen [5]. Therefore, the equation of motion is mostly referred to as BBO-equation. Numerous publications deal with the extension of the BBO equation for turbulent flows. The thesis of Tchen [6] was probably the first study on particle motion in turbulent flows based on the BBO equation. A rigorous derivation of the equation of motion for small particles in non-uniform flow has been performed by Maxey and Riley [7]. Inter-particle collisions, the method of the determination of the diffusion coefficient of the solid phases as well as the influence of the physical characteristics and velocity of the fluid on the bubble dynamics were investigated in the paper [8, 9, 10]. Effects of interaction of phases, stochastic nature of driving of a big assembly of particles can be considered within approach of Euler, in compliance with which the polyphase environment is considered as set multi-speed continuums (a carrying agent and various fractions of particles). For each of these continuums equations of motion in the form of Euler, and also equations of mass conservation of each considered continuums are written.

In case of particles with a small inertance this approach can be replaced with the models based on the concept of a drift of a disperse phase concerning a carrying agent. Thus the speed of a disperse phase is defined in the assumption of a smallness of inertial terms or, otherwise, dynamic balance of forces operating on particles. Thus, there is no need to solve the complete differential equations of driving, and it is enough to consider the equation of dynamic balance of forces.

Owing to the simplicity and profitability models of particles drift were widely adopted in engineering practice. Nevertheless, possibility of their application has to be defined not by intuitive estimates and simplicity reasons but by the corresponding quantitative assessments.

Both approaches of Lagrange and Euler are based on the assumption of a smallness of particle sizes d_p in comparison with the reference size of area of a current L for which within computing hydrodynamics it is convenient to take the size of a finite difference grid. Thus, the ratio has to be carried out:

$$Kn_p = d_p / L << 1. \quad (1)$$

In case of violation of conditions (1) within approach of Lagrange the particle cannot be taken for the material point, and it is necessary to investigate a picture of a flow of a particle. Non-performance of a condition (1) within approach of Euler is equivalent to impossibility to apply model interpenetrating continuums.

2. Mathematical model

The paper considers some volume of liquid corresponding to a finite difference cell. Carrying agent velocity in this volume $\vec{v}_l$ can be considered independent of coordinates. Entering the frame moving with a velocity $\vec{v}_l$, the carrying agent will be fixed in the frame.

At model operation of driving of a disperse phase in a finite difference cell it will be proceed from the following assumptions:

- driving of a particle is defined by Archimedes force and resisting strength;
- the particle of a disperse phase with initial velocity $\vec{v}_0$ gets to some volume of based liquid;
- the vector of velocity of a particle in an initial instant is parallel to a vector of the acceleration caused by Archimedes force;
- particles of a disperse phase are assumed as spherical;
- interaction between particles is not considered.

The last assumption enables to investigate driving of a single particle only.

Within these assumptions the equation of motion of a single particle can be presented in the following form [1 – 9]:

$$\frac{d\vec{v}_p}{dt} = -\frac{3}{4}\frac{\rho}{\rho_p}C_D d_p^{-1}|\vec{v}_p|\vec{v}_p + \frac{\rho_p - \rho}{\rho_p}\vec{g}, \tag{2}$$

where $\vec{g}$ – free fall acceleration, d_p - diameter of a disperse phase, ρ, ρ_p - density of a bearing and disperse phase $\vec{v}_p$ - the velocity of a disperse phase, drag coefficient C_D is function of a relative of Reynolds $Re = \rho|\vec{v}_p|d_p/\mu$.

For small Reynolds numbers ($Re < 0.5$) viscous effects dominate and no separation is observed. Therefore, an analytic solution for the drag coefficient is possible as proposed by Stokes. In the transition region ($0.5 < Re < 1000$) inertial effects become of increasing importance. Above a Reynolds number of about 24 the flow around the particle begins to separate. Initially this separation is symmetric. It becomes unstable and periodic above $Re \approx 130$. Above $Re \approx 1000$ the drag coefficient remains almost constant up to the critical Reynolds number, since the wake size and structure is not considerably changing. This regime is referred to as Newton-regime. At the critical Reynolds number $Re > 2 \cdot 10^5$ a drastic decrease of the drag coefficient is observed, being caused by the transition from a laminar to a turbulent boundary layer around the particle. This results in a decrease of the particle wake. In the super-critical region ($Re > 4 \cdot 10^5$) the drag coefficient again increases continuously. The reference drag curve can be approximated patch dependences of the form:

$$C_D = C'Re^n, \tag{3}$$

where values of coefficients are given in a formula (3) in Table 1.

The next step is defining the displacement velocity of a particle within drift model. In case of a gravitational sedimentation of particles the equation of dynamic balance of forces can be presented as follows:

$$\frac{3}{4}C_D\frac{\rho}{\rho_p}d_p^{-1}|v_s|v_s + \frac{\rho - \rho_p}{\rho}g = 0, \tag{4}$$

where v_s - the velocity of a stationary deposition of a particle (sedimentation velocity).

Table 1. Values of coefficients in resistance law.

Reynolds number Re	$Re < 1$	$1 < Re < 10$	$10 < Re < 800$	$800 < Re \le 2 \cdot 10^5$
C'	24	26.3	12.3	0.44
n	-1	-0.8	-0.5	0

At the driving of particles described by the law of resistance of Stokes speed of sedimentation it can be defined as:

$$v_s = \frac{\rho_p - \rho}{18\mu}gd^2. \tag{5}$$

In the transitional area the velocity of sedimentation will be defined as follows:

$$v_s = \left[\frac{4}{3}(C')^{-1}(\rho_p - \rho)\rho_p g\right]^{5/6}\frac{d_p^{3/2}}{\rho\mu^{4/6}} \quad \text{for } 1 < Re < 10, \tag{6}$$

$$v_s = \left[\frac{4}{3}(C')^{-1}(\rho_p - \rho)\rho_p g\right]^{2/3}\frac{d_p}{\rho\mu^{1/3}} \quad \text{for } 10 < Re < 800. \tag{7}$$

When driving particle the velocity of sedimentation will be equal in the Newtonian mode:

$$v_s = \sqrt{\frac{4}{3}\frac{\rho_p - \rho}{\rho}C_D^{-1}gd_p} \tag{8}$$

Velocity of sedimentation can be considered as the scale characteristic characterizing displacement velocity of a particle in the resisting environment. Other such scale characteristics include time of a sedimentation t_s and sedimentation length l_s. These scales can be defined as follows:

$$t_s = \frac{\rho}{\rho - \rho_p} \frac{v_s}{g}, \qquad l_s = v_s t_s . \tag{9}$$

For the further analysis it is convenient to present equations of motion in the dimensionless look. For this purpose we will enter the dimensionless velocities, coordinates and time:

$$\varphi = \frac{v_z}{v_s}, \qquad \xi = \frac{x}{l_s}, \qquad \tau = \frac{t}{t_s}. \tag{10}$$

Thus, particle driving within model of a drift will be described by the following dependences:

$$\varphi = \varphi_d = 1, \qquad \xi = \xi_d = \tau . \tag{11}$$

In need of the accounting of inertial properties of a particle it is necessary to solve differential equations of driving which in the dimensionless form are presented as follows:

$$\frac{d\varphi}{d\tau} = -|\varphi|\varphi^{1+n} + 1, \qquad \frac{d\xi}{d\tau} = \varphi, \tag{12}$$

where n - an exponent at a Reynolds number in resistance law.

Equations of motion of a particle (12) become isolated by the following starting conditions:

$$\tau = 0: \quad \varphi = \varphi_0, \qquad \xi = 0. \tag{13}$$

The solution of a set of equations (12) with starting conditions (13) depends on resistance law.

Let us consider at first driving of particles in a mode of resistance of Stokes: $n = -1$. Integration of the equations (12) with conditions (13) allows determining the velocity and particle movement for various instants. In the dimensionless look these dependences have an appearance:

$$\varphi = 1 + (\varphi_0 - 1)e^{-\tau}, \qquad \xi = \tau + (\varphi_0 - 1)\left[1 - e^{-\tau}\right]. \tag{14}$$

Movement of the particle moving in a mode of resistance of Stokes differs from the speed of a drift and the movement caused by a drift on the following values:

$$\Delta\varphi = (\varphi_0 - 1)e^{-\tau}, \qquad \Delta\xi = (\varphi_0 - 1)\left[1 - e^{-\tau}\right]. \tag{15}$$

Time for which the particle will move on the reference size of area of current L, within model of a drift will be defined as:

$$\Delta\tau = \frac{L}{l_s} = \lambda = \frac{18}{Kn_p Re_s}, \tag{16}$$

where $Re_s = \rho v_s d_p / \mu$ - a sedimentation Reynolds number, $Kn_p = d_p / L$ – Knudsen number. Thus, the relative accuracy in the particle shift, given by model of a drift can be written down as:

$$\frac{\Delta\xi}{\lambda} = \frac{Kn_p Re_s}{18}(\varphi_0 - 1)\left[1 - exp\left(-\frac{18}{Kn_p Re_s}\right)\right]. \tag{17}$$

The model of a drift of particles can be applied if $Kn_* = \Delta\xi/\lambda << 1$.

Limits of applicability of a formula (17) are defined by limits of applicability of the Stokes law of resistance:

$$Re_s \le 1, \qquad Re\varphi_0 \le 1 . \tag{18}$$

At the Newtonian mode of resistance change of speed and particle coordinate over time, and also the relative accuracy of model of a drift allows finding the solution of equations of motion of a particle:

$$\varphi = \frac{(1+\varphi_0) - (1-\varphi_0)exp(-2\tau)}{(1+\varphi_0) + (1-\varphi_0)exp(-2\tau)}, \tag{19}$$

$$\xi = ln\left|\frac{(1-\varphi_0)+(1+\varphi_0)exp(3/2\,C_D Kn_p)}{2}\right| - 2\tau, \tag{20}$$

$$\frac{\Delta\xi}{\lambda} = \frac{4}{3}C_D^{-1}Kn \cdot ln\left|\frac{(1-\varphi_0)+(1+\varphi_0)exp(3/2\,C_D Kn_p)}{2}\right| - 2. \tag{21}$$

Definition of the law of motion of a particle and definition of an error of model of a drift in the transitional area in an analytical form is, unfortunately, impossible. Therefore, for definition of characteristics of driving it is necessary to use numerical methods.

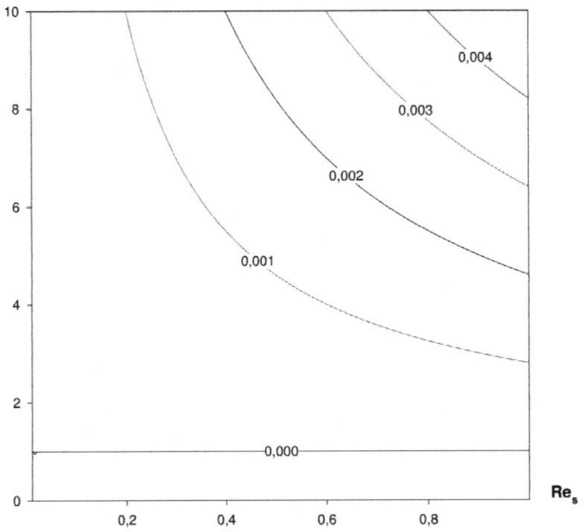

Figure 1. The relative error of the model drift $\mathrm{Re} < 1$

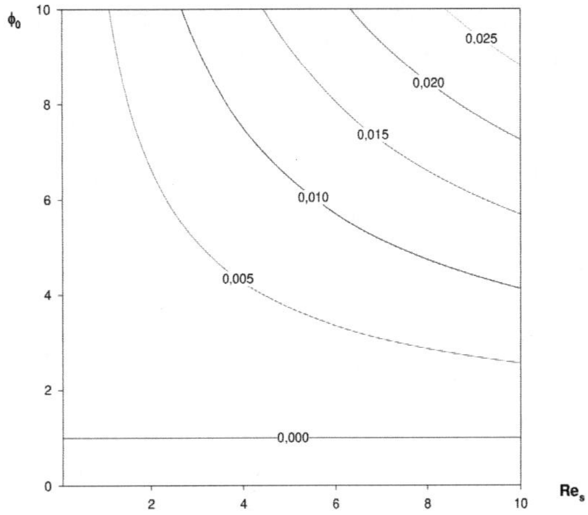

Figure 2. The relative error of the model drift $1 < Re < 10$

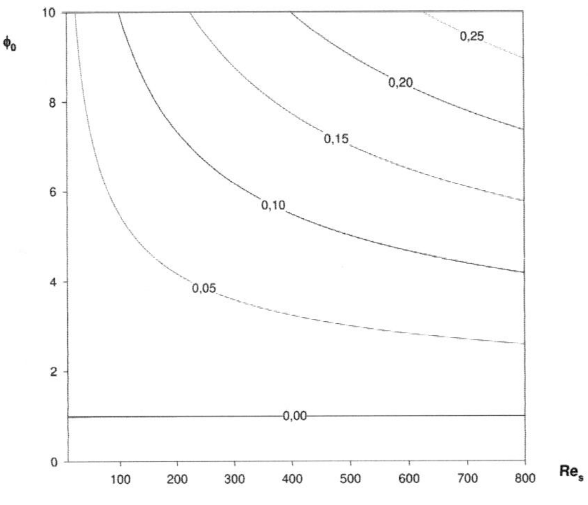

Figure 3. The relative error of the model drift $10 < Re < 800$

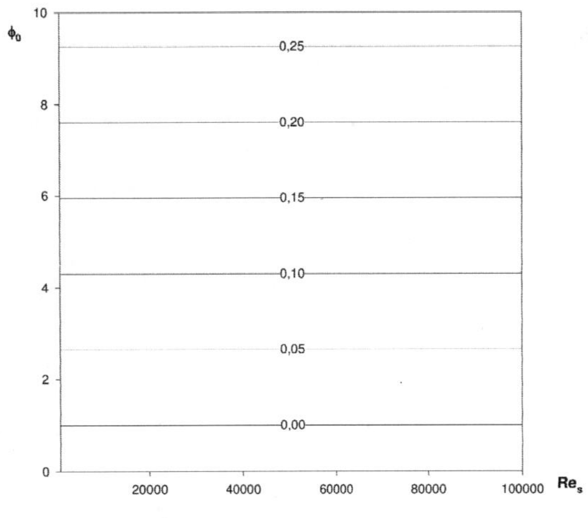

Figure 4. The relative error of the model drift $800 < Re$

In Figures 1-4 lines of equal relative accuracies $Kn_* = const$ for various values Re_s and φ_0, and also various laws of resistance are presented. The analysis of these drawings allows to come to

IOP Conf. Series: Materials Science and Engineering **71** (2015) 012033 doi:10.1088/1757-899X/71/1/012033

conclusion that at reasonably chosen parameters of a grid ($Kn_p < 0.1$) usage of model of a drift of a particle at $\mathrm{Re} < 1$ gives the error which is not exceeding 1%.

With increasing of a Reynolds number and initial velocity of a particle the relative accuracy increases in the transitional area. So, at $\mathrm{Re} = 800$ with increase from φ_0 2 to 5 Kn_* increases from 1% up to 10%. At $\mathrm{Re} > 800$ the size of the relative accuracy depends only on φ_0 size.

Conclusions

Theoretical analysis of the particle motion is carried out and the model of particle drift in the flow is proposed. When calculating disperse multi-phase flows with fine particle (for example dust), the relaxation time for the particles will be much smaller than the flow time scale. If disperse phase are at low volume fractions, it is unnecessary to solve momentum equations for the particle phase separately, as it can be assumed that the particles are always at their terminal velocity.

References

[1] Crowe C T, Sommerfeld M, Tsuji Y 1998 *Fundamentals of Gas-Particle and Gas-Droplet Flows* (CRC Press Boca Raton USA)
[2] Sommerfeld M 1992 *Int. J. of Multiphase Flow* **18** 905-926
[3] Boussinesq J V *1885 Sur la resistance d'une sphere solide* (Acad. Sci. Paris)
[4] Basset A B 1888 *Phil. Trans. Roy. Soc* **A179** 43-69
[5] Oseen C W 1927 *Hydromechanik* (Leipzig Akademische Verlagsgem)
[6] Tchen C-M 1947 *Mean value and correlation problems connected with the motion of small particles suspended in a turbulent fluid* Dissertation Technische Hochschule Delft
[7] Maxey M R, Riley J J 1983 *Phys. of Fluids* **26** 883-889
[8] Matvienko O V Ushakov V M Evtyushkin E 2004 *Vestn. TGPU* **6** (43) 50-53
[9] Matvienko O V Agafontseva M V Bazuev V P 2012 *Vestnlk of TSUAB* 144
[10] Matvienko O V Daneyko A M 2013 *Izvestia vuzov. Fizika* 9/3 190

IOP Conf. Series: Materials Science and Engineering **71** (2015) 012034 doi:10.1088/1757-899X/71/1/012034

Numerical investigation of Herschel Bulkley fluids mixing

O V Matvienko [1,2], V P Bazuev[3], V N Venik[3] and N G Smirnova[3]

[1]Tomsk State University of Architecture and Building, Department of Theoretical Mechanics, Tomsk, 634003, Russia

[2]National Research Tomsk State University, Department of Physical and Computational Mechanics, Tomsk, 634050, Russia

[3]Tomsk State University of Architecture and Building, Research Institute of Building Materials, Tomsk, 634003, Russia

E-mail: matvolegv@mail.ru

Abstract. Bitumen modification by means of introducing the components into bitumen swirling flows using an injector mixer has been investigated in this paper. The properties of mixing depending on geometry and operating conditions and the swirling intensity are presented herein. The results obtained have shown that the optimum mixing can be achieved within a strongly swirled flow when the injector mixer locates in close proximity to a swirler.

1. Introduction

A priority challenge in the field of highway engineering is to develop a set of measures and technologies for production of new high-quality bituminous binders to provide an improved type of pavement of the roads, which would increase their service life between the periods of road reconditioning. A way to solve it could be a development of new low-cost technologies and devices that would allow new asphalt binders to be produced via incorporation of new modifying components taking into account the conditions of road construction and operation [1]. In [2 - 7], the authors studied the preparation of a new bituminous binder using the principle of cavitation-mixing dispersion. In some cases, however, it is necessary to use other approaches, namely, to introduce the components into the swirling bitumen flow through an injector or to combine injection mixing with the cavitation-mixing dispersion to obtain new high-quality binders. A priority application of injection mixing combined with cavitation-mixing dispersion is the preparation of polymer-bitumen binders, since it is difficult to manufacture a new high-quality binder due to the fact that polymers have complex structural bonds and making the manufacturing process difficult.

2. Mathematical model

To describe the field of flow and mixing of flows, the following assumptions were taken. The fluid flows are assumed to be stationary, axis-symmetric and swirling, depending on the law of rigid body rotation at the entrance. The model included two-dimensional rheological equations written in cylindrical coordinates, which are the most suitable for description of the axisymmetric flow [8 - 11].

$$\frac{\partial \rho u}{\partial x}+\frac{1}{r}\frac{\partial \rho v r}{\partial r}=0 \qquad (1)$$

$$\frac{\partial \rho u^2}{\partial x}+\frac{1}{r}\frac{\partial \rho u v r}{\partial r}=-\frac{\partial p}{\partial x}+\frac{\partial \tau_{xx}}{\partial x}+\frac{1}{r}\frac{\partial r\tau_{xr}}{\partial r}, \qquad (2)$$

$$\frac{\partial \rho u v}{\partial x}+\frac{1}{r}\frac{\partial \rho v^2 r}{\partial r}=-\frac{\partial p}{\partial r}+\frac{\partial \tau_{xr}}{\partial x}+\frac{1}{r}\frac{\partial r\tau_{rr}}{\partial r}-\frac{\tau_{\varphi\varphi}}{r}+\frac{\rho w^2}{r}, \qquad (3)$$

$$\frac{\partial \rho u w}{\partial x}+\frac{1}{r}\frac{\partial \rho v w r}{\partial r}=\frac{\partial \tau_{x\varphi}}{\partial x}+\frac{1}{r^2}\frac{\partial r^2\tau_{r\varphi}}{\partial r}-\frac{\rho v w}{r}. \qquad (4)$$

$$T^2=\frac{\tau_{xx}^2+\tau_{rr}^2+\tau_{\varphi\varphi}^2}{2}+\tau_{x\varphi}^2+\tau_{r\varphi}^2+\tau_{xr}^2>\sigma^2:$$

$$\tau_{xx}=2\left(kU^{n-1}+\frac{\sigma}{U}\right)\frac{\partial u}{\partial x},\quad \tau_{xr}=\left(kU^{n-1}+\frac{\sigma}{U}\right)\left(\frac{\partial u}{\partial r}+\frac{\partial v}{\partial x}\right),\quad \tau_{rr}=2\left(kU^{n-1}+\frac{\sigma}{U}\right)\frac{\partial v}{\partial r},$$

$$\tau_{x\varphi}=\left(kU^{n-1}+\frac{\sigma}{U}\right)\frac{\partial w}{\partial x},\quad \tau_{r\varphi}=\left(kU^{n-1}+\frac{\sigma}{U}\right)\left(r\frac{\partial}{\partial r}\left(\frac{w}{r}\right)\right),\quad \tau_{\varphi\varphi}=2\left(kU^{n-1}+\frac{\sigma}{U}\right)\frac{v}{r},$$

$$U=\sqrt{2\left[\left(\frac{\partial u}{\partial x}\right)^2+\left(\frac{\partial v}{\partial r}\right)^2+\left(\frac{v}{r}\right)^2\right]+\left(\frac{\partial u}{\partial r}+\frac{\partial v}{\partial x}\right)^2+\left(\frac{\partial w}{\partial x}\right)^2+\left(r\frac{\partial w/r}{\partial r}\right)^2}. \qquad (5)$$

The constant k is called the fluid consistency index; the smaller its fluidity, the higher is k. The parameter n characterizes the degree of the non-Newtonian behavior of material; the more n differs from unity (either higher or smaller), the more distinct the viscosity anomaly and the nonlinearity of the flow curve reveal themselves. Corresponding to the values $0 < n < 1$ are pseudoplastic fluids, the apparent viscosity of which decreases with an increase in the shear velocities. Corresponding to the values $n > 1$ are dilatant fluids, the apparent viscosity of which increases with the shear velocities. Thus, the medium is determined by three parameters: consistency k, the nonlinearity index n and yield stress σ. The Shvedov-Bingham fluid is characterized by the parameters $n=1, \sigma>0$, the Oswald-de Weil fluid is characterized by the parameters $n \neq 1$, $\sigma=0$, and Newtonian fluid – $n=1$, $\sigma=0$.

By analogy with a Newtonian liquid, a Bulkley-Herschel liquid is conveniently characterized by the effective viscosity μ_{eff}, determined as $\mu_{\text{eff}}=kU^{n-1}+\frac{\sigma}{U}$. The dilatant properties of a liquid manifest themselves as an increase in its effective viscosity with increase in the rate of shear deformation of the liquid; the pseudoplastic and viscoplastic properties of a liquid manifest themselves as a decrease in its effective viscosity with increase in the rate of shear deformation of the liquid [12-15].

For description of the mixing process, the concentration transport equations were used.

$$\frac{\partial \rho u C}{\partial x}+\frac{1}{r}\frac{\partial \rho v r C}{\partial r}=\frac{\partial}{\partial x}\left[D\frac{\partial C}{\partial x}\right]+\frac{1}{r}\frac{\partial}{\partial r}\left[Dr\frac{\partial C}{\partial r}\right] \qquad (6)$$

2. Results

Below, the process of bitumen modification in an injector mixer is considered. It has a cylindrical case with a radial nipple for the modifier input, a nozzle for molten bitumen in the face end of the case and a mixing chamber. The mixing chamber communicates with the bitumen nozzle directly and with the radial nipple through a ring nozzle. The following parameters are calculated parameters of the device: $R=0.1\,\text{m}$, $L=1\,\text{m}$, $h=0.2\,\text{m}$, $d=0.05\,\text{m}$.

At first a flow with impenetrable walls is considered. Figure 2 shows the profile of the radial distribution of axial and tangential velocities in the swirling flow. In the absence of a flow swirl, the flow in the channel in the initial part is characterized by a plane profile of the axial velocity and a narrow zone of the boundary layer. In the swirling flow the axial velocity decreases in the axial zone and increases in the near-wall zone (Figure 1a).

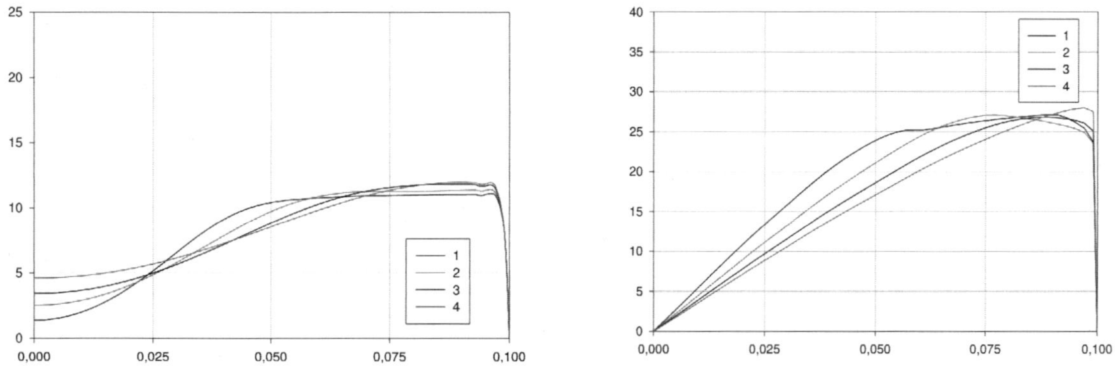

a b

Figure 1. Radial distribution of the axial (a) and tangential (b) velocities: $u_{in} = 10\text{m/s}$, φ = 50°, 1 – $x = 0.285\text{m}$, 2 – $x = 0.485\text{m}$, 3 – $x = 0.685\text{m}$, 4 – $x = 0.885\text{m}$.

Radial distribution of the tangential component of velocity is shown in Figure 1b. One can see three characteristic zones in the channel: near-wall zone of the boundary layer, axial zone, in which the tangential velocity varies by the law of rigid body rotation, and the central zone characterized by the constant value of the tangential velocity.

Figure 2 shows the downstream profile of the radial velocity. In the non-swirling flow and at small swirling angles ($\varphi < 50°$), the fluid flow lines are parallel to the axis of symmetry, which is characterized by nearly zero radial velocities.

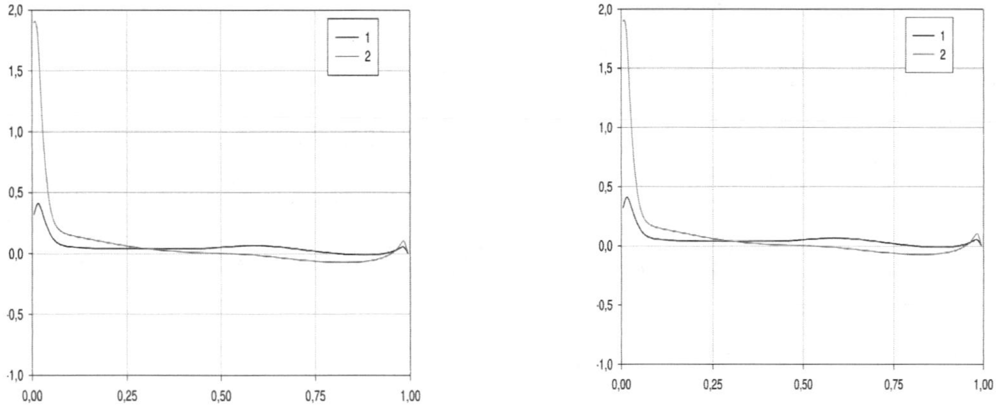

Figure 2. Downstream profile of the radial velocity: non-swirling flow (a) $u_{in} = 10\text{m/s}$, φ = 50°, 1 – $r = 0.076\text{m}$, 2 – $r = 0.036\text{m}$.

Only under conditions of intense swirling of the flow, we can observe a sharp increase in the radial velocity in the zone adjacent to the swirler owing to the flow edging by the field of centrifugal forces. Consider now the flow structure in a channel with an injector. Figure 3a shows the downstream profile of the radial velocity in the case without swirling at different distances from the flow axis. In the zones far from the injector, the radial velocity is close to zero and only in the injector zone we can see significant radial motions.

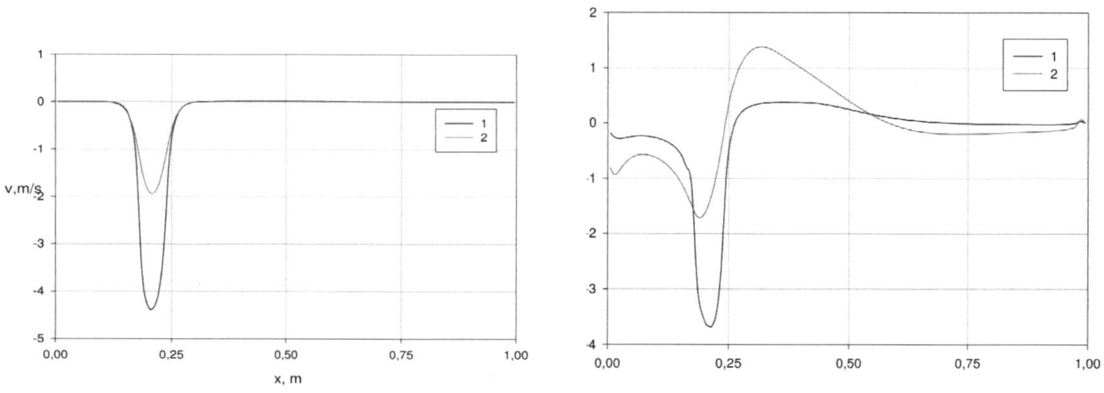

a b

Figure 3. Downstream profile of the radial velocity: $u_{in}=10$ m/s, $v_{in}=5$ m/s, a – non-swirling flow φ = 0, b – φ = 50°, 1 – $r=0.076$m, 2 – $r=0.036$m.

The highest value of the radial velocity is observed near the injector, while closer to the channel axis the radial velocity decreases. For weak ($\varphi=20°$) and moderate ($\varphi<50°$) swirls, there are no qualitative changes in the radial velocity observed. However, in the case of a strong swirl of the flow ($\varphi\geq 50°$), the plot $v(x)$ is characterized by the more complex behavior. In the case where the injector is rather far from the entrance, near the swirler one can observe a zone of positive values of the radial velocity owing to the flow edging by the field of centrifugal forces in the initial part. If the injector is located near the swirler, this feature is not observed. It should be noted that behind the injector there is a zone with positive values of the radial velocity. In this case, the smaller is the separation between the injector and the swirler, the more intense is the motion of fluid behind the injector toward the wall. Thus, the intense swirling of the flow intensifies the radial mass transfer.

Figure 4 shows the radial profile of the axial velocity (along the channel radius). It can be seen that upstream the injector the velocity profile is plane in the whole flow zone except for the near-wall area of the boundary layer. Downstream the injector, the radial distribution of the axial velocity is characterized by three zones: axial zone with high values of the axial velocity, zone of the central flow, in which the axial velocity is nearly constant, and zone of the boundary layer. It should be noted that the maximal value of the axial velocity at the channel axis decreases downstream.

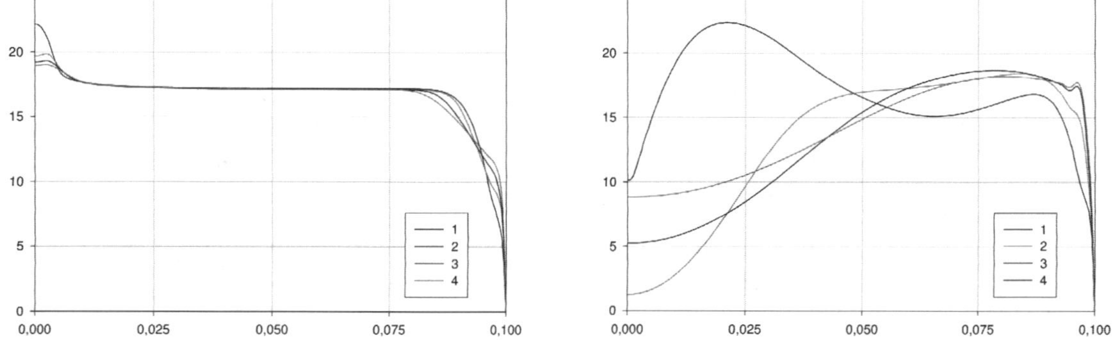

a b

Figure 4. Radial distribution of the axial velocity in the channel with an injector: $u_{in}=10$ m/s, $v_{in}=5$ m/s: a – φ = 0, b – φ = 50°,1 – $x=0.285$m, 2 – $x=0.485$m, 3 – $x=0.685$m, 4 – $x=0.885$m.

The flow swirl results in the formation of a zone of minimal axial velocity in the center of the channel in the section between the swirler and the injector. Behind the injector, the velocity on the channel axis increases sharply.

Figure 5 shows the radial distribution of the tangential velocity. The axial velocity includes a zone of quasi-solid rotation. In the central part of the flow, the rotation is characterized by a constant swirling angle, while near the wall the rotation is described by the potential vortex.

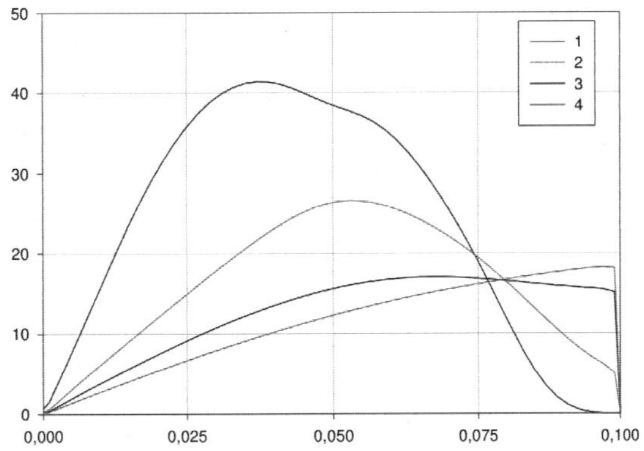

Figure 5. Radial distribution of tangential velocity in the channel with an injector: $u_{in} = 10\,\text{m/s}$, $v_{in} = 5$ m/s, φ = 50°, 1 – $x = 0.285$m, 2 – $x = 0.485$m, 3 – $x = 0.685$m, 4 – $x = 0.885$m.

From the practical point of view, the radial change in the concentration is of greatest interest (Figure 6).

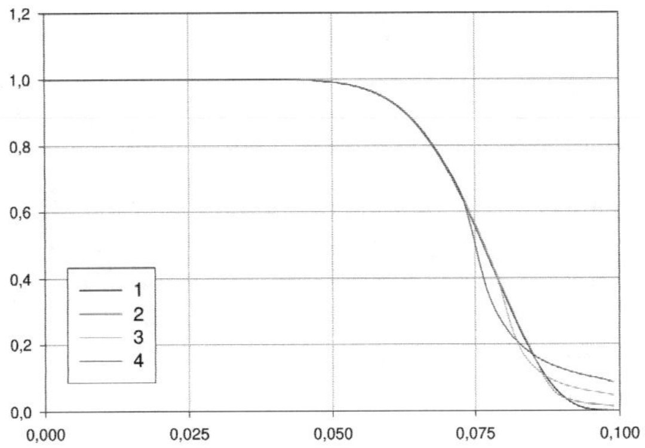

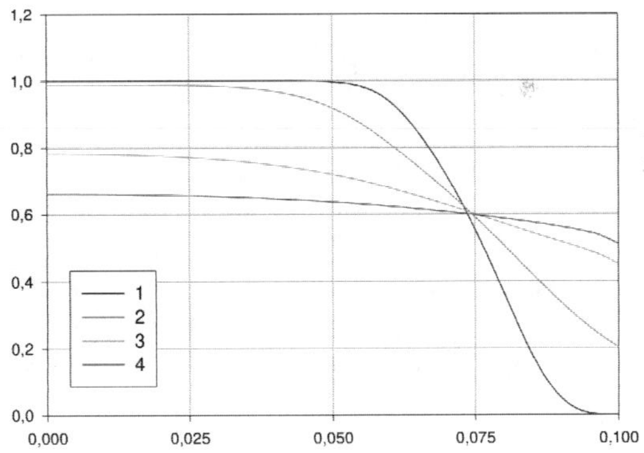

a b

Figure 6. Radial distribution of concentration in the channel with an injector: $u_{in} = 10\,\text{m/s}$, $v_{in} = 5$ m/s: a – φ = 0, b – φ = 50°, 1 – $x = 0.285$m, 2 – $x = 0.485$m, 3 – $x = 0.685$m, 4 – $x = 0.885$m.

The closer to the entrance cross section the injector is, the wider is the mixing zone. The flow swirl improves the mixing characteristics. In this case, intensive radial motions in the swirling flow result in a fast equalization of concentrations in the flow. This analysis shows that the best mixing can be achieved in a strongly swirling flow with the injector located in close proximity to the swirler.

3. Conclusions

The paper describes the results of investigations of a non-Newtonian fluid flow and mixing in a cylindrical channel with injector. The values of the effective viscosity of the viscoplastic Shedov - Bingham fluid in the axial zone of the channel are increased downstream. The hydrodynamic

IOP Conf. Series: Materials Science and Engineering **71** (2015) 012034 doi:10.1088/1757-899X/71/1/012034

stabilization of the flow leads to the conversion of the axial highly viscous region into a rigid zone of quasi-solid flow. The the hydrodynamically unstabilized flow of the dilatant Oswald - de Weil fluid in the peripheral region of the channel is characterized by the growth of the effective viscosity. Investigations of the dilatant Bulkley - Herschel fluid flow have shown that, for large values of Bulkley - Herschel number in the flow the dilatant properties are dominated and for small – viscoplastic ones. The minimum values of the effective viscosity are realized at moderate shear rates. With increasing of Bulkley-Herschel number with one and the same value of the nonlinearity parameter the effective viscosity in the axial region of the flow is reduced, and in the central and peripheral areas - increases. With increasing of the nonlinearity parameter with one and the same value of Bulkley-Herschel number the dilatant properties of the fluid are more pronounced. The flow swirl improves the mixing characteristics. In this case, intensive radial motions in the swirling flow result in a fast equalization of concentrations in the flow. This analysis shows that the best mixing can be achieved in a strongly swirling flow with the injector located in close proximity to the swirler.

References

[1] Karpeko F V and Gureev A A 1998 *Bitumen Emulsions. Principles of Physicochemical Production and Application* (Moscow: Khimiya)
[2] Wilkinson W L 1964 *Non-Newtonian Fluids. Fluid Mechanics, Mixing, and Heat Transfer* (Moscow: Mir)
[3] Klimov D M, Petrov A G, and Georgievskii D V 2005 *Viscoplastic Flows: Dynamic Chaos, Stability and Mixing* (Moscow: Nauka)
[4] Kutepov A M, Polyanin L D, Zapryanov Z D et al. 1996 *Chemical Hydrodynamics: Handbook* (Minsk: Byuro Kvantum)
[5] Ostrovskii G M 2000 *Applied Mechanics of Inhomogeneous Media* (St. Petersburg Nauka)
[6] Matvienko O V, Bazuev V P, Venik V N and Smirnova N G 2014 *Vestnik of TSUAB* **4** 231
[7] Matvienko O V, Bazuev V P, Dul'zon N K, Smirnova N G and Agafonova M V 2014 *Vestnik of TSUAB* **2** 80
[8] Matvienko O V and Evtyushkin E V 2011 *J Eng Phys and Thermophys* **84** 241
[9] Matvienko O V, Bazuev V P and Yuzhanova N K 2011 *J Eng Phys and Thermophys* **84** 58
[10] Matvienko O V, Bazuev V P and Dul'zon N K 2014 *J Eng Phys and Thermophys* **87** 200
[11] Matvienko O V, Bazuev V P and Dul'zon N K 2014 *J Eng Phys and Thermophys* **87** 1129
[12] Malkin A Ya, Isaev A I 2005 *Rheology: conception, methods, applicationsn* (Saint-Petersburg: Professia)
[13] Reiner M *Rheology* 1965 (Moscow: Nauka)
[14] Tanner R I 2000 *Engineering Rheology* (New York: Oxford University Press)
[15] Macosko C W 1994 *Rheology: Principle, Measurement and Application* (New York: VCH)

IOP Conf. Series: Materials Science and Engineering **71** (2015) 012035 doi:10.1088/1757-899X/71/1/012035

Mathematical investigation of ellipsoidal particles separation in the hydrocyclones

O V Matvienko [1,2]
[1]Tomsk State University of Architecture and Building, Department of theoretical mechanics, Tomsk, 634003, Russia

[2] National Research Tomsk State University, Department of Physical and Computational Mechanics, Tomsk, 634050, Russia

E-mail: matvolegv@mail.rul

Abstract. This paper presents the results of the numerical studies of the non-spherical particles separation in the hydrocyclones. The calculations are done for freely rotating particles, and particles with orientations fixed by means of an external torque exerted by a strong orienting field. The results demonstrate that separation characteristics of the hydrocyclone are strongly depends on the flow shear rate, particle size, and particle shape. The model is validated by the good agreement between the measured and predicted results.

1. Introduction

Hydrocyclones are used extensively for particle separation and classification in mineral, environmental and chemical engineering. The popularity of the hydrocyclone can be described to its simplicity, ease of operation, small physical size in proportion to the high volumetric throughputs that it can handle and the relatively low capital, operating and maintenance costs associated with its use.

Figure 1. General view of the hydrocyclone.

A typical hydrocyclone consists of a cylindrical section with a central tube connected to a conical section with a discharge tube (Figure 1). An inlet tube is attached to the top section of cylinder at the point of entry. Two exits, at the top (or overflow) and bottom (or underflow) of the cyclone yield the separation products together with a carried liquid. The fluid being injected tangentially into the hydrocyclone causes swirling within device, thus generating centrifugal force. Under the influence of different centrifugal forces, the heavy phase is pushed towards the wall and then leaves the hydrocyclone through the underflow; the light phase towards the cyclone axis, where it joins the up wards flow in a central vortex, leaves the cyclone through the overflow [1].

In recent years, alongside the traditional application of hydrocyclones in mineral processing, new applications, particularly in the field of environmental engineering, have opened up for these separators. Examples in this context include their application for gypsum separation in wet flue gas desulphurization processes, in washing plants for contaminated soils and in separation plants for tunnel driving projects. To ensure that the potential of hydrocyclone engineering is fully utilized, the optimum hydrocyclone geometry specific to the respective application, the correct combination of materials and appropriate operating parameters are essential. Extensive experiments in pilot scale are therefore required in the planning phase. The effort involved in this process can be reduced substantially with the help of modern computational engineering.

The flow behavior in hydrocyclone is quite complex. This complexity of flow processes has led designers to rely on empirical equations for predicting the equipment performance. These empirical relationships are derived from an analysis of experimental data and include the effect of operational and geometric variables. Different sets of experimental data lead to different equations for the same basic parameters. Empirical models correlate with classification parameter, such as the cut size with device dimensions and slurry properties [2].

The first successful work in predicting the fluid flow in hydrocyclones is the one of Pericleous and Rhodes [3], who used the PHOENICS computer code for the solution of the partial differential equations. Using the simple Prandtl mixing length model and the axisymmetry assumptions, the authors reported the velocity predictions in a 200-mm hydrocyclone. Later, Hsieh and Rajamani (1991) numerically solved the turbulent momentum equations to obtain the velocities and compared them with the Laser Doppler Velocimetry measurements in a 75-mm hydrocyclone. Numerical calculations of the separation of suspensions with different particle size distribution in the hydrocyclone show that feed solid concentration affects the separation parameters of the hydrocyclone [4].

The influence of the presence of an air column in a hydrocyclone and the regime of flow of particles through its discharge orifices on the separation characteristics of the hydrocyclone was investigated in paper [5, 6]. It was established that the flow of heavy particles through the lower discharge orifice of the hydrocyclone in the case where its upper discharge orifice is isolated from the atmosphere is larger than that in the case of free flow of particles to the atmosphere though both two orifices of the hydrocyclone. In the case where the lower discharge orifice of the hydrocyclone is isolated from the atmosphere, the involvement of light particles from the lower receiving bin into the reverse fluid flow in the axial zone of the hydrocyclone increases the resulting yield of particles through the upper discharge orifice of the hydrocyclone.

The majority of experimental and theoretical investigations use spherical particles. However, in a number practical situations particle shape deviates from spherical, either being irregular or having a well-defined geometry, such as granulates or fibers. The goal of this paper is to analyze the hydrodynamics, non-spherical particles motion and its separation in the hydrocyclone.

2. Mathematical model

The equations required for the description of the flow patterns in vortex devices express the time averaged fluid flow balance of mass and momentum. These equations are given here in cylindrical coordinates, which are best suited to the geometry of the conventional hydrocyclone. The parameters of the flow depend on the solid concentration and the density and viscosity of the slurry. It is assumed that particles concentrated suspension influence fluid velocities via density and viscosity. The turbulent Reynolds equations used for description of the flow fields, written in time-averaged variables have the following form:

$$\frac{\partial \rho u}{\partial x}+\frac{1}{r}\frac{\partial \rho v r}{\partial x}=0\,, \tag{1}$$

$$\frac{\partial \rho u^2}{\partial x}+\frac{1}{r}\frac{\partial \rho u v r}{\partial r}=-\frac{\partial p}{\partial x}+\frac{\partial}{\partial x}\left[\mu_{eff}\left(2\frac{\partial u}{\partial x}-\frac{2}{3}\left(\frac{\partial u}{\partial x}+\frac{1}{r}\frac{\partial v r}{\partial r}\right)\right)\right]+\frac{1}{r}\frac{\partial}{\partial r}\left[\mu_{eff}\, r\left(\frac{\partial u}{\partial r}+\frac{\partial v}{\partial x}\right)\right], \tag{2}$$

$$\frac{\partial \rho u v}{\partial x}+\frac{1}{r}\frac{\partial \rho v^2 r}{\partial r}=-\frac{\partial p}{\partial r}+\frac{\partial}{\partial x}\left[\mu_{\text{eff}}\left(\frac{\partial v}{\partial x}+\frac{\partial u}{\partial r}\right)\right]+$$
$$+\frac{1}{r}\frac{\partial}{\partial r}\left[\mu_{\text{eff}}\, r\left(2\frac{\partial v}{\partial r}-\frac{2}{3}\left(\frac{\partial u}{\partial x}+\frac{1}{r}\frac{\partial v r}{\partial r}\right)\right)\right]-2\frac{\mu_{\text{eff}}\, v}{r^2}+\frac{\rho w^2}{r}, \tag{3}$$

$$\frac{\partial \rho u w}{\partial x}+\frac{1}{r}\frac{\partial \rho v w r}{\partial x}=\frac{\partial}{\partial x}\left[\mu_{\text{eff}}\frac{\partial w}{\partial x}\right]+\frac{1}{r^2}\frac{\partial}{\partial r}\left[\mu_{r\varphi} r^3\frac{\partial}{\partial r}\left(\frac{w}{r}\right)\right]-\frac{\rho v w}{r}\,. \tag{4}$$

Here *u, v, w* are velocity components, p is pressure, x, r are the axial and radial co-ordinates, x, r are the axial and radial coordinate.

The turbulence characteristics were calculated on the basis of two-equation k - ε model (k being the turbulent energy, ε is the turbulence dissipation rate) with a Richardson numbers correction to the dissipation equation with the turbulent viscosity for the swirl equation reduced by a factor of $\sigma_{r\varphi}$:

$$\frac{\partial \rho u k}{\partial x}+\frac{1}{r}\frac{\partial \rho v k r}{\partial x}=\frac{\partial}{\partial x}\left[\frac{\mu_{\text{eff}}}{\sigma_k}\frac{\partial k}{\partial x}\right]+\frac{1}{r}\frac{\partial}{\partial r}\left[\frac{\mu_{\text{eff}}}{\sigma_k} r\frac{\partial k}{\partial r}\right]+G-\rho\varepsilon-D\,, \tag{5}$$

$$\frac{\partial \rho u \varepsilon}{\partial x}+\frac{1}{r}\frac{\partial \rho v \varepsilon r}{\partial x}=\frac{\partial}{\partial x}\left[\frac{\mu_{\text{eff}}}{\sigma_\varepsilon}\frac{\partial \varepsilon}{\partial x}\right]+\frac{1}{r}\frac{\partial}{\partial r}\left[\frac{\mu_{\text{eff}}}{\sigma_\varepsilon}\frac{\partial \varepsilon}{\partial r}\right]+(C_1-C_2\rho\varepsilon)\frac{\varepsilon}{k}+E\,, \tag{6}$$

$$G=\mu_t\left\{2\left[\left(\frac{\partial u}{\partial r}\right)^2+\left(\frac{\partial v}{\partial r}\right)^2+\left(\frac{v}{r}\right)^2\right]+\left(\frac{\partial u}{\partial r}+\frac{\partial v}{\partial x}\right)^2+\left(\frac{\partial w}{\partial x}\right)^2+\left(r\frac{\partial w/r}{\partial r}\right)^2\right\}, \tag{7}$$

$$D=2\mu\left[\left(\frac{\partial\sqrt{k}}{\partial x}\right)^2+\left(\frac{\partial\sqrt{k}}{\partial x}\right)^2\right], \tag{8}$$

$$E=2\frac{\mu_0\mu_t}{\rho}\left\{\left[\left(\frac{\partial^2 u}{\partial x^2}\right)+\frac{1}{r^2}\left(\frac{\partial}{\partial r}\left(r\frac{\partial u}{\partial r}\right)\right)\right]^2+\left[\left(\frac{\partial^2 v}{\partial x^2}\right)+\frac{1}{r^2}\left(\frac{\partial}{\partial r}\left(r\frac{\partial v}{\partial r}\right)\right)\right]^2+\right.$$
$$\left.+\left[\left(\frac{\partial^2 w}{\partial x^2}\right)+\frac{1}{r^2}\left(\frac{\partial}{\partial r}\left(r\frac{\partial w}{\partial r}\right)\right)\right]^2\right\}, \tag{9}$$

$$Ri_g=\frac{2\mu w}{G}\frac{\partial}{\partial r}\left(\frac{w}{r}\right), \qquad Ri_t=\frac{k^2}{\varepsilon^2}\frac{w}{r^2}\frac{\partial(wr)}{\partial r}\,. \tag{10}$$

Here the following designation is used: μ_{eff} is effective viscosity equal to the sum of the slurry μ and turbulent $\mu_t=C_\mu f_\mu \rho k^2\varepsilon^{-1}$ viscosities. The values of constants are: $C_\mu=0.09$, $C_1=1.44(1+C_3 Ri_g)$, $C_2=1.92(1-C_4\text{Ri}_t)$, $C_3=0.9$, $C_4=0.001$, $C_\mu=0.09$, $\sigma_k=1$, $\sigma_\varepsilon=1.3$, $\sigma_{r\varphi}=2.5$, $f_\mu=exp\left(-3.4/(1+0.02Re_t)^2\right)$, $Re_t=\rho k^2/\mu\varepsilon$.

From a mathematical point of view, the set of equations presented above is of an elliptical type and requires boundary conditions to be specified at all surfaces constraining the flow. Wall functions were employed to calculate the velocities and turbulence parameters in the vicinity of the wall. The boundary conditions at the inlet are specified once for all variables. To be able to treat the problem in two-dimensions, the fluid was assumed to be inlet into the cyclone chamber through a circumferential slot, the width of which was adjusted so that the flow rate and the inlet velocities matched the experimental conditions. The ratio of tangential to radial velocity component at the inlet was of the order 10. The inlet turbulent kinetic energy is assumed to be proportional to kinetic energy of the mean flow in the inlet tube. Zero gradient type conditions are assumed on the axis of symmetry of hydrocyclone for axial velocity, turbulent energy, dissipation and particles concentration, the tangential and radial components of velocity are zero by virtue of symmetry. On the walls of hydrocyclone velocity components equal zero. To calculate the turbulent parameters we assume the local equilibrium in the near-wall region. At the exit plane, for both overflow and underflow the gradients of the tangential velocity, turbulence parameters and particles concentrations where assumed to vanish in the axial directions, thus, having no upstream influence. For radial velocity here was assumed zero conditions. The pressure at the overflow exit was fixed according to the relation:

$$p(x_{of}, r) = \int_0^r \rho w^2 (x_{of}, \xi) d\xi \tag{11}$$

and at the underflow pressure was assumed equal the environmental pressure.

The set of the equations was solved with the use of an algorithm suggested by Patankar, where the finite difference equations where obtained by integrating the differential equations over scheduled volumes which incorporated points of a staggered finite-difference grids. The continuity equation was satisfied indirectly using an iterative method named PISO.

The trajectory of the discrete phase particle is obtained by integrating the force balance on the particle, which can be written in a Lagrangian reference frame. This force balance equates the particle inertia with the forces acting on the particle, and could be written as [7]:

$$\rho_p \frac{\pi d^3}{6} \frac{du_p}{dx} = F_{Dx} + \frac{\pi d^3}{6}(\rho_p - \rho)g, \tag{12}$$

$$\rho_p \frac{\pi d^3}{6} \frac{dv_p}{dx} = F_{Dr} + \frac{\pi d^3}{6}(\rho_p - \rho)\frac{w_p^2}{r}, \tag{13}$$

$$\rho_p \frac{\pi d^3}{6} \frac{dw_p}{dx} = F_{D\varphi} - \frac{\pi d^3}{6}(\rho_p - \rho)\frac{v_p w_p}{r}, \tag{14}$$

where F_{Dx}, F_{Dr}, $F_{D\varphi}$ – are the axial, radial and tangential components of the drag force.

We consider the ellipsoidal particle described by

$$\frac{\xi^2}{a_1^2} + \frac{\eta^2}{a_2^2} + \frac{\zeta^2}{a_3^2} = 1, \tag{15}$$

where a_1, a_2, a_3 denote the lengths of three semi-principal axes and ξ, η, ζ denotes the body-fixed coordinate system.

The drag force, acting on the particle, is related to the particle velocity by the resistance tensor **K** as: $F_{Di} = K_{ij}(v_j - v_{pj})$. For an axisymmetric particle with the symmetry axis given by the unit vector $\vec{n}$ this has the form $K_{ij} = K_1 n_i n_j + K_2(\delta_{ij} - n_i n_j)$.

Explicit results for the resistance coefficient K_1 for the motion parallel to the symmetry axis and K_2 for the motion transverse to the axis are given by the Happel and Brenner [11]. With $a_1 > a_2 = a_3$ and the aspect ratio $\lambda = a_1/a_2$, the resistance coefficients for a prolate spheroid ($\lambda > 1$) are

$K_1 = 8\pi\pi\mu_2\tau^3\left[(2\tau^2+1)\ln(\lambda+\tau)-\lambda\tau\right]^{-1}$, $K_2 = 16\pi6\pi_2\tau^3\left[(2\tau^2-1)\ln(\lambda+\tau)+\lambda\tau\right]^{-1}$,

where $\tau = \sqrt{|\lambda^2 - 1|}$.

For an oblate spheroid with $\lambda < 1$:

$K_1 = 8\pi\pi\mu_2\tau^3\left[\left(2\tau^2-1\right)\arctan(\lambda/\tau)+\lambda\tau\right]^{-1}$, $K_2 = 16\pi6\pi_2\tau^3\left[\left(2\tau^2-1\right)\arctan(\lambda/\tau)-\lambda\tau\right]^{-1}$,

The equation of the rotational motions for spheroid is:

$$I\frac{d\vec{\omega}_p}{dt}-(I\cdot\vec{\omega})\times\vec{\omega}=-\vec{T}. \quad (16)$$

The relation between the external torque $\vec{T}$, the angular velocity $\vec{\omega}_p$ of the spheroid and the ambient .ow is given by:

$$T_i=\mu\left[R_1\left(\omega_{pj}-\omega_j\right)n_in_j+R_2\left(\left(\omega_{pi}-\omega_i\right)-\left(\omega_{pi}-\omega_i\right)n_in_j\right)\right]-\mu R_2D\varepsilon_{ijk}E_{kl}n_jn_l \quad (17)$$

where $D=\left(\lambda^2-1\right)/\left(\lambda^2+1\right)$, ω is the locally uniform vorticity, and E_{kl} is the locally uniform rate of strain. For a prolate spheroid:

$$R_1=\frac{16}{3}\pi a_2^3\tau^3\left[\lambda\tau-\ln(\lambda+\tau)\right]^{-1}, \qquad R_2=\frac{16}{3}\pi a_2^3\tau^3\left[\left(2\tau^2+1\right)\ln(\lambda+\tau)-\lambda\tau\right]^{-1}. \quad (18)$$

For an oblate spheroid:

$$R_1=\frac{16}{3}\pi a_2^3\tau^3\left[\arctan(\lambda/\tau)-\lambda\tau\right]^{-1}, \quad R_2=\frac{16}{3}\pi a_2^3\tau^3\left[\left(2\tau^2-1\right)\arctan(\lambda/\tau)+\lambda\tau\right]^{-1}. \quad (19)$$

3. Results and Discussions

The calculations were performed for a hydrocyclone with a chamber of diameter 75 mm, and diameters of underflow and overflow 12.5 mm and 25.0 mm respectively. The lengths of the cylindrical and conical parts were 100 mm and 200 mm.

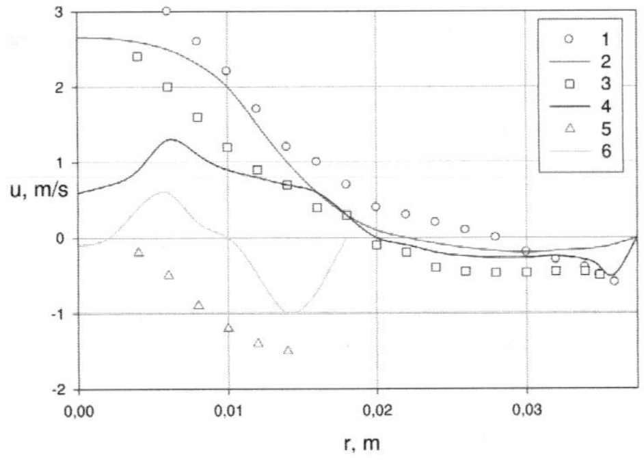

Figure 2. Radial distribution of axial velocity: 1, 3, 5 – experiment, 2. 4, 6 – calculations, 1, 2 – $x = 60$ mm, 2, 4 – $x = 120$ mm, 3, 4 – $x = 200$ mm.

On the Figure 2 one can see the calculated axial velocity distribution. It is obvious that axial velocity was positive in internal, meaning that fluid flow is upwards. In external part of the hydrocyclone axial velocity was negative, meaning that fluid flow is downwards.

The flow field in the cyclone indicates the expected forced/free combination of the Rankine vortex. The value of the tangential velocity equals zero on the wall and the centre of the flow field (air core is a forced vortex). The tangential velocity decreases as the flow spins down along the wall. Before it goes below the vortex finder, the fluid flow collides with the follow-up flow and forms a chaotic flow close to the vortex finder outside wall.

When calculating trajectories of non-spherical particles, both translational and rotational motions should be simultaneously considered, since these motions are coupled. The orientation distribution of non-spherical particles in flows is an important factor affecting the cumulative transport properties of the fluid–particle system. Lateral drift was found to occur for inertial particles with both fixed and free orientations. For a particle with a fixed orientation the transverse motion also strongly depends on the particle orientation. The fine particles flow rate through the underflow is equal approximately to the liquid flow rate, but almost all particles with diameter more then 20 microns leave the hydrocyclone through the underflow.

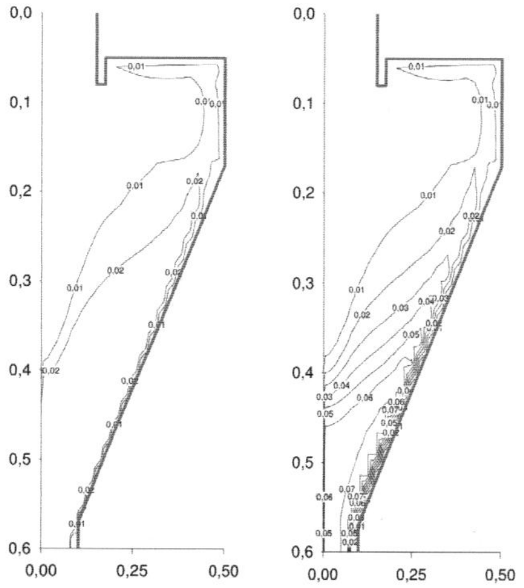

Figure 3. Particles concentration in the hydrocyclone:
$\rho_p = 1200$ kg/m^3, $d_p = 20\,\mu$m; a – $\lambda = 0.6$, b – $\lambda = 1.0$.

The decrease of the magnitude of the sphericity coefficient $\lambda = a_1/a_2$ leads to diminishing of the particles relative velocity. Therefore medium particles in the conical part of hydrocyclone cannot reach the vicinty of the wall and get with the upstream vortex in the overflow. The coarse nonspherical particles have a reduced radial displacement from the centerline. So the concentration of coarse non-spherical particles in the underflow decreasing (Figure 3).

This is mechanism for decrease the separation efficiency of the non-spherical particles. The calculations of the cut size in several hydrocyclone geometries and also particles trajectories are calculated for both inertial and inertialess particles, the orientation of which is fixed or free. The results demonstrate that separation characteristics of the hydrocyclone are strongly depends on the flow shear rate, particle size, and particle shape.

4. Conclusions

The flow in cyclones has been investigated k−ε model modified for streamline curvature effect and two Reynolds stress models. Using this model led to results that were in good agreement with experimental results. The results demonstrate that separation characteristics of the hydrocyclone are strongly dependent on the flow shear rate, particle size, and particle shape.

References

[1] Svarovsky L 1984 *Hydrocyclones* 1997-1998 (London: Technomic Publishing Co.)
[2] Fahlstrom, P H 1963 Studies of the hydrocyclone as a classifier *Proc. 6th Int. Miner. Process. Congr.* (Cannes) 211
[3] Pericleous K A, Rhodes N and Cutting G W 1984 A mathematical model for predicting the flow field in a hydrocyclone classifier *Proc. Int.Conf. Hydrocyclones* (Bath, England) 765
[4] Monredon T C, Hsieh K T and Rajamani R K 1990 Fluid flow model of the hydrocyclone: an investigation of device dimensions *Int. J. Miner. Process* **35** 143
[5] Dueck J, Matvienko O V and Neesse Th 2000 *Theor Found Chem Eng+* **34** 478
[6] Matvienko O Dueck J 2006 *Theor Found Chem Eng+* **40** 216
[7] Happel J and Brenner H 1986 *Low Reynolds Number* (Dordrecht: Martinus Nijhoff)

IOP Conf. Series: Materials Science and Engineering **71** (2015) 012036 doi:10.1088/1757-899X/71/1/012036

Study of the average heat transfer coefficient at different distances between wind tunnel models

A Gnyrya[1], S Korobkov[1], D Mokshin[1] and A Koshin[1]
[1]Tomsk State University of Architecture and Building, Construction Engineering Technology Department, Tomsk, 634003, Russia

E-mail: mokshin@sibmail.com

Abstract. The paper presents investigations of physical and climatic factors with regard to design and process variables having effect on heat transfer in the building model system at different distances between them in the airflow direction. The aim of this work is to improve energy efficiency of exterior walls of buildings. A method of physical simulation was used in experiments. Experimental results on the average values of the heat transfer coefficient in the building model system are presented herein. A series of experiments was carried out on a specific aerodynamic test bench including a subsonic wind tunnel, heat models and devices for giving thermal boundary conditions, transducers, and the record system equipment. The paper contains diagrams of the average heat transfer distribution at fixed Reynolds number and the airflow angle of attack; the average values of the heat transfer coefficient for each face and wind tunnel models as a whole at maximum, medium, and large distances between them. Intensification of the average heat transfer was observed on the downstream model faces depending on the distance between models.

1. Introduction

Researchers of the Construction Engineering Technology Department of Tomsk State University of Architecture and Building (Tomsk, Russia) are involved in a variety of problems dealing with the airflow structure [1], and the local and the average values of the heat transfer coefficient [2] with the aim to improve the design technique and energy efficiency of exterior walls.

In the literature, there are papers devoted to the study of heat transfer of bluff bodies [3–15]. Comparison of these data on the average heat transfer indicates a significant (more than 100%) variation in the given data.

All experiments on the average value of heat transfer coefficient were carried out using the subsonic open-return wind tunnel with air sucked through a duct. The experimental methodology and processing of wind tunnel test results are described in the work of Mokshin *et al.* [16].

In this study, the key parameter is calibration of wind tunnel models expressed by the distance ratio $L1/a$ between models in the light ($L1$) and the cross-sectional dimension of the wind tunnel model (a) equal to 50 mm.

2. Experimental

Figure 1 a, b shows schematic layouts of the test Model 2 relative to the test Model 1 at the airflow angle of attack $\varphi = 0^{\circ}$.

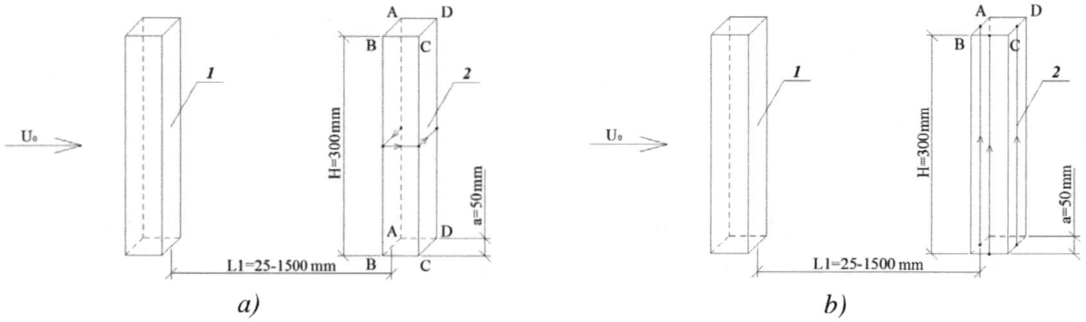

a) horizontal section; b) vertical section

Figure 1. Schematic layouts of the test Model 2 relative to the test Model 1 at the airflow angle of attack $\varphi = 0^{\circ}$.

Figures 2 – 4 contain plots of the average values of the heat transfer on faces depending on distance *Ll/a* between wind tunnel models at the airflow angle of attack $\varphi = 0^{\circ}$. Also, for comparison, these Figures present data for $Ll / a \to \infty$ in the plane case of the single-model airflow [2].

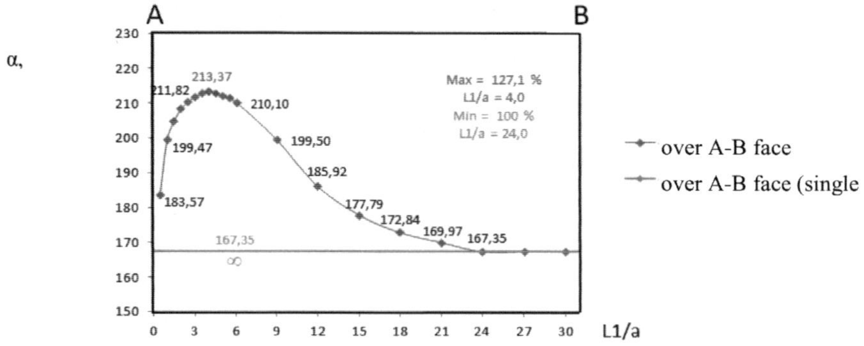

Figure 2. Average heat transfer distribution on *(A – B)* face of Model 2 at the increase of distance ratio *Ll/a* at $\varphi = 0^{\circ}$, $Re = 4{,}25 \cdot 10^{4}$.

At the increase of distance ratio *Ll/a* from 0,5 to 4,0 the average heat transfer on *(A – B)* face of Model 2 also increases. At a maximum short distance of *Ll/a* = 0,5 the average heat transfer on *(A–B)* face of Model 2 is 8,6 % higher than on the same face of Model 1. This is because the flow separation occurred at the top of *(A – B)* face of Model 2 and the arch-type vortex formed by Model 1. Their maximum effect on *(A – B)* face of Model 2 was recorded at *Ll/a* = 4,0 [1] when the value of the heat transfer coefficient achieved its maximum, that is 27,1 % higher than on *(A – B)* face of Model 1. At distance ratio ranging between 4,0 and 24,0 the flow separation force at the top of *(A – B)* face of Model 2 decreases as well as the effect from the arch-type vortex formed by Model 1. The average value of the heat transfer coefficient on *(A – B)* face of Model 2 is reduced and thereby approximates to that of *(A – B)* face of Model 1.

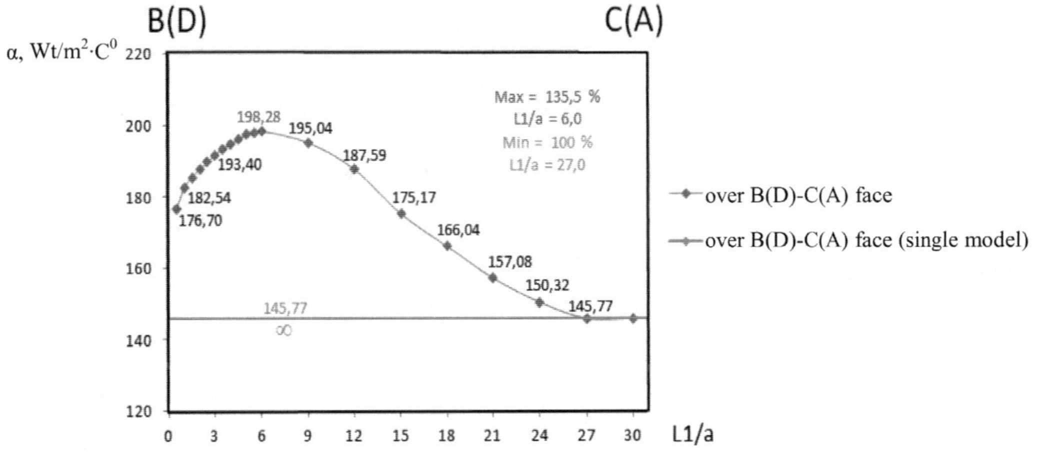

Figure 3. Average heat transfer distribution on *(B – C)* and *(D – A)* faces of Model 2 at the increase of distance ratio *L1/a* at $\varphi = 0^{\circ}$, $Re = 4{,}25 \cdot 10^4$.

Since in Figure 3 the average heat transfer distribution over faces (*B – C*) and (*D – A*) is identical, the description below will refer to face (*B – C*) only.

At distance ratio $L1/a$ ranging between 0,5 and 6,0 a gradual increase of the average heat transfer is observed on (*B – C*) face of Model 2. At $L1/a = 0{,}5$ the average value of the heat transfer coefficient on (*B – C*) face of Model 2 is 21,3 % higher than on the same face of Model 1. It can be explained by the fact that at a maximum short distance between the models, recirculation regions formed by Models 2 and 1 and the flow separation formed at the top of *(A – B)* face of Model 2 noticeably affect the bottom of (*B – C*) face. At distance ratio $L1/a = 6{,}0$ these forces have maximum effect on (*B – C*) face of Model 2. Average values of the heat transfer coefficient on (*B – C*) face of Model 2 increase by 35,5 % as compared to that ones obtained from (*B – C*) face of Model 1. At a distance between the wind tunnel models ranging from 6,0 to 27,0 the effect from recirculation regions and the flow separation on (*B – C*) face of Model 2 is reduced. The average value of the heat transfer coefficient on (*B – C*) face of Model 2 also decreases, thereby approximating to that of (*B – C*) face of Model 1.

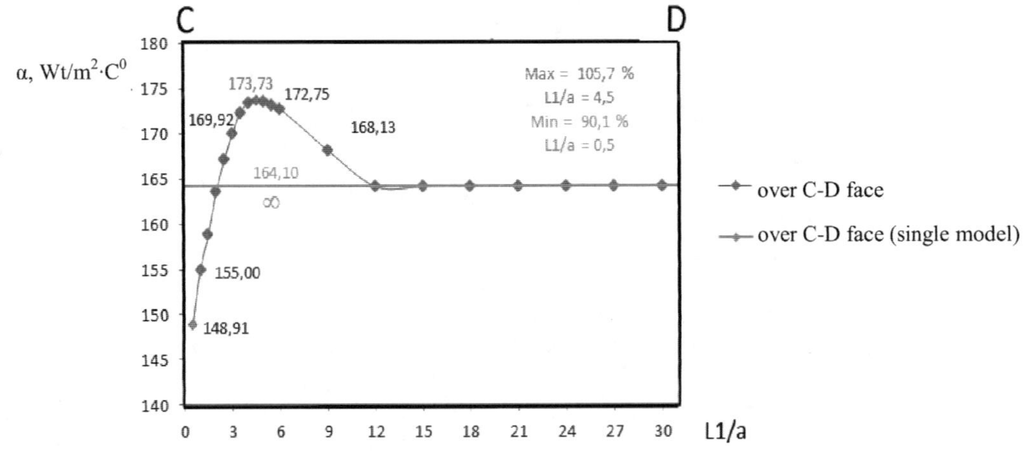

Figure 4. Average heat transfer distribution on *(C – D)* face of Model 2 at the increase of distance ratio *L1/a* at $\varphi = 0^{\circ}$, $Re = 4{,}25 \cdot 10^4$.

IOP Conf. Series: Materials Science and Engineering **71** (2015) 012036 doi:10.1088/1757-899X/71/1/012036

At a distance between the wind tunnel models ranging from 0,5 to 2,0 a sharp increase of the average value of the heat transfer coefficient on $(C-D)$ face of Model 2 is observed, the value of which is lower than on the same face of Model 1. At distance ratio $L1/a = 0{,}5$ the heat transfer coefficient of $(C-D)$ face of Model 2 is 9,9 % lower than on the same face of Model 1. At a maximum short distance, the arch-type vortex formed by Model 2 is less intensive than that one formed by Model 1. In increasing the distance ratio up to 2,0 the arch-type vortex formed by Model 2 promotes a stronger effect on Model 2. At that, average values of the heat transfer coefficient on $(C-D)$ face of Model 2 are practically the same (0,46 % difference) as on $(C-D)$ face of Model 1.

At a distance between the wind tunnel models ranging from 2,0 to 4,5 average values of the heat transfer coefficient on $(C-D)$ face of Model 2 are being sharply increased, and become higher than on the same face of Model 1. At distance ratio $L1/a = 4{,}5$ the heat transfer coefficient of $(C-D)$ face of Model 2 is 5,7 % higher than on the same face of Model 1. At this distance, the intensity of the arch-type vortex formed by Model 2 exceeds the similar arch-type vortex formed by Model 1.

At a distance between the wind tunnel models ranging from 4,5 to 12,0 the intensity of the arch-type vortex formed by Model 2 on its $(C-D)$ face becomes lower. The average value of the heat transfer coefficient on $(C-D)$ face of Model 2 is reduced, thereby approximating to that of $(C-D)$ face of Model 1.

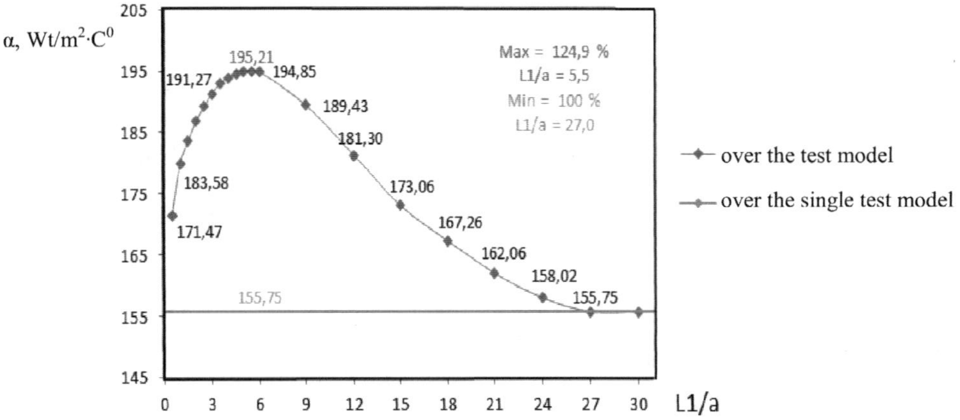

Figure 5. Average heat transfer distribution over Model 2 at the increase of distance ratio $L1/a$ at $\varphi = 0^{\circ}$, $Re = 4{,}25 \cdot 10^{4}$.

At a distance between the wind tunnel models ranging from 0,5 to 5,5 the average value of the heat transfer coefficient increases on the entire Model 2. At a maximum short distance $L1/a = 0{,}5$ the heat transfer coefficient on Model 2 is 9,7 % higher than on Model 1. This is because Model 2 is affected by arch-type vortices formed by Model 1 and the strong flow separation occurred at the top of Model 2. The ultimate effect from these phenomena is recorded at a distance ratio $L1/a = 5{,}5$ when the average value of the heat transfer coefficient achieves its maximum, i.e. 24,9 % higher than on Model 1.

At a distance between the wind tunnel models ranging from 5,5 to 27,0 these forces become lower, and the average value of the heat transfer coefficient on Model 2 is reduced, thereby approximating to that of Model 1.

The diagrams of the average heat transfer for each face of Model 2 are similar to those of Model 1 at different model calibrations, namely: $L1/a = 24{,}0$ for $(A-B)$ face; $L1/a = 27{,}0$ for $(B-C)$ face; $L1/a = 12{,}0$ for $(C-D)$ face. Average values of the heat transfer coefficient on Model 2 and Model 1 are identical at a distance ratio of 27,0 and larger.

This identity is being observed at the subsequent increase of the distance ratio up to $L1/a = 30{,}0$.

Figure 6 demonstrates the summary plot of the average heat transfer distribution over Model 2 faces at the increase of distance ratio $L1 / a$ at $\varphi = 0^{\circ}$, $Re = 4{,}25{\cdot}10^{4}$.

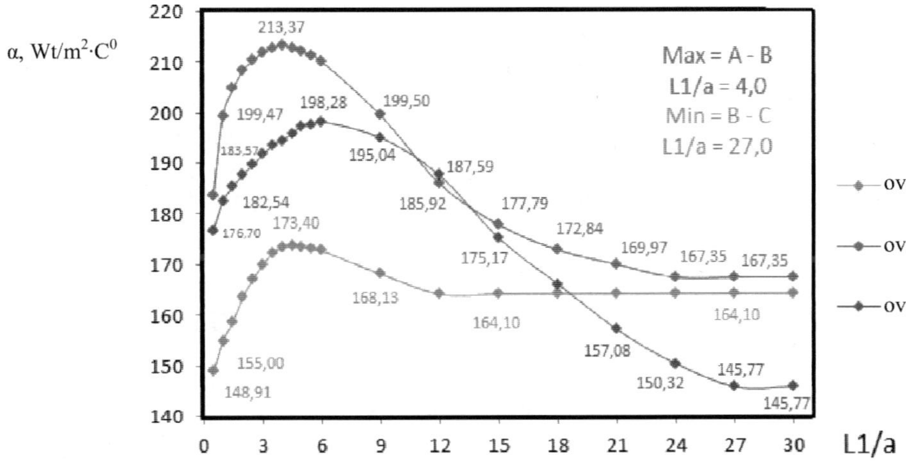

Figure 6. Average heat transfer distribution on Model 2 faces at the increase of distance ratio $L1 / a$ at $\varphi = 0^{\circ}$, $Re = 4{,}25{\cdot}10^{4}$.

As shown in Figure 6, at the increase of distance ratio $L1 / a$ from 0,5 to 30 the airflow behavior is changed between the wind tunnel models and on lateral faces of Model 2. This leads to the intensification of the average heat transfer on Model 2 faces, while the distance between wind tunnel models increases. At the same time, in varying the distance ratio $L1/a$ not more than by 10%, heat transfer on $(C - D)$ face of Model 2 changes insignificantly.

3. Conclusions

1. At a maximum short distance ranged from 0,5 to 6,0 the clear-cut distinction between the average values of the heat transfer coefficient on Model 2 faces was observed. The maximum average heat transfer was on $(A - B)$ face of Model 2, while the minimum was on its $(C - D)$ face. As for lateral faces $(B - C)$ and $(D - A)$ of Model 2, the heat transfer value varied between those on $(A - B)$ and $(C - D)$ faces of Model 2.
2. At a medium distance $L1 / a$ ranged between 6,0 and 21,0 the average heat transfer was more intensive on Model 2 faces. At first, the heat transfer on $(B–C)$ and $(D–A)$ faces of Model 2 at distance ratio $L1 / a$ = 6,0 was average as compared to faces $(A–B)$ and $(C–D)$ of Model 2, however, at $L1 / a$ increase up to $12{,}0 \div 15{,}0$ it achieved its relative maximum. Then, at $L1 / a$ increase from 15,0 to 21,0 the average value of heat transfer lowered sharply, and its contribution to the total heat transfer on the entire Model 2 became minimum.
3. At a maximum distance $L1 / a$ ranged from 21,0 to ∞, the clear-cut distinction between the average values of the heat transfer coefficient was observed on Model 2 faces. The maximum average heat transfer was on $(A - B)$ face of Model 2, while the minimum was on its lateral faces $(B - C)$ and $(D - A)$. On face $(C - D)$ of Model 2, it was lower than on face $(A - B)$ of Model 2 (some 3,0 % difference).

Acknowledgements
The authors acknowledge the financial support from the Grant Council of the Russian Foundation for Basic Research within the Project N 13-08-00505-a. The authors also acknowledge Mr. Terekhov V.I. and Mrs. Yarygina N.I. from Kutateladze Institute of Thermophysics, Siberian Division of the Russian Academy of Sciences, Novosibirsk, Russia, for great assistance and helpful comments.

References

[1] Гныря А И, Коробков С В, Кошин А А, Мокшин Д И и Терехов В И 2011 Комплексные экспериментальные исследования аэродинамики и теплообмена моделей зданий и сооружений *Вестник ТГАСУ* **4** 113–126

[2] Мокшин Д И и Коробков С В 2014 Исследование среднего теплообмена отдельностоящих квадратных призм *Фундаментальные и прикладные исследования: проблемы и результаты* **13** 196–202

[3] Meinders E R, Van Der Meer T H and Hanjalic K 1997 Local convective heat transfer from an array of wall-mounted cubes *Int. J. Heat Mass Transfer* 335–346

[4] Meinders E R and Hanjalic K 1999 Vortex structure and heat transfer in turbulent flow over a wall-mounted matrix of cubes *Int. J Heat Mass Transfer* 255–267

[5] Nakamura H, Igarashi T and Tsutsui T 2001 Local heat transfer around a wall-mounted cube in *the turbulent boundary layer Int J Heat Mass Tran **44** 3385–3395*

[6] Meinders E R and Hanjalic K 2002 Experimental study of the convective heat transfer from in-line and staggered configurations of two wall-mounted cubes *Int J Heat Mass Tran* 465–482

[7] Yakhot A, Liu H and Nikitin N 2006 Turbulent flow around a wall-mounted cube: A direct numerical simulation *Int J Heat Fluid Fl* **27** 994−1009

[8] Rathnam G S and Vengadesan S 2008 Performance of two equation turbulence models for prediction of flow and Heat transfer over a wall mounted cube *Int J Heat Mass Tran* **51** 2834−2846

[9] Defraeye T, Blocken B and Carmeliet J 2010 CFD analysis of convective heat transfer at the surface of a cube immersed in a turbulent boundary layer *Int J Heat Mass Tran* **53** 297−308

[10] Terekhov V I, Kalinina S V, Mshvidobadze Yu M and Sharov K A 2009 Impingement of an impact jet onto a spherical cavity. Flow structure and heat transfer **52** 2498-2506

[11] Богатко Т В, Терехов В И и Халатов А А 2012 Структура течения и теплообмен при турбулентном обтекании одиночных преград различной формы в трубе *Тепловые процессы в технике* **4** 146-155

[12] Terekhov V I and Pakhomov M A 2009 Predictions of turbulent flow and heat transfer in gas-droplets flow downstream of a sudden pipe expansion *Int. J. Heat Mass Transfer* **52** 4711-4721

[13] Terekhov V I , Yarygina N I and Zhdanov R F 2003 Heat transfer in turbulent separated flows in the presence of high free-stream turbulence *Int. J. Heat Mass Transfer* **46** 4535-51

[14] Терехов В И, Смульский Я И и Шаров К А 2012 Интерференция отрывных потоков за обратным уступом при наличии пассивного управления *Письма в Журн. техн. физики* **38** 46-53

[15] Богатко Т В и Терехов В И 2011 Влияние тепловой предыстории на турбулентное отрывное течение при внезапном расширении трубы *Теплофизика и аэромеханика* **18** 225-232

[16] Мокшин Д И и Коробков С В 2014 Методика проведения и обработки экспериментов по исследованию локальной и средней теплоотдачи зданий и сооружений *Наука и современность* **31** 112–122

Strength and deformability of compressed concrete elements with various types of non-metallic fiber and rods reinforcement under static loading

A V Nevskii[1], I V Baldin[1] and K L Kudyakov[1]
[1]Tomsk State University of Architecture and Building, Department of Reinforced Concrete and Masonry Structures, Tomsk 634003, Russia

E-mail: andrenevsky@gmail.com

Abstract. Adoption of modern building materials based on non-metallic fibers and their application in concrete structures represent one of the important issues in construction industry. This paper presents results of investigation of several types of raw materials selected: basalt fiber, carbon fiber and composite fiber rods based on glass and carbon. Preliminary testing has shown the possibility of raw materials to be effectively used in compressed concrete elements. Experimental program to define strength and deformability of compressed concrete elements with non-metallic fiber reinforcement and rod composite reinforcement included design, manufacture and testing of several types of concrete samples with different types of fiber and longitudinal rod reinforcement. The samples were tested under compressive static load. The results demonstrated that fiber reinforcement of concrete allows increasing carrying capacity of compressed concrete elements and reducing their deformability. Using composite longitudinal reinforcement instead of steel longitudinal reinforcement in compressed concrete elements insignificantly influences bearing capacity. Combined use of composite rod reinforcement and fiber reinforcement in compressed concrete elements enables to achieve maximum strength and minimum deformability.

1. Introduction

Development of construction industry involves introduction of new technologies and modern materials in design of new-built buildings and structures of various purposes. Studying possibility of integration of modern high-strength materials in structures is one of the important issues to improve design solutions. Non-metallic materials based on fibers are of particular interest as their use is already proved in the military, aerospace engineering and in the manufacture of aircrafts, ships, cars and other industrial products. The advantages of such materials include high strength properties and low weight, corrosion resistance, chemical and magnetic inertia.

Several types of raw building materials are produced based on fibers: fiber, obtained by cutting the filament used as the fiber (dispersed) reinforcement for the manufacture of fiber-reinforced concrete; composite rebar, which is non-metallic rods of continuous filament, bonded by polymer adhesive composition, which may be used as a concrete rod reinforcement; fibrous webs, which are used for enhancing various concrete structures. The production volume of such materials is increasing in different countries while scientists conduct studies to gather new data on the materials properties [1]. Research [2] has shown areas of their effective applicability in construction.

However, there are still many open issues in the field of application of these materials in manufacture of building structures on the basis of concrete [3].

2. Materials

For experimental research the following raw materials, based on non-metallic fibers were selected: basalt and carbon fiber for manufacture of fiber-reinforced concrete and composite rebar based on glass and carbon fibers used as rod reinforcement.

Preliminary testing of composite reinforcement under axial compression demonstrated that its compressive strength may reach values up to 650 MPa. Moreover, different types of composite reinforcement have high relative strain, higher values of deformations than of the steel reinforcement. Composite rebar has relatively low modulus of elasticity under compression compared with steel reinforcement. Fracture behavior of composite rebar based on carbon fibers is "bulging" of external fibers from the central compressive load and the scheme of destruction of composite reinforcement based on glass fibers has a shape of the shear.

Preliminary tests of fiber-reinforced concrete on basis of non-metallic fibers under axial compression showed that fiber non-metallic reinforcement of concrete increases its compressive strength and reduces deformability. After adding basalt or carbon fiber into concrete compressive strength increased up to 51%. This result was obtained using the developed technology of producing fiber-reinforced concrete [4,5].

3. Experimental procedure

The stress-strain state of compressed concrete elements with various types of non-metallic rod and fiber reinforcement has been studied experimentally. Program of experimental research included design, manufacture and compression testing of five concrete samples under static loading. For fiber reinforcement two types of non-metallic fibers were used: basalt and carbon. Composite reinforcement on the basis of glass and carbon fibers was used as non-metallic rod reinforcement. Experimental samples are concrete and fiber reinforced concrete elements with geometric dimensions of 100*100*1000 mm, reinforced with four longitudinal rods of metal or non-metallic reinforcing:

1) Concrete sample with metal longitudinal rod reinforcement, diameter of rods 10 mm Class A400 (standard compressive strength of 400 MPa) [6];

2) Concrete sample with non-metallic longitudinal reinforcement with rods of composite fiberglass reinforcement (GFRP) brand MONSTEROD (production of holding company "Composite", Moscow) nominal diameter of 9,4 mm;

3) Concrete sample with non-metallic longitudinal reinforcement with rods of composite carbon fiber reinforcement (CFRP) brand FibARMRebar (production of holding company "Composite", Moscow) nominal diameter of 9,6 mm;

4) Basalt fiber-reinforced concrete (BFRC) sample with non-metallic longitudinal reinforcement with bars of the CFRP brand FibARMRebar nominal diameter of 9,6 mm;

5) Carbon fiber-reinforced concrete (CFRC) sample with non-metallic longitudinal reinforcement with bars of the CFRP brand FibARMRebar nominal diameter of 9,6 mm.

Reinforcement of experimental samples in the transverse direction was made of rebar class Bp500 [4], diameter of 3 mm, located in pace of 75 mm. In abutting areas there were grids of confinement reinforcement arranged at a pitch of 20 mm to the height size of the sample bearing surface area. Fine grained concrete was used. The marking of the experimental samples was done in accordance with their design as shown in Table 1. Test program is shown in Figure 1.

Table 1. Marking of the experimental samples.

Identifier of the sample	Number of samples, units	Quantity, type and nominal diameter of the longitudinal bars	Percentage of cross-sectional reinforcement, %	Type of fiber, which used for fiber-reinforcement of concrete	The percentage of fiber reinforcement (by weight of binder in concrete), %
CS-1	1	4xØ10 mm A-400	3,14	-	-
CS-2	1	4xØ9,4 mm GFRP	2,89	-	-
CS-3	1	4xØ9,6 mm CFRP	2,77	-	-
CS-4	1	4xØ9,6 mm CFRP	2,77	Basalt fiber	0,5
CS-5	1	4xØ9,6 mm CFRP	2,77	Carbon fiber	0,2

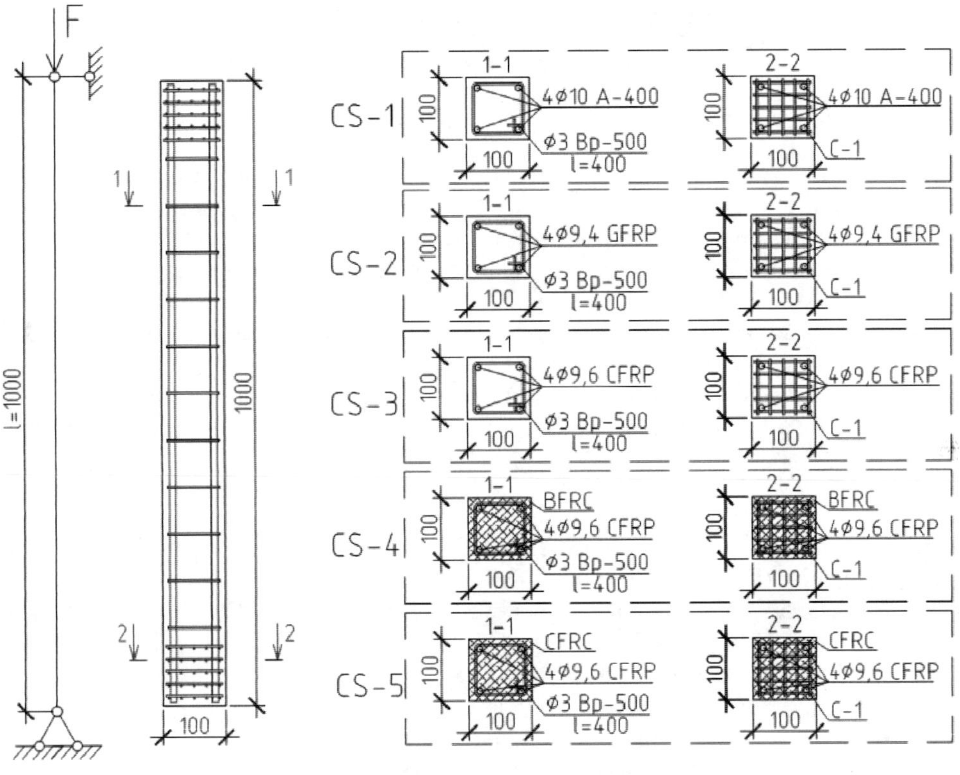

Figure 1. Design of experimental research.

Experimental samples were tested in accordance with the requirements of National State Standard GOST 8829-94 by static compressive load up to a limiting condition. Tests were carried out on an automated press PMM-250, equipped with advanced displacement sensors and electronic data recording. The loading was performed by the stages of 20 kN. After application of the load each stage was followed by a pause of 8 ... 10 minutes, during this pause redistribution of the internal stresses in the samples occurred. At this time the sample was examined, the formation and development of cracks was recorded. After holding the structure under the load the readings of devices were taken. The compressed samples were destroyed, while the value of the maximum efforts was recorded. General view of testing samples of compressed elements is given in Figure 2.

Figure 2. General view of compression testing of concrete elements.

The following measurements were taken during tests. Longitudinal strain of reinforcement was recorded by resistance-strain gauge mounted at the stage of preparing samples on each of the four longitudinal bars in the middle of their length. Longitudinal and transverse deformations of concrete and fiber-reinforced concrete were also recorded by resistance-strain gauges mounted on the body of each of the sample pairs in mutually perpendicular directions along two adjacent sides in the middle of them. General longitudinal deformations of the samples were recorded by inductive displacement sensors mounted on two opposite sides of each of the samples.

The testimonies of electronic sensors were read through the measuring and computing complexes MIC-400D and MIC-300M. Force at each stage was recorded by an electronic display of the pressure machine.

4. Results

As a result of the tests performed schemes of crack formation and destruction of samples under static loading were obtained (Figure 3). The destruction of the samples was accompanied by crushing of concrete in all tests. Breaking load for samples were respectively: for CS-1 (A400) 360 kN, for CS-2 (GFRP) 339 kN, for CS-3 (CFRP) 347 kN, for CS-4 (BFRC, CFRP) 413 kN for CS-5 (CFRC, CFRP) 428 kN. Also, as a result of statistical treatment of readings of devices for each of the sample, the graph of longitudinal deformation of concrete was obtained (Figure 4).

The experimental data analysis showed that the bearing strength at the compression of concrete samples with longitudinal reinforcement with composite fiber reinforced glass and carbon rods (CS-2 and CS-3) is smaller than the bearing capacity of the sample reinforced with steel reinforcement rods class A400 (CS-1) by 5.8% and 3.6% respectively. The bearing capacity of samples of the basalt fiber-reinforced concrete (CS-4) and carbon fiber-reinforced concrete (CS-5) with a longitudinal reinforcement of composite carbon rods is more than load-carrying capacity of the sample of concrete (CS-3) and the remaining samples (CS-1 and CS-2).

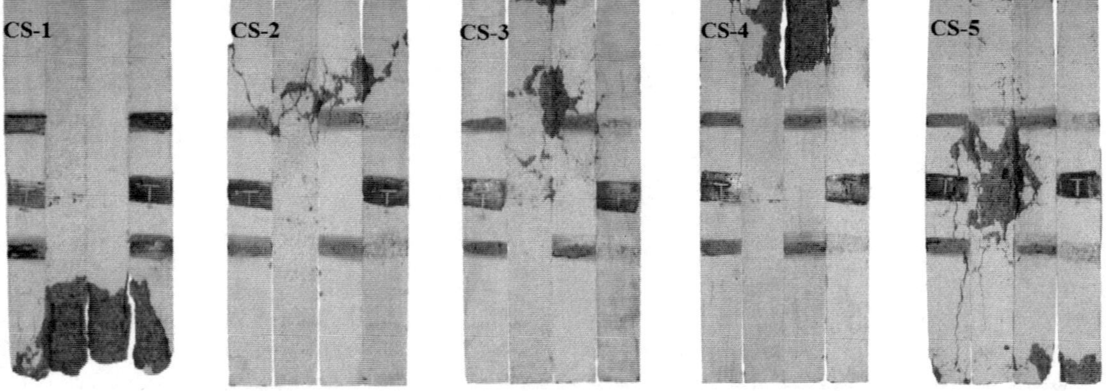

Figure 3. Schemes of destruction and crack formation of compressed concrete elements.

Using basalt fiber for disperse reinforcement of compressed elements with carbon rod longitudinal reinforcement allowed to increase the carrying capacity by 19%, and with the use of carbon fiber it allowed to increase the carrying capacity by 23%. The reinforced-concrete sample of compressed element with composite rod reinforcement showed higher deformability: for sample with fiberglass reinforcement with longitudinal rod 142%, for the sample with carbon fiber-reinforced with longitudinal rod 133%.

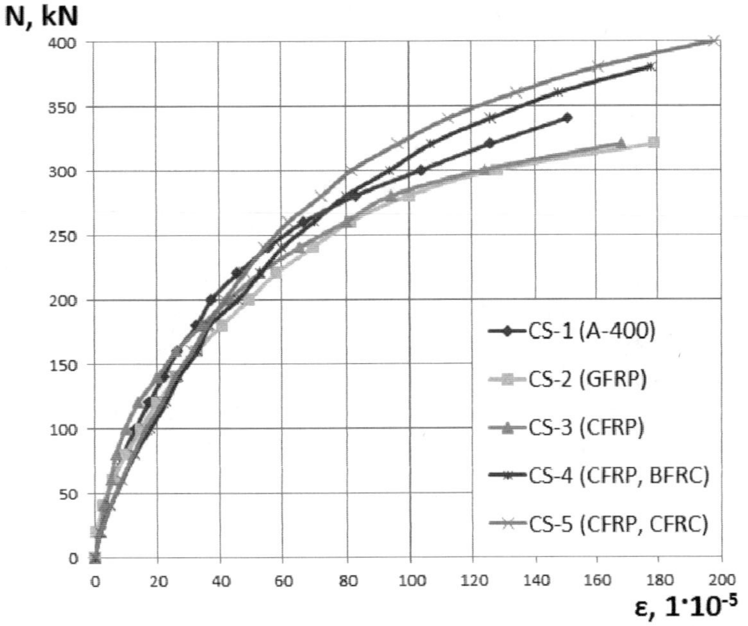

Figure 4. Graph of longitudinal deformation of the compressed concrete elements.

Disperse reinforcement of concrete by basalt and carbon fiber contributed to reduction of deformability of the compressed concrete elements as a whole. It was also noted that the composite rebar and concrete (fiber-reinforced concrete) operate under strain compatibility.

5. Conclusion

According to the research, it was found that the fiber reinforcement of concrete enables increase in the carrying capacity of the compressed concrete elements and reduces their deformability. Despite the high rate of tensile strength of composite reinforcement its strength properties under compression are not realized completely due to the high marginal relative deformations in compression. In this regard, the composite rod reinforcement of compressed elements is advisable to use together with the dispersed fiber-reinforcement.

Further scientific research is required to study the possibility of joint application of rod composite reinforcement and fiber dispersed reinforcement in production of building structures.

Acknowledgement

The research performed with the financial support of the Ministry of Education and Science of the Russian Federation within Governmental Decree No.218 of “Measures of state support for cooperation of Russian higher education institutions and organizations, for implementation of complex projects of high-tech production”, integrated project No. 02.G25.31.0022, “Production technologies of energy- and resource-efficient housing of economy class based on the universal framed prefabricated architectural and construction system”.

References

[1] Almerich Chulia A, Martin Concepcion P, Molines Cano J M and Rovira Soler J 2012 New GFRP bars as internal reinforcement in concrete Structures *Proc. 15th European Conf. on Composite Materials* (Venice)

[2] De Luca A, Matta F and Nanni A 2010 Behavior of full-scale glass fiber-reinforced polymer reinforced concrete columns under axial load *ACI Structural Journal* **107** 589-596

[3] Степанова В Ф, Степанов А Ю и Жирков Е П 2013 *Арматура композитная полимерная* (Москва: АСВ) 200

[4] Кудяков К Л, Невский А В и Ушакова А С 2013 Влияние дисперсного армирования базальтовым волокном на прочностные свойства бетона *Труды X Международной конференции студентов и молодых учёных Перспективы развития фундаментальных наук* (Томск: ТПУ) 708-710

[5] Кудяков К Л, Невский А В и Ушакова А С 2014 Влияние дисперсного армирования углеродными волокнами на прочностные свойства бетона *Труды XI Международной конференции студентов и молодых учёных Перспективы развития фундаментальных наук* (Томск: ТПУ) 799-802

[6] СП 63.13330.2012 Бетонные и железобетонные конструкции. Основные положения Актуализированная редакция СНиП 52-01 (Москва: НИИЖБ)

IOP Conf. Series: Materials Science and Engineering **71** (2015) 012038 doi:10.1088/1757-899X/71/1/012038

Creation of functional solid-state composites based on black peat

L B Naumova[1], V N Batalova[1], N P Gorlenko[2], Yu S Sarkisov[2] and A A Kartashova[3]

[1]National Research Tomsk State University, Analytical Chemistry Department, Tomsk, 634050, Russia

[2]Tomsk State University of Architecture and Building, Chemistry department, Tomsk, 634003, Russia

[3]National Research Tomsk Polytechnic University, graduate student of the chemistry department, Tomsk, 634050, Russia

E-mail: naumovalb@mail.ru

Abstract. The paper presents investigations of composite materials based on black peats of Barabinskoe and Taganskoe deposits of the Tomsk region and carboxymethyl cellulose, both modified with iron (III) and copper (II) chloride solutions. In order to improve hydrophobic properties of compositions, optimum salt concentrations are detected. Water sorption and desorption isotherms are obtained for modified specimens. It is suggested to employ synthesized solid-state compositions as insulators in the capacity of both humidity controller and indoor-contaminant absorber.

1. Introduction

Peat based compositions are widely used as heat insulation and exterior walls [1,2] in construction material production and also as sorbents and geochemical barriers in zones of contaminant accumulation [3,4]. Naumova *et al* [5-8] stated the improvement of hydrophobic properties of iron-containing peat compositions, and showed their effective use in sorption of various ions, radionuclides, and other contaminants from man-made solutions. A high sorption capacity of peat allows the effective use of modifiers of diverse natures when controlling the quality characteristics of materials. For the process development of peat-based composites and their effective use, it is necessary to establish a correlation between technological properties and the field of utilization of these materials, their composition, structure, and production conditions.

The aim of this work is to ascertain physicochemical parameters of creating solid-state composites based on black peats at introduction of inorganic modifiers.

The subject of research is black peat from Barabinskoe and Taganskoe deposits of the Tomsk region. Iron (III) and copper (II) chloride solutions are selected as modifying agents.

2. Materials and methods

Metal-containing peat composites were prepared by iron (III) and copper (II) chloride treatment of peat with 0.25 mmol/l concentration in static conditions with intensive shaking for 7-10 days up to maximum saturation. Control of the amount of iron ions before and after modification was carried out using complexometry. Metal content in the peat phase was detected by this difference. Modified specimens were washed with distilled water up to a complete elimination of relevant ions. To prevent leaching, composites were granulated with carboxymethyl cellulose and processed in a drying chamber at 105 °C. Bulking of the solid occurred due to sorption (swelling) of liquid produced by the original and modified specimens, was detected using the technique described in the work of Markhol [9]. At first, to calculate the amount of the modifier, acid-base properties of peats (complete exchange capacity, ionization constant, p*K*) were investigated by potentiometry [9]. Sorption activity of peats in relation to iron (Ш) and copper (II) ions was calculated by distribution coefficients *D* obtained by the method described in [9]. The structure and the qualitative and quantitative composition of specimens were studied with QUANTAX 70 energy dispersive spectroscopy on the scanning electron microscope Hitachi TM-300. Water vapour adsorption was studied using the gravimetrical static method described in the work [10].

3. Results and discussion

Table 1 presents results of investigation of acid-base properties of black peats.

Table 1. Values of swelling and complete exchange capacity of acid and basic groups and ionization constant of original black peats.

Peat types	*CEC^{H+} (mmol/g)	*pK_a	CEC^{OH-} (mmol/g)	pK_b	*W(%)
Barabinskoe deposit	5.40	9.40	2.10	12.50	80.10
Taganskoe deposit	4.90	10.80	1.30	12.30	79.10

Note: CEC is complete exchange capacity; p*K* is ionization constant; W is swelling.

From this Table, specimens under research are high-capacity amphoteric electrolytes, while in their acid-base and swelling properties they are not essentially different. Later on, the amount of the modifier was calculated using values of the complete exchange capacity.

Figure 1 illustrates the peat structure from Barabinskoe deposit and Fe (III) ion distribution in it. Fe (III) ions as shown in this Figure are uniformly distributed in the bulk of the peat-based composite. The average content of iron in modified specimens ranges between 3.8 and 5 %. Compressive strength of these specimens is (8-11) MPa.

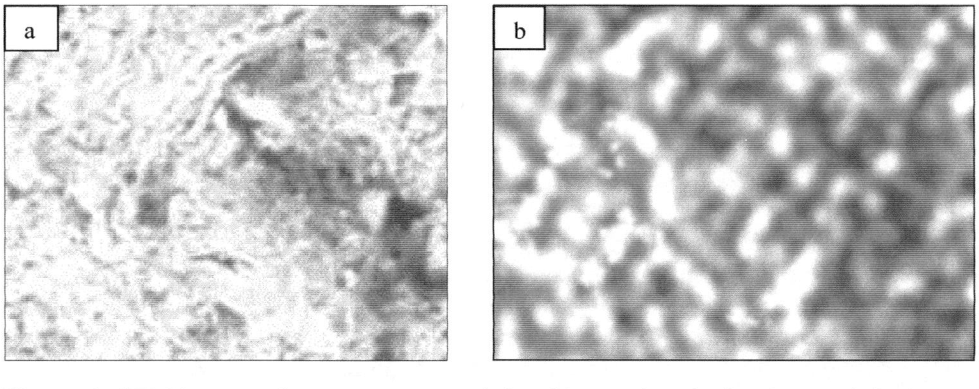

Figure 1. SEM images of peat structure and the elemental analysis of: a – original peat; b – peat modified with iron (III) ions.

Table 2 presents the dependence between distribution coefficients, relative humidity of peats and $C(FeCl_3)$. According to this Table, swelling of modified specimens is considerably lower than that of the original ones. At the same time, minimum of water absorption was observed in peats modified with $FeCl_3$ and $CuCl_2$ solutions with 0,25 mmol/l concentration, and specimens obtained after drying were of hard consistency, with node hydration upon repeated contact with water. Data obtained prove those described in the work of Lych [11] who observed internal restructuring processes in the peat phase at absorption of polyvalent metal ions with certain concentrations. This is due to clustering of associates leading to the primary and secondary structure formations, and a change of water holding capacity and strength properties of the peat-based composite.

Table 2. Distribution coefficients and relative humidity of peats depending on C_{FeCl3} at pH = 1.5 and C_{CuCl2} at pH = 4.3.

Peat types	C_{FeCl3} (mmol/l)	*W (%)	*D (ml/g)	C_{CuCl2} (mmol/g)	*W (%)	D (ml/g)
Barabinskoe	0.50	69.7±0.6	49.2±2,4	0.5	79.6±0,7	48.5±2,1
deposit	0.25	52.2±0.3	61.1±2.7	0.25	62.4±0,6	59.8±2,0
	0.10	75.1±0.5	51.1±2,8	0.1	72.3±0,6	47.9±1,9
Taganskoe	0.50	56.6±0.3	44.8±2,1	0.5	73.1±0,6	49.7±2,3
deposit	0.25	48.7±0.3	26.1±1.5	0.25	56.8±0,5	24.2±1,4
	0.10	74.6±0.6	52.9±3,1	0.1	68.8±0,5	51.8±2,3

Note: D is distribution coefficient; W is relative humidity.

Taking into account high values of distribution coefficients, one can state that the composite material will be the efficient contaminant absorber.

To study the moisture exchange in the peat-based composition, isotherms of sorption (1) and desorption (2) were constructed for water vapour in peat specimens taken from Taganskoe deposit. Experimental data are shown in Figure 2.

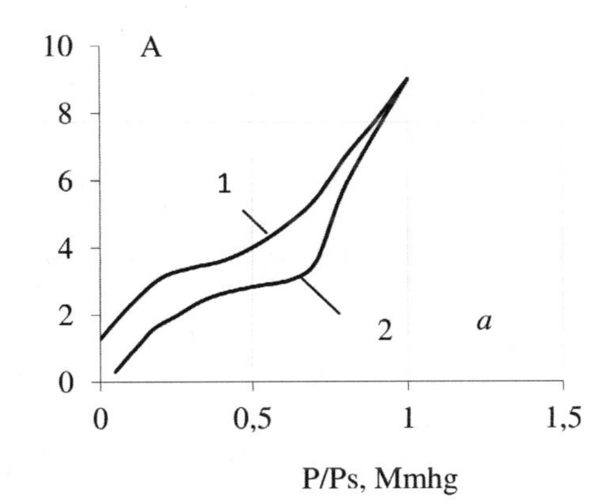

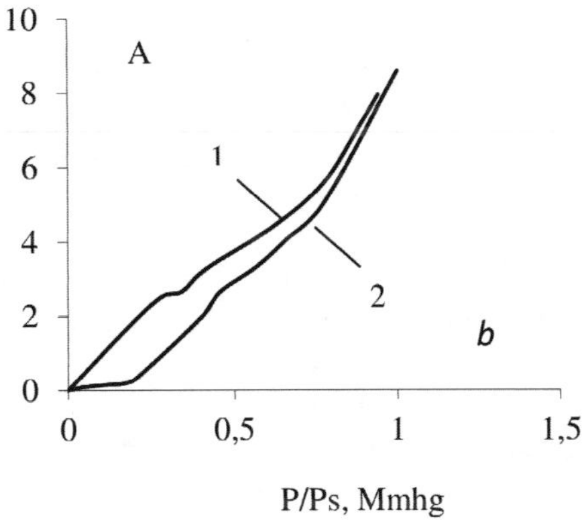

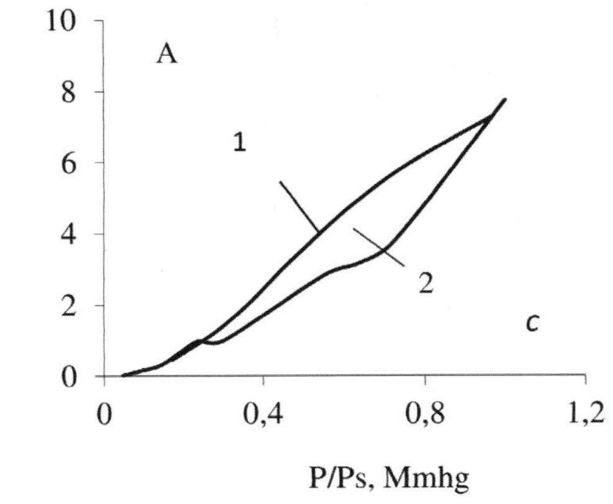

Figure 2. Sorption and desorption isotherms of water vapour in:
a - original peat;
b - peat modified with Fe^{3+};
c - peat modified with Cu^{2+}.
A - the amount of absorbed humidity, mmol·g^{-1};
P/P_s–pressure ratio, Mmhg.
1- humidity sorption curves,
2- humidity desorption curves.

As characterized by Figure 2, in case of original peat the sorption isotherm falls under sigmoidaltype II according to Brunauer's classification. At low pressure ratios, this isotherm is convex relative to X-axis, lies higher the adsorption isotherm, and does not return to a zero position within the whole interval of pressure ratio that proves a strong interaction between primary portions of water and the surface. At the same time, adsorption exhibits its irreversible character and sorption and desorption isotherm types indicate different-size pores presenting in the composite material. As for modification, it considerably changes isotherm types: they fall under type III (J-shaped) and show hysteresis extending up to the pressure ratio of 0 and 0.15 mm Hg for iron and copper respectively. The initial section of the sorption isotherm indicates the degree of water repellency of both specimens. The latter is connected with blocking of active centres on the peat surface due to their interaction with absorbed ions. At that, porosity calculation showed that all modified specimens possess basically large pores.

4. Conclusions

1. Solid-state compositions are suggested based on black peats of Barabinskoe and Taganskoe deposits of the Tomsk region as well as carboxymethylcellulose modified with iron (Ш) and copper (II) chloride solutions. It is shown that treatment with iron (Ш) and copper (II) chloride of 0.25 mmol/l concentrations reduces the moisture absorption by composites.
2. The laws of moisture exchange were established as a result of moisture sorption isotherms describing modified peat specimens.
3. Results obtained show the prospectivity of creation of efficient construction materials based on solid-state peat-based compositions that can be used as new generation heat insulators capable to function both as humidity controller and indoor contaminant absorber.

References

[1] Касицкая Л В, Саркисов Ю С, Горленко Н П и др. 2007 *Торфяные ресурсы Томской области и пути их использования в строительстве* (Томск: STT Publ.)
[2] Копаница Н О, Кудяков А И, Копаница Н О и Саркисов Ю С 2013 Строительные материалы и изделия на основе модифицированных торфов Сибири Томск (Томск: Издательство ТГАСУ
[3] Белькевич П И и Чистова Л Р 1979 *Торф и проблема защиты окружающей среды* (Минск: Наука и техника)
[4] Разворотнева Л И, Богуславский А Е, Коваль В П и Бабушкин А Н 2007 Экология промышленного производства **3** 33
[5] Naumova L B, Gorlenko N P and Otmakhova Z I 1995 *J. Physical Chem.* **68** 1461
[6] Наумова Л Б, Горленко Н П, Отмахова З И и Мокроусов Г М 1997 Химия в интересах устойчивого развития **5** 609
[7] Naumova L B, Chashchina O V and Gorlenko N P 1994 *J. Physical Chem.* **66** 688
[8] Naumova L B, Kiseleva L A and Egorova L A 2013 *Vestnik of TSUAB* **3** 243
[9] Мархол М 1985 *Ионообменники в аналитической химии* (Москва: Мир)
[10] Naumova L B, Minakova T S, Chernov E B, Gorlenko N P and Ekimova I A 2011 *J. Physical Chem.* **84** 762
[11] Лыч П И 1991 *Гидрофильность торфа* (Минск: Наука и техника)

IOP Conf. Series: Materials Science and Engineering **71** (2015) 012039 doi:10.1088/1757-899X/71/1/012039

Numerical simulation of wooden beams fracture under impact

P Radchenko[1], S Batuev[2] and A Radchenko[3]
[1]Tomsk State University of Architecture and Building, Department of Applied Mathematics, Tomsk, 634003, Russia
[2]Tomsk State University of Architecture and Building , Department of Geoinformatics and Cadastre,Tomsk, 634003, Russia
[3]Tomsk State University of Architecture and Building, Institute of Cadastre, Economics and Engineering Systems in Construction, Tomsk, 634003, Russia

E-mail: andrey-radchenko@live.ru

Abstract. In this paper the fracture of wooden (pine) beams under low-speed impact of steel ram tester is numerically investigated, considered speed is ranging between 10-30 m/s. The behavior of wood is described within phenomenological approach taking into account anisotropy of elastic and strength properties. The behavior of the steel ram tester is modelled by the elastoplastic medium. Numerical modeling is carried out by a finite elements method in three-dimensional statement. Fracture propagation in wood is under study.

1. Introduction

Impact action of solid bodies in a wide range of kinematic and geometric conditions is a complex task of mechanics. Difficulties connected with theoretical studies of the process of materials destruction and deformation under impact action using analytical approaches makes one to introduce a number of simplifying hypotheses that cannot always reflect the real situation. Therefore it should be considered that the leading role in studying events connected with high-speed interaction of solid bodies is currently given to experimental and numerical studies.

Research in materials destruction while impact actions demonstrate that due to change in impact conditions the fracture mechanisms are also changed [1]. Experiments demonstrate that in number of cases final destruction is defined by a combination of several mechanisms. However the experiments do not allow tracing the sequence, time of impact and contribution of the different fracture mechanisms. Furthermore, fractures obtained at the initial stages of process cannot always be identified while analysis of the final materials destruction. Therefore numerical simulation is relevant when studying impact interaction [2].

2. Major equations of mathematical model

The system of equations describing the time-dependent adiabatic motion of a compressible medium in an arbitrary coordinate system ($i=1,2,3$) includes the following equations:

- equation of continuity

$$\frac{\partial \rho}{\partial t} + div \rho \vec{v} = 0 \quad (1)$$

- equation of motion

$$\rho a^k = \nabla_i \sigma^{ki} + F^k, \tag{2}$$

where

$$a^k = \frac{\partial \upsilon^k}{\partial t} + \upsilon^i \nabla_i \upsilon^k \tag{3}$$

$$\nabla_i \sigma^{ki} = \sigma^{ki}_{,i} + \sigma^{mi} \Gamma^k_{mi} + \sigma^{ki} \Gamma^m_{mi} \tag{4}$$

- equation of energy

$$\frac{dE}{dt} = \frac{1}{\rho} \sigma^{ij} e_{ij}, \tag{5}$$

where ρ - density of the media; $\vec{\upsilon}$ - velocity vector; a^k - components of acceleration vector; F^k - components of the nodal forces vector; Γ^k_{ij} - Christoffel symbols; σ^{ij} - contravariant components of symmetric stress tensor; E - specific internal energy; e_{ij} - components of the symmetric strain velocity tensor:

$$e_{ij} = \frac{1}{2}\left(\nabla_i \upsilon_j + \nabla_j \upsilon_i\right). \tag{6}$$

3. Constitutive equations of isotropic metal materials

The behavior of isotropic metal materials is considered within elastoplastic model.

The stress tensor is represented in the form of the sum of deviatoric S^{ki} and spherical part P: $\sigma^{ki} = -P\delta_{ki} + S^{ki}$, where δ_{ij} - Kronecker symbol.

Assuming that for media true is the principle of minimum work of true stresses on increments of plastic deformations, the connection of tensor component of strain rate and stress deviator will be recorded in the form of:

$$2G\left(e_{ij} - \frac{1}{3} e_{kk} \delta_{ij}\right) = \frac{DS^{ij}}{Dt} + \lambda S^{ij}, (\lambda \geq 0). \tag{7}$$

Here

$$\frac{DS^{ij}}{Dt} = \frac{dS^{ij}}{dt} - S^{ik}\omega_{jk} - S^{jk}\omega_{ik},$$

where $\omega_{ij} = \frac{1}{2}\left(\nabla_i \upsilon_j - \nabla_j \upsilon_i\right)$, G - shear modulus.

Parameter $\lambda = 0$ under elastic deformation and under plastic deformation ($\lambda > 0$) determined by the help of a condition of Mises: $S^{ij} S_{ij} = \frac{2}{3}\sigma_d^2$, σ_d – dynamic yield stress. Pressure in the material was calculated from the Mie-Grüneisen equation as a function of the specific internal energy E and density ρ:

$$P = \sum_{n=1}^{3} K_n \left(\frac{V_0}{V} - 1\right)^n + K_0 \rho E, \tag{8}$$

where, K_0, K_1, K_2, K_3 - material constants.

4. Constitutive equations of anisotropic materials

To describe the behavior of anisotropic (orthotropic) body the following conditions and restrictions are accepted:

1. The body is solid (continuous media);
2. The cross dimensions of structural (reinforcing) elements are small in comparison with the body sizes, i.e. the medium is quasi-homogeneous;
3. Relationships between increments of stress tensor components and tensor components of strain rate s linear.

Therefore the relations between components of stress tensor and strain rate tensor before fracture to orthotropic body can be given as:

$$\frac{d\sigma_{11}}{dt} = C_{11}e_{11} + C_{12}e_{22} + C_{13}e_{33}, \quad \frac{d\sigma_{12}}{dt} = C_{44}e_{12}, \tag{9}$$

$$\frac{d\sigma_{22}}{dt} = C_{12}e_{11} + C_{22}e_{22} + C_{23}e_{33}, \quad \frac{d\sigma_{13}}{dt} = C_{66}e_{13}, \tag{10}$$

$$\frac{d\sigma_{33}}{dt} = C_{13}e_{11} + C_{23}e_{22} + C_{33}e_{33}, \quad \frac{d\sigma_{23}}{dt} = C_{55}e_{23}, \tag{11}$$

where, C_{ij} – elastic constants.

Elastic constants C_{ij} can be expressed by means of shear modulus G_{ij}, Young's modulus: E_i and Poisson's ratios ν_{ij}:

$$C_{11} = \frac{1}{E_2 A}\left(\frac{1}{E_3} - \frac{\nu_{23}^2}{E_2}\right); \quad C_{12} = \frac{1}{E_3 A}\left(\frac{\nu_{31}\nu_{23}}{E_3} + \frac{\nu_{12}}{E_1}\right);$$

$$C_{13} = \frac{1}{E_2 A}\left(\frac{\nu_{12}\nu_{23}}{E_1} + \frac{\nu_{31}}{E_3}\right); \quad C_{33} = \frac{1}{E_1 A}\left(\frac{1}{E_2} - \frac{\nu_{12}^2}{E_1}\right);$$

$$C_{22} = \frac{1}{E_3 A}\left(\frac{1}{E_1} - \frac{\nu_{31}^2}{E_3}\right); \quad C_{23} = \frac{1}{E_1 A}\left(\frac{\nu_{12}\nu_{31}}{E_3} + \frac{\nu_{23}}{E_2}\right);$$

$$C_{44} = G_{12}; \quad C_{55} = G_{23}; \quad C_{66} = G_{13}; \tag{12}$$

$$A = \frac{1}{E_1 E_2 E_3}\left(1 - 2\nu_{12}\nu_{23}\nu_{31} - \frac{E_1}{E_3}\nu_{31}^2 - \frac{E_2}{E_1}\nu_{12}^2 - \frac{E_3}{E_2}\nu_{23}^2\right); \tag{13}$$

Knowing nine sizes of technical constants in the main axes of symmetry of orthotropic material, one can determine the value of any constant in any direction.

5. Fracture model of anisotropic material under dynamic loadings

Defining the transition of the barrier material to complete fracture the Hoffman criterion was used [3] that takes into account the different strength structural properties of the material in compression and tensile actions:

$$f(\sigma) = C_1(\sigma_2 - \sigma_3)^2 + C_2(\sigma_3 - \sigma_1)^2 + C_3(\sigma_1 - \sigma_2)^2 +$$
$$C_4\sigma_1 + C_5\sigma_2 + C_6\sigma_3 + C_7\sigma_4^2 + C_8\sigma_5^2 + C_9\sigma_6^2 \geq 1, \tag{14}$$

where coefficients C_{1-9} are determined from the following equations:

$$C_1 = [(Y_t Y_c)^{-1} + (Z_t Z_c)^{-1} - (X_t X_c)^{-1}]/2;$$
$$C_2 = [(X_t X_c)^{-1} + (Z_t Z_c)^{-1} - (Y_t Y_c)^{-1}]/2;$$
$$C_3 = [(X_t X_c)^{-1} + (Y_t Y_c)^{-1} - (Z_t Z_c)^{-1}]/2;$$
$$C_4 = (X_t^{-1} - X_c^{-1});\ C_7 = S_{yz}^{-2};$$
$$C_5 = (Y_t^{-1} - Y_c^{-1});\ \ C_8 = S_{zx}^{-2};$$
$$C_6 = (Z_t^{-1} - Z_c^{-1});\ \ C_9 = S_{xy}^{-2}. \tag{15}$$

Here X_t, X_c - ultimate tensile and ultimate compressive strength along the X-axis respectively, Y_t, Y_c - ultimate tensile and ultimate compressive strength along the Y axis , Z_t, Z_c - ultimate tensile and ultimate compressive strength along the Z axis, S_{xy}, S_{yz}, S_{zx} - ultimate shear strength along the corresponding axes. It is supposed that fracture of anisotropic materials in the conditions of intensive dynamic loadings occurs as follows [4]:

1. If failure criterion is broken under compression $(e_{kk} \le 0)$, it is considered that material loses properties of anisotropy, and its behavior is described by hydrodynamic model, and the material retains only compressive strength ; stress tensor becomes in this case spherical $\sigma = -P$ and its value is determined from the relation [5]:

$$P = \left[\exp\left(4b\frac{V_0 - V}{V_0}\right) - 1\right]\frac{\rho_0 c_0^2}{4b}, \tag{16}$$

where ρ_0 - the initial density, c_0 and b - the coefficients of the shock adiabat;

$$D = c_0 + bu_m,$$

u_m - mass velocity, V_0 and V - initial and current specific volumes.

2. If failure criterion is broken under tension $(e_{kk} > 0)$, the material is considered completely destructed, and components of stress tensor are set equal to zero $(\sigma_{ij} = 0)$. Numerical simulation is carried out by finite element method [6].

6. Results and discussion

The process of interaction of steel (ρ_0=2710 kg/m^3, σ_d=310 MPa) ram tester is considered, corresponding 400g in weight, speed range of 10-30 m/s (ρ_0=2710 kg/m^3, σ_d=310 MPa) with a barrier from glued wood (pine) thickness of 60 mm. The density of the barrier material ρ_0=580 kg/m^3, the elastic and strength properties have the following values: E_x=11700 MPa, E_y=500 MPa, E_z=620 MPa, X_t=84 MPa, X_c=46 MPa, Y_t=30 MPa, Y_c=46 MPa, Z_t=30 MPa, Z_c= 46 MPa, S_{xy}=30 MPa, S_{yz}=30 MPa, S_{zx}=30 MPa.

Figures 1 - 2 show the calculated configuration of the ram tester and the barrier and gradation of grey color show values of relative level of fracture V_r/V_t in barrier. The values V_r and V_t are given to the nodes of the computational grid, V_r – is a volume of the elements connecting in node in which the

condition of fracture (14) was satisfied, V_t – the total amount of the elements forming this node. Value $V_r/V_t=1$ corresponds to complete fracture of material in the node of the computational grid.
Resulting from the impact action on the barrier annular area of fracture occurs on a front surface along the perimeter of the ram tester. This destruction area spreads to the back surface along the barrier thickness. Figure 3 demonstrates changes in the time of speed of the mass center of ram tester for various initial speeds of interaction which illustrate intensity of ram tester braking. Penetration of a barrier isn't observed up to speed 30m/s.

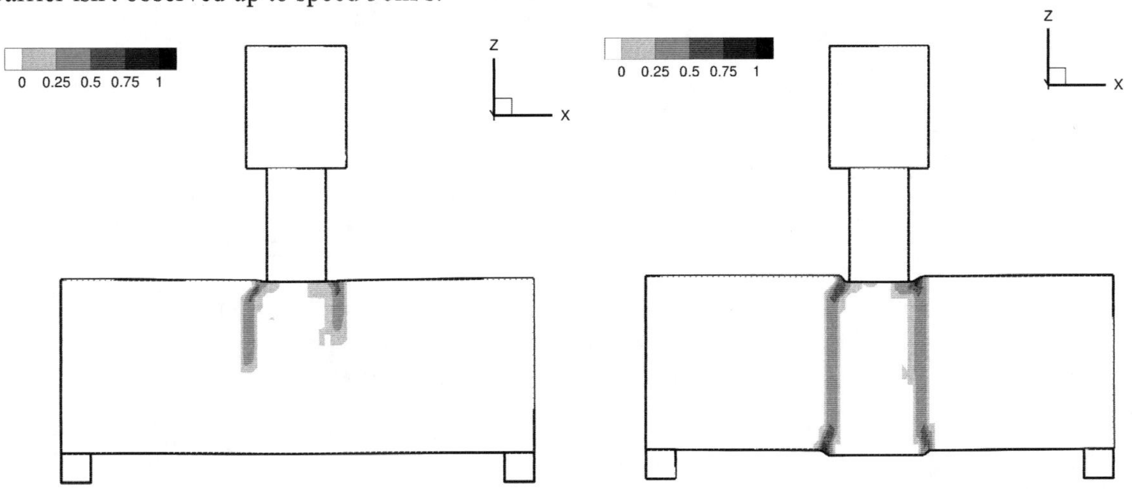

t=60 мкс, t=200 мкс

Figure 1. Configuration of the projectile and target, isolines of relative volume of fracture in barrier. $\upsilon_0=15$ m/s.

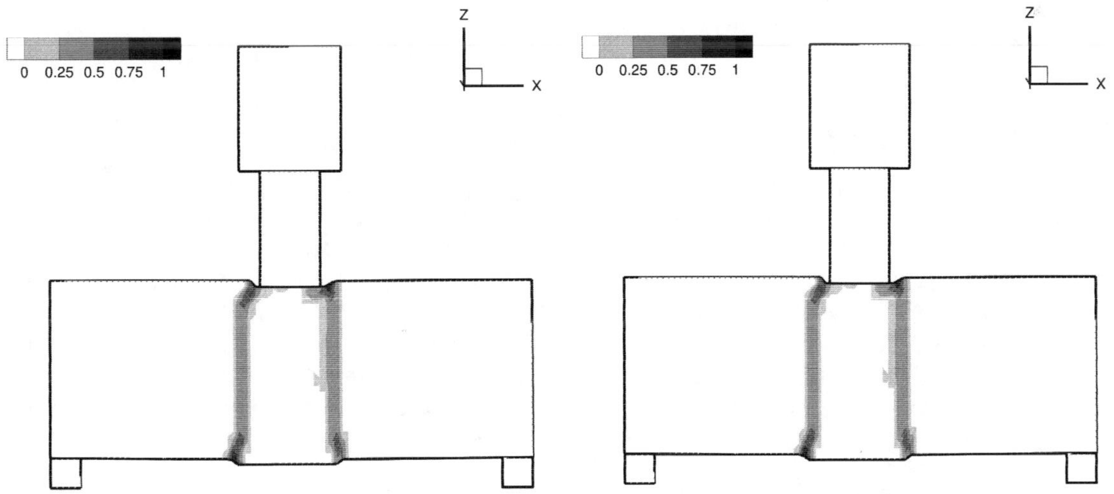

t=60 мкс, t=200 мкс

Figure 2. Configuration of ram tester and barrier, isolines of relative volume of fracture in barrier $\upsilon_0=20$ m/s.

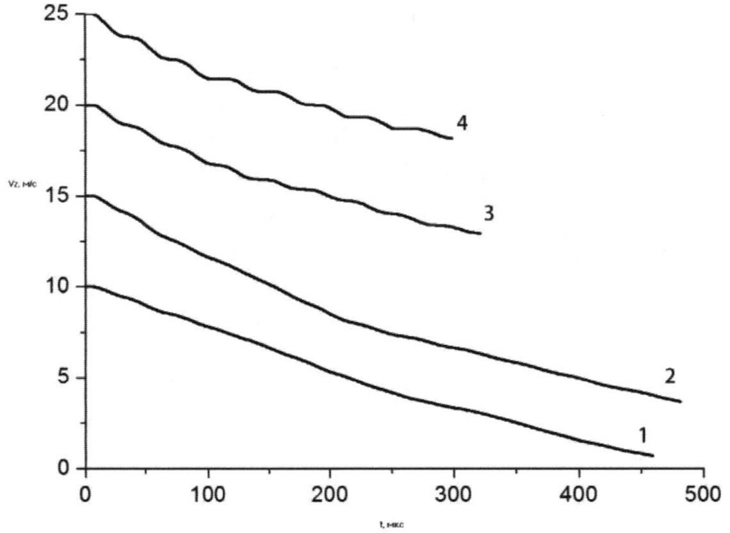

Figure 3. Change in time of velocity of the center of mass of ram tester.
1 – for initial velocity 10 m/s, 2 – 15 m/s, 3 – 20 m/s, 4 – 25 m/s.

7. Conclusion

The developed model describes behavior of brittle-fractured orthotropic material, transferring into model of brittle-fractured transtropic or isotropic material at equality of properties in the directions X and Y, or in all directions. It allows investigating the behavior of a wide class of materials with various properties using the same model. Wood and structural elements obtained from it can also be described well by the suggested model thus allowing its broad application in studying the behavior of similar materials under dynamic loading. The model does not restrict the criterion of strength, its choice is defined by appropriateness and given strength properties for particular material.

References

[1] Канель Г И, Разоренов С В, Уткин А В и Фортов В Е 1996 *Ударноволновые явления в конденсированных средах* (Москва: Янус-К) 407
[2] Фомин В М, Гулидов А И, Сапожников Г А и др. 1999 *Высокоскоростное взаимодействие тел* (Новосибирск: Издательство СО РАН) 600
[3] Ву Э 1978 Феноменологические критерии разрушения анизотропных сред *Механика композиционных материалов* **2** 401
[4] Radchenko A and Radchenko P 2011 Numerical modeling of development of fracture in anisotropic composite materials at low-velocity loading *J. Mat. Sci.* **46** 8
[5] Канель Г и Щербань В 1980 Пластическая деформация и откольное разрушение железа "Армко" в ударной волне *ФГВ* **16** 93
[6] Johnson G 1977 *J. Appl. Mech.* **6** 771

Numerical simulation of composite fracture under dynamic loads

P A Radchenko[1]
[1]Tomsk State University of Architecture and Building, Department of Applied Mathematics, Tomsk, 634003, Russia

E-mail: radchenko@live.ru

Abstract. This paper presents methodology devloped to describe numericaly deformation and fracture of composite materials under various dynamic loadings. The task is solved in full three-dimensional approach. The offered model allows describing well the behavior of composites in a wide range of initial and boundary conditions. On the basis of the offered model program complexes were developed, considering computing capacity of modern computer architecture.

1. Introduction

During last decades the application of anisotropic materials significantly increased in different engineering fields and particular in civil engineering. Development and improvement of production technologies of materials with the set properties that allowed their optimum use in various structures, made application of such materials even wider. It mainly refers to the structures whose effective operation is possible only by optimization of wide number of parameters of their elements, it is in turn is possible only by use of materials of primary orientation with physical and mechanical properties.

The performance of dynamic structures, used in aviation, power and space engineering can be enhanced by increasing the structural performance of materials. If direction of the force is known, one of the approaches to increase strength is orientation of properties, i.e. ordering the material structure, the other approach would be reinforcment by strengthening elements.

Behavior of such anisotropic materials under deformation significantly differs from behavior of the isotropic ones: the spherical body subjected to all-round compression turns into an ellipsoid and bending of anisotropic beam is accompanied by twisting. Velocity of stress waves distribution in anisotropic materials depend on the direction of wave distribution and on the stress level when loading is beyond the elastic limit. For example, in orthotropic materials in some directions transverse wave is faster, and longitudinal wave is slower. It causes significant change in the nature of interaction of the reflected and refracted waves in anisotropic materials. As at impulse impact on material the area of energy concentration is limited by velocity of waves distribution. Changing the degree of anisotropy of the material can directly change the distribution of the stress waves in the area surrounding the impulse loading zone. Due to changes in the form of anisotropic bodies under the impact of hydrostatic pressure it is necessary to consider the impact of hydrostatic pressure on the strength of materials. The condition of equally dangerous states can only be given for static short-term loading at the set temperature mode. Influence of time-dependent factors can be introduced in the parametrical way.

Most of the composite materials refer to anisotropic materials by particular mechanical properties: thermal, electric, magnetic, elastic, strength, plastic, etc. Four classes of fibers are considered appropriate for space structures: glass, boric, carbon and organic. This paper is focused on modeling the fracture of elastic-brittle composite materials at impulse loadings.

2. Fracture modeling of anisotropic materials

Fracture of anisotropic material was described within phenomenological approach. Phenomenological approach has to meet strength condition within the following requirements:

1. Failure criterion has to be invariant regarding choice of system of coordinates.
2. It has to possess sufficient flexibility to model any kinds of destruction without any special limiting assumptions regarding the mechanism of failure.
3. Any given radial load path must have only one point of intersection with the surface of strength.
4. The failure criterion should allow its easy action depending on the desirable level of accuracy.
5. It has to be mathematically convenient and allow possibility of the formulation both in space of tension, and in space of deformations at biunique compliance between these formulations; the criterion has to be applicable for structure calculation.
6. Definition of the main axes of durability has to be single valued.

Geometrically the failure criterion can be interpreted as some limiting surface in stress space i.e. the condition of destruction is satisfied while the set stress vector crosses this surface of strength.

Tsai and Wu [1] considered the general failure criterion of anisotropic materials. This criterion is written in terms of scalar functions of the components of the stress tensor as follows:

$$f(\sigma_{ij}) = F_{ij}\sigma_{ij} + F_{ijkl}\sigma_{ij}\sigma_{kl} + ... \geq 1, \quad \text{i, j, k, l} = 1, 2, 3. \tag{1}$$

Components F_{ij} and F_{ijkl} are components of tensor of the 2nd and 4th rank respectively, and are submit to the following laws of transition:

$$F'_{ab} = F_{ij}q_{ia}q_{jb}, \tag{2}$$

$$F'_{abcd} = F_{ijkl}q_{ia}q_{jb}q_{kc}q_{ld}, \tag{3}$$

where q_{ij} – a cosine of the angle between the corresponding axes. Destruction tensor components in square form are determined by the following ratios:

$$F_{ii} = \frac{1}{X_{ii}} - \frac{1}{X'_{ii}} \quad F_{iiii} = \frac{1}{X_{ii}X'_{ii}}, \tag{4}$$

$$F_{ij} = \frac{1}{2}\left(\frac{1}{X_{ij}} - \frac{1}{X'_{ij}}\right) \quad F_{ijij} = \frac{1}{4X_{ij}X'_{ij}} \quad i \neq j, \tag{5}$$

where X_{ii}, X'_{ii} - ultimate strength at tension and at compression in the direction i; X_{ij}, X'_{ij} - shear strength in two opposite directions at i≠j. Coefficients F_{1122}, F_{2233}, F_{3311} are determined in experiments on biaxial tension in planes 1–2, 2–3, 3–1 respectively. The remained coefficients are defined similarly at the combined loading in the corresponding planes.

The accuracy of this failure criterion is determined by the highest rank tensors F. To set the highest order of items that have to constitute the tensor polynomial formulation of failure criterion of Wu it is possible as follows: if the value F_{ijklmn} measured in experiment is more than maximum permissible accuracy, the accounting of tensors of this rank (and, probably, higher) can lead to increase in accuracy of criterion. In practice the accuracy of failure criterion represents a compromise between requirements of engineering practice and time available for the experimenter.

The phenomenological failure criterion can be built without numerical model, using a large number of experimental data and fracture of material can be predicted on the basis of nomograms. Numerical model allows reducing number of experimental data and facilitates by mathematical modeling of structural behavior under loading. Advantages of failure criterion of Wu are the maximum simplicity of operation, considering strength parameters in tension and compression in one equation at development of a surface of strength. Another feature of the Wu failure criterion is the dependence of the surface strength tensor components on the technical properties of strength. Therefore, it is possible to reach any necessary accuracy of the description by consecutive inclusion of units of the highest degrees without recalculation of the constants found in the previous calculations.

Some failure criteria of anisotropic environments actually represent special and limited cases of this tensor-polynomial formulation. The failure criteria offered by Hoffman is as follows:

$$\begin{gathered} C_1(\sigma_{22}-\sigma_{33})^2+C_2(\sigma_{33}-\sigma_{11})^2+C_3(\sigma_{11}-\sigma_{22})^2+ \\ +C_4\sigma_{11}+C_5\sigma_{22}+C_6\sigma_{33}+C_7\sigma_{12}^2+C_8\sigma_{23}^2+C_9\sigma_{31}^2\geq 1 \end{gathered}, \quad (6)$$

where C_i – material constants.

The presence of linear terms in Hoffman criterion significantly increases its flexibility, but compared to the tensor-polynomial formulation of Hoffman's criterion still has some disadvantages, including: (i) it is applicable only to orthotropic materials as it is accepted that coefficients defining interaction of shear tension equal to zero; (ii) coefficients F_{1122}, F_{2233}, F_{3311} that characterize mutual influence of normal tension are not independent constants, so reducing flexibility of the criterion.

The failure criterion offered by Ashkenazi [2] at complex stress states contains tension in the second and in the fourth degree. The fourth order of polynom not only allows better approximation of experimental data, than the second one, but also meets the phenomenon nature for strongly anisotropic solids: surface durability for highly anisotropic materials may contain both convex and concave sites due to different character of the dangerous condition in these areas. The polynomial criterion of the fourth degree in the reduced tensor record is as follows:

$$\alpha_{ik\ell m}\sigma_{ik}\sigma_{\ell m}-\left[\frac{\left(\sigma_{ik}\delta_{ik}\right)^2+\sigma_{ik}\sigma_{ik}}{2}\right]^{\frac{1}{2}}\geq 0. \quad (7)$$

The first term ($\alpha_{ik\ell m}\sigma_{ik}\sigma_{\ell m}$) represents a joint invariant of stress tensor and strength tensor, and the second term expresses dependence of strength of anisotropic bodies on two invariants – J_1 and J_2 of the stress tensor. The first term of this equation in expanded form is "plastic potential". The invariant equation of equally dangerous states can be considered in physical aspect as generalization of "the plastic potential" of Mises for anisotropic bodies in case when there is an obvious dependence of limit state on the first invariant of stress tensor (from the hydrostatic pressure of J_1). The expanded form of the criterion represents polynom of the fourth degree respecting to six component of the operating tension. Therefore the equation is called as polynomial criterion of the fourth degree. The constants α_{iklm} are in formula components of symmetric tensor of the fourth rank - strength tensor, and the size σ_{ik} and σ_{lm} - components of stress tensor (field tensor of second rank). In this criterion the symmetry of coefficients is accepted answering to orthogonal symmetry of material and considered in the present work in details. To reveal the coefficient α_{iklm}, you must specify all the stresses σ_{xx}, σ_{yy}, σ_{zz}, σ_{xy}, σ_{xz}, σ_{yz} comprising the formula of values corresponding to particular cases of stress states. For example, at unaxial tension or compression these values will be: σ_{zz}-σ_{bz} - tensile strength or compression along the axis z, while all other components of the stress will equal to zero, i.e. $\sigma_{xx}=\sigma_{yy}=\sigma_{xy}=\sigma_{xz}=\sigma_{yz}=0$. Thus we will receive that:

$$\alpha_{1111}=\frac{1}{\sigma_{bz}}. \quad (8)$$

Similarly also the sense of other coefficients is investigated. Also easily the sense is investigated the component $\alpha_{ik\ell m}$ for which turns out: $\alpha_{ikik} = \frac{1}{\tau_{bik}}$, where τ_{bik} - tensile strength in pure shear, at which there is a change of an initial right angle between axes of symmetry of i and k, etc.

The polynomial criterion of the fourth degree of Ashkenazi for orthotropic material in the Cartesian system of coordinates of XYZ is the following:

$$\frac{\sigma_{xx}^2 + c\sigma_{yy}^2 + b\sigma_{zz}^2 + d\tau_{xy}^2 + p\tau_{yz}^2 + r\tau_{xz}^2 + s\sigma_{xx}\sigma_{yy} + t\sigma_{yy}\sigma_{zz} + f\sigma_{xx}\sigma_{zz}}{\sqrt{\sigma_{xx}^2 + \sigma_{yy}^2 + \sigma_{zz}^2 + \tau_{xy}^2 + \tau_{yz}^2 + \tau_{xz}^2 + \sigma_{xx}\sigma_{yy} + \sigma_{yy}\sigma_{zz} + \sigma_{xx}\sigma_{zz}}} \geq [\sigma_0], \tag{9}$$

where

$$[\sigma_0] = \frac{\sigma_{bx}}{k_b};\ c = \frac{\sigma_{bx}}{\sigma_{by}};\ b = \frac{\sigma_{bx}}{\sigma_{bz}};\ d = \frac{\sigma_{bx}}{\tau_{bxy}};p = \frac{\sigma_{bx}}{\tau_{byz}};r = \frac{\sigma_{bx}}{\tau_{bxz}};$$

$$s = \frac{4\sigma_{bx}}{\sigma_{bxy}^{(45)}} - c - d - 1;\ t = \frac{4\sigma_{bx}}{\sigma_{byz}^{(45)}} - c - b - p;\ f = \frac{4\sigma_{bx}}{\sigma_{bzx}^{(45)}} - b - r - 1. \tag{10}$$

Here σ_{xx}, σ_{yy}, σ_{zz} and τ_{xy}, τ_{yz}, τ_{xz} – normal and shear components of the stress tensor, the letter σ with the lower ''b'' index denotes the value of the yield strength in tension or compression in the direction of the axis corresponding to the second lower index; letter τ - the same as in pure shear, in which the right angle between the axes designated in an index changes. The top index (45) at the letter σ means tensile strength in diagonal direction (at an angle of 45° to the axes of symmetry), lying in the plane corresponding to the lower indexes. k_b – safety factor which relied identical to all directions, in this case to k_b=1. As a rule, materials have different ultimate compressive and tensile strength, and the used model allows considering it.

If at least one of the failure criteria was carried out, it is suggested that the environment is damaged, i.e. its mechanical properties have changed. In areas where the criterion is carried out under compression ($e_{kk} < 0$), it is considered that material loses properties of anisotropy and for the description of such environment the model of liquid is used, resisting volume compression and not resisting shear and tension. Pressure in such material was calculated from the state equation offered by Kanel G. I. [3]:

$$P = \left[\exp\left(4\beta\frac{V_0 - V}{V_0}\right) - 1\right]\frac{\rho_0\alpha^2}{4\beta}, \tag{11}$$

where ρ_0 - the initial density, α and β - the coefficients of the shock adiabat $D=\alpha+\beta u_m$, u_m - the mass velocity, V_0 and V - initial and current specific volumes. In areas, where the criterion is carried out under tension ($e_{kk} > 0$) the material is considered completely destroyed, and components of stress tensor are set equal to zero.

3. Calculation procedure

Since the development of high-quality finite element mesh is generally a non-trivial task, free software products were used to its realization, based on an open program code. Tetrahedrons having the highest resistance to large deformations are used as finite elements of computational space. The program complex allows considering unlimited number of the bodies with arbitrary geometry. For each body it is possible to set both various initial, and boundary conditions. The problem was solved completely in three-dimensional position.

After building the numerical grid for bodies and initial conditions the integration block starts working in the following order [4]:

1. Strain rates are calculated;
2. Tension and pressure are defined;
3. Stress of fluidity or fracture stress is updated;
4. The concentrated forces in nodes, new nodal velocities and nodal coordinates are calculated;
5. Velocities and coordinates of nodes in the area of contact boundaries are updated.

At each calculation step Courant stability condition is calculated. Application of the coefficient of stability that close to 1 for quasi-static problems allows significantly increasing the step value in time, significantly reducing the overall solution time.

The program complex uses possibility of parallelization of computational process. OpenMP API realized for systems with the general memory is applied for this purpose. The program complex doesn't impose restrictions on number of processors and shows a good ratio of the used computing cores and time of the solution of a task. Also there are no restrictions on quantity of finite elements if the 64-bit system possesses enough random access memory.

4. Conclusion

In this article the numerical model of fracture of composites under dynamic loadings is offered. The model allows describing behavior of a wide class of elastic-brittle materials taking into account complex structure of internal reinforcing both with metal, and nonmetal reinforcement. While studies two program complexes were developed: for quasi-homogeneous (certificate on the state registration of the computer program No. 2011615952) and heterogeneous materials (certificate on computer program registration No. 2014614671).

References

[1] Tsai S Wand Wu E M 1971 A general theory of strength for anisotropic materials *J. Compos. Mat.* **5** 58-80
[2] Ашкенази Е К 1966 Прочность анизотропных древесных и синтетических материалов *Лесная промышленность* 167
[3] Канель Г И и Щербань В В 1980 Пластическая деформация и откольное разрушение железа "Армко" в ударной волне 16493-103
[4] Johnson G R 1977 High velocity impact calculations in three dimension *J. Appl. Mech.* 95-100

The use of the chrysotile cement waste as the secondary aggregate for the concrete

V Semenov[1], A Pligina[1] and T Rozovskaya[1]
[1]Moscow State University of Civil Engineering, Department of Building Materials, Moscow, 129337, Russian Federation

E-mail: science-isa@yandex.ru

Abstract. The article presents the results of research on the effective concrete with secondary chrysotile cement aggregate. One of the important problems of modern science of construction materials is the use of secondary resources for the production of construction materials, and a considerable part of them are the chrysotile cement waste and scrapped chrysotile cement products. The aim of presented research is the development of effective concrete for the production of foundation wall blocks with the use of crushed chrysotile cement products as a secondary aggregate. The main characteristics of the secondary chrysotile cement aggregate have been determined. The concrete with different compositions and with different content of secondary chrysotile cement rubble has been studied. The dependences of the strength and the specific strength of concrete with a constant W/C ratio and constant binder consumption on the consumption of the secondary aggregate have been obtained. It is stated that the introduction of secondary chrysotile cement aggregate does not significantly effect the water resistance and frost resistance of the concrete. It is shown that the variation of the fractions of secondary aggregates and the binder makes it possible to obtain the effective concrete with a wide range of strength values.

1. Introduction

One of the important trends in the resource economy in construction is the use of recycled resources in the production of building materials [1]. This allows us to expand the resource supplies for the production of building materials, and it is one of the solutions to important national economic problems caused by the environmental pollution.

2. Review of published works

The chrysotile cement wastes are a considerable part of secondary resources [2-3]. The chrysotile cement industry has a significant impact on the environment. The chrysotile cement plants produce specific waste: chrysotile cement dust, wet waste, crushed sheets and pipes, trash from the use of paper bags [2]. Each chrysotile cement plant has a waste storage. According to [4], the waste output in the chrysotile cement industry is 7 – 10 % of the weight of the products, including the dry waste (3 – 5 %) and the wet waste (4 – 5 %). According to the "Chrysotile Association" Society, there were produced about 1.26 million tons of chrysotile cement products in 2010 [5]. Thus, the waste output was equal to about 126 000 tons per year. In this regards the use of chrysotile cement waste is a quite acute question. In addition, the construction waste landfills have accumulated a large amount of overage chrysotile cement products (pipes, sheets, slate). Reliable methods of disposal of chrysotile cement waste are costly; therefore the development of technologies of the use of chrysotile cement waste in the building materials industry is promising. Disposal of waste will free up their disposal space and solve important environmental and economic problems [2, 4].

There are various methods of disposal of dry chrysotile cement waste, i.e. crushed sheets, scraps, and overage chrysotile cement products. The simplest of these is the crushing of waste and their following use as a secondary aggregate for concrete. For example, in the middle of the last century, some chrysotile cement plants in our country periodically manufactured the masonry units from a mix of chrysotile cement waste, slag and cement. They have been used for the erection of rural low-rise buildings [4]. In this paper, it was suggested to use crushed chrysotile cement sheets in the production of foundation wall blocks.

The safety of chrysotile cement products has been confirmed by many years of research done by both Russian and foreign scientists, the results were adopted by the World Health Organization (WHO) [6].

It is known that the asbestos minerals include serpentine group minerals (for example, chrysotile) and amphibole minerals [5]. The structure and properties of serpentinite and amphibole differ much. Currently, the amphibole asbestos is recognized by the International Agency for Research on Cancer (IARC) and the WHO as the 1st group carcinogens and is forbidden to be used [4]. In the construction they use exclusively the chrysotile asbestos, which has a less biological activity. Combining these minerals under generally accepted term "asbestos" caused in its time the asbestos phobia and led to the restriction of the use of materials containing chrysotile asbestos.

In [2], it is shown that the destruction of corrugated chrysotile cement sheets (roofing slate) may occur inside the cement matrix or inside the chrysotile fiber. In this system, the contact zone "fiber – cement matrix" has the highest strength, which eliminates a spontaneous emission of fibers from the material. The fibers can leave the products only in case of their mechanical damage [2]. Asbestos-containing serpentine rock is widely presented in nature, therefore the air contains a certain amount of asbestos fibers (natural asbestos background) [7]. There is a threshold dose of asbestos in the air, below which it does not have any negative impact on human health.

In [8-9], it was shown that all emitted chrysotile fibers from the chrysotile cement surface were almost completely covered with the hydration products and the subsequent carbonization clinker phases. In [8], an active sorption of the products of Portland cement hydration by chrysotile fibers is shown. It is found that all chrysotile fibers in the cement matrix chemically react with the cement hydration products, changing their composition and structure. Thus, it is proved that the surface of chrysotile cement products cannot extract the fibers with their initial properties. According to [10], the chrysotile fibers coated with cement hydration products have a reduced biological activity.

The environmental-and-health studies [11] showed that chrysotile cement materials are not any significant sources of air pollution by chrysotile asbestos fibers inside or outside buildings. Even the process of machining of products by high-speed equipment, producing some dust does not make the content of chrysotile fibers higher than the maximum permissible concentration (MPC).

Indirectly, the safety of chrysotile cement products is proved by current hygiene regulations, according to which it is allowed to apply pressure asbestos cement pipes in the networks of hot and cold potable water supply [6].

Thus, it was assumed that the use of crushed chrysotile cement sheets as an aggregate for the concrete completely eliminates the emission of fibers from the concrete due to a double "conservation" of fibers in the concrete. And the use of this concrete for the foundation blocks completely eliminates the negative impact of asbestos fibers on humans.

The use of secondary aggregates in the concrete technology has been the subject of numerous studies. Considerable experience of using concrete and reinforced concrete waste as aggregates for concrete was accumulated [12], which is one of the most important reserves of saving material and energy in the production of building materials [13]. The question of the use of crushed chrysotile cement products as a secondary aggregate for concrete has not been studied yet.

The aim of the present research is the development of effective concrete for the production of foundation wall blocks with the use of crushed chrysotile cement products as a secondary aggregate.

3. Materials and methods

According to the State Standard of Russian Federation GOST 13579-78 "Concrete fundament wall blocks. Specifications", the strength class for the fundament wall concrete blocks is C 12/15. The composition of the concrete was taken with respect to the required strength of 19.5 MPa, with consideration of the coefficient of the required strength equal to 1.3 (GOST 18105-10 "Concrete. Rules for monitoring and evaluation of strength") with a coefficient of variation of the concrete strength equal to 13.5 %.

According to the recommendations [14], it is advisable to take the Portland cement Holcim CEM II/B-S 32.5 N with slag (GOST 31108-2003 "Cements. Specifications") as the binder. As a fine aggregate, the medium size silica quarry sand (GOST 8736-93 "Sand for construction works. Specifications") was used. The properties of the sand were determined according to the standard procedures in accordance with GOST 8735-88 "Sand for construction works. Test methods". The bulk density of the sand was equal to 1596 kg/m^3, the true density was equal to 2650 kg/m^3. The air volume coefficient for the sand was determined analytically and was equal to 39.8 %. The sand satisfies the requirements of the State Standard of Russian Federation with respect to the grain composition. The gradation factor of the sand was equal to 2.1, the water demand was equal to 7.5 %.

As a coarse aggregate, the rubble granite (fraction 5-20 mm) according to GOST 8267-93 "Rubble and gravel from dense rocks for construction work. Specifications" was used. The properties of the rubble were determined by standard methods in accordance with GOST 8269.0-97 "Rubble and gravel from dense rocks and industrial waste products for the construction work. Test methods". The grain composition of the rubble satisfies the requirements of the State Standard of Russian Federation. The average rubble density (in a piece), determined by hydrostatic weighing, was equal to 2640 kg/m^3, the bulk density was equal to 1437 kg/m^3, the air volume coefficient was equal to 45.6 %.

As a secondary aggregate, the crushed chrysotile cement flat sheets were used. The properties of chrysotile cement rubble are determined primarily by the properties of the original chrysotile cement products. The chrysotile cement material is dispersed reinforced material with anisotropic properties [3]. The strength of the chrysotile cement along the rolling direction is 1 to 1.25 times higher than the strength across the rolling direction and is significantly higher than the strength in the direction perpendicular to the layers. The average density of the chrysotile cement is 1870...1900 kg/m^3. The compressive strength of the chrysotile cement products is more than 24.5 MPa [15]. The tensile bending strength of chrysotile cement sheets is 17...35 MPa. The porosity is equal to 20...25 %, the pores are mostly open. The water absorption is 20...24 %. A great open porosity determines a low frost resistance of the chrysotile cement products. According to [15], the strength of chrysotile cement on average is reduced by 10 % after 25 freeze-thaw cycles for the average density of 1570 kg/m^3 after 50 freeze-thaw cycles for the average density of 1670 kg/m^3 and after 100 freeze-thaw cycles for the average density of 1800 kg/m^3.

The mineral composition of the chrysotile cement wastes, according to [16], is presented by hydrosilicates, hydroaluminates, calcium hydroxide and calcium carbonate, and chrysotile asbestos. The content of clinker minerals hydrates is 53...58 %, that of calcium hydroxide and magnesium is 27...35 %, that of calcium carbonate is 15...20 % and that of asbestos is 6...20 %.

The chrysotile cement flat sheets were crushed using a jaw crusher. The derived secondary rubble was fractionated for the production of rubble with a grain structure similar to the used granite rubble. Due to the laminate structure of asbestos sheets, the rubble obtained from them was characterized by a higher content of grains with elongate form, which determined its higher air volume coefficient. The average density of chrysotile cement rubble was determined by hydrostatic weighing and was equal to 2028 kg/m^3, the bulk density was equal to 843 kg/m^3, the air volume coefficient was equal to 58.4 %. The water absorption by mass of the rubble was equal to 19.2 % that by volume was equal to 38.9 %. The strength characteristics of the rubble were determined by the compression in the cylinder in accordance with GOST 9758-86 "Aggregates porous inorganic for construction work. Test methods". The value of squeezing strength is 3.9 MPa, the softening coefficient is 0.57. The frost resistance grade is 50 cycles.

As a plasticizer, we used the superplasticizer C-3; the dose was equal to 0.75 % by mass of the cement in terms of dry matter.

The composition of the concrete was taken by the conventional method [14] with the use of the Bolomey-Skramtaev formula, the equation for the absolute volume and the equation for the filling of voids in the coarse aggregate. The secondary chrysotile cement rubble in the concrete mix is characterized by the water demand by 30...50 % less than the water absorption of the rubble in the water, which was considered in the process of calculation.

4. Results and discussion

The properties of concrete with different content of second aggregate have been studied. The mobility of all concrete mixtures was equal to 2…4 cm. The following concrete compositions have been studied: the control one (without secondary aggregate), the one with the content of chrysotile cement rubble equal to 10 %, 30 %, 50 % and 100 % of the mass of granite rubble (9 %, 23 %, 33 % and 50 % of the mass of coarse aggregate), and the one with secondary rubble without granite rubble. The calculation of the concrete composition in this case was carried out in order to ensure the conditions of constant water-to-cement ratio of the concrete mix, which, with the consideration of the water demand of secondary rubble, led to an increase of up to 50 % of the cement consumption in the process of increasing the proportion of the secondary aggregate. The increase of the cement consumption also contributes to a higher air volume coefficient of the secondary rubble.

The properties of the concrete and concrete mixes were determined by standard methods in accordance with the State Standard of Russian Federation. The compressive strength of the concrete was determined on the cube samples with the edge length of 10 cm with actuation by means of a scaling factor 0.95 to the strength of standard samples with the edge length of 15 cm. The results are shown in Table 1.

Table 1. The properties of concrete with secondary chrysotile cement rubble.

Characteristics, dimension	The values of characteristics: Control sample (0 % of chrysotile cement)	The share of chrysotile cement rubble with respect to granite rubble,%: 10 %	30 %	50 %	100 %	100 % of chrysotile cement
Compression strength, MPa	19.5	22.6	21.5	22.2	26.8	27.3
Average density of concrete, kg/m^3	2425	2395	2280	2310	2235	2120
Specific strength, MPa	8.0	9.4	9.4	9.6	12.0	12.9

As one can see from Table 2 that average density of the concrete decreases but its strength increases with the increase of the content of porous secondary aggregate in the concrete; and the higher the content of the secondary aggregate in the concrete is, the greater is the strength (Figure 1). That increase is up to 40 %. Accordingly, the specific strength of the concrete with secondary aggregate increases up to 60 % compared with the control composition (Figure 2). This effect is due to increased cement consumption with the water-to-cement ratio in the concrete being unchanged. In addition, it is known that porous aggregates, such as crushed chrysotile cement, play a positive role in the concrete structure formation by absorbing the water at the initial stage of hardening, thereby compacting the contact zone "cement stone - aggregate", and then giving out the absorbed water during the concrete strength set contributing to a more complete hydration of the binder.

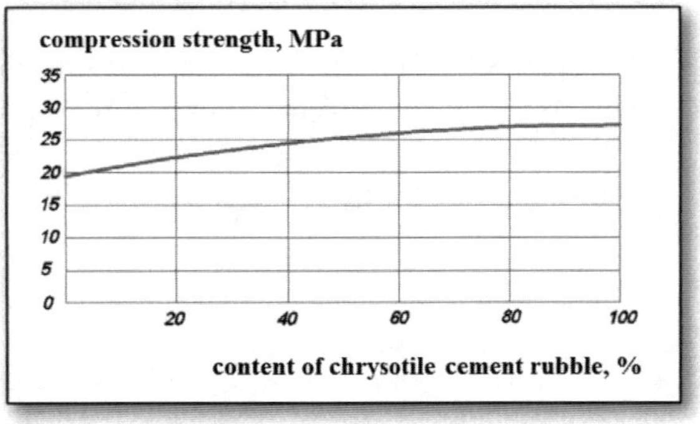

Figure 1. The dependence of the compression strength on the content of chrysotile cement rubble.

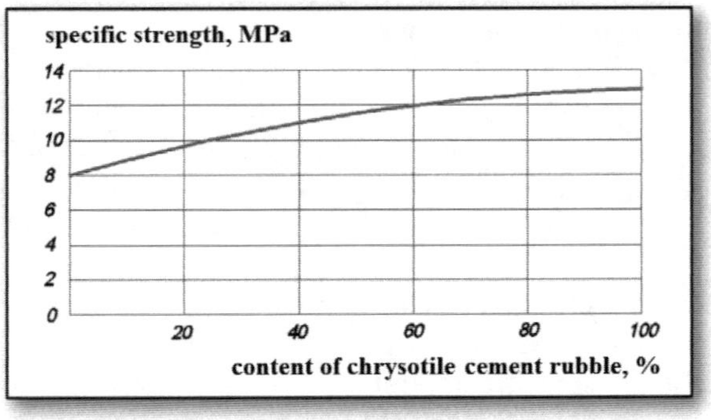

Figure 2. The dependence of the specific strength on the content of chrysotile cement rubble.

It has been found that the introduction of secondary chrysotile cement aggregate into the concrete does not effect the water resistance and frost resistance of tested concrete samples.

The composition of the concrete, in order to maintain the constancy of the water-to-cement ratio in this case is not entirely correct, since the excess water in the concrete mixture is absorbed by the porous aggregate, and so it has no significant effect on the strength of the cement matrix. The strength of concrete with various contents of secondary aggregate at a constant consumption of Portland cement has been determined; it turned out to be the same as for the control composition. The dependence of the concrete strength on the content of secondary rubble is shown in Figure 3.

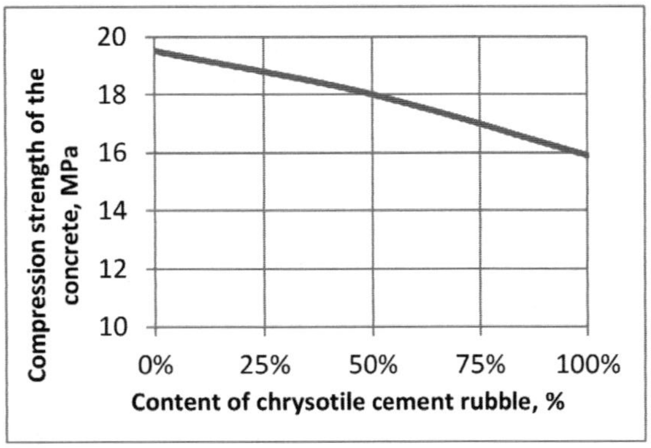

Figure 3. The dependence of the compression strength of the concrete on the content of chrysotile cement rubble at the constant cement consumption.

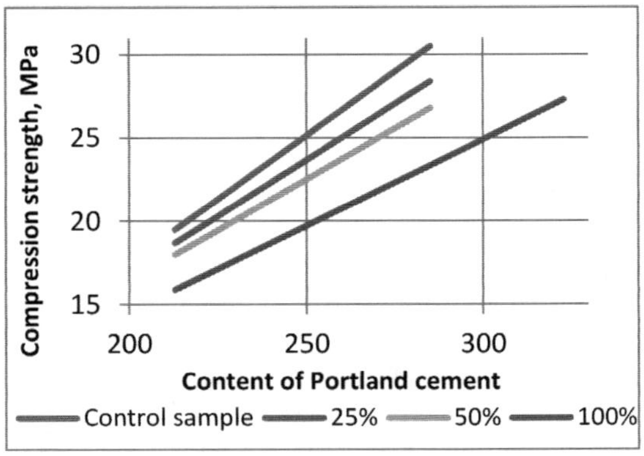

Figure 4. The dependence of the strength of the concrete with different contents of secondary chrysotile cement rubble on the content of Portland cement.

Figure 3 shows that the strength of the concrete decreases with the increase of the content of secondary aggregate. But this is due not only to a low strength of the aggregate, but also to the shortage of cement paste in these compositions, since the secondary aggregate has a higher air volume coefficient (58.4 %), which implies an increase in the sand consumption.

5. Conclusions

The studies confirmed the possibility of using the crushed chrysotile cement products as a secondary aggregate for concrete production; the effective concrete for the production of foundation wall blocks with compressive strength values up to 15 MPa has been developed. The economic and ecological efficiency is achieved through the use of secondary chrysotile cement aggregate. Figure 4 shows the dependence of the strength of the concrete with different contents of secondary chrysotile cement rubble on the content of Portland cement CEM II/B-S 32.5 N. The graphs show that, through the variation of the proportion of the secondary aggregates and the binder, it is possible to obtain the effective concrete with a wide range of strength values. There exists a reserve for further widening of the range for the strength values of the concrete with such aggregate due to the use of the cement with higher strength values. In the process of mass production of some types of products, it is possible to replace completely the granite rubble by the secondary rubble without decrease in physical, mechanical and maintenance characteristics of the products as well as without any increase in the products cost.

References

[1] Oreshkin D 2010 Problems of Building Materiology and Production of Building Materials *Construction Materials* **11** 6-8

[2] Kozlov V, Popov K, Mezhov A and Liljak A 2011 Ways of using chrysotile cement indusrtry waste *Proc. Moscow State University of Civil Engineering* **1(2)** 284-287

[3] Oreshkin D, Popov K, Liljak A and Mezhov A 2011 Utilization of asbestos cement waste in the building materials *Proc. Moscow State University of Civil Engineering* **1(2)** 296-298

[4] Neiman S, Bagautdinov A and Bondarenko M 1992 The use of asbestos waste in the production of building materials *Information center of technical, economic and marketing studies in the construction industry*

[5] Mezhov A 2012 The effective concrete with the use of modified chrysotile cement waste *Cand. of Sc. Thesis*

[6] Chesnokov V and Babich V 2006 Asbestos cement pipes is a reliable tool real housing and utilities reform *Construction Materials* **4** 35-37

[7] Steblyuk A 2012 The study of raw materials for road concrete *Bulletin of SibSIU* **1** 44-46

[8] Neiman S, Luginina I, Vezentcev A, Turskii V, Naumova L, Nesterova L, Gorshkov A and Sivtcov A 2002 Modification of the surface of asbestos fibers in asbestos-cement under the influence of cement hardening products *Construction Materials* **4** 30-31

[9] Luginina I, Vezentcev A, Neiman S, Turskii V, Naumova L and Nesterova L 2001 Changing the properties of chrysotile asbestos cement products under the action of the cement stone and weather factors *Construction Materials* **9** 16-18

[10] Pylev L, Vasil'eva L, Vezentcev A, Gudkova E, Naumova L, Neiman S and Smirnova O 2007 Effect of surface modification of chrysotile fibers on their biological activity *Hygiene and sanitation* **2** 77-80

[11] Manakova N, Kashanskii S and Plotko E 2001 The use of asbestos: ecological and hygienic aspects *Construction Materials* **9** 19-20

[12] Pilipenko A 2012 The effective decorative concrete products for facades with crushing concrete scrap sitted *Cand. of Sc. Thesis*

[13] Zaurbekov Sh, Murtazaev S, Saidumov M and Abdullaev M 2011 Utilization of concrete and reinforced concrete scrap Ecology and Industry of Russian Federation **2** 26-28

[14] Bazhenov Yu 2011 The technology of concrete *Publisher Association Building universities*

[15] Betnei I and Kolbasov V 1985 The technology of asbestos products *Stroyizdat*

[16] Ved' E and Tereshenko L 1963 The binding properties of asbestos and its use for the production of thermal insulation products *Construction Materials* **9** 35

Properties of modified dry masonry mixtures for effective masonry units

V Semenov[1] and T Rozovskaya[1]
[1]Moscow State University of Civil Engineering, Department of Building Materials, Moscow, 129337, Russian Federation

E-mail: tamara.roz@yandex.ru

Abstract. The paper is devoted to the problem of the development of dry light-weight mixtures with hollow ceramics microspheres (CMS) for masonry works. For the one-layer fencing structures including effective masonry units, the use of “warm” masonry mortars is necessary. The used light-weight masonry mortars do not provide the brand strength and thermal uniformity of the fencing structures because of high average density. The CMS are effective light-weight aggregate for such mortars. The influence of the dosage of CMS on the physics-and-mechanics parameters and the technological properties of the masonry mortars has been studied. The optimal mixture compositions have been obtained and their main properties have been determined. The influence of an air-entraining admixture and redispersible polymer powders on the average density and physics-and-mechanics parameters of the masonry mortars have been studied. The optimal compositions of light-weight dry masonry mixtures with CMS have been suggested. It has been established that the mortars, obtained from such mixtures, have the requisite average density, the water retention capacity more than 95 %, high homogeneity and high strength characteristics. The application of the proposed mixtures enables to reduce the construction material costs and to improve the energy efficiency of the fencing structures.

1. Introduction

At present many countries, and particularly Russia, pay great attention to the question of energy efficiency. This is due to several factors such as the limited energy resources, high cost of energy and the negative environmental impact caused by the energy production. Today, according to the Institute of Building Physics, up to 45 % of the energy produced in Russia is consumed by buildings [1]. Herewith, the considerable part of energy is spent on maintenance of the required indoor climate parameters. The primary task of saving energy is to reduce heat loss, including the heat loss through the fencing structures. Studies show that during the operation of a multistory apartment building the heat loss through walls is within the range from 18 to 40 %, depending on the materials used and design solutions applied [2, 3]. The possibility of application of either type of fencing structures is caused by the necessity to provide the normative requirements for thermal protection.

The standard requirements for thermal protection of buildings in Russia are one of the highest in the world. It is confirmed by the fact that the reduced heat-resistance value, which considers the effect of thermal conductive inclusions, is standardized [4]. The fencing structures can be single- or multi-layer ones. However, the use of multilayer structures reduces the coefficient of thermal uniformity and, consequently, increases the wall thickness. In addition, multi-layer structures include thermal insulation made of mineral wool or polystyrene plates. Warranty lifetime of such thermal insulation materials is not more than 15...25 years. Thus, the single-layer fencing structures with the combination of structural and thermal protective qualities in one material, meeting the requirements for thermal protection, are the most technological ones, and they give a cost-effective solution. Such materials used for the erection of single-walled structures include foam ceramics concrete, cellular concrete, polystyrene concrete, etc. [5-7].

The use of cement and sand masonry mortars in one-layer wall structures including effective masonry units leads to the formation of structural cold bridges causing significant heat loss and condensation on interior wall surfaces. To exclude such negative effects, it is necessary to have the average density of masonry mortar not more than the average density of masonry unit. On the other hand, the masonry mortar strength should be at least by 10 to 15 % higher than the strength of the wall material to ensure the required values of the wall design strength. For example, in case of use of the foam ceramics concrete with the average density value of 900 kg/m^3 and the compression strength value of 12.5 MPa, the brand strength value of the mortar should be not less than 14 MPa. The brand strength of the light-weight dry masonry mixtures containing such light-weight fillers as expanded perlite and vermiculite sands, granulated foamed glass and foamed polystyrene does not exceed 5 MPa, while their average density values are not less than 700 kg/m^3. The most widely used foam and cellular concrete blocks have the average density of 600 kg/m^3. Thus, the light-weight masonry mortars available at the building materials market do not provide the design strength of ceramics walls and so do not guarantee the thermal uniformity of a cellular concrete block wall.

The hollow ceramics microspheres are effective light-weight aggregates for the construction and special mortars [8-19]. Previously suggested masonry mortars containing hollow glass microspheres (HGMS) [8-12]. The application of HGMS as a light-weight filler reduces the average density of the mortar down to 600 kg/m^3, but the compression strength value of such compositions is not greater than 2.5 MPa, moreover, the hollow glass microspheres are expensive. The hollow ceramics (alumosilicate) microspheres (CMS) are more effective light-weight fillers, they allow producing mortars with a high strength value and a considerably lower cost [14-19].

The purpose of this study was to develop effective light-weight dry cement mixtures with hollow ceramics microspheres for masonry work, meeting the modern requirements to masonry mortars for effective masonry units.

2. Materials and testing methods

The Portland cement CEM II/A-W 42.5 H (Holcim) with the addition of 9.8 % of granulated slag in accordance with GOST 31108–2003 was used as the binder. The hollow ceramics microspheres with the fraction of 100 to 500 μm "INOTEC Kuznetsk Siberia" were used in this study. The bulk density of the microspheres is 370 to 390 kg/m^3. The average density of the microspheres is 800 kg/m^3. The density of the microsphere wall material is 2450 kg/m^3. The thickness of the microsphere sheath is 1 to 5 μm. The coefficient of thermal conductivity of the microspheres is 0.08 W/m•°C at the temperature of 20 °C, the softening point is over 1000 °C. The minimum compressive strength for a hydrostatic compression (10 % destruction) is 15 to 28 MPa. The gas phase inside the spheres is a mixture of CO_2 (70 %), N_2 (30 %). The photos of CMS microstructure are shown in Figure 1.

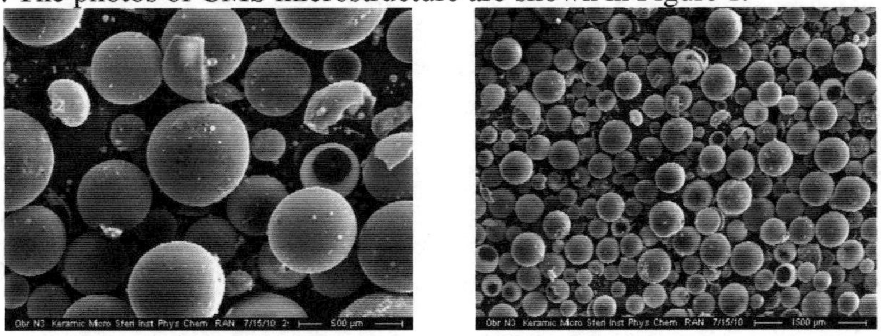

Figure 1. The photos of CMS microstructure

As a plasticizing additive for dry mixtures, the supersoftener PERAMIN SMF 10 (0.4 % of the Portland cement mass) was used which is a sulfonate melamine powder.

In this work the standard testing methods through GOST 5802-86 "Mortars. Test methods", GOST 28013-98 "Mortars. General specifications", GOST 31356-2007 "Dry cement mixtures. Test methods", GOST 31357-2007 "Dry cement mixtures. General specifications" were used. The mobility of all masonry mortars corresponds to the immersion value of a standard cone of 4 to 8 cm. The average density of mortar was determined by a steel pycnometer with the capacity of 100 cm^3. The setting time of the mortar was determined with the Vicat apparatus. The determination of the water-retention capacity of masonry mortar was performed with the help of the DW-RC device. The mortar bending and compression strength values were determined for the prismatic samples with the dimensions of 4x4x16 cm.

3. Results and discussion

The compounds of the mortars with the CMS dosage of 10 %, 20 %, 30 %, 40 %, 50 % and 60 % from the cement mass were studied. The strength of the mortar was determined at the age of 28 days under the conditions, regulated by GOST 5802-86. The compositions and properties of light-weight masonry mortars with CMS are shown in Table 1.

Table 1. The composition and properties of light-weight masonry mortars with hollow ceramics microspheres.

№	Composition, mass %				W/C	Average density of mortar, kg/m^3	Average density of dry cement stone, kg/m^3	Setting time, h-min		Strength of masonry mortar at the age of 28 days, MPa	
	PC	CMS	Water	SS				start	end	compression	bend
1	100	10	31	0,4	0,31	1728	1543	4-00	4-45	48,1	4,89
2	100	20	39	0,4	0,39	1535	1367	4-20	5-00	37,2	4,15
3	100	30	44	0,4	0,44	1386	1277	4-30	5-10	31,3	3,69
4	100	40	48	0,4	0,48	1253	1107	4-35	5-15	24,86	3,38
5	100	50	58	0,4	0,58	1195	1008	4-50	5-45	20,56	3,21
6	100	60	64	0,4	0,64	1104	837	5-15	5-55	17,14	2,84

Notes: PC = Portland cement, CMS = hollow ceramics microspheres, SS = supersoftener, W/C = water/cement ratio.

As we can see from Table 1, the average density of the mortar is 1100 to 1800 kg/m^3 depending on the dosage of the microspheres, the average density of dry mortar is 950 to 1650 kg/m^3. The compressive strength of the mortar is 17.0 to 48.0 MPa, the bend tension strength of the mortar is 2.8 to 4.9 MPa. The average density and the strength of the mortar are reduced when increasing the dosage of the microspheres (Figure 2-4). The developed compositions meet the GOST requirements. The setting time of mortar is great enough for masonry work; the water-holding capacity of mortars is over 95 % for all compositions.

The developed dry masonry mixture can be used for the foam ceramics concrete with the average density of 1000 to 1400 kg/m^3. However, for some other masonry units the average density of the dry stone should be reduced from 400 to 1000 kg/m^3. It has been found that the increase of the dosage of the ceramics microspheres in the mortar to the value of over 60 % does not provide an effective reduction of the average density of the cement stone, causing the unjustifiable rise in the cost of dry mixture [19]. Therefore, to reduce the average density of the mortar, we introduced an air-entraining admixture into the composition. In this study we used the air-entraining admixture ASCO 93 representing an anionic surfactant based on a high-molecular olefin sulfonate. The composition and properties of light-weight masonry mortars with hollow ceramics microspheres and an air-entraining admixture are shown in Table 2.

Table 2. The composition and properties of light-weight masonry mortars with hollow ceramics microspheres and air-entraining admixture.

№	Composition, mass %					W/C	Average density of mortar, kg/m^3	Average density of dry cement stone, kg/m^3	Strength of masonry mortar at the age of 28 days, MPa	
	PC	CMS	Water	SS	AFA				compression	bend
1	100	30	49	0,4	0,03	0,49	1093	887	8,44	2,41
2	100	40	54	0,4	0,03	0,54	965	708	7,21	2,32
3	100	50	64	0,4	0,03	0,64	892	622	6,54	2,01
4	100	60	71	0,4	0,03	0,71	802	576	5,01	1,67

Notes: AFA = air-entraining admixture.

As we can see from Table 2 and Figures 2-4, the introduction of an air-entraining admixture significantly reduces both the average density and the strength of the stone: the average density of dry stone is reduced by 30 %, the compressive strength of the mortar is by 70 to 75 % lower, and the bending strength is by 35 to 40 % lower. Therefore the introduction of the admixture enhancing the physics-and-mechanics properties of the mortar is advisable. The redispersible polymer powders (RPP) which are dry polymer powders obtained through spraying the dry latex emulsion (dispersion of polymer particles in water), belong to this category of admixtures. When mixing with water, RPP again form aqueous polymer dispersions. The RPP improve the adhesive strength with the base and both the tension strength and the bending one [20]. In this work we used admixtures Vinnapas 4023 N, (a copolymer of ethylene and vinyl acetate) and Vinnapas H 8034 (a terpolymer of ethylene, vinyl laurate, and vinyl chloride). The consumption of RPP was 3 % by mass of the Portland cement. The results are shown in Table 3 and Figures 2-4.

Table 3. The composition and properties of light-weight masonry mortars with hollow ceramics microspheres, air-entraining admixture and redispersible polymer powders.

№	Composition, mass %							W/C	Average density of mortar, kg/m^3	Average density of dry cement stone, kg/m^3	Strength of masonry mortar at the age of 28 days, MPa	
	PC	CMS	Water	SS	AFA	V 4023 N	V 8034 H				compres sion	bend
1	100	30	51	0,4	0,03	3	–	0,51	1233	1042	19,72	3,21
2	100	30	53	0,4	0,03	–	3	0,53	1142	951	14,95	2,91
3	100	40	56	0,4	0,03	3	–	0,56	1072	986	15,12	2,81
4	100	40	58	0,4	0,03	–	3	0,58	982	798	12,31	2,59
5	100	50	65	0,4	0,03	3	–	0,65	1003	916	12,58	2,63
6	100	50	69	0,4	0,03	–	3	0,69	928	729	10,42	2,17
7	100	60	73	0,4	0,03	3	–	0,73	907	811	10,68	2,43
8	100	60	76	0,4	0,03	–	3	0,76	831	598	8,38	1,89

Notes: V 4023N – Vinnapas 4023 N, V 8034H – Vinnapas 8034 H.

Analysis of the Table 3 shows that the introduction of RPP into the mortar essentially reduces the effect of air entrainment. However, the strength of the mortar increases, i.e. the expected effect is achieved. Thus, the addition of Vinnapas 4023 N increases the average density of the dry mortar with an air-entraining admixture by 17 to 47 %, the compressive strength becomes 1.9-fold to 2.3-fold greater and the bending strength is by 30 to 40 % higher. The addition of Vinnapas 8034 H does not significantly increase the average density of dry mortar, it is only by 4 to 17 % higher, the compressive strength is by 60 to 80 % higher, the bending strength is by 20 to 80 % higher.

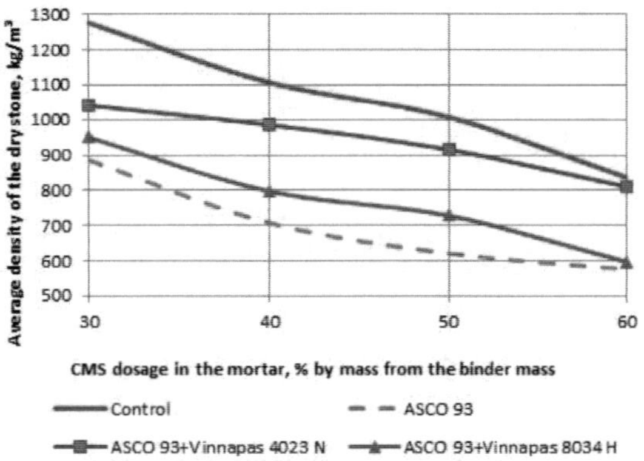

Figure 2. The average density of masonry mortars with CMS, with an air-entraining admixture and redispersible polymer powders depending on the CMS dosage

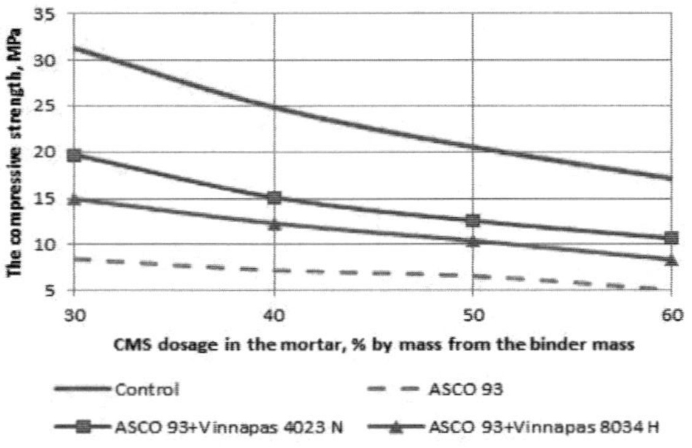

Figure 3. The compressive strength of masonry mortars with CMS, with an air-entraining admixture and redispersible polymer powders depending on the CMS dosage

The actual influence of an air-entraining admixture and redispersible polymer powders can be evaluated in terms of specific strength of the mortar (Figure 5). Thus, for the control composition with the dosage of microspheres of 30 to 60 % by mass of the binder, the specific strength is of 20 to 24 MPa, for the mortar with the air-entraining admixture it is of 8.5 to 10.5 MPa, for the mortar with the air-entraining admixture and Vinnapas 4023 N it is of 13 to 19 MPa, for the mortar with air-entraining admixture and Vinnapas 8034 H it is of 14 to 16 MPa. That is, in terms of the specific strength of the mortars, the used RPP show approximately the same results.

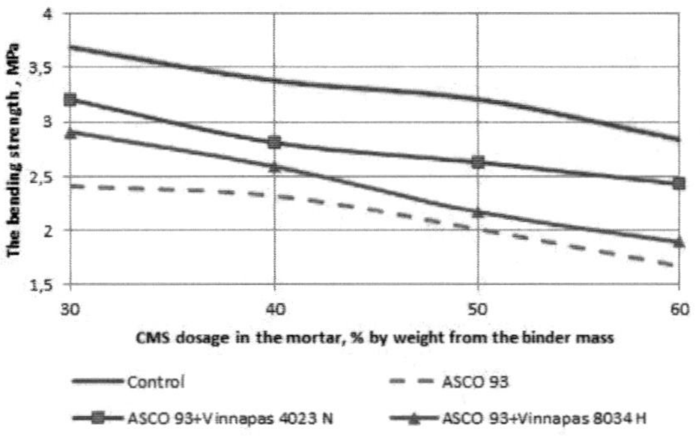

Figure 4. The bending strength of masonry mortars with CMS, with an air-entraining admixture and redispersible polymer powders depending on the CMS dosage.

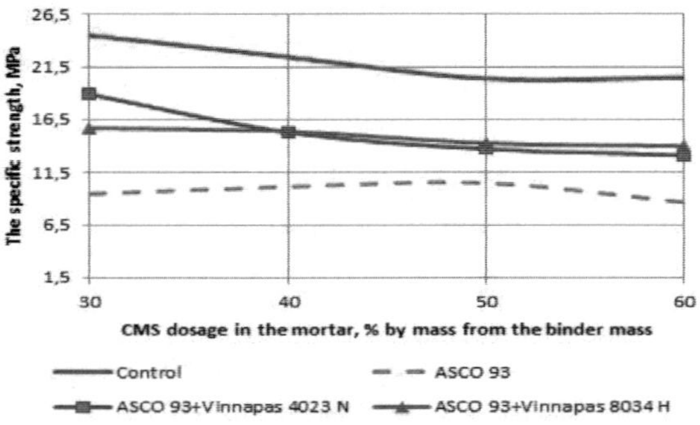

Figure 5. The specific strength of masonry mortars with CMS, with an air-entraining admixture and redispersible polymer powders depending on the CMS dosage.

4. Conclusion

Thus, the use of hollow ceramics microspheres as fillers for masonry mortars, allows producing mortars with an average density of 1000 to 1400 kg/m^3 and with the strength values of 15 MPa to 30 MPa at the dosage of CMS of 20 to 50 % by mass of the binder, designed for efficient ceramics units. Introduction of the air-entraining admixture and RPP into the dry mixture allows getting a mortar with average density of 600 to 1000 kg/m^3 and with the strength values of 5 MPa to 15 MPa for masonry works with cellular concrete blocks.

The developed dry masonry mixtures for effective masonry units provide a high thermal uniformity of the fencing structures and allow us to reduce the masonry seam thickness value to 4…5 mm. It is established that the mortars obtained from such dry mixtures have the required average density value, the water-holding capacity over 95 %, a high homogeneity and high strength parameters. The use of such mortars allows us to reduce the construction cost and to increase the energy efficiency of the fencing structures.

References

[1] Matrosov Yu 2008 The energy-saving in buildings. The problem and the ways for its solution *The Institute of Building Physics (Moscow)*

[2] Gagarin V and Dmitriev K 2013 Accounting heat engineering heterogeneities when assessing the thermal protection of enveloping structures in Russia and European Countries *Construction Materials* **6** 14–16

[3] Oparina L 2013 Results of Calculation of Energy Consumption of a Building Life Cycle *Housing Construction* **11** 50

[4] Gagarin V and Kozlov V 2011 Prospects of increase of power efficiency of residential buildings in Russia *Moscow State University of Civil Engineering Bulletin* **3 (1)** 192–200

[5] Sirazin M 2006 The warm ceramics is a promising material for residential construction in Russia *Construction Materials* **4** 18–19

[6] Gudkov U and Akhundov A 2004 Wall materials based on cellular concrete *Construction Materials* **1** 9–10

[7] Katsynel R 2013 Cellular Concrete and Energy Efficient Construction *Housing Construction* **4** 24–26

[8] Oreshkin D, Belyaev K and Semenov V 2010 High-qualitative building and plugging mortars with hollow glass microspheres *Industrial and Civil Engineering* **10** 56–58

[9] Semenov V, Oreshkin D and Rozovskaya T 2013 Properties of light-weight masonry mortars with hollow ceramics microspheres and antifreeze additives *Industrial and Civil Engineering* **11** 9–11

[10] Semenov V 2011 Efficient light-weight masonry and plugging mortars for inclement climatic conditions *Cand. of Sc. Thesis*

[11] Klochkov A, Pavlenko N, Strokova V and Belentsov Yu 2012 On the use of hollow glass microspheres in heat-insulating structural masonry mortars *Belgorod State Technical University Bulletin* **3** 64–66

[12] Oreshkin D, Belyaev K and Semenov V 2010 Thermophysical Properties, Porosity and Vapour Permeability of Light-Weight Cement Mortars *Construction Materials* **8** 51–54

[13] Oreshkin D, Belyaev K and Semenov V 2010 The overall diagram of obtaining the lightened and super-lightweight cement mortars *Construction of oil and gas wells on land and at sea* **11** 32–33

[14] Inozemtsev A and Korolev E 2013 Hollow microspheres as an effective aggregate for high-strength lightweight concretes *Industrial and Civil Engineering* **10** 80–83

[15] Korolev E, Inozemtcev A 2013 Preparation and research of the high-strength lightweight concrete based on hollow microspheres *Advanced Materials Research* **746** 285–288

[16] Blanco F, García P, Mateos P, Ayala J 2012 Characteristics and properties of lightweight concrete manufactured with cenospheres *Cement and Concrete Research* **30** 1715–1722

[17] Danilin L, Drozhzhin V, Kuvaev M, Kulikov S, Maksimova N, Malinov V, Pikulin I, Redyushev S and Khovrin A 2012 The hollow microspheres from the flue ash as a multi-functional aggregate for composition materials *Cement and its use* **4** 100–105

[18] Teryaeva T, Kostenko O, Ismagilov Z, Shikina N, Rudina N and Antipova V 2013 The physical and chemical properties of alumosilicate hollow microspheres *Kuzbass State Technical University Bulletin* **5** 86–90

[19] Semenov V and Rozovskaya T 2013 Dry masonry mixtures with ceramic hollow microspheres *Science Review* **9** 195–199

[20] 2013 Modern trends in development of the consumption market for complex and multi-functional additives for concrete. Discussions *Concrete Technologies* **7** 8-15

IOP Conf. Series: Materials Science and Engineering **71** (2015) 012043 doi:10.1088/1757-899X/71/1/012043

Modification of cement systems with oxalic aldehyde

N V Subbotina[1], N P Gorlenko[1], Ju S Sarkisov[1], L B Naumova[2] and T S Minakova[3]

[1]Tomsk State University of Architecture and Building, Chemistry Department, Tomsk, 634003, Russia

[2]National Research Tomsk State University, Analytical Chemistry Department, Tomsk, 634050, Russia

[3]National Research Tomsk State University, Physical and Colloid Chemistry Department, Tomsk, 634050, Russia

E-mail: subnv@sibmail.com

Abstract. The experimental results of physical-chemical properties of composite materials on the basis of cement and wood waste modified by an aquatic solution of oxalic aldehyde are presented in this paper. The injection of a chemical addition agent being in optimal concentration is shown to result in the increase of compressive strength of a cement stone by 30%, that of wood-cement composition – in 7 times. IR spectroscopy investigations, microphotographs of structures, kinetics of samples strength changes are shown.

1. Introduction

The injection of chemical addition agents in the compositions based on cement is one of the most effective methods to increase building materials operational properties. For example, the injection of oxalic aldehyde having optimal concentration into a concrete mixture allows increasing the compressive strength, regulate the duration of concrete composition setting [1]. Modification of wood with the help of chemical additional agents of organic origin is of great importance now. The usage of low quality technogenic wood raw material as one of the components for the building materials will allow solving not only the problem of industrial wastes utilization but the ecological task dedicated to the environment conservation. During the last decades wood-cement materials such as fibreboard, woodcrete, and wood-chip rafts are becoming widely used in construction. The pace of development of materials production in this branch of industry, however, is limited by quite low factors of operational characteristics, such as compressive strength, water absorption and others [2-5]. One of the most effective methods of increasing the quality of building materials based on wood-cement compositions application is the modification of the system by means of chemical additional agents. Oxalic aldehyde refers to the perspective additional agent while solving the problems of building materials quality improvement due to the manifestation of different properties this compound possesses.

Oxalic aldehyde is a high active non-toxic substance easily undergoing the reactions of oxidation, reduction and polymerization. It possesses the properties of a crosslinking agent that defines its application for polymer cellulose-containing materials production. It is capable for suppressing water and returning it into the volume of composition during the late periods of structure formation that must lead to the increase of hydration degree for the grain of cementing agent. These properties permit to confirm the fact that oxalic aldehyde is a perspective chemical additional agent in cement compositions especially when the wood is used.

The aim of the paper is to study the processes of structure formation and to determine the physic-chemical properties for composite materials based on cement and wood waste by means of their modification with oxalic aldehyde.

2. Materials and methods

The processes of structure formation of the system "cement - water" at the initial time of hardening was estimated according to the values of setting duration with the help of Vica apparatus GOST 310.3-76, and at the final time after 28 days of hardening in wet conditions the samples sized $(5 \times 5 \times 5) \times 10^{-2}$ m were studied to clear up the limit of compressive strength. For preparation of wood-cement mixture the pine sawdust or chip was used as a wood aggregate. The mixtures obtained by such methods were prepared in the following way: a wood aggregate was treated with an aquatic solution of oxalic aldehyde having different concentrations and Portland cement was injected in the process of agitation. Then this mixture was placed into forms-blocks sized $(5 \times 5 \times 5) \times 10^{-2}$ m, and the impaction was carried out. After stripping of the samples the experiments on compressive strength at different time of composition setting were fulfilled. In both cases cement of brand 400 was used.

3. Results and discussion

The experimental data are presented in Table 1 and Figure 1.

Table 1. The values of the limits for compressive strength, the beginning and the end of cement stone samples setting after 28 days of hardening for the system "cement - water" in accordance with the concentration of oxalic aldehyde solution.

Number of the sample	Concentration of oxalic aldehyde (mass. %)	The beginning of the system "cement – water" aquatic solution setting (min)	The end of the system "cement – water" aquatic solution setting (min)	The limit of compressive strength (MPa)
Control sample	0.00	210	250	40.3
Sample № 2	0.16	180	390	5.4
Sample № 3	0.33	90	360	47.0
Sample № 4	0.67	-	-	0

The results of the experiment show (Table 1) that the compressive strength of modified samples in the system "cement - water" was increased approximately by 30%, the initial setting time of the cement composition was decreased and the final one was increased in comparison with the control samples at the optimal concentration of oxalic aldehyde solution which equals to 0,16 % mass. The effect obtained can be explained with the fact that oxalic aldehydes suppresses water and gradually returns it into the system of structural new compounds during the later setting time that can increase the degree of cement particles hydration and, hence, results in increasing the compressive strength of the samples.

In the system of "wood – cement – water" (WCW) the compressive strength in the final time of structural formation is increased in comparison with control samples by 7 times (Figure 1). In this case the significant dependency of samples strength from the concentration of oxalic aldehyde solution injected into the system is observed. Such essential deviations of samples strength compared with the wood-cement composition without injecting of any chemical additional agent is probably connected with the chemical interactions of oxalic aldehyde with wood. Therefore, for the interpretation of obtained results IR- spectra for wood samples and WCW transmission were made (Figure 2).

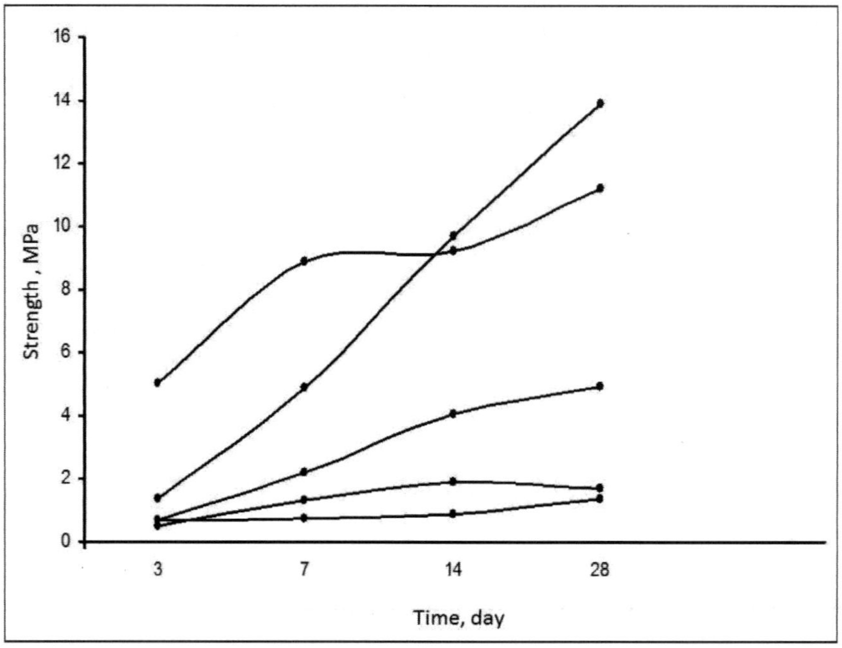

Figure 1. Kinetics of the limits for compressive strength changes which were detected for wood-cement composition and a control sample modified by an aquatic solution of oxalic dehyde in comparison with the control sample: a – a control sample, b – WCW tempered by 3% solution of oxalic dehyde, c - WCW tempered by 0.2 solution of oxalic dehyde, d – WCW tempered by 1% solution of oxalic dehyde, e – WCW tempered by 2% solution of oxalic aldehyde.

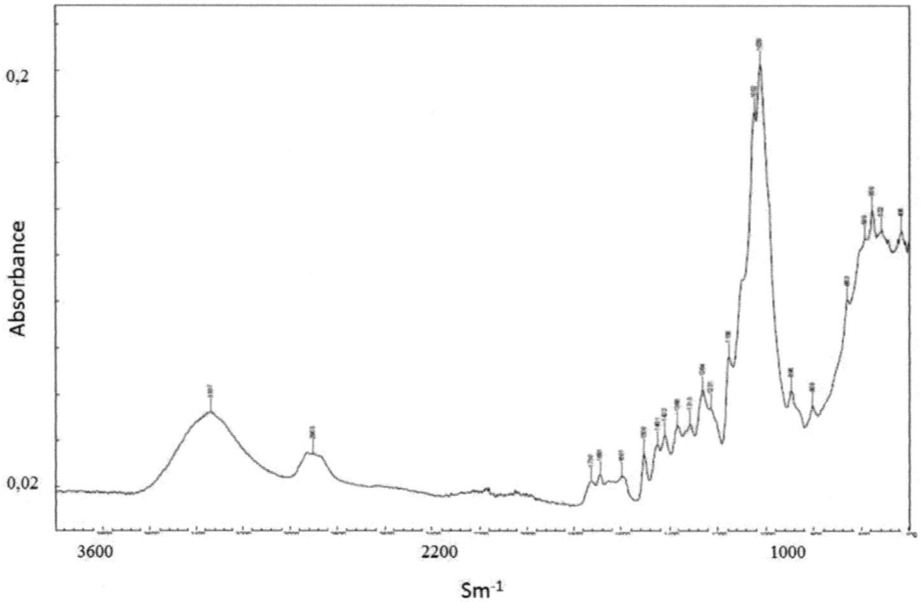

Figure 2. IR- spectrum of pine wood.

It can be concluded from the analysis of IR-spectra presented in Figure 2 that the line of valence vibration of O-H bond (3200 sm^{-1}) is shifted in a high energetic part of the spectrum 3339 sm^{-1} that indicates the appearance of free hydroxide groups. The band in the range of (2945-2845) sm^{-1} characterizes the valence vibrations of CH- and CH_2- groups. The absorption which occurs at the wave length of 1740 sm^{-1} designates tightly associated water that belongs to carbohydrate and lignin components. The bands of absorption in the region of (1507, 1451, 1421) sm^{-1} point at the presence of aromatic compounds in the content of wood, and they belong to the skeletal vibrations of aromatic ring. Very wide and intensive band at (1190-900) sm^{-1} refers to the vibrations of simple aliphatic ethers and to carbohydrate skeletons [6, 7].

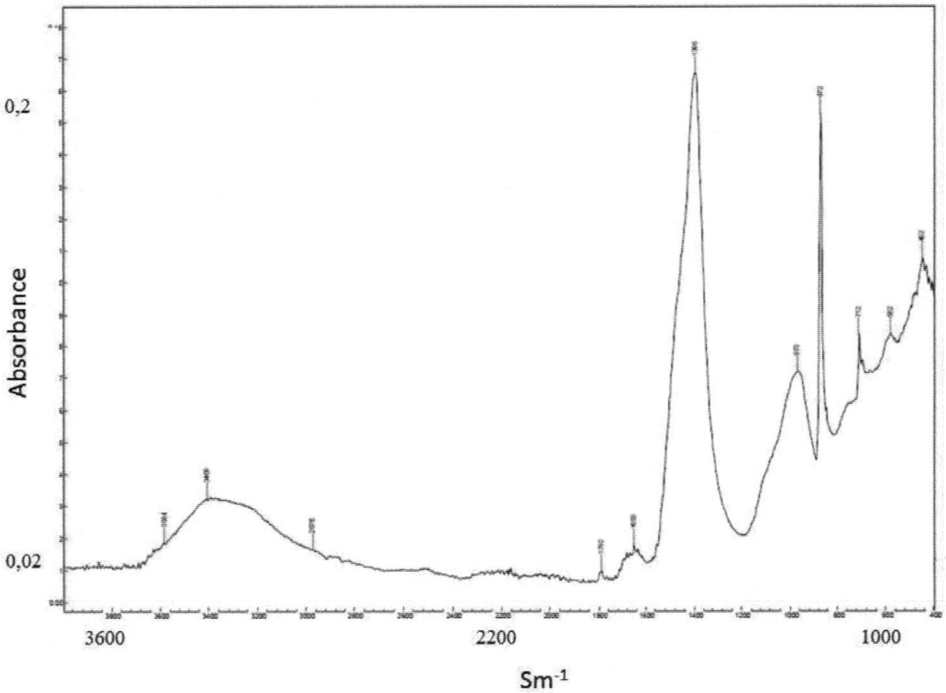

Figure 3. IR-spectrum of wood-cement composition modified by oxalic aldehyde after 28 days of samples hardening.

The main differences of IR-spectra of wood-cement composition (Figure 3) from IR- spectra of wood (Figure 2) are the following. The shift of valence vibrations peak for O–H bond in the region of a longer wave's length is observed. The bands of absorption in the region of aromatic compounds vibrations are practically absent. The narrow and intensive bands in the region of 1396 and 872 sm^{-1} are revealed. The region of cellulose spectra (1500–900) sm^{-1}characterizes different vibrations of C–H, C–O and O–H - bonds, the vibrations of glycoside bond and glucopyranose ring of cellulose, skeletal vibrations of guaiacil ring C_{Ar}–O–C. Except that the intensity of bands in this region is sensitive to the nature of substituting components in the structure of chemical compounds [8-11].

The essential changes in IR-spectra of the samples speak for the complicated chemical and structural interactions in the system of WCW modified with oxalicaldehyde. Taking in to consideration the fact that processes of chemical interaction occur in the region of alkaline condition (pH=12), these changes are connected with lignin, cellulose and other wood components destruction.

Microphotographs (Figure 4) of the initial sawdust and of that modified with oxalic aldehyde, and also of wood-cement composition being compared with the compressive strength of the samples show that the largest strength of WCW samples is achieved in the case when the surface of wood is partially covered with oxalic aldehyde or with the product of its interaction (Figure 4b). These conditions are reached at the concentration of oxalic aldehyde that equals to up to 2 mass., % from the mass of water.

IOP Conf. Series: Materials Science and Engineering **71** (2015) 012043 doi:10.1088/1757-899X/71/1/012043

It should be noticed that the optimal concentration of oxalic aldehyde for a cement stone equals to only 0,16 mass. % (Table 1), that indirectly confirms the processes of ethandial with chemical components of sawdust interaction. During the injection of 3 and more mass. % of oxalic aldehyde the strength of modified samples becomes comparable with that of control samples (Figure 1b). Taking into consideration the fact that cement grout fully covers the surface of wood aggregate (Figure 4c) the contact of cement with the surface of pure wood is considered to be an important moment for strength of WCW structures improvement.

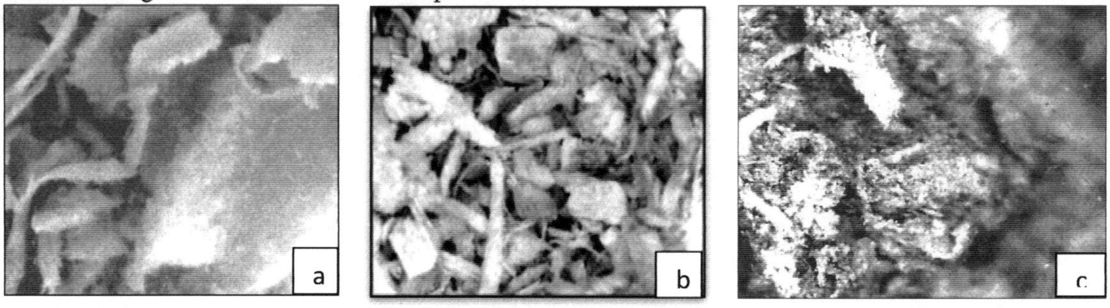

Figure 4. Microphotographs of structures: a – wood; b – wood modified with oxalic aldehyde (2 mass.%); c – wood-cement composition modified with oxalic aldehyde.

On the basis of obtained experimental results the following reasons for the increase of WCW strength modified with oxalicaldehyde can be suggested.

1. It is known that bakelite is formed at the room temperature and condensation of phenol with formaldehyde in the presence of alkaline condensing substances [12]. Bakelite is a solid substance possessing strength. Since oxalicaldehyde is the analogue of formaldehyde, and the wood comprises phenols and aromatic rings in the structures of lignin and cellulose, and also the alkalinity of the medium (pH = 12) is created as a result of cement hydrolysis, one of possible chemical transformations can occur according to the scheme presented in Figure 5. The data of IR-spectra confirm this fact indirectly.

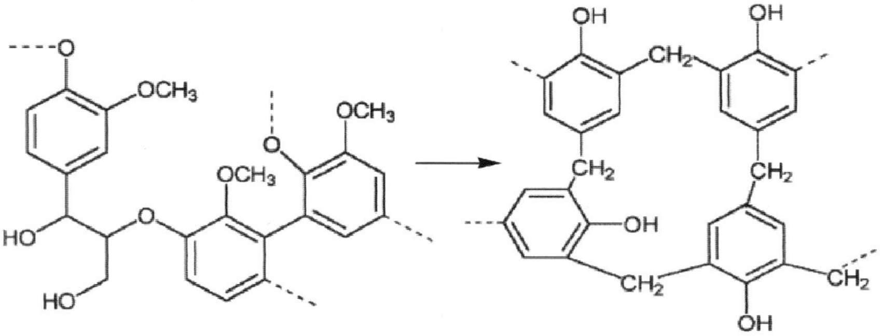

a fragment of lignin structure a fragment of bakelite structure

Figure 5. The simulated scheme of lignin structure transmission into bakelite structure.

2. Oxalic aldehyde suppresses water and gradually returns it in the system of structural new compounds in the later setting time that can increase the degree of cement particles hydration. Even after 28-day hardening the characteristic frequencies which correspond to the water with a normal web of hydrogen bonds and to that with the destructed structure are presented on the obtained spectrograms (Figure 3).

IOP Conf. Series: Materials Science and Engineering **71** (2015) 012043 doi:10.1088/1757-899X/71/1/012043

4. Conclusion

The effectiveness of oxalic aldehyde aquatic solution application in the systems based on cement was shown in the paper. The injection of a chemical additional agent having optimal concentration was determined to regulate the time of cement stone setting, to increase the compressive strength of the samples. The effect of compressive strength of the samples improvement is clearly revealed for the system of wood-cement composition modified with oxalic aldehyde. The optimal concentrations of a chemical additional agent to the system based on cement were detected.

Some possible interactions of ethandial with components of wood were studied by means of IR spectroscopy. It was shown that in alkaline conditions created by the products of cement minerals hydrolysis the destruction of wood components are fulfilled. Some possible mechanisms of the strength improvement for the modified wood-cement compositions were proposed.

References

[1] Gorlenko N P, Sarkisov Yu S and Volkov V A *Bulletin of Higher Institutions "Physics"* 127
[2] Lotov V A 2008 *Problems of Geology and Expansion of Resources: col. of scientific papers of XIII International Symposium* (Tomsk: Tomsk politechnical university) p 819
[3] Pichugin A P, Denisov A S and Khritankov V F 2005 *Building Materials* **5** 2
[4] Nelson M L and O'Connor R T 1964 *Journal of Applied Polymer Science* **8** 1311
[5] Nizina T A 2012 Influence of Mineral Additives on Rheological and Strength Characteristics of Cement Composites *Vestnik of TSUAB* **2** 148
[6] Belamy L 1963 *Infrared spectroscopy of complicated molecule* (Moscow: Publishing house of foreign literature)
[7] Дехант И, Данц Р, Киммер В и Шмонсе Р 1976 *Инфракрасная спектроскопия полимеров* (М.: Химия)
[8] Hergert G L 1960 Infrared Spectra of Lignin and Related Compounds. II. Conifer Lignin and Model Compounds *Journal of Organic Chemistry* **25** 405
[9] Zhbankov R G 1964 *Infrared spectra of cellulose and its derivatives* (Minsk: Nauka i Tekhnika) edited by Golman L P and Reznikov V M 1973 *Journal of Applied Spectroscopy* **19** 494
[10] Pilipchuk Yu S, Pen R Z and Phinkelshtein A V 1965 *Journal of Physical Chemistry* **39** 1768
[11] Chemical Encyclopedia 1988 edited by I L Knuyantsa (Moscow: Publishing house Soviet Encyclopedia) **1**

IOP Conf. Series: Materials Science and Engineering **71** (2015) 012044 doi:10.1088/1757-899X/71/1/012044

Active mineral additives of sapropel ashes

V A Khomich[1], E V Danilina[1], O I Krivonos[2] and G V Plaksin[2]
[1] Siberian State Automobile and Highway Academy, the Department of Engineering Ecology and Chemistry, Omsk, 644050, Russia
[2] Institute of Hydrocarbons Processing of the Siberian Branch of the Russian Academy of Sciences (SB RAS), Laboratory of Catalytic Transformation of Hydrocarbons, Omsk, 644040, Russia

E-mail: vera_khomich@mail.ru

Abstract. The goal of the presented research is to establish a scientific rational for the possibility of sapropel ashes usage as an active mineral additive. The research included the study of producing active mineral additives from sapropels by their thermal treatment at 850-900 °C and afterpowdering, the investigation of the properties of paste matrix with an ash additive, and the study of the ash influence on the cement bonding agent. Thermogravimetric analysis and X-ray investigations allowed us to establish that while burning, organic substances are removed, clay minerals are dehydrated and their structure is broken. Sapropel ashes chemical composition was determined. An amorphous ash constituent is mainly formed from silica of the mineral sapropel part and alumosilicagels resulted from clay minerals decomposition. Properties of PC 400 and PC 500A0 sparopel ash additives were studied. Adding ashes containing Glenium plasticizer to the cement increases paste matrix strength and considerably reduces its water absorption. X-ray phase analysis data shows changes in the phase composition of the paste matrix with an ash additive. Ash additives produce a pozzolanic effect on the cement bonding agent. Besides, an ash additive due to the alumosilicagels content causes transformation from unstable calcium aluminate forms to the stable ones.

1. Introduction

More than a half of all the Russian lacustrine sapropel reserves are available on the territory of West Siberia. As of 2013, 174 deposits were found in Omsk region. The volume of the sapropel reserves is 156...186 million tons. Lake Puchai of the Tyukalinsk district is the main sapropel deposit, with the annual production up to 20 thousand tons. As sapropel belongs to the renewable natural resources its extraction, processing and application may be predicted [1].

Active mineral additives of sapropel ashes are obtained when sapropelic raw materials are burnt at a certain temperature. Sapropelic raw materials consist of water, mineral (silica, clay, carbonates, iron compounds, etc.) and organic parts. While burning the organic part is removed.

The mineral part of sapropels is formed from skeletal sediments of aquatic vegetation (diatoms) and microorganisms. The genesis of this mineral constituent predetermines the amorphous silica content. Amorphous silica SiO_2 is an ingredient of active mineral (pozzolanic) cement additives [2]. Besides silica, there are some other reactants of a pozzolanic reaction, these are alumosilicagels ($nAl_2O_3 \cdot mSiO_2$). They are obtained by burning of sapropelic raw materials, resulted from clay minerals decomposition.

2. The study of obtaining active mineral additives from sapropel

Thermogravimetric analysis and X-ray investigations with the Shimadzu DTG-60 analyzer were used for studies of sapropel thermal transformation. Effects of the exothermic reactions leading to organic substance decomposition, oxidation and carbonization are registered on the DTA curves of the Puchai sapropels at 305-315 and 490-500 °C as well as at the temperature range of 655 – 815 °C. A weak endothermic effect at 540-560 °C occurs when aluminosilicates lose their hydroxyl groups. At 875-885 °C an endothermic effect without an obvious loss of weight is registered. It is connected with mica (muscovite) structure breaking and with melt formation due to alumosilicagels.

X-ray investigations were carried out in the D8 Advance (Bruken) X-ray powder diffractometer with the CuK_α-radiation. Puchai sapropels contain phases of silica SiO_2, albite $Na(AlSi_3O_8)$, muscovite $KAl_2(AlSi_3O_{10})(OH)_2$ and hydrosilicate $H_{10}Si_{10}O_{15}$. Phases of muscovite and hydrosilicate disappear after sapropel burning and additional phases Fe_2O_3 and $Na(AlSi_2O_6)$ are formed in the ash.

On the basis of the experimental results it is determined that organic substances are removed while burning; clay minerals are dehydrated, their structure is broken and alumosilicagels are formed. It leads to increase of the reacting capacity of the material amorphous constituent. However, if the material is burnt longer, the ashes reaction activity starts decreasing since the amorphous components are turned into new crystalline phases.

Ashes obtain pozzolanic properties at the temperature of raw material burning which is higher than those of organic substances removal and of clay minerals dehydration and decomposition. Heat treatment has a significant influence on ash activity. Burning time depends on the type of sapropel, the origin and composition of sapropelic raw materials.

Sapropel ashes were produced by burning at 850 - 900 °C and after powdering. The quantitative analysis of the ashes was conducted on the Varian 710-ES «Agilent Technologies» (Al, Fe, Ca, Mg) atomic emission spectrometer and the AA-6300 «Shimadzu» (Na, K) atomic absorption spectrometer, with the silica being determined by a gravimetrical method, GOST 2642.3-97, Table 1.

Table 1. Chemical composition of sapropel ashes, %.

Sapropel deposit	SiO_2	Al_2O_3	Fe_2O_3	CaO	Na_2O	K_2O	MgO
Puchai	67,72	14,54	5,74	3,37	1,17	2,23	2.33
Zhiloy Ryam	67,30	12,03	5,91	2,67	1,37	2,45	1,92
Gorkoe	63,40	7,08	3,79	8,82	2,41	1,76	5,55
Gorchakovskoe	56,59	13,92	6,69	7,11	2,17	2,30	9,32

Ashes amorphous constituent reactivity was determined by the pozzolanic activity. Y.M. Butt and V.V. Timashev's method was used for determining the pozzolanic activity, in %. Its value for the ash of lake Puchai was 47, and 46 and 34 for the ashes of lake Zhiloy Ryam and lake Molokovskoe respectively.

3. The study of the paste matrix properties with sapropel ashes additives

The ashes influence on the set cement strength properties was the determinative factor for evaluating the efficiency of the ashes pozzolanic activity (GOST 5802-86). Cement mixes were prepared with Portland cement PC 400 and PC 500A0 (additives = 0) produced at Topkinsky cement factory. The Glenium 115 superplasticizer of the "BASF Construction Systems" company was applied for water reducing.

According to the analysis of strength characteristics of the PC 400-based paste matrix (after 28 days of near-water hardening at 20 ± 5 °C) it was found that ashes (8…10 % of the cement weight) with the plasticizer increase the strength by 42…56 % [3]. Compounds with the sapropel ashes of lake

Puchai reveal the highest strength. The PC 500A0-based paste matrix characteristics are given in Figure 1 and Table 2.

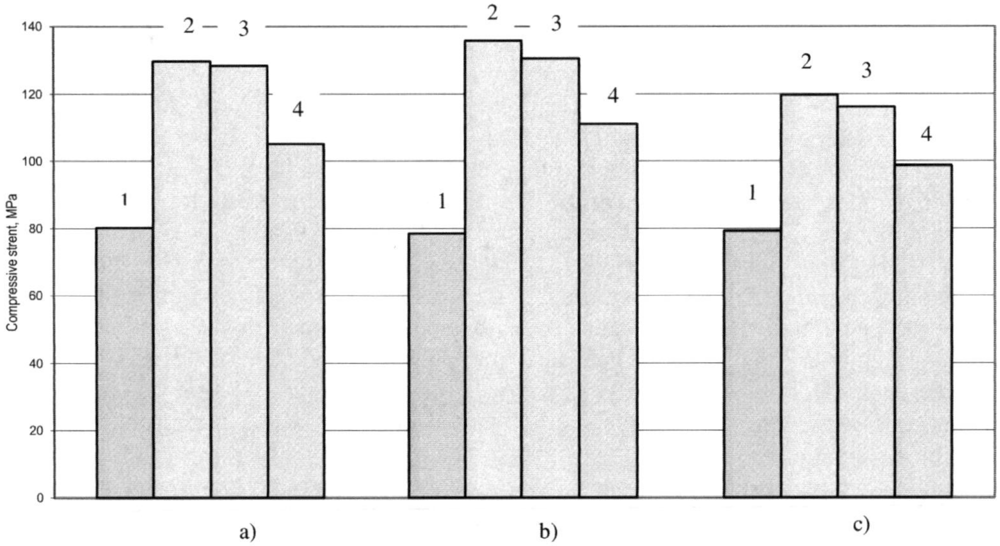

Figure 1. Compressive strength of the PC 500A0-based paste matrix with Glenium (Gl) and sapropel ash: a – 5 %, b – 10 %, c – 15 %; 1 – composition without additives, compositions with additives Gl and ashes of the lakes: 2 – Puchai, 3 – Zhiloy Ryam, 4 – Gorkoe.

Table 2. Paste matrix characteristics.

Additive availability	Compressive strength, MPa	Water-cement ratio, W/C	Water saturation, %	Deflection strength, MPa
Without additives	65,1	0,27	17,6	7,05
Glenium (Gl)	89,8	0,20	-	7,57
Ash	39,4	0,27	-	-
Gl and ash 6 %	98,7	0,21	-	11,3
Gl and ash 8 %	110,5	0,23	3,0	13,47
Gl and ash 10 %	98,4	0,23	-	7,93
Superplasticizer-1 and ash 8 %	81,8	0,28	9,2	-

A PC 500A0-based construction mix additive containing 8 % of the Puchai ash and 1 % of the Glenium 115 superplasticizer increases strength of paste matrix by 70 % and reduces water absorption from 18 to 3 %. Temperature changes during ash production and the degree of ash grinding made it possible to increase paste matrix strength by 94 %.

4. The study of the sapropel ash action mechanism in concrete

To study the influence of an ash additive it was decided to explore the paste matrix phase composition (Figure 2). On the X-ray diagrams of the paste matrix with an ash additive peaks corresponding to the Portlandite (d/n = 4,89; 1,97; 1,76) go down. New peaks corresponding to stable crystallized cubic hydrated calcium aluminates C_3AH_6-type (d/n = 5,14; 4,4; 3,35; 2,78;.1,95; 1,66) appear. Besides pozzolan activity, an ash additive causes transformation from unstable aluminate forms to the stable ones which provide a steady durability during the period of paste matrix use.

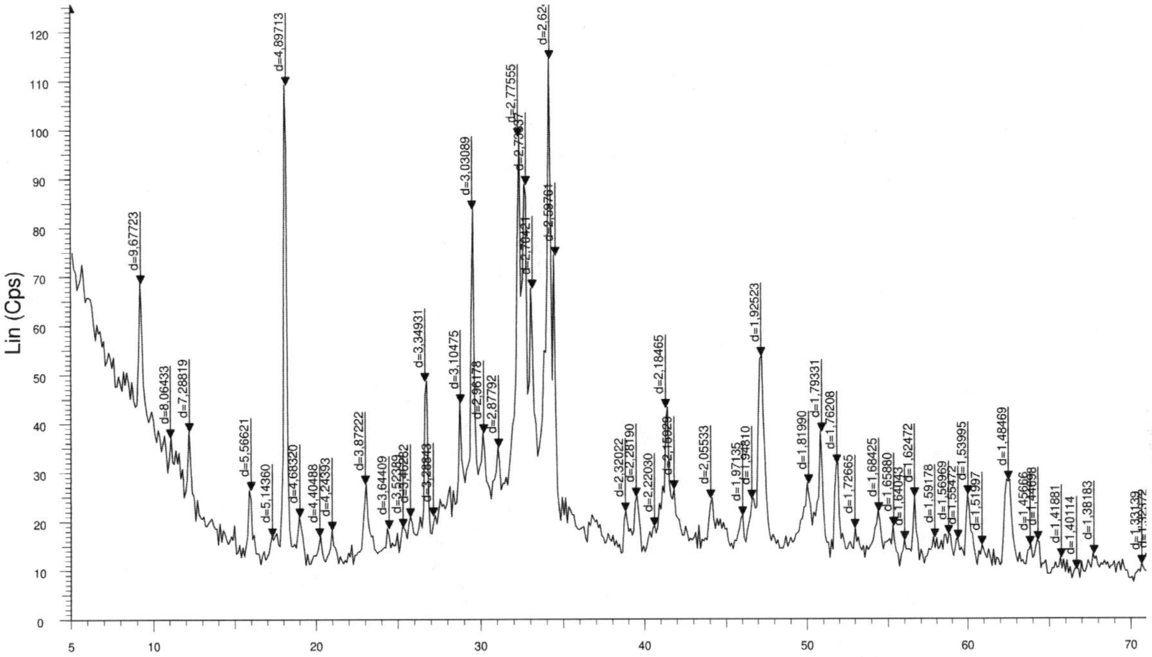

Figure 2. X-ray diagram of the paste matrix with an additive sapropel ash.

5. Conclusion

Thus, adding sapropel ashes to the cements increases paste matrix strength and reduces its water absorption. It has been found that sapropel ashes function as active mineral additives due to the reactive amorphous component, containing SiO_2 and Al_2O_3 in a certain ratio.

Both negative (raw material) sapropels and solid wastes after removal of the organic part including biologically active substances during thermal and chemical processing can be used as raw materials for producing active mineral additives. Waste petroleum and carbon-mineral sorbents, sorbent carriers and catalysts derived from sapropels can be used as raw materials [4].

Producing cement active mineral additives from sapropel ashes is a separate trend in complex processing of Omsk region sapropelic raw materials.

References

[1] Shmakov P F, Tretyakov A G and Levitsky V A 2005 *Sapropel resources of Omsk region lakes and their conservation* (Omsk) 51-70

[2] Kramar L Y, Trofimov B Y, Gamaliy E A, Chernykh T N and Zimich V V 2012 *Modifiers of cement concretes and mortars* (Chelyabinsk: Llc "Iskra-Profi") 202

[3] Khomich V A and Danilina E V 2013 Sapropel ashes as Portland cement additives *Bulletin of the South Ural State University: Series "Construction Engineering and Architecture"* **13** 41 – 44

[4] Krivonos O I 2012 *New approach to integrated sapropel processing*: *Abstract of a Cand. Sc. (Chemistry) Thesis* (Omsk) 18

Dolomite magnesium oxychloride cement properties control method during its production

T N Chernykh[1], A V Nosov[1] and L Ya Kramar[1]
[1]South Ural State University (National Research University), Chelyabinsk, 454080, Russia

E-mail: chernyh_tn@mail.ru

Abstract. The work considers the possibility of reducing the decomposition temperature of $MgCO_3$ in dolomite rock, provides the results of studies of the effect of various additives and enhancers on the decomposition of magnesium and calcium components of dolomite. Chlorides additives are the most promising for dolomite rocks roast intensification. They allow shifting the $MgCO_3$ decomposition to lower temperatures, without exerting a significant influence on the decomposition of $CaCO_3$. Introduction of additives-enhancers is found to be an effective method of controlling the properties of dolomite MOC during roasting, producing high-strength dolomite magnesium oxychloride cements with change in volume during solidification.

1. Introduction

The main features of magnesium oxychloride cements (MOC), including dolomite ones, are setting time, strength and evenness in volume change during solidification. For dolomite MOC these properties are directly dependent on presence and the percentage of free calcium oxide they contain. In case of its presence the strength decreases and it may lead to uneven change in volume during solidification.

It was investigated that in order to obtaining high-quality MOC of dolomite $CaMg(CO_3)_2$ one should ensure that formation of CaO is excluded at the most complete decomposition of the magnesium component. This provides MOC with high strength and evenness in volume change [1-3].

Another and more efficient way of dolomite MOC properties control is a directed formation of a given quality material by controlling the processes of decomposition and crystallization of the original rocks using furnace-charge compilation of several components that work together to form the finished product of specified quality at lower energy costs [4-8].

2. Materials and methods

To research the initial rock decomposition processes authors resorted to a thermal method, the analysis was performed under nitrogen Netzsch Luxx STA409 derivatograph, the heating rate was 10 % min. Binder properties were determined in accordance with All Union State standard and technical specifications. The binder were produced by roasting in the furnace chamber for 2 hours, followed by grinding in a ball mill up to 008 sieve residue of 10-12 %. Satka deposit dolomite rock with the dolomite content of 95-98 % was used as a source raw material; the impurities comprised serpentin-like minerals.

IOP Conf. Series: Materials Science and Engineering **71** (2015) 012045 doi:10.1088/1757-899X/71/1/012045

3. Results

In the study of various dolomite rocks roasting process, it was noted that the intervals of magnesium and calcium components of dolomite decarbonization overlap. This is illustrated by derivatographic analysis (Figure 1). For Satka dolomite magnesium component decarbonization occurs at 680-850 °C, and calcium - at 850-950 °C. However, there is no clear boundary between the magnesium carbonate decomposition end and calcium carbonate dissociation start and at some point two dolomite constituents dissociation processes occur simultaneously. By means of XRF (X-ray fluorescence) authors conducted the study of the dolomite rocks phase composition changes during roasting, and noted the emergence of reflections corresponding to the formation of CaO at a temperature of 750°C. Therefore, the decomposition of the calcium component begins prior to 750 °C, which causes significant difficulties in obtaining MOC with a maximum MgO content and CaO lack.

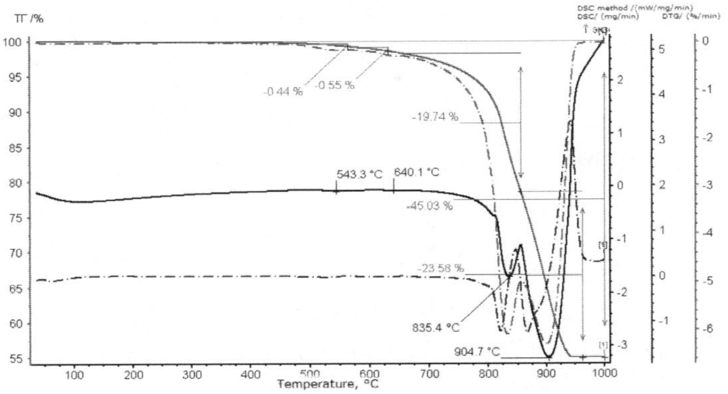

Figure 1. DTA of dolomite rock.

These revealed features of dolomite decomposition during roasting allow us to conclude that to obtain MOC from dolomite the rock magnesium carbonate temperature decomposition decrease is required.

Earlier studies [4] have revealed that chloride additives are the most effective for intensifying roasting dolomite rocks. They allow shifting $MgCO_3$ decomposition to lower temperatures, without entering into the ion-exchange reactions with magnesium and calcium carbonate, and without significant influence of $CaCO_3$ on the decomposition.

Dolomite rock DTA with the addition of NaCl (Figure 2) shows that under the action of the additive intensifier the main peaks of $MgCO_3$ and $CaCO_3$ decomposition to lower temperatures (100 °C and 20 °C respectively) have shifted. Decarbonization of the magnesium component occurs at 620-780 °C, and calcium - at 780-950 °C.

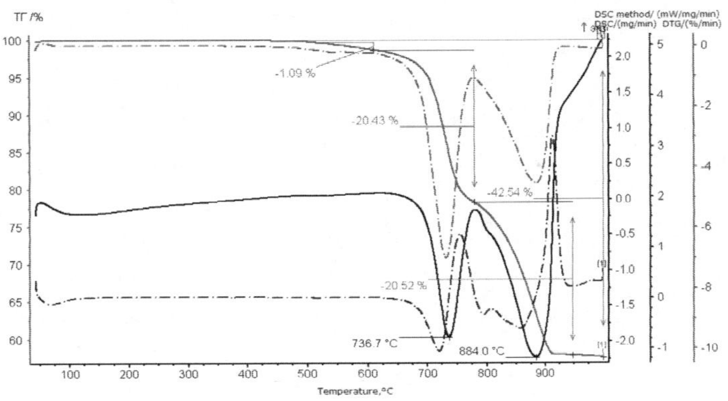

Figure 2. DTA of dolomite rock with NaCl.

When introducing bishofite ($MgCl_2$) additives the main peaks of dolomite endothermic effects shift to lower temperatures by 150 °C for $MgCO_3$ and by 25 °C for $CaCO_3$ (Figure 3). In the course of dolomite roasting with this additive at a temperature of 420-480 °C, a partial dissociation of anhydrous bishofite takes place leading to destabilization of its crystalline grid and reduces the initial decomposition temperature of $MgCO_3$ (endoeffect at 480-600 °C). But active bishofite melting occurs at 600-650 °C, which changes the rate of dissociation of $MgCO_3$ (endoeffect 620-720 °C).

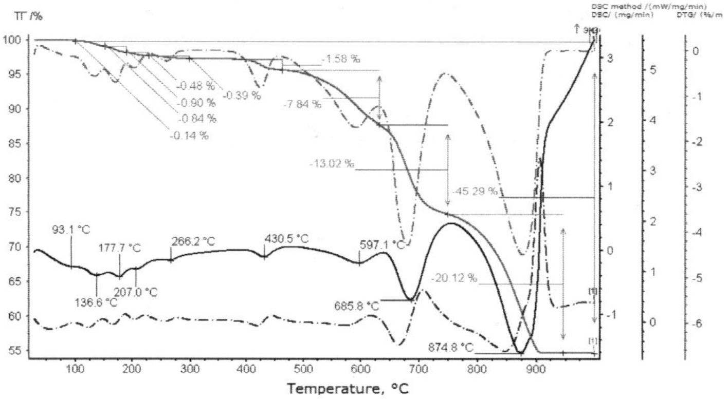

Figure 3. DTA of dolomite rock with $MgCl_2$.

With the introduction of carnallite (KCl·MgCl2·6H2O) the maximum shift of the main endothermic effect occurs corresponding to MgCO3 decomposition - by 200 °C to lower temperatures. Lowering the dissociation temperature is related to the melting of the carnallite, which occurs at 485 °C. Periclase phase, as in the presence of bishofite, already start to form at 500 °C, i.e. 200 ° C lower than in straight dolomite therefore the presence of carnallite encourages a complete decomposition of the magnesium component. The formation of calcium oxide begins at 670 °C.

The studies revealed that the optimum roasting temperature of dolomite with additives is in the range of 550-650 ºC. Table 1 shows the properties of the obtained MOC agents depending on the roasting temperature and the presence of additives.

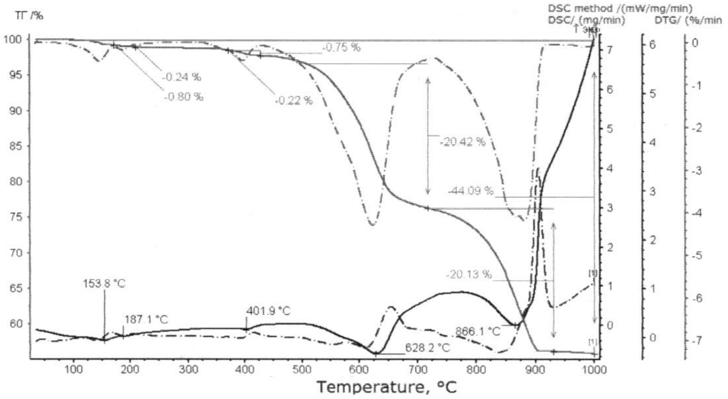

Figure 4. DTA of dolomite rock with $KCl·MgCl_2·6H_2O$.

Table 1. Properties of the resulting MOC.

	Without additives		With carnallite	
Roasting temperature, ºC	750	850	550	650
Free magnesium oxide content, %	10	22	21	22
Free calcium oxide content, %	4	15	0	prints
Endurance limit during pressing at the point of 28 days of solidification, MPa	10	0	68	72
Evenness in volume change during solidification	Net of cracks produced	Completely destroyed to separate units	Without cracks	Without cracks

As can be seen from the results presented in Table 1, the use of additives-enhancers allows obtaining high-strength binder with a uniform change in volume during solidification.

4. Conclusions

Chlorides additives are the most promising for dolomite rocks roast intensification. They allow shifting the $MgCO_3$ decomposition to lower temperatures, without exerting a significant influence on the decomposition of $CaCO_3$.

Introduction of additives-enhancers is an effective method of controlling the properties of dolomite MOC during roasting; it can produce high-strength dolomite MOC with even change in volume during solidification.

References

[1] Бутт Ю М, Богомолов Б Н и Дворкин Л И 1970 Высокопрочный магнезиально-доломитовый цемент *Вяжущие материалы Сибири и Дальнего востока* (Новосибирск: Академия наук СССР, Сибирское отделение, Издательство Наука) 179

[2] Шелихов Н С и Рахимов Р З 2008 Особенности формирования активной фазы MgO в доломитовом цементе *Строительные материалы* **10** 32-33

[3] De Silva P, Bucea L and Sirivivatnanon V 2009 Chemical, microstructural and strength development of calcium and magnesium carbonate binders *Cement Concrete Res* **39** 460-465

[4] Орлов А А и Крамар Л Я 2010 Низкообжиговое магнезиальное вяжущее из бруситовых пород *Вестник ЮУрГУ Серия «Строительство и архитектура»* **11** 25-28

[5] Kolovos K, Loutsi P, Tsivilis S and Kakali G 2001 The effect of foreign ions on the reactivity of the CaO–SiO2–Al2O3–Fe2O3 system: Part I. Anions *Cement Concrete Res* **31** 425-429

[6] Kolovos K, Tsivilis S, Kakali G 2001 The effect of foreign ions on the reactivity of the CaO–SiO2–Al2O3–Fe2O3 system: Part II: Cations *Cement Concrete Res* **32** 463-469

[7] Zhang Y and Ding S 2008 Study on impact of thermal decomposition of dolomite on MgO activity *Industrial Minerals & Processing* **4**

[8] Zhang Y and Li J 2008 Influence of Initial Water Amount on Production of Magnesium Oxychlorride Cement *Liaoning Chemical Industry* **4**

IOP Conf. Series: Materials Science and Engineering **71** (2015) 012046 doi:10.1088/1757-899X/71/1/012046

Processing of materials of laser scanning of roads on the basis of recursive cubic splines

E A Esharov[1] and B M Shumilov[1]

[1]Tomsk State University of Architecture and Building, Department of Applied Mathematics, Tomsk, 634003, Russia

E-mail: sbm@tsuab.ru

Abstract. The problem of approximation of discrete dependence (t_i, y_i) by means of continuous function $f(t)$ arises in tasks of the analysis and data processing of measurements y_i at successive time points of t_i. The solution of the presented task becomes significantly complicated in cases when measurements form a continuous stream, and the results of processing are required to be used before full reception of measurements. The computational schemes based on recurrent spline of degree 3 depths 1 and 2 which provide accuracy for polynomials of the third degree were used in the work. The main results consist in elaboration of mathematical formulas and the proof of stability of the constructed computing schemes. Relevance of the work consists in possibility of application of the constructed splines, for example, for processing materials of laser scanning of roads. Advantages of the constructed splines are the piecewise cubic character of the recovering of the simulated smooth spatial curve and possibility of non-uniform sampling on the route when the quantity of spline knots are less than the number of samples of a function. Examples of calculation of parameters of computing schemes, and also results of numerical experiments are presented.

1. Introduction

Samples of experimental data are often represented as an array consisting of pairs of numbers (t_i, y_i). Therefore, in order to further use them for analysis and processing, the problem of approximating a discrete dependence $y(t_i)$ by continuous function $f(t)$ arises. Function $f(t)$, depending on the specific tasks, may respond to different requirements [1]:

- $f(t)$ should pass through the points (t_i, y_i), i.e. $f(t_i) = y_i$, $i = 1,2, ..., n$. In this case it should be said about the *interpolation* function $f(t)$ at the interior points between t_i or *extrapolation* outside the range that contains all the t_i – forecasting (prediction);
- $f(t)$ should in *some* way (for example as determined according to analytical dependence) approximate y_i, not necessarily pass through the points (t_i, y_i). This is the statement of the problem of *regression*, which in many cases can also be called *smoothing* data;
- $f(t)$ should *approximate* the experimental dependence $y(t_i)$ at the given points considering that the data (t_i, y_i) are obtained with an error, which expresses the measurement noise component. The function $f(t)$ using a particular algorithm reduces the error in the data (t_i, y_i). This type of problem is called *filtering*. *Smoothing* – a special case of filtering;
- $f(t)$ is used for more efficient use of storage and data transfer. This allows the data compression. *Data compression* process is recoding data produced in order to reduce their storage.

The solution of the presented problems becomes significantly complicated in cases when measurements form a continuous stream, so results of processing are required to be used before full reception of these measurements. Such situation arises, for example, at laser scanning of roughness of a roadbed [2].

2. Recursive formulas of approximation compressing splines of degree 3

Let the measured process *f(t)* is registered in the equally spaced time points t_i for measuring steps $\frac{\Delta t}{m}$, and *f(t)* allows for adequate representation in the form of an expansion in *B*-splines of the third degree and between neighboring nodes are $m-1$ measurements. Then the measured values $y_{i,k}$ in the intervals $[t_k, t_{k+1}]$, $k = 0, 1, \ldots$, are [3]:

$$y_{i,k} = \frac{1}{6}\left(1-\frac{i}{m}\right)^3 y_{k-1} + \left[\frac{2}{3}-\left(\frac{i}{m}\right)^2+\frac{1}{2}\left(\frac{i}{m}\right)^3\right]y_k + \\ +\frac{1}{2}\left[\frac{1}{3}+\frac{i}{m}+\left(\frac{i}{m}\right)^2-\left(\frac{i}{m}\right)^3\right]y_{k+1}+\frac{1}{6}\left(\frac{i}{m}\right)^3 y_{k+2}+\xi_{i,k}, \tag{1}$$

where $\xi_{i,k}$ – the measurement errors.

1.1. *Recursive formulas of approximation compressing splines of degree 3 depth 1*

In this case to find estimates $\hat{y}_k$ of expansion coefficients of the cubic spline we will use a recursive filter of depth 1 with the pattern averaging width $m + 1$:

$$\hat{y}_k = \lambda_1 \hat{y}_{k-1} + \sum_{i=0}^{m} a_i\, y_{i,k+1}, \quad k = -1, 0, 1, \ldots, \tag{2}$$

where the initial condition $\hat{y}_k$, $k = -2$, determined by the measured values $y(t)$ at the initial segments, for example, by the least squares method, $m + 1$ – quantity of measurements on each segment of spline, $y_{i,k+1}$ – observation coming from the $(k + 1)$-th segment, the coefficients λ_1, a_i – free parameters of the recursive averaging algorithm.

Theorem 1. Let for a given $m > 2$ the coefficients λ_1, a_i of algorithm (2) satisfy the following conditions

$$\sum_{i=0}^{m} a_i = 1-\lambda_1,\ \sum_{i=0}^{m} a_i\left(\frac{i}{m}\right) = 2\lambda_1 - 1,\ \sum_{i=0}^{m} a_i\left(\frac{i}{m}\right)^2 = \frac{2-11\lambda_1}{3},\ \sum_{i=0}^{m} a_i\left(\frac{i}{m}\right)^3 = 6\lambda_1. \tag{3}$$

$$|\lambda_1|<1 \tag{4}$$

Then the coefficients $\hat{y}_k$ of the 3rd degree spline approximation of depth 1, defined by the algorithm (2), are stable calculated. Spline, constructed by Theorem 1, provides accuracy up to third-degree polynomials.

Proof. Verify the conditions of the accuracy of third-degree polynomials [4]. For that spline, in order to exactly reproduce the measurement $y_{i,k}$ $\forall i, k$ for the corresponding $y_{i,k}$ from the set of values of any polynomial with a degree not higher than 3, it is necessary and sufficient that the flowing equalities valid

$$\begin{cases} \sum_{i=0}^{m} a_i + \lambda_1 = 1, \\ \sum_{i=0}^{m} a_i t_{i,k+1} + \lambda_1 t_{k-1} = t_k, \\ \sum_{i=0}^{m} a_i t_{i,k+1}^2 + \lambda_1\left(t_k^2 - 2t_k\Delta t + \frac{2\Delta t^2}{3}\right) = t_k^2 - \frac{\Delta t^2}{3}, \\ \sum_{i=0}^{m} a_i t_{i,k+1}^3 + \lambda_1\left(t_k^3 - 3t_k^2\Delta t + 2t_k\Delta t^2\right) = t_k^3 - t_k\Delta t^2. \end{cases}$$

Here $t_{i,k} = t_k + \frac{i}{m}\Delta t,\ \ i = \overline{0,m}$. Performance the first of (2) is obvious, from the first equality we obtain $\sum_{i=0}^{m} a_i = 1 - \lambda_1$.

From the second equality, substituting $t_k + \frac{i+m}{m}\Delta t$ for $t_{i,k+1}$ and substituting $t_k - \Delta t$ for t_{k-1}, we obtain $\sum_{i=0}^{m} a_i \left(t_k + \frac{i+m}{m}\Delta t \right) + \lambda (t_k - \Delta t) = t_k$. Removing the brackets and taking t_k out, we have

$$t_k \left(\sum_{i=0}^{m} a_i + \lambda_1 - 1 \right) + \sum_{i=0}^{m} a_i \frac{i}{m}\Delta t + \sum_{i=0}^{m} a_i \Delta t - \lambda_1 \Delta t = 0 .$$

Since the first equality has been proved, the expression in the brackets is zero. Hence, we obtain $\sum_{i=0}^{m} a_i \frac{i}{m}\Delta t + \sum_{i=0}^{m} a_i \Delta t - \lambda_1 \Delta t = 0$.

Reducing both sides of this equality by $\Delta t \neq 0$, we come to the second of the conditions being proved: $\sum_{i=0}^{m} a_i \frac{i}{m} = 2\lambda_1 - 1$.

The third and fourth equalities of system are similarly considered.

Denote

$$\tilde{y}_k = \sum_{i=0}^{m} a_i y_{i,k+1} . \tag{5}$$

Then (2) can be rewritten in the following form

$$\hat{y}_k = \lambda_1 \hat{y}_{k-1} + \tilde{y}_k . \tag{6}$$

Equation (6) is a 1st order linear inhomogeneous differential equation, with the following form of characteristic equation

$$z - \lambda_1 = 0.$$

The equation root equals λ_1, and it is known [5] that condition (4), i.e. $|\lambda_1|<1$, makes the solution of the equation (6) be stable to the initial condition $\hat{y}_{-2}$. Theorem 1 is proved.

1.2. Recursive formulas of approximating compression spline of degree 3 depth 2

Unlike to the previous section, to find the estimates $\hat{y}_k$ of expansion coefficients of the cubic spline will be used the recursive filter of depth 2, with predetermined averaging width $m + 1$:

$$\hat{y}_k = \lambda_1 \hat{y}_{k-1} + \lambda_2 \hat{y}_{k-2} + \sum_{i=0}^{m} a_i\, y_{i,k+1},\quad k = 1, 2, ..., \tag{7}$$

where $m + 1$ – the quantity of measurements at each spline segment, $y_{i,k+1}$ are the observations coming from the $(k + 1)$-th segment and the coefficients λ_1, λ_2, a_i – free parameters of the recursive averaging algorithm. Here, the initial conditions $\hat{y}_k$, $k = -3, -2, -1, 0$, are determined from the measured values $y(t)$ at the initial segments, for example, by the least squares method.

Theorem 2. Let for a given m coefficients λ_1, λ_2, a_i of algorithm (7) satisfy the following conditions

$$\sum_{i=0}^{m} a_i = 1 - \lambda_1 - \lambda_2,\quad \sum_{i=0}^{m} a_i \frac{i}{m} = 2\lambda_1 + 3\lambda_2 - 1,$$
$$\sum_{i=0}^{m} a_i \left(\frac{i}{m} \right)^2 = \frac{2 - 11\lambda_1 - 26\lambda_2}{3},\quad \sum_{i=0}^{m} a_i \left(\frac{i}{m} \right)^3 = 6\lambda_1 + 24\lambda_2. \tag{8}$$

$$\lambda_1^2 + 4\lambda_2 > 0;\quad |\lambda_1| + |\lambda_2| < 1. \tag{9}$$

Then the coefficients $\hat{y}_k$ of the spline approximation of degree 3 depth 2, defined by the algorithm (7), are stable calculated. Spline, constructed under the conditions of Theorem 2, provides accuracy to third-degree polynomials.

Proof is carried out similarly, beginning with the conditions of the accuracy for third-degree polynomials:

$$\begin{cases} \sum_{i=0}^{m} a_i + \lambda_1 + \lambda_2 = 1, \\ \sum_{i=0}^{m} a_i t_{i,k+1} + \lambda_1 t_{k-1} + \lambda_2 t_{k-2} = t_k, \\ \sum_{i=0}^{m} a_i t_{i,k+1}^2 + \lambda_1 (t_k^2 - 2t_k \Delta t + \frac{2}{3}\Delta t^2) + \lambda_2 (t_k^2 - 4t_k \Delta t + \frac{11}{3}\Delta t^2) = t_k^2 - \frac{\Delta t^2}{3}, \\ \sum_{i=0}^{m} a_i t_{i,k+1}^3 + \lambda_1 (t_k^3 - 3t_k^2 \Delta t + 2t_k \Delta t^2) + \lambda_2 \left(t_k^3 - 6t_k^2 \Delta t + 11 t_k \Delta t - 6\Delta t^3\right) = t_k^3 - t_k \Delta t^2. \end{cases}$$

3. Numerical examples and results of the experiments

3.1. Numerical examples

3.1.1. Case p = 1. When $m < 2$ solution does not exist, because the quantity of parameters is less than the amount of conditions (3). For $m = 2$ the measuring points are conveniently placed at the nodes of the spline and at one point midway between the nodes of the spline. Then the values $\frac{i}{m} = 0, 1/2, 1,$ where i – numbers of measurements between the nodes of the spline. As a result, we have 4 unknown values λ_1, a_0, a_1, a_2 to determine which is required to solve a system of linear algebraic equations of the form

$$\begin{cases} a_0 + a_1 + a_2 + \lambda_1 = 1, \\ \frac{1}{2}a_1 + a_2 - 2\lambda_1 = -1, \\ \frac{1}{4}a_1 + a_2 + \frac{11}{3}\lambda_1 = \frac{2}{3}, \\ \frac{1}{8}a_1 + a_2 - 6\lambda_1 = 0. \end{cases}$$

The determinant of the matrix of the system is –25 / 8, so there is a unique solution

$$a_0 = \frac{271}{75}, a_1 = \frac{-296}{75}, a_2 = \frac{91}{75}, \lambda_1 = \frac{3}{25}.$$

Because the absolute value of $\lambda_1 = 3/25$ is <1, the obtained solution fully satisfies the conditions of Theorem 1. The next version – when the measurement points are located at the nodes of the spline and other 2 points for measurement are spaced evenly between the nodes of the spline. If the calculation of spline coefficients takes the points corresponding $\frac{i}{m} = 1/3, 2/3, 1$ (shift of averaging pattern), the only solution exists:

$$a_0 = \frac{721}{96}, a_1 = \frac{-127}{12}, a_2 = \frac{1123}{288}, \lambda_1 = \frac{25}{144}.$$

Since in this case $\lambda_1 = \frac{25}{144} < 1$, the resulting solution fully satisfies the conditions of Theorem 1. However, it is not entirely comfortable with the point of view of prediction properties of the resulting spline solution. Nevertheless, there is an acceptable option, corresponding to the points $\frac{i}{m} = 0, \frac{1}{3}, \frac{2}{3}$, in which the only solution exists,

$$a_0 = \frac{1123}{182}, a_1 = \frac{-732}{91}, a_2 = \frac{39}{14}, \lambda_1 = \frac{8}{91},$$

and it completely satisfies the conditions of Theorem 1.

Let now $m = 3$. As a result, we have 5 unknowns λ_1, a_0, a_1, a_2, a_3, to determine which is required to solve the under determined system of 4 linear algebraic equations, depending on the parameter λ_1,

$$\begin{cases} a_0 + a_1 + a_2 + a_3 = 1 - \lambda_1, \\ \frac{1}{3}a_1 + \frac{2}{3}a_2 + a_3 = 2\lambda_1 - 1, \\ \frac{1}{9}a_1 + \frac{4}{9}a_2 + a_3 = \frac{2 - 11\lambda_1}{3}, \\ \frac{1}{27}a_1 + \frac{8}{27}a_2 + a_3 = 6\lambda_1. \end{cases}$$

The determinant of the matrix of the system is equal to 4/243, so there is a total solution

$$a_0 = -72\lambda_1 + \frac{25}{2},\ a_1 = \frac{363}{2}\lambda_1 - 24,\ a_2 = -156\lambda_1 + \frac{33}{2},\ a_3 = \frac{91}{2}\lambda_1 - 4.$$

3.1.2. Case $p = 2$. If $m < 1$ there is no solution, since the quantity of parameters is less than the amount of conditions (8). In the case of $m = 2$ points of measurement are conveniently placed at the nodes of the spline and at one point midway between the nodes of the spline. Then the values $\frac{i}{m} = 0, \frac{1}{2}, 1$. Thus result we have five unknowns λ_1, λ_2, a_0, a_1, a_2, to determine which is required to solve a system of linear algebraic equations of the form

$$\begin{cases} a_0 + a_1 + a_2 + \lambda_1 + \lambda_2 = 1, \\ \frac{1}{2}a_1 + a_2 + \lambda_1 = 2, \\ \frac{1}{4}a_1 + a_2 + \frac{2}{3}\lambda_1 - \lambda_2 = \frac{11}{3}, \\ \frac{1}{8}a_1 + a_2 - 6 = 0. \end{cases}$$

The determinant of the matrix of the system is equal to 3/8, so there is a common solution, depending on the parameter λ_2,

$$a_0 = \frac{1}{3} + \frac{\lambda_2}{3}, a_1 = \frac{-8}{3} - \frac{8\lambda_2}{3}, a_2 = \frac{19}{3} + \frac{\lambda_2}{3}, \lambda_1 = -3 + \lambda_2.$$

Graphs of received coefficients values depending on λ_2 at $|\lambda_2|<1$ shown in Figure 1.

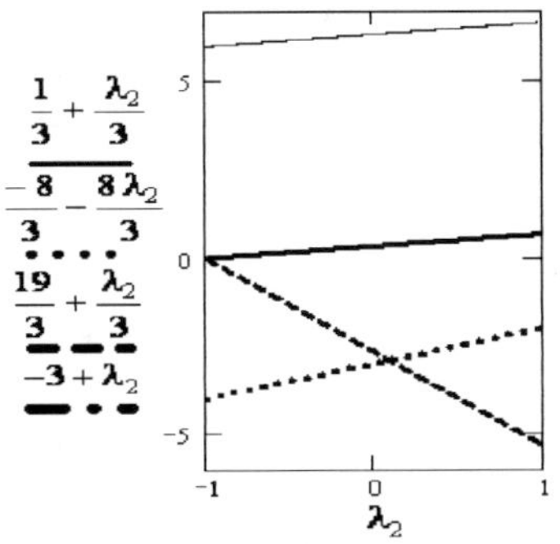

Figure 1. Graphs of received coefficients values a_0, a_1, a_2, λ_1 depending on λ_2.

Thus, the solutions satisfying the conditions of the Theorem 2 do not exist.

In the case $m = 3$ we have 6 unknown values λ_1, λ_2, a_0, a_1, a_2, a_3, to determine which is a general solution, which depends on the parameters λ_1 and λ_2:

$$a_0 = -4 - \frac{3}{2}\lambda_1 + 2\lambda_2, a_1 = \frac{33}{2} + 6\lambda_1 - \frac{15}{2}\lambda_2, a_2 = -24 - \frac{15}{2}\lambda_1 + 6\lambda_2, a_3 = \frac{25}{2} + 2\lambda_1 - \frac{3}{2}\lambda_2.$$

For example, in the case of $\lambda_1 = \lambda_2 = 0$ is obtained one of the possible local approximation formulas:

$$\hat{y}_k = \frac{1}{2}\left(-8\, y_{k-1} + 33 y_{k-\frac{2}{3}} - 48 y_{k-\frac{1}{3}} + 25 y_k\right).$$

3.2. The results of the experiments

3.2.1. The case p = 1. Using the obtained above stable recursive formulas for spline prediction a series of numerical experiments on the approximation of monomials of degree less than or equal to the degree of the spline 3 was done.

1) It was found that at $\sigma = 0$ these formulas provide a zero error of approximation of functions $f(t) = t^l$, l=0, 1, 2, 3. Thus received plots are practically identical.

2) The numerical experiments were carried out with functions interfered with normally distributed noise with zero mean and variance σ^2, $\sigma = 1$, 5, 10, 15. It turned out that for $\sigma \le 9$ properties of accurately reproducing polynomials of sufficiently high degree met, for $\sigma \ge 10$ filters cannot repay so large error.

3) At the last part of the experiments was considered on approximating polynomial of degree $3 + l$, $l > 1$, i.e. greater than the degree of the spline estimates. The resulting error is called monospline and in most cases has a characteristic graph. As known, monosplines are the main error term for the approximation of sufficiently smooth functions. For $l = 1$ the result is shown in Figure 2.

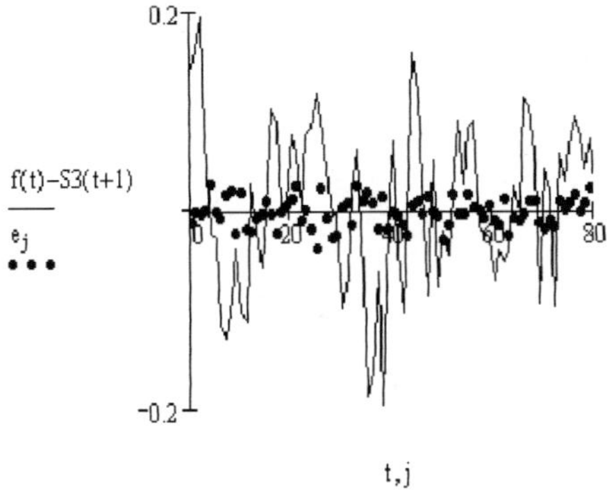

Figure 2. The error $S_3\,(t^{3+l};\, t) - t^{3+l}$ for spline of degree 3 depth 1, e_j – error (random numbers that are normally distributed with mean zero and standard deviation σ =10).

IOP Conf. Series: Materials Science and Engineering **71** (2015) 012046 doi:10.1088/1757-899X/71/1/012046

From comparison of the received results with [1] it is possible to draw the following conclusion: increase of degree of the filter favorably affects an error of the received predicting values on condition of smoothness of approximated dependence.

References

[1] Shumilov B M, Esharov E A and Arkabaev N K 2010 *Numerical Analysis and Applications* **3** 186–198
[2] Shumilov B M, Baigulov A N and Abdykalyk kyzy Zh 2014 *Vestnik of Tomsk State University of Architecture and Building* **1** 142-152 (In Russian)
[3] De Boor C 1978 *A Practical Guide to Splines* (Springer-Verlag: New York)
[4] Shumilov B M 1996 Recurrent approximation by splines *Russian Math. (Izvestiya VUZ. Matematika)* **40** 78–80
[5] Gel'fond A O 1971 *Calculus of finite differences* (Hindustan Publ. Corp.)

Feasibility study of building envelopes selection for low-rise construction

N V Gusakova[1] and A M Gusakov[1]
[1]Tomsk State University of Architecture and Building, Department of Economics and Construction Organization, Tomsk, 634003, Russia

E-mail: Gusakovanata@mail.ru

Abstract. Optimal selection of different types of building envelopes for low-rise construction is a key issue in effective implementation of Regional Programs of Low-rise Housing. When studying this issue the area of construction and ways of transportation should be taken into consideration. In the present paper calculation of minimum thickness and cost per square meter of different variants of load bearing building envelopes was performed at the same thickness of insulant and minimum heat transfer resistance. Comparative analysis was performed providing data on choice of materials for different types of construction.

1. Introduction

In 2011 Ministry of Regional Development of Russian Federation approved the Development Strategy for building materials industry and industrial housing up to 2020. The development and implementation of the Strategy is determined by the necessity of better quality and wider range of products of construction industry at the local market [1].

Low-rise construction is currently spreading over the territory of Tomsk Region. In 2013 85.3 thousand square meters of low-rise housing has been launched. In 2014 60 Ha of land area will be tendered for individual housing.

According to construction codes 50.13330.2012 “Heat protection of buildings” (the updated edition of construction codes SNIP 23-02-2003) the construction of buildings should be performed in accordance with the requirements to heat protection of buildings to provide the necessary microclimate inside the building, the necessary reliability and durability of structures, climatic conditions for operation of technical facilities under minimal energy consumption for heating and ventilation during heating season [2].

Currently the designed building should meet the energy efficiency class A, B^{++}, B^{+}, B and C.

The low-rise building can be considered energy-efficient when while design stage they use building materials enabling to reduce heat losses through exterior walls during operation and resulting in lower energy consumption. In conditions of Tomsk region the problem of optimal choice of materials for the exterior wall is relevant [3].

Modern solutions make it possible to use building materials in different combinations. Therefore the thickness of exterior wall can be decreased up until minimum carrying capacity while keeping thermal and physical properties of the house and also reducing the time of construction [4].

Building materials for building envelopes should meet the following requirements:
1. Energy efficiency;
2. Eco-friendliness and safety for people;
3. Cost-effectiveness;
4. Durability;
5. Fire safety;
6. Aesthetic view.

2. Materials and methods

Thermal-technical calculation of the exterior wall will be held in accordance with Russian construction codes SNIP 23-02-2003 "Heat protection of buildings".

The exterior envelope is calculated as a flat wall that divides the air environment with different temperature and humidity; it is bordered by parallel surfaces and is perpendicular to heat flow. The required heat transfer resistance of building envelopes meeting the sanitary and comfort requirements is calculated by construction codes SNIP 23-02-2003.

Degree day of the heating season:

$$D_d = (t_{\text{int}} - t_{ht}) \cdot z_{ht} = (21 - (-8{,}4) \cdot 236 = 6938{,}4^0 C \cdot \text{ñóò}\ , \quad (1)$$

where $t_{\text{int}} = 21\ ^0$C – design temperature of inside air; t_{ht}=-8,4 – the average temperature of heating season; z_{ht} – duration of heating season, equals to 236 days.

Standard value of heat-transmission resistance of building envelopes (walls) is given in Table 4 Construction codes SNIP 23-02-2003.

$$R_{reg} = a \cdot D_d + b = 0{,}00035 \cdot 6938{,}4 + 1{,}4 = 3{,}83\text{ì}\ ^{2} \cdot {}^{0}\text{Ñ} / \text{Âò}\ . \quad (2)$$

The reduced resistance of building envelope:

$$R_0 = R_{si} + R_k + R_{se}\,, \quad (3)$$

where:

R_{κ} – thermal resistance of building envelope, m^2·°C/W;

$R_{si} = 1/\alpha_{\text{int}}, \alpha_{\text{int}}$ – heat transfer coefficient of the internal surface of building envelope, W / (m^2·°C), is accepted as in Table 7 SNIP 23-02;

$R_{se} = 1/\alpha_{ext}, \alpha_{ext}$ – heat transfer coefficient of the external surface of building envelope for cold climate conditions, W / (m^2·°C), is accepted as in Table 8 SNIP 23-101-2004.

Thermal resistance R, m^2·°C/W, of homogeneous layer of multi-layered building envelope should be defined by the formula:

$$\text{R} = \delta / \lambda\,, \quad (4)$$

where:

δ - thickness of layer, m;

λ - designed coefficient of heat transfer of layer, $\text{Âò} / \text{ì}\ ^{2} \cdot {}^{0}\text{Ñ}$.

Heat transmission resistance:

$$R_o = \frac{1}{\alpha_{\hat{a}}} + \frac{\delta_1}{\gamma_1} + \frac{\delta_2}{\gamma_2} + \frac{\delta_3}{\gamma_3} + \frac{\delta_4}{\gamma_4} + \frac{1}{\alpha_{\acute{\imath}}}. \quad (5)$$

Note that:

1. The thickness of insulant is accepted as 100 mm for all of the variants of building envelope;
2. For all the structures of building envelope it is accepted that R_0^{min}= 3,83 $m^2 \cdot °C/W$;
3. The thickness of the bearing layer of building envelope is accepted from the condition of providing the carrying capacity.

Table 1. Calculation of heat transfer coefficient of building envelope.

Structure of building envelope	Calculation of the thickness of bearing layer of building envelope	R_o^{actual}, $m^2 \cdot °C/W$	Design of building envelope
Ceramic brick Foam polystyrene PENO-PLEX-35 Plaster mortar	$\delta_3 = \left(3{,}83 - \left(\frac{1}{8{,}7} + \frac{0{,}1}{0{,}03} + \frac{0{,}015}{0{,}76} + \frac{1}{23}\right)\right) \cdot 0{,}64 = 0{,}21\text{ì}$ δ=380 mm	4,26	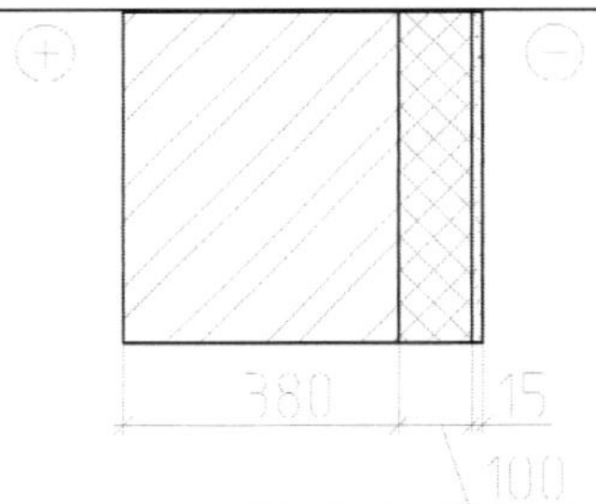
Large-size conventional brick PORO-THERM 25 Foam-polystyrene PENO-PLEX-35 Plaster mortar	$\delta_3 = \left(3{,}83 - \left(\frac{1}{8{,}7} + \frac{0{,}1}{0{,}03} + \frac{0{,}015}{0{,}76} + \frac{1}{23}\right)\right) \cdot 0{,}24 = 0{,}08\text{ì}$ δ=250 mm	4,43	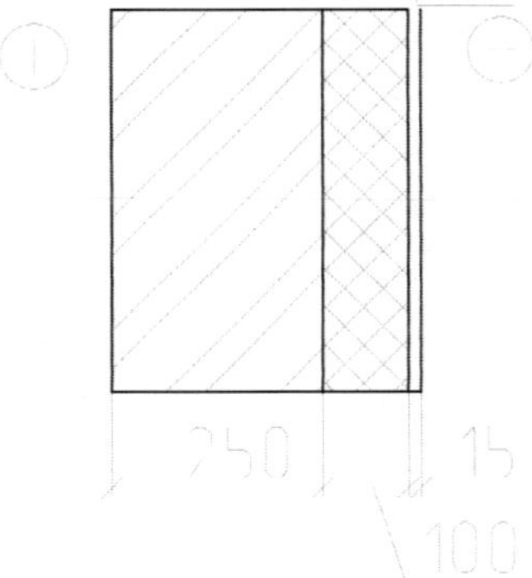
Gas-concrete block Foam polystyrene PENO-PLEX-35 Plaster mortar	$\delta_3 = \left(3{,}83 - \left(\frac{1}{8{,}7} + \frac{0{,}1}{0{,}03} + \frac{0{,}015}{0{,}76} + \frac{1}{23}\right)\right) \cdot 0{,}4 = 0{,}169\text{ì}$ δ=300 mm	4,41	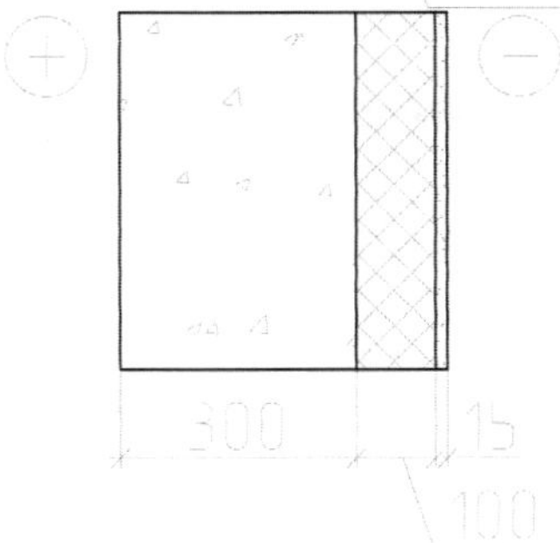

Bar of square section (pine-tree) Foam polystyrene PENO-PLEX-35 Plaster mortar	$\delta_3 = \left(3{,}83 - \left(\frac{1}{8{,}7} + \frac{0{,}1}{0{,}03} + \frac{0{,}015}{0{,}76} + \frac{1}{23}\right)\right) \cdot 0{,}18 = 0{,}058ì$ δ=200 mm	4,62	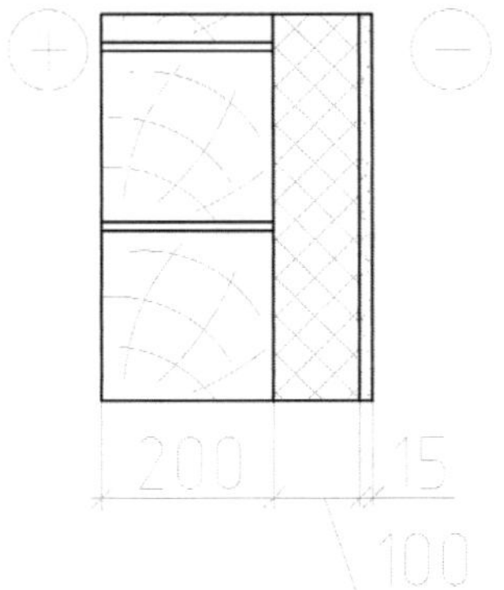
Wall panel	$R_ò = \frac{1}{8{,}7} + \frac{0{,}12}{0{,}67} + \frac{0{,}2}{0{,}03} + \frac{0{,}08}{0{,}67} + \frac{1}{23} = = 7{,}13ì^2 \cdot {}^\circ Ñ/Âò \succ R^o_{min}$ δ=400 mm	7,13	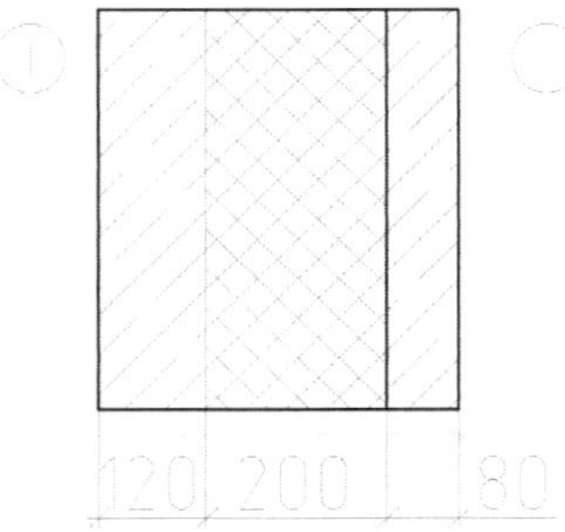

All the variants under consideration correspond with the accepted minimum heat transfer coefficient of building envelopes R_0^{min}= 3,83 $m^2 \cdot °C/W$. This in its regard will allow construction of low-rise buildings in Tomsk region with the highest energy-efficiency class A.

To select the optimal material of the bearing building envelope for low-rise construction we will perform cost calculation for 1 square meter of living floor area (Table 2).

Table 2. Total cost of building envelope counting 100 m^2 of residential floor area.

Material of building envelope	Amount	Cost of material per cubic meter, roubles	Cost of material, roubles	Cost of insulant + finishing, roubles	Total cost, roubles	Cost of material per square meter, roubles
Ceramic brick, m^3	31,92	3200	102144	37800+2898	142842	1428
Large-size conventional brick PORO-THERM 25, m^3	21,0	4100	86100	37800+2898	126798	1267
Gas-concrete block, m^3	25,2	3600	90720	37800+2898	131418	1314
Bar of square section (pine-tree), m^3	16,8	6300	105840	37800+2898	146538	1465
Wall panel	33,6	8000*	-	-	268800	2688

Note: Total cost includes 1 m^3 of reinforced concrete and the chosen insulant.

All the data provided can be applied for construction of different types of low-rise settlements. Variability of choice is defined by the area for construction and the way of materials transportation.

Taking for instance industry or field-based villages the most appropriate is construction of fast erected structures on a lightweight foundation. For construction in the countryside where there are companies-manufacturers of ceramic brick, ceramic large-scale conventional brick or gas concrete block, the construction of these materials will be more reasonable. In case the countryside lacking such companies the choice of materials for building envelopes should be selected regarding the minimal cost.

Table 3. Variability of materials selection for different types of construction.

Type of construction / Material of building envelope	Construction within the city line	Marginal development	Countryside construction	Industry-based villages (agricultural, gas or oil-industry-based villages)
Ceramic brick	+	+	–	–
Large-size conventional brick	+	+	+	–
Gas concrete block	+	+	–	–
Bar of square section (pine-tree)	+	+	+	+
Wall panel	+	+	–	–

3. Conclusions

Building materials under study are reliable and durable; they correspond to the climate conditions of Tomsk Region, while having minimum energy consumption for heating and ventilation of buildings during heating season.

The optimum choice of material use for the load bearing building envelope was performed for low-rise construction in Tomsk and Tomsk Region. The minimum thickness of building envelope and its cost per square meter was estimated for Tomsk and Tomsk region. The table of materials choice variability was offered for different types of construction.

The given data can be used in Development Programs of low-rise construction providing the economical feasibility of investment projects.

References

[1] Министерство регионального развития Российской Федерации Стратегия развития промышленности строительных материалов на период до 2020 года
[2] СНиП 23-02-2003 «Тепловая защита зданий»
[3] Минаев Н Н, Филюшина К Э, Колыхаева Ю А 2014 Исследование закономерностей процессов энергосбережения и повышения энергетической эффективности в регионах России *Сибирская финансовая школа* **2 (103)** 55-60
[4] Овсянников С Н 2009 Повышение энергоэффективности жилых зданий в процессе реновации жилищного фонда *Academia. Архитектура и строительство* **5** 313-318
[5] Гусакова Н В 2014 Инновационные технологии энергообеспечения малоэтажных поселений в инвестиционных проектах малоэтажного строительства *«Проблемы экономики и управления строительством в условиях экологически ориентированного развития»*

IOP Conf. Series: Materials Science and Engineering **71** (2015) 012048 doi:10.1088/1757-899X/71/1/012048

Technical regulation in low-rise construction in the context of heat supply energy efficient projects

G I Prokofyeva[1] and N V Gusakova[1]
[1]Tomsk State University of Architecture and Building, Department of Economics and Construction Organization, Tomsk, 634003, Russia

E-mail: Gusakovanata@mail.ru

Abstract. The present article is devoted to the problem of energy saving and energy efficiency in construction industry and solving the issues of technical regulation and development of the sufficient regulatory basis. Economic calculations have been performed to study the influence of energy efficient heat sources and application of energy saving ventilation systems on the total operation costs. The efficiency of their use has been proved. The use of energy efficient heat sources and energy saving ventilation systems allowed significant reduction of utility expenses providing comfortable living conditions for the residents.

1. Introduction

Reduction of energy intensity of the national economics is one of major issues that Russia faces today. Construction industry in Russia consumes 40 – 45% of the entire energy output. Currently heating of 1 m^2 of living quarters costs Russia 13 liters of standard fuel. At the same time it costs Canada similar to Russia in climatic conditions only 3,5 – 4 liters per year [1]. These conditions determine the relevance of studying energy efficiency in construction in Russia. The amount of energy consumed is planned to be decreased by 30% up to 2016 and by 40% up to 2020 in comparison with basic level of consumption [2] which is accomplishable providing there is substantial development of legislative and technical standards basis. For this purpose it is essential to make arrangements in technical regulation and standardization of energy efficiency.

Federal law No.261 "On energy efficiency and increase in energy performance and on introduction of amendments to several legal acts of the Russian Federation" can be considered the beginning of technical standards basis development [3].

Modern means of technical regulation in construction in Russia is Standards of National Association of Builders (NOSTROY). Integrated programme of development of regulatory documents on energy efficiency and increase of energy performance of buildings and structures has been designed as the result of scientific and development research of the Association. The programme is intended to promote more than 200 regulatory documents on the problem.

Energy efficiency involves systematic approach not only in delivery of energy resources for buildings already being in operation but also application of energy efficient building materials and heat supply system at the point of initial design of the building [4].

2. Experimental

Experience has proven that low-rise construction is most susceptible to application of innovative heat-saving materials and equipment. In this regard, regional investment project of low-rise construction in Tomsk implies:

1. Maximum application of innovative technologies in heat supply for minimal operating costs.
2. Comfortable conditions inside the buildings through installation of most energy efficient and cost efficient ventilation systems.

As the result, the project plans to utilize heat pumps as a basis of heating system and variable air delivery ventilation system that are supposed to ensure up-to-date quality of climatisation of buildings as well as decrease its maintenance costs.

Heat pump utilized in the project is not only a source of heating but air conditioning as well. In winter heat pump accumulates environmental heat to further employ it in the heating process and cold (7 – 9% of which is used for creation of micro-climate in buildings) from driven well in summer. Heat pump allows obtaining of 75 – 84 % of renewable thermal energy free of charge.

Let us consider how exactly application of heat pumps influences utility expenses in low-rise construction. In the table below there is a comparison of utility expenses of a townhouse with heat pump and an apartment with central heating.

Table 1. Utility expenses of accommodation with local and central heating systems.

No.	Index	Townhouse (local heating system)	Apartment (central heating system)
1.	Total area, m^2	140	66,3
2.	Heating expenses per year, RUR	6101	23984
3.	Heating expenses per $1m^2$, RUR	44	361
4.	Heating expenses per month, RUR	762	2998

Calculation shows that the cost of $1m^2$ of area covered by central heating is 8,3 times more than the one covered by local heating based on heat pumps.

In the table below there is a comparison of maintenance costs of buildings with local and central heating systems.

Table 2. Maintenance expenses of accommodation with local and central heating systems.

No.	Index	Townhouse (local heating system)		Cottage (central heating system)
1.	Total area, m^2	140		140
2.	Heating expenses per year, RUR	6101		50646
3.	Saving through heat pump utilization, RUR		44544	
4.	Hot water expenses, RUR	1373		13515
5.	Saving through heat pump utilization for water heating, RUR		10383	
	Total saving (heating, hot water), RUR		54928	

Therefore, total annual cost advantage of local heating system based on heat pumps per building is equal to 54928 RUR, which is equivalent to 9,2 times reduction in operating costs.

Moreover, it is essential to notice that nowadays more and more attention is being paid to two interconnected problems: energy supply and microclimate of buildings. Basing on what has been said previously, traditional approach to climatisation of buildings cannot be acknowledged in economic way.

Traditional systems are rather energy-consuming which means large expenditure of electricity against small refrigeration capacity.

The project is aimed at working out solutions providing decrease in energy consumption and reduction of maintenance costs. Such solutions include installation of ventilation system with recuperation of energy. The system is most fit for kindergartens, schools, cottages which can accommodate system of air supply pipes, recuperator and exhausting system. Major disadvantage of ventilation system with recuperation of energy is imbalance in air supply and exhaust currents leading to loss of efficiency of recuperation.

We applied another solution – variable air volume ventilation system (VAV-system). This system allows regulating air supply in each room independently. VAV-system with $550m^2/h$ performance is sufficient for a middle-sized cottage. Ventilation system is split into 2 sections with separate control shutter.

System's cost including equipment and installation is inconspicuously higher but it allows saving up to 50% of electricity spent for ventilation.

Let us compare maintenance cost of ventilation system with recuperation of energy and variable air volume ventilation system.

Table 3. Calculation of cost efficiency of maintenance expenses.

No.	Index	Ventilation system with recuperation of energy	Variable air volume ventilation system
1	Total thermal energy ventilation expenditure, kW per year	8190	7793
1.1	System of heat disposal (degree of efficiency – 70%) kW per year	-	5455
1.2	Heating with central heat supply system, kW per year	8190	2337
1.3	Ventilation electricity expenditure, kW per year	-	431
2	Total thermal expenditure, kW per year	4550	2338
3	Total annual electricity expenditure, kW per year	-	431
4	Maintenance costs, RUR per year	7772	3202
4.1	Including electricity by 1,82 RUR per kW	-	784
4.2	Thermal energy by 0,949 RUR per kW	7772	2218
5	Total saving provided by VAV-system	-	4570

Therefore, total saving of VAV-system utilization is 4570.5 RUR which exceeds saving of ventilation system with recuperation of energy by 2,4 times.

3. Conclusions

The calculation results showed that when applying heating pump systems heating cost per 1m^2 of the total floor area is reduced by 8,3 times in comparison with central heating system.

The application of local heating system and hot water supply system would lead to economy per one house of total surface area 140 m^2 and will comprise 54,928 RUR per year.

Application of economic ventilation systems for the same house makes the economy of 4570 RUR and enables to reduce the utility expenses by 2,4 times per year.

In conclusion, application of ecologically efficient heating system and modern sources of ventilation allow:

- Reducing total expenditure of thermal energy spent for heating, ventilation and hot water supply compared to annual expenditure in accordance with basic level of energy efficiency requirements.
- Reducing emissions of pollutants into the atmosphere supporting environmental protection.
- Reducing utility payments, transform them into autonomous ones which means independent of energy resources price rise.
- Creating efficient microclimate conditions simultaneously with reducing maintenance costs.

Reference list

[1] ЭСКО Электронный журнал энерго-сервисной компании «Экологические системы» 2009 **11**
[2] Государственная программа Российской федерации «Энергосбережение и повышение энергетической эффективности на период до 2020 года»
[3] Федеральный закон от 23 ноября 2009 г. N 261-ФЗ «Об энергосбережении и о повышении энергетической эффективности и о внесении изменений в отдельные законодательные акты Российской Федерации» (с изменениями и дополнениями)
[4] Колыхаева Ю А и Филюшина К Э 2012 Комплексная оценка эффективности функционирования системы теплоснабжения *ПСЭ* **1** 322-325
[5] Прокофьева Г И и Гусакова Н В 2014 Экономическое обоснование использования технологий энергосбережения в строительстве малоэтажной жилой и социальной инфраструктуры *Вестник ТГАСУ* **1** 42

Application of information systems in road-climatic zoning

V Efimenko[1], S Efimenko[1], A Sukhorukov[1] and A Yankovskaya[2]
[1]Tomsk State University of Architecture and Building, Department Automobile Road, Tomsk, Russia
[2] Tomsk State University of Architecture and Building, Department of Applied Mathematics, Tomsk, 634003, Russia

E-mail: svefimenko_80@mail.ru

Abstract. The article is devoted to the problem of organization and integration of information streams when making territorial road-climatic zoning taking into account geocomplexes of zonal and intrazonal character. The scheme of data acquisition and processing at the stages of allocation of zones, subzones, and road districts is shown. The relevance of the article is due to insufficient consideration of peculiarities of climate conditions while designing of highways in newly reclaimed regions of Russia, for example, Siberia and the Far East.

1. Introduction

The regulatory documents of road industry provide insufficient consideration of climate conditions of the West Siberian region and other Russian regions. Therefore it leads to reduced reliability of highways while commissioning for operation. This also results in significant material and economic losses during their life cycle, associated with bringing of transport facilities to the required conditions.

Nowadays, the tasks of consideration of geographical complexes while designing and construction of roads are mainly solved based on the method of road-climatic zoning of territories. The "zone" was taken as a unit of road-climatic zoning already in the middle of the last century [1, 2]. The "zone" is usually defined as a broad strip of land characterized by a specific combination of heat and moisture, which determines the development of specific and interrelated types of soils and vegetation.

The results of research carried out by famous Russian specialists at Moscow and Leningrad road junctions, [3, 4] (beginning of the thirties of the last century) were taken as a basis when composing standards for designing transport infrastructure facilities. The standards are currently valid in Russia. The findings were extended to other areas, including Western Siberia, and documented in the form of requirements and recommendations. However, the qualitative verification of these results is not always possible on the newly reclaimed areas.

In some regions of Russia, current research is aimed at specification of the boundaries of road-climatic zones. Zoning of separate areas includes overlapping of distribution schemes of zonal, azonal, intrazonal, and regional nature geocomplexes, as well as applying novel solutions – mathematical methods in particular. Considering foreign practice of territorial zoning one should note the experience of scientists from the USA [5]. The effective application of new solutions for road construction industry cannot be implemented due to insufficient development of homogeneity and integrity criteria of the allocated areas, the rules for selection of observation elements, designation of geographical boundaries, as well as the diversity of methods for collecting and processing of initial data. Today, related research in Russian regions is of local nature. It relates only to particular administrative entities due to complexity of the executive organization of work and significant labor and financial costs at their implementation. As there are no unified requirements to work performance on road zoning, the results presented by different researchers are difficult to consider and apply. The aforementioned determines the relevance of the present work. Territorial distribution of boundaries of road-climatic zones considering peculiarities of geocomplexes of zonal, intrazonal, and regional nature should be performed because of absence of a unified methodology for performance of zoning works (within Russian Federation). The methodology should take into account peculiarities of the water-and-thermal regime of roadbeds formed under the influence of various climatic conditions.

2. Problem statement

The methodology of road-climatic zoning should consider additional structuring of territories [6, 7] occupied by zones with allocation of subzones and road districts, for example, on the basis of the taxonomic scheme: "zone – subzone – district" [8, 9].

In this system, the taxon "road district" corresponds to a genetically homogeneous territory characterized by geocomplexes (climate, geology, terrain and other conditions) typical and peculiar only to this territory. Within the territory of the road district, the same-type road structures, especially the roadbed and pavement, should be characterized by approximately same strength and resistance.

The main geocomplex feature of subzones is the terrain. According to the nature of the relief, the terrain is divided into flat, undulated, and mountainous. The guiding criterion at allocation of subzones is the morphostructure significantly affecting design, construction, and usage of auto roads.

The taxon "zone" combines subordinated terms "region" and "subzone" into a subsystem that characterizes the earth's surface with a uniform distribution of heat and moisture, determining the development of certain types of soils and vegetation.

2.1. Brief information on the problem area

The methodical scheme for specification of territorial dislocation of boundaries of road-climatic zones in the system "zone - subzone – district" includes three stages of research [7]. The first stage is aimed at building of information base of geocomplex indicators of zonal and intrazonal nature. Zonal features include climatic conditions that determine the flow of the water-and-thermal regime of roadbeds in the region (average, maximum, and minimum air temperatures, the amount and seasonal distribution of precipitations, evaporation from the land surface, snow height, the depth and the rate of roadbed freezing, the moisture content of the territory). Intrazonal natural factors can vary significantly within the limits of the territory of each zone. Such features can include: terrain, granulometric composition of soils, and other. Indicators considered at zoning can be assigned on the basis of field and laboratory research, consideration of peculiarities of the water-and-thermal regime of roadbeds [10], as well as on the basis of the reference source in relation to control points (settlements) located within the limits of the territorial distribution of zones near hydrometeorological stations.

The second stage of the research on road zoning of the Western Siberia territory consists in component-wise overlapping of geocomplex element distribution schemes, with the involvement of mathematical methods of characteristics processing included in the data base [11]. Features of the solution of the second stage of the work on specification of geographical location of the boundaries of road areas within territorial distribution of zones are discussed in detail in previous works [10].

The third stage of work focuses on positioning of boundaries of districts and zones located within the limits of adjacent administrative entities geographically situated in large regions, for example, in Western Siberia.

2.2. Information systems in road-climatic zoning

Mathematical algorithms for specification of dislocation boundaries of zones, subzones, and road districts consider vector operations, therefore all the initial modelling data is represented in the form of information matrix [9, 12]:

$$\begin{pmatrix} I_1 \\ I_2 \\ \dots \\ I_i \\ \dots \\ I_n \end{pmatrix} = \begin{pmatrix} X_1 & X_2 & \dots & X_j & \dots & X_m \end{pmatrix} = \begin{pmatrix} x_{11} & x_{12} & \dots & x_{1j} & \dots & x_{1m} \\ x_{21} & x_{22} & \dots & x_{2j} & \dots & x_{2m} \\ \dots & \dots & \dots & \dots & \dots & \dots \\ x_{i1} & x_{i2} & \dots & x_{ij} & \dots & x_{im} \\ \dots & \dots & \dots & \dots & \dots & \dots \\ x_{n1} & x_{n2} & \dots & x_{nj} & \dots & x_{nm} \end{pmatrix} \quad (1)$$

The row of the matrix represents a uniform sequence of characteristics of analyzed attributes, the vertical column – a fixed sequence of considered controlled points. Where x_{ij} is the value of the j- th attribute in the controlled point with a number i; m is the number of attributes (the number of columns) characteristic to controlled points; n is the number of controlled points; X_j is the j- th column of the matrix; I_i is the i-th row of the matrix in which the information on the i-th controlled point is recorded. The initial data is standardized by presenting them in dimensionless random values with a zero mathematical expectation and a unit variance.

One major factor that describes the road-climatic zone in an integrated manner is found with the use of the method of principal components of the factor analysis. This factor is a linear combination of parameters and serves as a criterion for classifying the studied territory to a particular road-climatic zone.

Based on the known factor values a polynomial $f(x, y)$ is built for reference points, where for x, y are coordinates of the earth's surface point. The polynomial coefficients are determined by the least-squares method and lines of the polynomial level are built, which define the boundaries of the road and climatic zones, subzones, and districts by groups of intrazonal and regional parameters. Then, an imposition of the obtained results is carried out, and homogeneous territories are allocated by the accepted features.

Mapping of boundary positions for adjacently located zones, for example, administrative entities, as a rule, does not coincide at border segments. Therefore, for the obtained expressions $F1(x, y)$ and $F2(x, y)$ of two neighboring administrative entities, for connecting borders it is possible to use the following solution.

If there is an intersection point, then we shall assume that at this point one line merges into another. If the intersection point in regions is absent, then:

$$F3(x, y) = \frac{F1(x, y)(b-x)(d-y) + F2(x, y)(x-a)(y-c)}{(b-a)(d-c)}, \quad a \le x \le b, \quad c \le y \le d, \quad (2)$$

where (a, c) and (b, d) on lines F1(x, y)=C and F2(x, y)=C, respectively.
The points (a, c) and (b, d) can be assumed, for example, as intersection point of lines F1(x, y)=C and F2(x, y)=C with a common boundary of administrative entities, or points near this boundary point, if the lines do not cross the border.

Division of the studied territories into zones is carried out using the lines of the expression level F11 (x, y), F21(x, y), F31(x, y) (and so on for all administrative entities). Division of the studied territory into subzones is carried out using the lines of the expression level F12(x, y), F22(x, y), F32(x, y) (and so on) corresponding to the second principal factor. Overlaying of the obtained results together with the data on regional factors allows allocating homogeneous territories designated in accordance with the taxon system: road area - subzone - district [13].

3. Conclusion

Among the issues associated with the implementation of the proposed methodical scheme of structured execution of works on road-climatic zoning it is possible to distinguish:

- criterial substantiation of requirements to formation of an information base to prove the territorial distribution of boundaries of road areas, subzones and zones within regionally unified administrative entities;

- development of algorithms for determination of various types of regularities (uninformative, alternative, dependent, binding , pseudobinding, etc.);
- development of a logical-combinatorial optimizing of transformation in the space of optimal subset of attributes and the use of original means of cognitive graphics.

The essence of the latter position is that in accordance with existing representations [14] on the territory, for example Western Siberia, 4 road-climatic zones (RCM) are allocated: I - with distribution of permafrost soils; II - with overmoistened soils (Selyaninov hydrothermal moistening coefficient HTC>1,40); III - with significant moistening of soils in separate years (HTC = 1.00 - 1 40); and IV - with insufficient moistening of soils (HTC = 0.50 - 1.00). The considered territory is characterized by three subzones according to the terrain – flat, undulated, and mountainous. The preliminary analysis showed that within the limits of administrative entities on the territory I RCZ it is possible to allocate three road districts. Within the limits of administrative entities on the territory of II RCZ, according to terrain features and distribution of vegetation, it is expedient to allocate 5 road districts. On the territory III RCZ, according to the same features, it is expedient to conditionally allocate 5 road districts, and on the territory IV RCZ - 2 road districts.

The proposed methodological approach to specification of the boundaries of road-climatic zones in the vast territory of Russia can contribute to a better consideration of road-climatic conditions in the design, construction, repair, and maintenance of highways in regions with poorly developed road network and, therefore, to ensure a decrease in transport and production costs over their life cycle.

The work is supported by Russian Foundation for Basic Research grant (project № 14-07-00673 A).

References

[1] Бабков В Ф 1956 *Основы грунтоведения и механики грунтов* (Москва: Автотрансиздат) 308 с
[2] Преферансова Л А 1953 *Влияние природных условий местности на устойчивость грунтовых оснований усовершенствованных дорог.* Проектирование грунтовых оснований усовершенствованных покрытий с учетом их работы в зимних условиях (Москва: Дориздат) с 5-39
[3] Тулаев А Я 1938 К вопросу характеристики участков, подверженных пучинообразованию *Строительство дорог* **8-9**
[4] Пузаков Н А 1935 *Исследование дорожного полотна* (Ленинград: Ленгострансиздат)
[5] Richlinien fur die Standartisierung des Oberbaues von Verkehrsfiuchen RStO 01 Ausgabe 2001 [online source] http://tu-dresden.de/die_tu_dresden/fakultaeten/vkw/ivs/gsa/dateien/Auszug_RStO_2001.pdf
[6] Ярмолинский А И 2005 *Проектирование конструкций автомобильных дорог с учетом природно-климатических особенностей Дальнего Востока* (Хабаровск: Издательство Тихоокеанского государственного университета) 197 с
[7] Efimenko V N 2002 Methodical fundamentals of road-climatic zoning of the South-East of Western Siberia *Bulletin of higher educational institutions Construction* **10** 87-90
[8] Russam K and J D Coleman 1969 The effect of climatic factors on subgrade moisture conditions *Geotechnique.* **XI** 22–28
[9] Zapata C E and Houston W N 2008 *Calibration and validation of the enhanced integrated climatic model for pavement design* (Washington, D.C.: Transportation Research Board) 62
[10] Efimenko V N 1987 The influence of the water-and-thermal regime on variability of road service quality *Abstracts of the XI All-Union Scientific-Research Conf. "Perspective, economical, and durable highway constructions and the technology of their construction"* 13
[11] Efimenko S V, Efimenko V N and Afinogenov A O 2014 *Vestnik of TSUAB* **1** 125-134
[12] Kupriyanova T P 1977 *Principles and methods of physical-and-geographical zoning using computers* (Moscow: Science) 126
[13] Efimenko S V and Cherepanov D N 2013 *Bulletin of MSUCE* **6** 214-22
[14] SNiP (Russian Construction Codes) 2.05.02 – 85* 2004 *Highways* 56

The concept of increasing energy efficiency of low-rise construction in the context of technical regulation

N Minaev[1,2], K Filushina[1] and E Jarova[1]
[1]Tomsk State University of Architecture and Building, Department of Economics and Urban Management, Tomsk, 634003, Russia
[2]National Research Tomsk Polytechnic University, Department of Natural Resources Economics, Institute of Natural Resources, Tomsk, 634050, Russia

E-mail: nickolian@mail2000.ru

Abstract. Intensive development of low-rise construction and modern trends directed at enhancing energy efficient construction considering regulatory requirements determine the necessity to develop the concept based on variable control models managing the processes of low-rise construction. However effective process management cannot be performed without correlation of sectoral planning documents of federal and regional level also considering the technical regulations requirements. Development of concept to increase energy efficiency of low-rise construction on the basis of governmental housing policy should be built within integrated and system approach. This approach should take into account the factors and objectives of social and economic regional development and its peculiar features, as well as complex urban planning, transport and engineering conditions of low rise housing development. The program and goal-oriented approach used in the present study enables to provide integrity and interrelation with the technical requirements of energy efficiency.

1. Introduction

Efficient management of the processes in low-rise construction cannot be put into practice without harmonization of documents of federal and territorial sector planning with technical regulation requirements. As a result, the concept development of increasing energy efficiency of low-rise construction on the base of state housing policy should be realized within the framework of complex and system approach, including factors and purposes of social and economic regions development and a specific character of social and economic development, architectural, transport and engineering conditions developing territories of low-rise construction. Program-oriented approach with the realization of low-rise construction development provides complexity and interconnection with technical energy efficiency requirements [1].

It was established that low-rise construction has a range of substantial advantages over multi-storey buildings construction, being as follows: mobility, ease of transportation, quickness and easiness of installation, security and reliability, esthetics, comfort, identity, operating ability, cheaper autonomous housing and utility infrastructure and flexible use of energy-saving technologies. However models of interaction between subjects are still not worked out at the federal and regional levels. These models include effective mechanisms and instruments, providing stability of low-rise construction processes development. It is also necessary to improve all requirements to energy efficiency of low-rise construction.

The necessity and significance of developing the concept to increase energy efficiency and integrate energy saving technologies for low-rise construction is defined by high development rates of low-rise housing. It also depends on new tendencies which provide moving from point-like construction inside the settlement to long-distance structures of low-rise settlements around large settlements. They are forming both along engineering utilities and transport hubs and also in areas demanding establishment of stations of energy generation while executing the project of comprehensive development in towns and villages. Therefore arises a whole range of problems connected with development of new energy supply schemes that influences economic efficiency of construction.

2. Experimental

For the last 20 years input of new low-rise housing in the Russian Federation increased by 7 times. Since the period of 2008 to 2013 activation of processes in low-rise construction has been observed. In 2012 about 34,9 mln. m^2 of new low-rise housing has been put into commission (the interest of low-rise construction by total volume of housing equals 53,6 %), and in 2013 in Russia 37 mln. m^2 of low-rise housing has been commissioned which amounts 53,3 % by total volume. From the period of 2008 to 2013 more than 183 mln. m^2 of low-rise housing was commissioned (Table 1), which amounts more than 50 % by total volume on the territory of the Russian Federation. Increase in the level of new housing commissioning, including low-rise construction should make 60 % or around 54 mln. m^2 up to 2015 in accordance with Federal Target Program "Housing"[11]. However, the mentioned Program does not include requirements to energy efficiency of new housing and does not consider its necessity to comply with new technical regulations.

Table 1. New housing commissioning from 2008-2013 thous. m^2.

	2008	**2009**	**2010**	**2011**	**2012**	**2013**
Low-rise housing (total square houses), mln m^2:	27,2	28,5	25,5	30,3	34,9	37

The analysis [2] estimated that the process of low-rise construction development has a range of problems, on the one hand connected with absence of essential regional regulatory and legal framework along with the changes to active land and town-planning legislation and on the other hand connected with the absence of territories provided with access to utilities including the development of communal, transport and social infrastructure for the low-rise construction [3].

Another issue in low-rise construction that is more specific for developed social [4] and economic territories is a high cost of land. So, around 25 % of the housing cost is production cost, 15 % is finishing cost, around 30 % is cost of settling the utility lines of traditional technologies to towns and inside it and 30 % is land cost due to design expenses. In regard to the low income level of social groups (young professionals, public sector workers and other benefit- entitled citizens) needing low-cost comfortable housing, effective demand is essentially lower than volume of buildings planned for commissioning. In this case, financial and credit institutions, reluctantly give loans for low rise structures thus not allowing attracting of loan resources forming other loan programs. The situation deteriorates, because using of traditional low energy effective building technologies of low-rise housing makes high further operational expenses.

Finally, the problem of technical regulation and the development of national standards in the aspect of energy efficiency of low-rise construction must be emphasized. The standards, developed by National Association of Builders (NOSTROI) are established by assembly as private and become the ones to apply to by all the members. However, according to the first half of 2014 in the field of energy efficiency, one standard operated was developed "R NOSTROI 2.15.6-2013 "Guidelines for arranging energy efficient ventilation and air conditioning systems" (Resolution of the Board of National Association of Builders, protocol of 2003, 19 September, No.46) which underwent publisher's editing.

It is worth noting, that the acceptance of technical regulations, according to section 6 of Federal Law "About technical regulation" [12] is providing energy efficiency and resource conservation.

Thus, processes of developing energy efficient low-rise construction are sufficiently restrained by the absence of the corresponding regulations. Consequently, the low-rise housing supply cannot be corresponded to modern regulations and requirements. For this reason, resolving the estimated problems, it requires the active transfer towards low-cost energy efficient technologies of fast erected buildings with necessary functional performance (sustainability, durability, comfort, heat stability), and, therefore, developing regional market of construction materials and structures.

Solving the estimated problems is based on developing of interaction models of management entities motivated to active integration of energy efficient technologies and increasing energy efficiency of low rise construction regarding the requirements of technical regulation.

It is defined by the governmental decision – Regulation No. 1 of 24.01.2014, which demands making energy performance certificate and assign energy efficiency class for every building and structure before its commissioning or after reconstruction. Energy performance certificate is a document necessary for commissioning as it proves correspondence of building parameters with the design documentation. There are also some tax advantages: tax free are newly built buildings with high energy efficiency class (B, B+, B++ and A) in accordance with section 21 Article 381 Tax Code of the Russian Federation, during 3 years from the date of personal property registration.

The model implementation should as well contribute to solving problems of old and failing housing stock.

In that way, a development concept of low-rise construction with account of energy efficiency requirements in the regions of Russia over a long period should follow:

– developing institutional environment, being formed under influence of regional operators of low-rise construction, coordinating interaction between subjects of low-rise construction and providing mutual guarantees and forming energy efficient environment;

– improving of the regulatory environment, stimulating development of energy efficient low-rise construction considering “Comprehensive program of regulatory documentation development in the field of energy-saving and increasing energy-efficiency of buildings and structures for the period until 2015”, that was considered and approved by the Ministry of Regional Development of the Russian Federation;

– developing interaction model between subjects of low-rise construction, taking into account energy efficiency requirements;

– forming the concept of integrated development of low-rise housing (including procedures for the provision of land plots for low-rise construction, developing social and energy efficient engineering and communal infrastructure [5];

– carrying out measures to analyze modern energy efficient technologies of low-rise construction according with the system of technical regulation (it is the system of NOSTROI standardization, which provides for those included in self-regulating organizations an opportunity to impose, on the base of NOSTROI, the uniform requirements in the sphere of construction).

– developing financial arrangements of low-rise construction, with the application of funds “Support Fund for the Low-Rise Construction Development”, “Housing and Utility Reform Foundation”, loan procedures, maternity capital and other mechanisms to provide land properties and construction materials, etc. [6,7];

– cataloging of the available low-rise construction, including defining standard design solutions of energy efficient construction, providing cost reduction at the design stage [8].

The above mentioned considers application of program- and goal-oriented approach which enables to perform the following:

– complex approach of resolving the problems including all aspects of low-rise construction development in the regions, in the context of energy efficiency and funding streams, defining priority ranking of measures and events;

– delegation of power and responsibility between the authority levels, builders, investors, etc. therefore increasing efficiency of programs and events performance;

– efficient planning and monitoring of operational results, where it's necessary to elaborate system of indicators, which allows annual estimation of results of program actions realization.

3. Results

To increase availability of the modern energy efficient low-rise housing, it should be emphasized that financial resources will be primarily directed at development and modernization in accordance with energy efficiency requirements, engineering and communal infrastructure of the low-rise towns. Land use planning and management will be based on the principles of governmental and private partnership and include supplying of accessible properties for the benefit-entitled citizens [9].

It should be considered, that the sufficient demand for the low-rise construction in suburban zones of big cities [10], deficit of free territories and environmental requirements restrict the average size of land by the low-rise construction. In addition, the main reserves of land properties are agriculture territories (not being used for designated purpose) of a municipal formation, where the subtraction is the reasonable decision.

The complex of events includes the measures of governmental and regional support of organizations dealing with construction complex, population, non-profit association of citizens, directed at demand and supply promotion at the housing market, in accordance with the energy efficiency requirements, market balance and development of low-rise construction and building industry in the region.

4. Conclusions

The developed concept realization should be carried out within the following directions:

– multipurpose projects of low-rise buildings (cottage settlements, including economy class), their construction is sometimes performed with the support of projects of integrated land use and development;

– low-rise construction for rural areas, in order to improve living conditions of young specialists and members of their families in the line of Ministry of Agriculture of the Russian Federation;

– collective and individual low-rise construction (Housing Association).

The mentioned measures will enable to influence directly the final effectiveness of the projects of low-rise construction where effect should be built of the two components:

1. Economic effects occurring during the construction process starting from the new technology and building materials integration.

2. Economic effects occurring while long-term operation of low-rise construction objects.

References

[1] Форрестер Дж 1974 *Динамика развития городов* (Москва: Прогресс)

[2] Филюшина К Э 2012 *Управление рисками при реализации инвестиционно-строительных проектов в регионе на основе государственно-частного партнерства (на примере Томской области)* (Санкт-Петербург: Санкт-Петербургский государственный архитектурно-строительный университет)

[3] Шодонере Жанн 2000 *Основные участники современного градостроительного процесса во Франции. Генеральное управление городского планирования, жилищного и капитального строительства Франции* (Москва)

[4] Бейер Э и Корзелиус П 2002 *Социальная жилищная сфера в Нидерландах: что может быть полезно для России* (Москва: Недвижимость)

[5] Минаев Н Н 2010 *Управление жилищно-коммунальным комплексом города: территориально-отраслевой аспект* (Челябинск: Южно-Уральский государственный университет) с 12-14

[6] Филюшина К Э 2012 *Формирование модели государственно-частного партнерства в строительном комплексе региона* (Москва: Региональная экономика: теория и практика) с 41-44

[7] Марку Лаппалайнен 2009 *Энергия, экология и деревянное строительство* (Санкт-Петербург: НР АДД)
[8] Тед Кесик, Майкл Лио 2004 *Строительство деревянных каркасных домов по каркасной технологии* (Канадская ипотечная и жилищная корпорация. пер. с англ.: СМСН) с 311
[9] Минаев Н Н и Филюшина К Э 2011 *Методика анализа и оценки рисков в строительном комплексе региона* (Москва: Интеграл) с 158-159
[10] Марку Карьялайнен, Рику Патокоски 2009 *Урбанизированные деревянные поселки* (Хельсинки: Ракеннустието) с 168
[11] Постановление Правительства РФ от 17 декабря 2010 г. № 1050 «О федеральной целевой программе «Жилище» на 2011 - 2015 годы" (Система ГАРАНТ: http://base.garant.ru/12182235/#ixzz3KvLn8E6Y)
[12] Федеральный закон от 27.12.2002 № 184-ФЗ (ред. от 23.06.2014) «О техническом регулировании» (КонсультантПлюс: http://www.consultant.ru/popular/techreg/)

Quality assessment of urban environment

T Y Ovsiannikova[1] and M N Nikolaenko[1]
[1]Tomsk State University of Architecture and Building, Department of Expertize and Real Estate Management, Tomsk, 634003, Russia

E-mail: tatovs@mail.ru, preobrazhenskaya.m.n@gmail.com

Abstract. This paper is dedicated to the research applicability of quality management problems of construction products. It is offered to expand quality management borders in construction, transferring its principles to urban systems as economic systems of higher level, which qualitative characteristics are substantially defined by quality of construction product. Buildings and structures form spatial-material basis of cities and the most important component of life sphere – urban environment. Authors justify the need for the assessment of urban environment quality as an important factor of social welfare and life quality in urban areas. The authors suggest definition of a term "urban environment". The methodology of quality assessment of urban environment is based on integrated approach which includes the system analysis of all factors and application of both quantitative methods of assessment (calculation of particular and integrated indicators) and qualitative methods (expert estimates and surveys). The authors propose the system of indicators, characterizing quality of the urban environment. This indicators fall into four classes. The authors show the methodology of their definition. The paper presents results of quality assessment of urban environment for several Siberian regions and comparative analysis of these results.

1. Introduction

The quality management has recently become a crucial component of economic systems management. The sphere of quality management mainly includes production systems with the quality of processes, goods or services as the objects of management. It is valid for construction industry as well where practically all large enterprises have created corporate systems of quality management. It allows them increasing competitiveness of construction production in the market. Nevertheless, problems of quality management in construction continue to be extremely relevant because characteristics of construction product often do not satisfy the requirements of consumers. This is not only due to the level of design solutions or quality of materials and structures, but this is also connected with following regulatory requirements under work execution. In particular this is proved by the inspections of the state construction supervision in the Tomsk region (in 2011-2013 every fifth or sixth investigated object in the construction or reconstruction of the Tomsk Region has detected violations) [1].

Therefore research of quality management problems of construction products still takes an important place in the Russian economic science. However, in our opinion, it is necessary to expand quality management borders, transferring it principles to economic systems of higher order, in particular, on urban systems and on their separate elements. Substantially qualitative characteristics of urban system are defined by quality of the construction products which form space-material basis of the cities and the most important component of life environment – the urban environment.

2. Urban environment as an element of urban system

The global urbanization demands more attention to the urban areas, where concentrates not only considerable part of the population of planet, but also the main part of productive forces [2, 3]. The modern cities are polyhierarchical urban systems, which include social, economic and ecological sub-systems. One of the most important function of urban system is formation of the life environment. Socio-political, demographic and migratory processes, investment appeal of the territory and, eventually, dynamics of city development substantially depend on quality of the life environment.

The quality of life environment is studied by many Russian and foreign researchers, but from viewpoint of ecology and human hygiene predominantly. In particular researches of scientists from the A.N. Sysin Institute of Human Ecology and Hygiene of the Environment are devoted to these problems. Some scientists use such parameters as percent of built-up area and open spaces, population and household density, accessibility to roads, quality of air and noise pollution [4], some scientists consider additionally coastal pollution, slums [5] or surface temperature [6]. Experts of United Nations Environment Programme, the Cities Alliance and the International Institute for Environment and Development show that environmental resources are the "strategic assets" for the cities [7]. However dominating of artificial urban environment and increasing its impact on all spheres of human life becomes more and more evident in modern urban systems.

Authors state that the urban environment is anthropogenous part of environment formed by set of construction objects (residential, public, industrial buildings and structures, commercial facilities, objects of transport and municipal infrastructure) and adapted natural elements of the territory. They all function in interrelation as uniform urban system and people directly interact with this unique urban system while living, forming and changing the environment. The urban environment is an element of urban system that defines the quality of the life environment, and, eventually, quality of life of the population living in urban areas. Confirmation of it is that the United Nations, estimating an index of city prosperity (City Prosperity Index) [8], as one of private indicators, considers city infrastructure which is the most important element of the urban environment.

3. Methodology

Quality management of construction products is carried out at the level of the construction organization. The basic of methodology is object-process approach. However this approach is not applicable when assessing the quality of urban environment as a result of the whole construction sector activity. It is mainly defined by the quality of building materials, the units, the performed works, individual buildings and their complexes appearing integrally and mattering exactly in that measure as it provides favorable, comfortable conditions for life environment. The peculiarity of quality assessment of urban environment is determined by the urban environment being the emergent system with its quality depending not only on quality of its individual elements, but also on level of interrelations between components. In this regard, along with the quality assessment of individual elements it is necessary to estimate and control quality of all urban system and the urban environment as set of construction projects in particular.

The quality of the urban environment is latent property, i.e. it cannot be measured directly and is developed as a result of a certain combination of a number of private characteristics which this property depends on. The quality of the urban environment is a category difficult for assessment, which is characterized not only by objective indicators, which are defined by normative documents and technical rules, but also by many subjective characteristics depending on preferences of the individual. In such a way, a basis of methodology for the quality assessment of urban environment should be based on an integrated approach proposed by the authors [9], which is suggesting a system analysis of all factors and use of both quantitative methods of assessment (calculation of particular and integrated indices) and qualitative (expert assessments, and surveys).

4. Methods of urban environment quality assessment

Quality of the urban environment is a correspondence of the urban environment with modern requirements of the individual to comfortable, favorable and safe conditions of life where humans satisfy their needs for housing, access to health care and education, movement, physical, social and cultural development, accumulation and generation of the human capital provided.

When considering the urban environment as uniform object, indicators of it quality can be divided into four categories (Table 1).

Table 1. Indicators of quality of the urban environment.

Category of quality of the urban environment	Indicator set of quality of the urban environment	Indicators
	Housing per capita	$H = \frac{S_h}{N}$, S_h – total residential space, m^2; N – number of inhabitants, pl.
	Providing with social facilities	$P_{S_i} = \frac{Q_{S_i}}{\left(\frac{N}{1000}\right)}$, P_{S_i} – provision of type i of social facilities; Q_{S_i} – total capacity of type i of social facilities: in nursery schools; in schools; in hospitals; N – number of inhabitants, pl.
Providing with basic life facilities	Providing with sports facilities	$P_{sp} = \frac{Q_{sp}}{\left(\frac{N}{1000}\right)}$, Q_{sp} – quantity of sports facilities; N – number of inhabitants, pl.
	Providing with cultural establishments	$P_{C_j} = \frac{Q_{Cj}}{\left(\frac{N}{1000}\right)}$, P_{C_j} – provision of type j of cultural establishments; Q_{Cj} – quantity of type j of cultural establishments: museums, theatres, club-type cultural institutions, parks for cultural activities and rest, zoological gardens; N – number of inhabitants, pl.
	Providing with facilities of the communal infrastructure	$P_{CI_k} = \frac{L_k}{\left(\frac{N}{1000}\right)}$, P_{CI_k} – provision of type k of facilities of the communal infrastructure; L_k – length of type k of communal objects, km: water supply lines, sewerage and steam network, heating network;

Category of quality of the urban environment	Indicator set of quality of the urban environment	Indicators
		N – number of inhabitants, pl.
	Providing with transportation routes	$P_{TI_t} = \frac{M_t}{\left(\frac{N}{1000}\right)}$, P_{TI_t} – provision of type t of transportation routes; M_t – mileage of type t of transportation routes, km: automobile roads, railways; N – number of inhabitants, pl.
State of capital construction project	Housing improvements	$W_f = \left(1 - \frac{S_f}{S_h}\right) \cdot 100\%$, W_f – part off housing stock which is unequipped with any type of service f (%): cold-water supply system, hot-water supply system, canalization, heating system, baths (shower), gas-supply system, electric cookers; S_f – housing stock which is equipped with a type of service f, m^2; S_h – total area of housing stock, m^2; BD – percent of amount of slum and dangerous dwelling, %; EE – percent of building with class of energy performance "C" and higher, %
	State of facilities of the communal infrastructure	D_c – deterioration of facilities of the communal infrastructure, %
Sanitary-ecological characteristics of building	Degree of green planting	$D_{gz} = \frac{S_{gz}}{N}$, S_{gz} – total area of green planting, m^2; N – number of inhabitants, pl.
	Built-upness	$K_{b\text{-}u} = \frac{S_b}{S_l}$, S_b – total area of buildings, m^2; S_l – total area of territory, $м^2$
	Environmental friendliness of construction materials	Expert estimates in scale from 0 to 1 inclusive
	Recycling of construction materials	Expert estimates in scale from 0 to 1 inclusive
Architectural characteristics of building	Expressive elements of architecture	Expert estimates in scale from 0 to 1 inclusive
	Proportion of buildings and structures	Expert estimates in scale from 0 to 1 inclusive
	Color matching and	Expert estimates in scale from 0 to 1 inclusive

Category of quality of the urban environment	Indicator set of quality of the urban environment	Indicators
	facture of materials	

The first two categories are the basic characteristics of urban environment quality, as they reflect providing with basic facilities for the life and the state of these objects. These indicators are quantifiable and allow for inter-regional comparisons. Tables 2 and 3 show the results of the quality assessment of urban environment made for Siberian regions. As indicators have different units of measurement, comparability was provided by normalizing according to the formula of linear scaling.

Table 2. The map of region positioning in terms of indicators of providing with the main capital construction objects.

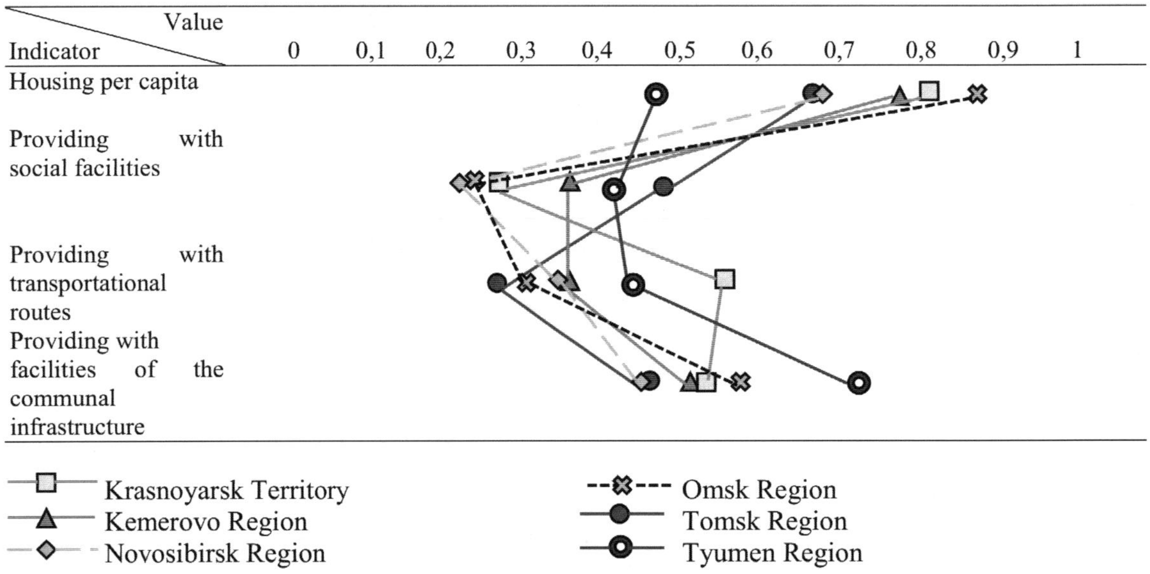

The analysis of results allows revealing the most and the least developed elements of the urban environment in each region, and, according to authors, it is important for determination of priority of the budgetary financing. For example, represented data shows that Omsk Region is one of leaders of housing per capita and providing with facilities of the communal infrastructure, Tomsk Region is the leader of providing with social facilities. The total deviation of providing lines to the left shows that urban areas have deficit of social facilities and transportational routes against the background of growing housing space. Indicators of the state of urban environment's objects (Table 3) is far less in Omsk region than in the neighboring regions, and Tyumen region is one of the leaders. Tomsk region has average values of indicators of urban environment quality.

5. Conclusion

The urban environment defines conditions of person's daily life and quality of life of the population on the urban area. Development of methods to assess urban environment quality is relevant, as the urban environment is the most important characteristic of urban system. The quality of urban environment is a multifactor category, which is subject to integrated assessment by authors system of the indicators. The analysis of assessment allows positioning the regions in terms of urban environment quality, to make interregional comparisons and to correct urban and budgetary activities.

Table 3. The map of region positioning in terms of indicators of housing facilities state.

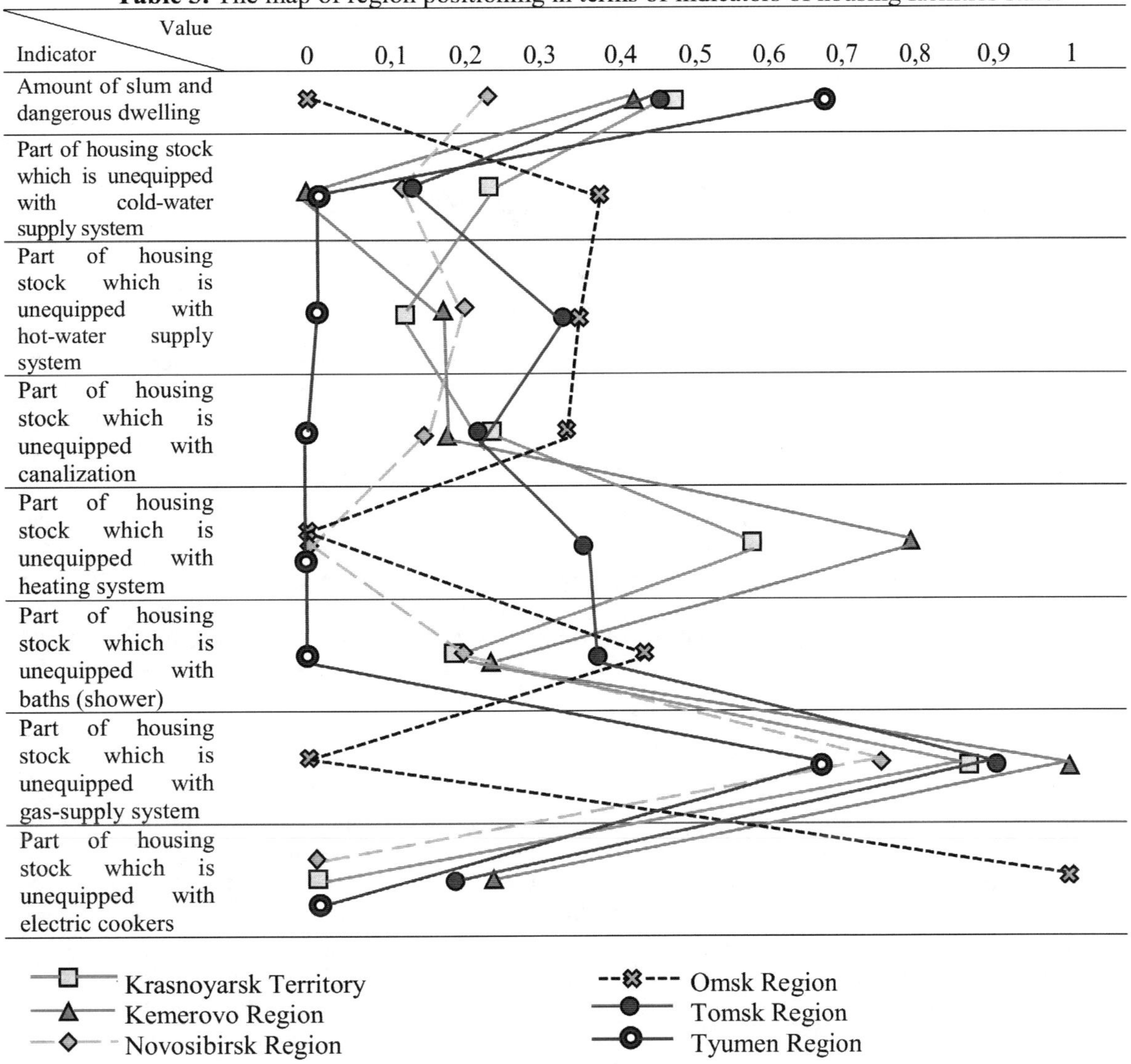

References

[1] Отчет об осуществлении государственного строительного надзора в 2014, 2013, 2012 и 2011 году Электронный ресурс [http://gsn.tomsk.gov.ru/деятельность-инспекции/отчеты]

[2] Bharat Dahiya 2012 *Cities* **29** 44–61

[3] Перелыгин Ю А 2012 *Управление развитием территорий* **1** 30–33

[4] Rahman A, Kumar Y, Fazal S and Bhaskaran S 2011 *Journal of Geographic Information System* **3** 62–84

[5] Joseph M, Wang F and Wang L 2014 *Habitat International* **41** 33–40

[6] Nichol J, Wong M S 2005 *Landscape and Urban Planning* **73** 49–58

[7] United Nations Environment Programme 2007 *Liveable cities. The benefits of urban environmental planning* (Washington: The Cities Alliance)

[8] UN–Habitat (2013) *State of the world's cities 2012/2013 Prosperity of cities* (USA, Canada: Routledge)

[9] Ovsiannikova T Yu and Preobrazhenskaya M N 2014 *Vestnik of Tomsk State University of Architecture and* Building **3** 191–200

Geoecological aspects of comprehensive development of Tomsk left-bank area

S Seryakov[1], R Seryakova[2], A Baranova[3] and A Potseluyev[4]
[1]National Research Tomsk Polytechnic University, Institute of Natural Resources, Tomsk, 634050, Russia

E-mail: svs_tpu@mail.ru, kh.rozali@gmail.com

Abstract. The article deals with the unique geoecological characteristics of the left bank of the river Tom' within Tomsk area (Russia). The authors suggest methodological approaches to land management in regions with intensive interaction of ecosystems and technogenic complexes. It proves the necessity to use the systems approach and estimating method to solve the tasks of urban development and land management. The authors make a conclusion about the most rational land use of the studied area in terms of ecologized land management.

1. Introduction

Nowadays the Russian scientific society is changing the paradigm of environmental perception. There comes the epoch of research based on systems approach and focusing on preserving favorable living environment as long as possible, which replaces the "technocentric" knowledge of the XXth century [1, 3]. These changes are a positive result of the global process of ecologization that has grown out of natural science and embraced all the aspects of the society ranging from urban planning to school education.

This process is manifested at a regional level in developing programs aiming at preserving and improving the environmental characteristics. The main principle of the process is the axiomatic necessity to protect natural areal complexes (NAC) and their constituent elements – unique landscapes under the conditions of growing industrial needs of the society.

Tomsk has been extending its boundaries for more than 400 years, and even more active for the last decade. At the moment it is possible to state the existence of Tomsk agglomeration. It needs a united systems development concept that would ensure the most effective use of NAC and the urban areas within its boundaries [7].

The use of the unique NAC as a site for urban development makes it necessary to take into account the whole complex of functional characteristics of the land. Comprehensive study of urban areas implies all-around consideration of interdependent influence of urban agglomerations on geoecological characteristics of local geo systems.

Land misuse caused by urbanization is the problem that can be solved by rational local land use and urban planning policy in particular areas. Taking into consideration the global concept of sustainable development the important task is to preserve undisturbed or slightly disturbed natural areas, their values increasing in the long-term.

In the frame of the development of sustainable land use planning, the fundamental approaches are those that ensure maximal optimization of resource use [4]. Complexity of urban development problems makes it necessary to apply the systems approach to design choice and to the design scientific rationale mostly by means of geoecological methods.

2. Basic data

A lot of developed countries have a great experience in natural area conservation while developing urban and industrial areas. This issue is being actively discussed in Russia though without being provided with sufficient practical and regulatory framework. The undisturbed natural areas of Russia are an ecological buffer. Being basic living environment, local urban areas should be paid special attention to.

Tomsk land reserves are changed every year due to the urban boundary extension by incorporation the nearest rural settlements. Thus, there appeared recreation areas within the boundaries of the fully developed and built-up city.

At the moment, the most promising direction of Tomsk comprehensive development is Kirovskiy district, the left bank of the river Tom', in particular. It is planned by the analogy with Novosibirsk and other big cities situated on both banks of rivers. It should be noted that the documents (draft area plan and parcel plan) regulating functional use of the area have not been adopted yet. Thus, its future use has not been definitely determined. The future image of Tomsk as well as its long-term prosperity depends on what kind of engineering and planning decisions will be taken.

The right-bank area of Tomsk is being actively modernized. Typical urban districts of the 1980s gradually change its architectural and infrastructural image that meets modern requirements.

The objects of the present research are natural objects that have spatial and functional connections with the objects of engineering infrastructure situated in the Ob'-Tom' interfluves near Tomsk. The aim of the work is to develop methodological approaches to the rational comprehensive development of Tomsk left-bank area in terms of underflooding threat.

The left-bank area of Tomsk is located in the Ob'-Tom' interfluvial plain. It is confined to the West Siberian folded margin platform, but at the same time has characteristics of a typical platform with its folded basement lying at low depth (up to 1 km). The platform cover embraces loose Mesozoic and Cenozoic sediments and consists of clay, sands and clay loams. This area is adjacent to the left-bank floodplain often flooded in spring and the terrace above the floodplain. All the flood plain lakes of the area are morphologically identical. Their origin is conditioned by flood waters of the river Tom'. The area basically consists of clay loams, from solid to liquid, the thickness of which is from 1.5 to 6.5 m. Also, sand loams ranging from solid to liquid are found, their thickness is up to 5.5 m. The area is well drained with a dense network of small rivers and lakes.

The combination of soil-formers makes grey forest soil and ashen-gray soil predominating in the area. The mother rocks are of different genesis: lacustrine, lacustrine-alluvial, alluvial and aeolian in places.

Ecological research conducted by Tomsk hydro geologists [2, 6] reveals a unique characteristic of the left-bank area: there are plenty of drinking water springs. Plicative tectonics and low dipping bedding of hydrogeological complexes and the above ground platform horizons condition lateral groundwater flow and springs in river valleys of the quaternary complexes. But the absence of folding or bench alluvial horizons that would work as a buffer in the places of the ground water discharge specifies high rate of surface water infiltration into deep soil horizons. This is also a water-collecting area. There are ground water wells of the first stage. They supply the city with drinking and domestic water from Palaeogene aquifer system. Nevertheless, the activity of Tomsk water intake led to strong landscape transformation.

Construction of anthropogenic objects within natural complexes with neglect of geoecological, social and economical risks doesn't meet the requirements of sustainable development of the region. Thus, the ecologized development of the left-bank area of Tomsk is undoubtful. The geoecological structure of the investigated area includes natural underground water outlets along the terrace of the flood-plain that comes out as springs identified in the road embankment near the bayou lakes. The largest concentrations of such springs are observed near settlements Timiryazevo and Eushta, they form the left-bank spring mega zone (Figure 1) [6, 11].

IOP Conf. Series: Materials Science and Engineering **71** (2015) 012052 doi:10.1088/1757-899X/71/1/012052

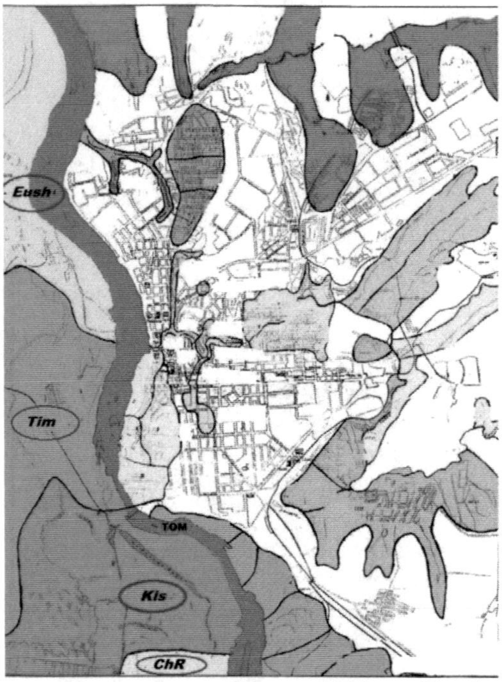

12 – The left-bank landscape-spring mega-zone:
Eush – Eushtinskaya;
Tim – Timiryazevskaya;
Kis – Kislovskaya;
ChR – Cherno-rechnskaya

Figure 1. Map of landscape and spring macro-zones of Tomsk.
©Nazarov A and Vertman E 2004

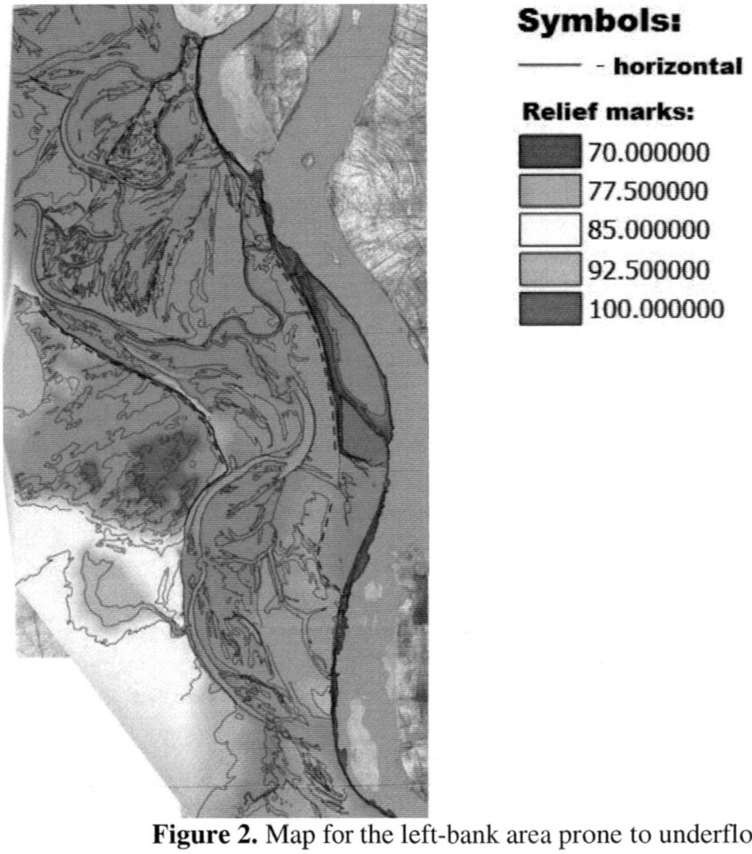

Figure 2. Map for the left-bank area prone to underflooding with sub-zones.

However, the left-bank area is not thoroughly investigated in terms of geoecology and geography. Our research of the unique geoecological features of the area resulted in compiling a map for area prone to flooding that manifest ground water-the river Tom' hydraulic connection (Figure 2) [8,9]. Flooding of the left-bank area is a negative process that leads to soil degradation due to erosion and suffusion processes, i.e. deterioration of engineering and geological conditions.

While developing the planning design of the left-bank area it is necessary to take into account the existing natural underflooding process and the technogenic underflooding attributed to the construction of water facilities engineering infrastructure. Thus, while developing the area it is necessary to apply the principle of sustainable and ecologized land use that implies continuous assessment and planning the whole "lifecycle" of anthropogenic objects. It will ensure further development of comfortable ecologized living environment [5, 10].

The left-bank area development will primarily influence the areas that are most attractive in terms of recreational and tourist activities in the valley of the river Tom'. Thus, planning design should include the whole complex of civil infrastructure (sewerage, water supply, waste disposal, traffic network etc.). While being designed and sited the objects of real estate are referred to land points therefore it is necessary to take into account the role of soil absorption capability and its buffer property. Potential environmental damage threatens the lands with the following functions: lands with water bodies, agricultural lands, special use zones, etc. [10]. Design and construction of civil infrastructure objects can result in the following negative consequences presented in Figure 3.

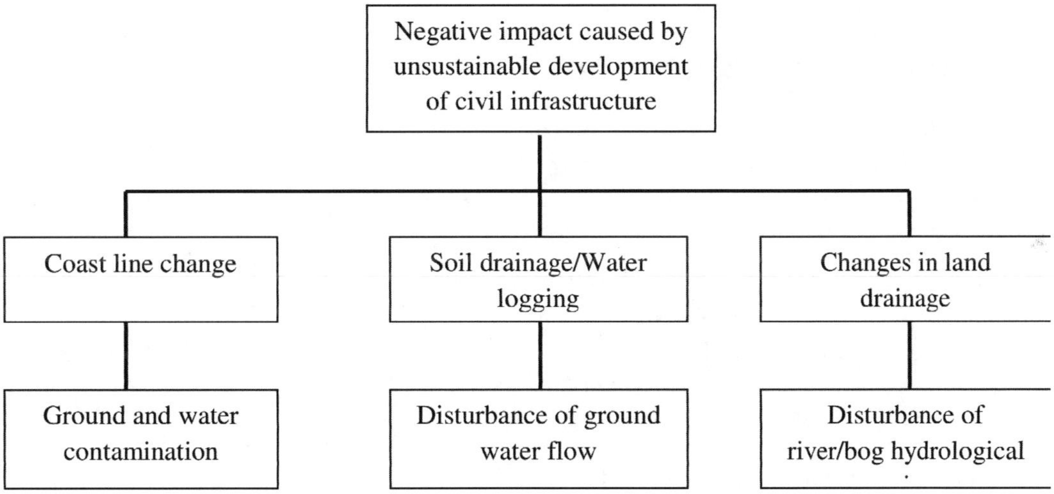

Figure 3. Basic negative consequences of unsustainable development of civil infrastructure on the left-bank area of Tomsk.

Civil infrastructure development will inevitably disturb the environment, thus, while developing the area it is necessary to apply ecologized approaches with regards to city development regulations. It is necessary to note that more and more local governments in Russian cities pay much attention to the problems of underflooding both of natural and technogenic genesis. The infiltration recharge value for the left-bank area of Tomsk is 95-197 mm/year. The technogenic factor leakage in water pipe systems can double this value.

3. Conclusions

1. The investigation of geoecological and geographical characteristics of Tomsk left-bank area allows identifying close interaction of these aspects with the technogenic factor. This interaction is manifested in the underflooding of the area caused by both natural and anthropogenic processes.

IOP Conf. Series: Materials Science and Engineering **71** (2015) 012052 doi:10.1088/1757-899X/71/1/012052

2. The research resulted in mapping for the left-bank area prone to underflooding. It is based on geoecological and geographical characteristics of the area.

References

[1] Петров К 2004 *Общая геоэкология* (Санкт-Петербург: Санкт-Петербургский государственный университет) 440

[2] Доклад о состоянии и использовании земель Томской области в 2011 году (Управление Федеральной службы государственной регистрации, кадастра и картографии по Томской области Томск 2012) 66

[3] Хромов А 2004 Современное природоохранное землеустройство и приоритетные направления его оптимизации (на примере Астраханской области) Автореферат на соискание степени кандидата географических наук (Астрахань, Россия) 24

[4] Шульгин А 2011 Совершенствование инновационной деятельности на основе государственно-частного партнерства Автореферат на соискание степени кандидата экономических наук (Воронеж, Россия) 24

[5] Попов В, Серяков С и Хафизова Р 2013 Экологизация землеустройства урбанизированных территорий в городе Томске: состояние и проблемы *Вестник ТГАСУ* **2** 337-345

[6] Вертман Е и Назаров А 2004 *Изучение гидродинамического и гидрогеохимического режима родников г. Томска* (Томск: Томский политехнических университет) 199

[7] Vernon Henderson J and Jacques-François Thisse 2004 Handbook of Regional and Urban Economics *Cities and Geography* **4** 2063-3073

[8] Shengguo Gao, Zhongli Zhu, Shaomin Liu, Rui Jin, Guangchao Yang and Lei Tan *Estimating the spatial distribution of soil moisture based on Bayesian maximum entropy method with auxiliary data from remote sensing* 54-66

[9] Mokhamad Yusup Nur Khakim, Takeshi Tsuji and Toshifumi Matsuoka 2014 Lithology-controlled subsidence and seasonal aquifer response in the Bandung basin, Indonesia, observed by synthetic aperture radar interferometry *International Journal of Applied Earth Observation and Geoinformation* **32** 199-207

[10] Doxani G, Karantzalos K and Tsakiri- Strati M 2014 Monitoring urban changes based on scale-space filtering and object-oriented classification *International Journal of Applied Earth Observation and Geoinformation* **15** 38-48

[11] Kasper Cockx, Tim Van de Voorde and Frank Canters 2014 Quantifying uncertainty in remote sensing-based urban land-use mapping *International Journal of Applied Earth Observation and Geoinformation* **31** 154-166

IOP Conf. Series: Materials Science and Engineering **71** (2015) 012053 doi:10.1088/1757-899X/71/1/012053

Factors affecting the use of modern methods and materials in construction

P Mesároš[1] and T Mandičák[1]
[1] Technical University of Košice, Faculty of Civil Engineering, Košice, 04200, Slovakia

E-mail: peter.mesaros@tuke.sk

Abstract. Sustainability of construction attracts much attention in construction industry. One of the factors driving this requirement is application of materials and components through modern methods and technologies. Modern methods of construction can be the way to obtain buildings assisting in minimizing the negative impact of construction industry on the environment. Article defines the factors affecting the use of these modern methods and materials of construction. At the same time it defines modern construction methods and materials that can be considered progressive in the construction process.

1. Introduction
The continued population growth affects the amount of construction. At the same time the need for housing and services is also increasing. Therefore increases the need to find new modern methods and materials for construction. Together with the needs of the population their demands for the development of technical and technological requirements steadily increases in all areas [1]. This results in the increasing demands for quality housing, which in turn also requires finding new ways and opportunities applicable in construction. One possibility includes modern methods and materials of construction. Prefabricated production construction represents mass production, which takes place in factories while greatly shortening the construction time and having positive impact on the environmental aspect. Modern construction methods and materials have very broad application in the construction industry. Their biggest advantage is ability to be implemented on site and also in the manufacture off-site, allowing shortening the time and increasing precision of construction. Usability of these methods is an important issue in regards of the ever increasing needs of the construction of buildings and structures.

2. Modern methods and materials of construction
Modern methods of construction can be defined as a method in which all building blocks (which are the walls, floor slabs, beams, columns and staircases) are mass-produced with high quality in the plant or in a place under the strict supervision [2]. The term "modern methods of construction 'is launched in the UK as a common designation for off-site construction methods, using prefabricated elements or parts of structures, constructed off-site and then transported and assembled on-site and on-site construction methods when making building blocks and parts of structures take place directly on the site [1].

Esa and Nurudin [3] argued that the MMC is a continuous system used by craftsmen at each construction activity that uses production output to minimize wastage of resources and increased value to end users. MMC in construction includes an industrial process in which parts of the building are conceived, planned, contrived, transported and erected on site. This system is a balanced combination of software and hardware components [4]. Between software features include a design that is a complex process of studying the requirements of end users, market analysis, development of standardized components, determination of the production and mounting arrangements, the process of allocating resources and materials, and definitions building designer conceptual framework. Software elements provide prerequisite to creating a favorable environment for the expansion of this industry.

Modern methods of construction are defined as those that ensure efficient management process providing more products of better quality with shorter time. The aim is to improve business efficiency, quality, customer satisfaction, environmental impact and sustainability. Modern materials are those that are used in modern methods of construction. Within the construction solutions are different types of materials whose main purpose is to speed up the construction process and minimize costs. The choice of material depends on the method of construction.

2.1 Classification of modern methods of construction

Modern construction methods can be classified in several aspects. At present, several authors start addressing this issue and the classifications in the literature is varying. The most common factor in the breakdown of modern methods of construction is production site. These methods are subdivided as follows [1] [5]:

- *Off-site production* - the production process takes place in a factory, off site. These include [6]:
  - *Systems of prefabricated houses* - consist of walls, floors and roofs in the form of prefabricated panels that are built at the place where they form a structure which then requires different levels of editing. The most commonly used open panels or frames that consist only of a skeleton and still need finishing work such as insulation and outer casing to be mounted directly on the spot. Another system which also used panels is closed. This is a more complex system more difficult to manufacture factory, which may include isolation and cladding material.
  - *Modular system* (volume / volumetric) - modules are composed and together constitute the structure of the building. This system allows the connection of a larger number of compatible modules on the spot, so often used to bigger, more standardized structures. Prefabricated modules are often referred to as cells, which are 80 to 95% complete internally in the factory, including mechanical and electrical services, and then transported to the final assembly site. They are mainly in specialized areas with typing and repetition, as kitchens and bathrooms.
  - *Hybrid* - This method combines a modular and panel system. Hybrid approach is used for the greater development opportunities while limiting uniformity of design.
  - *Sub-assemblies and components* - these methods can often be traditional, despite the use of factory assemblies or components. Examples include floor or roof boxes and kits prefabricated concrete foundations.
- *On-site production* - this approach involves technology and design systems that cannot fall into the category of off-site production. One characteristic of these methods is innovation. Illustrative examples are:
  - *"Tunnel form method"* - it is one of the most commonly used methods of cell construction. Characterized by cost efficiency, productivity and quality. This method is suitable for projects with more repeated components such as hotels, apartment buildings, student hostels. Using this technique it is possible construction of 40 floors and more. Construction process is simple and fast resulting in a 15% reduction in toll costs and 25% time savings. Construction of "tunnel" is formed by pouring concrete into the formwork, which forms the floor and walls. Every 24 hours, the formwork is moved so that it is possible to create another "tunnel". After completion of the deck, repeat this process to the next level.

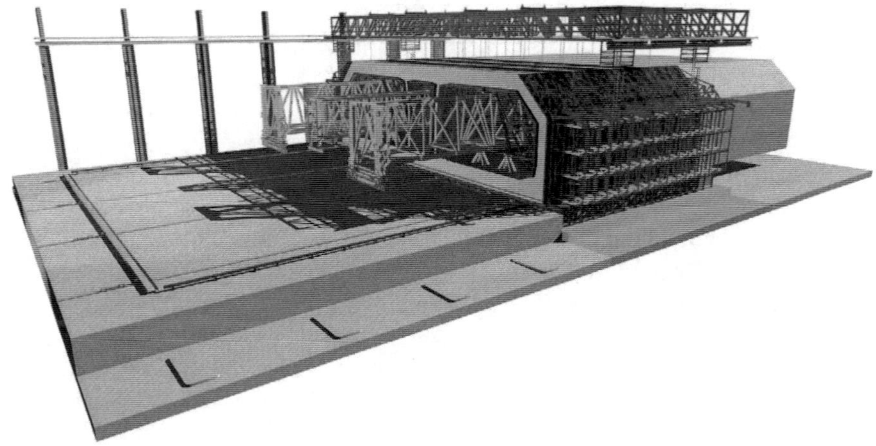

Figure 1. Tunnel form method [7].

- *Thin-joint masonry* (easily jointed masonry) - method is characterized by rapid, clean and precise use of aerated concrete blocks, which are essential size and shape produced under strict standards associated with a thin layer mortar. This method combines shortening construction time with greater flexibility aerated concrete structures. It is characterized by the easy handling.

The following breakdown of MMC provides the classification of systems and their associated materials. As shown in Table 1, it is possible to distinguish three basic systems (frame, panel and modular system).

Table 1. Overview of construction systems and materials.

System	Category	Material
frame system	lightweight frames	wood
	mediumweight frames	metal, reinforced plastic, laminated wood
	hardweight frames	steel, concrete
panel system	light and medium heavy panels	wood, metal, composite material
	heavy panels (manufactured off site)	concrete
	heavy panels (manufactured on site)	concrete
modular system	medium weight modules (mobile)	wood, steel, composite material
	medium weight modules (fixed deposited)	wood, steel, composite material
	heavyweights modules (made on site)	concrete
	heavy modules (tunnel production on site)	concrete

IOP Conf. Series: Materials Science and Engineering **71** (2015) 012053 doi:10.1088/1757-899X/71/1/012053

1.1. Properties and characteristics of the use of MMC affecting on key factors

For successful implementation of construction using modern methods and materials one needs to examine their desired properties, which are briefly described below.

- Mass production

Investment in equipment and human resources associated with industrialization can only be justified in cases where the recorded volume production. Distribution of this size allows fixed costs for large quantities, without unduly increasing the final cost.

- Flow production

Production of large amounts of standardized prefabricated elements allows a high level of production process. This process can be divided into a larger number of larger tasks. In such system work, workers perform repetitive activities while increasing their productivity.

- Production Equipment

Capital employed for the initial investment is a very important factor. Plant, equipment and skilled workers must be obtained before they start production. Such huge investments may be used only if there is sufficient demand for the product. On the other hand, the production of the product may be carried out directly on the construction site, thus minimizing transportation costs.

- Equipment site

Construction and installation of prefabricated panels and their location on their position requires different equipment. In the case of construction of such a system it is important to incorporate sufficient operating costs. A typical device is a crane that is used in the construction of multi-storey buildings.

- Construction time

The use of different systems of modern methods of construction can significantly shorten the construction period. In a large extent this is influenced by the fact that these methods require less amount of work on the construction site. This is illustrated in Figure 2.

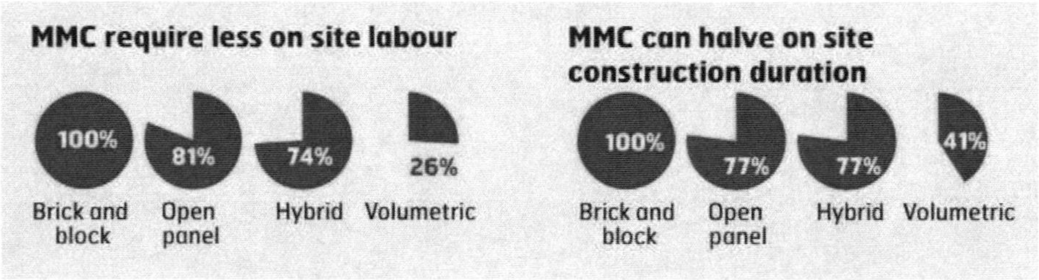

Figure 2. Degree of labor and construction time by using MMC [8].

- Modular coordination

Modular coordination is a unified management system for sizing space and components so that all elements fit into one another without any modifications, such as the trimming or lengthening, even if they are supplied by different manufacturers.

- Integration

Optimum results require a high degree of cooperation between the different parties, which are the designer, manufacturer, customer and supplier. This can only be achieved through an integrated system.

- Transport

Panels can reduce production costs by up to 30%. However, this saving is partially offset by shipping costs. Transport of large panels is also subject to certain restrictions. These limitations should be taken into account even for a friend that method of construction.

3. Factors and barriers affecting the use of MMC

These are not all advantages of using modern methods and materials in construction. They are closely related to factors that influence the use of MMC. Several authors have attempted to define the key factors. These factors include the following [1] [8]:

- political context,
- construction skills,
- quality of construction,
- costs,
- pace of construction and construction time,

Sustainability of construction is one of the main reasons to use modern progressive methods and materials of construction. Currently, the pressure on the sustainability of construction is increasing. There are several perspectives on sustainable construction. One is the economic view. Important is the social aspect and the environmental aspect also. It also has an impact on creating the environment in the form of legislation. Process automation, emphasis on environmentally friendly materials and technologies in the construction process is large.

- Construction skills

Construction skills are an important factor in choosing construction methods and materials. Progressive materials are in many cases difficult to integrate into the construction process because of the lack of knowledge about them. Construction companies should seek training programs for employees. Emphasis is on knowledge about new materials and work with them.

- Quality of construction

Quality construction and final construction work has a significant impact on the decision on the selection of materials and methods of construction. Perceived quality construction also plays a big role. Progressive materials should not only be flexible in terms of time and cost of construction sites, but also quality.

- Costs of construction

Perception of costs is very sensitive. Economic assessment of projects is often the basis for decision making. In general, it is the pressure to develop economically available materials. Modern methods often offer more efficient use of materials. When using modern methods of construction, costs are minimized, which has a positive effect on their use. [9]

- Pace of construction and construction time

For decision making, the choice of materials and methods used in construction is also important for planning the duration of construction time. The use of MMC has positive impact on the construction period. This is one of the biggest advantages of using these methods.

4. Conclusion

The use of MMC has indisputable advantages. Several studies declare an economic advantage and shorten the construction period. Similarly, the use of progressive materials has a positive impact on the sustainability of construction. Defining of these benefits is very important for several reasons. One of them is the interconnectedness of the key factors in the use of MMC and advanced materials in the construction with their advantages. These are characterized by multiple authors. Together with the economic efficiency of the solution the important factors are also building skills. The need for education and dissemination of material knowledge is the basis for their use.

The use of new materials and MMC has the potential in the future. In view of the growing need for sustainable construction projects is the assumption that they will benefit from an even larger extent than at present.

Acknowledgement

The paper presents the partial research result of project "VEGA - 1/0677/14 Research of construction efficiency improvement through MMC technologies".

IOP Conf. Series: Materials Science and Engineering **71** (2015) 012053 doi:10.1088/1757-899X/71/1/012053

References

[1] Bašková R and Kyjaková L 2013 *Parametrizácia konštrukčných riešení stavby pri uplatnení moderných metód výstavby* In: Multi-dimenzionálne prístupy podporujúce integrované navrhovanie a riadenie stavebných projektov

[2] Richard R B 2006 Industrialized, flexible and demountable building systems: Quality, economy and sustainability *Proc. of the CRIOCM 2006 Int. Sympos. on "Advancement of Construction Management and Real Estate"*

[3] Esa H and Nurudin M M 1998 Policy on industrialised building systems *Report on Colloquim on Industrialised Construction System Kuala Lumpur*

[4] Jaillon L and Poon C S 2009 *The evolution of prefabricated residential building systems in hong kong: A review of the public and the private sector*

[5] Bura Steering and Development Forum Report October 2005 *Modern Methods of Construction - Evolution or Revolution*

[6] Gibb A G F & Pendlebury M C 2005 *Buildoffsite Glossary of Terms*

[7] Incremental launching forming machine for the world´s longest underwater tunnel [online resource] http://www.peri.com/en/press/press/press-releases/pm2013/hzmb_china.cfm

[8] Lane A 2006 *Barriers and solutions to the use of modern methods of construction*

[9] Kozlovska M and Zupova L 2013 Modern methods of construction as a challenge for energy efficiency buildings *Proc. of 13th Int. Multidisciplinary Scientific Geoconf.: Nano, bio and green - technologies for a sustainable future* (Albena, Bulgaria) 677-684

IOP Conf. Series: Materials Science and Engineering **71** (2015) 012054 doi:10.1088/1757-899X/71/1/012054

Information systems for material flow management in construction processes

P Mesároš[1] and T Mandičák[1]
[1] Technical University of Košice, Faculty of Civil Engineering, Košice, 04200, Slovakia

E-mail: peter.mesaros@tuke.sk

Abstract. The article describes the options for the management of material flows in the construction process. Management and resource planning is one of the key factors influencing the effectiveness of construction project. It is very difficult to set these flows correctly. The current period offers several options and tools to do this. Information systems and their modules can be used just for the management of materials in the construction process.

1. Introduction

All resources are finite in financial terms, including materials. Their effective management, particular planning, is a must for sustainable construction. Material flow management is a challenging process and requires systematic approach. The human factor often fails to provide optimal use of material resources required in the construction process. Therefore numerous tools were developed in the digital time that offers a number of solutions. One of the available solutions is the use of information technology to support management of material flows. Information systems and modules facilitate efficient management of material utilization. This leads to optimal utilization of available resources in any form. The result is material saving and sustainable construction processes.

Information systems and especially Enterprise Resource Planning systems (ERP) are used in various fields. Construction is an area where innovative technologies are increasingly introduced. It includes not only process innovation by means of technology introduction in construction, such as new machinery and equipment, but also information and communication technology providing support for the management of the processes.

2. Information systems in construction

Construction field is characterized by fragmentation which exists both within individual phases as well as across project phases [1]. Participants from various organizations who are involved in a project phase or in different project phases are facing ineffectiveness and inefficiency in their coordination, collaboration and communication [2]. Powerful project management software has become a prerequisite to manage the projects more efficiently and effectively, and help the project managers in their decision-making [3]. The job is done easily, quickly and accurately thus constituting a competitive advantage. This may also be in the use of technologies for material flows management of construction production.

Progress of information and communication technologies (ICT) is undeniable in every field. Currently, the concept of IT (Information Technologies) covers both hardware and software which contains various essential software, application software and development resources [4]. There are all information resources used in the creation and use of IS performance. Information and communication technologies are the methods, procedures and methods of collection, storage and processing, evaluation, selection, distribution and current receipt of the necessary information in the required format and quality [5]. Information and communication technologies (ICT) present a way that organizations can use to meet their needs related to everyday operations, which are supported by modern information systems. Alone information systems pose to the organization, in addition to supporting and management activities, they often help to differentiate from other enterprises.

In construction projects and companies various types of information systems (IS) exist, such as construction management or business software. They have been developed, applied, and widely used. The basic question is what an information system in the construction enterprise is and what features should include the effective material management. Information system represents a set of people, processes, technical and software tools providing collection, transfer, storage and processing of data for distribution and presentation of information [6]. Information systems are currently based on the use of information and communication technology. The opinions of information systems are different. In its narrowest perception IS is a hardware and software solution for processing of information for decision support and control [7].

In many construction enterprises ERP systems are used as a system designed for enterprise resource planning. Information Systems ERP represent the core of enterprise information systems and provide the greatest range of its functions and processes. ERP is a system that incorporates internal management information with the external management information that covers the sales, accounting and finance, manufacturing and customer care [8].

3. Information systems for material flow management

Construction projects are very complex in nature and integrate numerous resources. The material flows are generated between the principal contractor of the construction project and subcontractors that represent smaller companies also involved in realization of a construction project, and some of them responsible for materials supply. Management of materials flows is essential to provide their most effective use, the information systems are of great help in this regard.

ERP systems are based on a modular system where each module has a specific function (Figure 1). In Figure 1 the ERP system is made up of production, finance, human resources modules. This system is not sufficient for all needs; especially it does not address the issue of management of material flows.

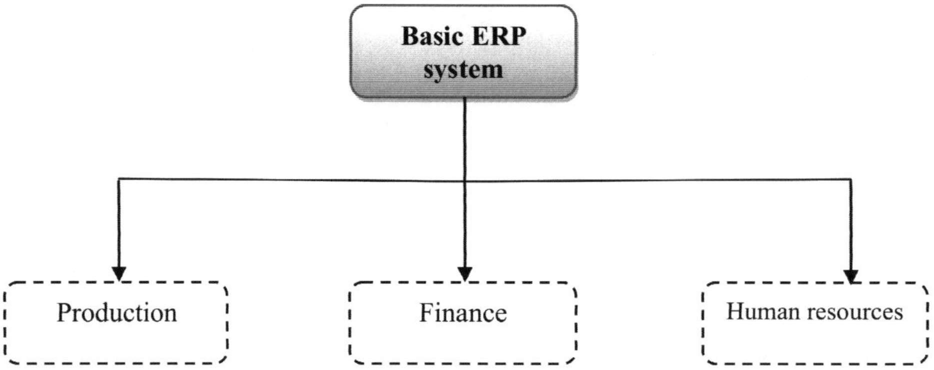

Figure 1. Basic ERP system and modules [6].

Supporting modules are important for planning and material flows management. It includes Customer Relationship Management (CRM), Supply Chain and Vendor Management (SCM), Human Resource (HR) modules and module for Finance and Accounting (Figure 2). Several sources report that the core ERP system creates only the production, finance and human resources [9]. The widespread perception of ERP also includes additional modules that extend the functionality and potential uses. It is Business Intelligence (BI), Production, Engineering, Sales and so on.

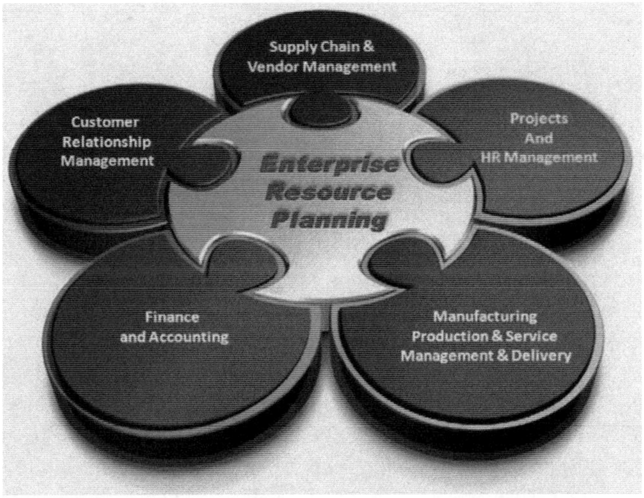

Figure 2. Extended ERP system and modules.

In terms of material flow management the most important modules are Supply Chain Management or Supply Chain and Vendor Management to management the entire supply chain.

4. Module SCM (Supply Chain Management)

The classic supply chain was essentially linear and consisted in the implementation of the basic bond:

Figure 3. The classic supply chain (material flow) [6].

In the Figure 3 the material flow direction is from the left to the right while the information flow and money (payments made for materials and services) are the opposites. Today's web-based businesses interconnect structures into more complex and more difficult ones to create mutually networked communities whose main common objective is to offer faster and cheaper desired and competitive product. It leads to exclusion (outsourcing) of a series of actions. Make use of the possibility of specialized companies, which have effective use of technology or know-how.

The supply chain today is no longer linear and covers more subjects. It can be defined as a combination of material, information and financial flows needed to ensure the development, manufacture, supply goods and services to customers - users. SCM is the management of this combined, complex flow. As SCM works well, ongoing transition from one phase to another chain is smooth. It does not include the production of unnecessary stock as well.

SCM is a set of tools and processes that optimize the management and maximum operational efficiency of all elements (cells) throughout the supply chain with respect to a customer. In case of construction it is a main contractor or investor [4]. SCM systems are a specific example of the interconnection of suppliers with subcontractors and investor, that is based on information and communication technologies. Through interconnection and exchange of information can be partners in the supply chain work to share information, plan and coordinate the overall process together in order to increase the responsiveness of the entire chain. The objective of SCM is mainly shortening production and delivery cycles, increasing flexibility, better communication with partners involved in the chain, decrease inventory through production according to specific orders.

The basic functions of SCM solutions mainly include the following [10]:

- ensure reducing the time and cost of the material requirements,
- support planning activities,
- focus on increasing customer satisfaction (in construction it is investor or main contractor).

Management of logistics processes and the supplier relationship gradually developed, and has its history. Therefore, SCM cannot be associated only with the application of Internet technology. According to several authors five developmental stages (phases) can be distinguished in the development of SCM from the simplest forms to virtual networks [11]. The fourth level represents SCM extended supply chain in which the possibility for data exchange in the form of EDI (Electronic Data Interchange) or the internet there is an increase communication between businesses in order to reduce the time and cost of material realization. Process design and material realization gradually engages customer as well. The fifth step of SCM is a virtual network with flexible arrangement of the partners. Measure of the value that this arrangement provides is each partner in the network.

ERP systems are often supplied for a particular purpose. For the needs of the construction industry they are often modified. They often contain other specific modules to manage material and its optimum use. Setting these material flows effectively is often only possible combinations of a number of modules. Information systems containing both modules to manage material flows in construction will likely accrue [12].

Conclusion

Use of available resources is crucial to sustainability of the construction. Set these activities optimally is one of the basic conditions for successful and efficient construction project. One of these resources is also material. Setting material flows is a necessity for efficient use of materials in the construction process [13]. At the moment a number of tools exist to automate this process. These tools include the progressive tools increasingly used in construction. One option is to use information systems for material flow management in the construction process. Information systems are based on the use of ICT. ERP systems are at the core of information systems, which consist of individual modules. Information systems focused on the needs of businesses and construction are different from others. What is important is functionality management of material flows. These systems help to manage and configure the material flows.

This phenomenon is also confirmed by several studies. The use of information and communication technologies is increasing also in construction [14]. Their implementation expects a positive impact on the efficiency and coordination of construction management processes. Overall sustainability of construction and efficiency in the building process also affects the use of material resources. Therefore, one of the main functionalities of these technologies and systems is the management of material flows between the parties. This trend will likely continue. Pressure grows for effective use. Similarly, the pressure to implement information systems is increasing. This results in automated material flow management.

Acknowledgement
The article presented is a partial research result of project "VEGA - 1/0677/14 Research of construction efficiency improvement through MMC technologies".

The paper presents the results of the project "Identification of key competencies of university students for the needs of knowledge society development in Slovakia", which is supported by the Ministry of Education, Science, Research and Sport of the Slovak Republic in supplying incentives for research and development from the state budget in accordance with Act no. 185/2009 Z. z. on incentives for research and development and on supplementing Act. 595/2003 Z. z. Income Tax, as amended, by Act no. 40/2011 Z. z.

References

[1] Howard C. et al. 1989 Computer integration: reducing fragmentation in AEC industry *Journal of computing inp* 18-32
[2] Nitithamyong P and Skibniewski M 2004 WEba-based construction project management systems: how to make them successful *Automation in construction* 491-506
[3] Lee S and Yu J H 2012 Success model of project management information system in construction *Automation in construction* 82-93
[4] Mesároš P 2013 *Manažment znalsotí v praxi slovenských stavebných podnikov* (Košice: VÚSI, spol. s r.o.) 182
[5] Stoffová V and Stoffa J 1999 Základné termíny z informačných, multimediálnych a didaktických technológií *Zborník MEDACTA '99*: Zväzok 1 (Nitra: Slovdidac) 64-69
[6] Čarnický Š and Mesároš P 2009 *Informačné systémy podnikov* (Bratislava: Ekonóm) 233
[7] Basl J 2008 *Podnikové informační systémy (podnik v informační společnosti)* 2. vyd. (Praha: Grada Publishing) 283
[8] Kikoy H 2012 *What is an ERP software and how can it help your business* [online source], Available: http://www.herbertkikoy.info/2012/05/erp-software-and-your-business/
[9] Thomas W S 2007 *Achieving success through adoption of prise resource planning (A quantitative analysis of sap users in North and South America)* (California: ProQuest Information and Learning Company)
[10] Mesároš F and Mesároš P 2009 *Controlling nákladov na stavebné procesy*. Košice: VÚSI spol. s r.o., 220 s. ISBN 978-80-89383-02-3
[11] Eckert C, Clarkson J and Stacey M 2011 Information flow in engineering companies: problems and their causes *Culley S (ed) International conference on engineering design* (Glasgow: ICED 01.Wiley) 43–50
[12] Mcintosh B and Sloan 2001 The potential impact of electronic procurement and global sourcing within the UK construction industry *Proc. ARCOM 17th Annual Conference* **1** (University of Salford, UK) 232–240
[13] Spišáková M and Kozlovská M 2013 Lean production as an innovative approach to construction *Selected Scientific Papers: Journal of Civil Engineering* **8** 87-96
[14] Mesároš F and Mesároš P 2010 Digital competencies in process of creating the knowledge company in construction sector. *ISARC 2010: The 27th International Symposium on Automation and Robotics in Construction: proceedings* (Bratislava, Slovakia)

Distribution of energy storage rate in area of strain localization during tension of austenitic steel

W Oliferuk[1,2], M Maj[2] and K Zembrzycki[2]
[1]Białystok University of Technology, Białystok, 15-351, Poland
[2]Institute of Fundamental Technological Research, Warsaw, 02-106, Poland

E-mail: wolif@ippt.gov.pl

Abstract. The present work is devoted to experimental determination of the energy storage rate in the area of strain localization. The experimental procedure involves two complementary techniques: i.e. infrared thermography (IRT) and visible light imaging. The results of experiments have shown that during the evolution of plastic strain localization the energy storage rate in some areas of the deformed specimen drops to zero. To interpret the decrease of the energy storage rate in terms of micro-mechanisms, microstructural observations using electron back scattered diffraction (EBSC) were performed.

1. Introduction

When a material deforms plastically, a part of the mechanical energy w_p expended on plastic deformation is converted into heat q_d while the remainder e_s is stored within the material as an energy related to defects of crystal lattice, mainly dislocation structures,

$$e_s = w_p - q_d. \tag{1}$$

The stored energy e_s represents a change in internal energy of the deformed material and it is an essential measure of its cold-worked state. This energy was discovered in calorimetric tests performed by Taylor and Quinney [1]. Since then the part of plastic work stored in the metallic material was considered to be a constant value of about 10% of the whole plastic work. Nevertheless, further experimental studies have shown that the ratio of the stored energy to the plastic work is not constant and depends on deformation level of the tested material [2-14]. Thus a measure of energy conversion at each instant of the plastic deformation process was introduced. It is the rate of energy storage Z defined as the ratio of the stored energy increment Δe_s to the plastic work increment Δw_p:

$$Z = \frac{\Delta e_s}{\Delta w_p}. \tag{2}$$

The stored energy increment is equal to the difference between Δw_p and the increment of energy dissipated as a heat Δq_d ,

$$\Delta e_s = \Delta w_p - \Delta q_d.$$

Therefore:

$$Z = \frac{\Delta e_s}{\Delta w_p} = 1 - \frac{\Delta q_d}{\Delta w_p}. \tag{3}$$

The entire deformation process, from initial state to the fracture of the tested specimen, can be divided into two stages: homogeneous deformation and heterogeneous one. Two indicators of appearance of plastic strain localization are usually applied. They are: non-uniform temperature distribution on the specimen surface and non-uniform strain field on the opposite surface of this specimen.

In the previous works it has been shown that during heterogeneous deformation (localization of plastic strain) of polycrystalline material, the energy storage rate rapidly decreases reaching the 0 value and then becomes negative [13, 15]. But up to now, only the average value of the energy storage rate for the gauge part of deformed specimen was estimated. The questions appear: a) What is the energy storage rate distribution along the gauge length of the strained specimen during development of plastic strain localization? b) What is relation between distribution of energy storage rate and microstructure evolution of deformed material? The purpose of the present work is to answer these questions. To reach the purpose, the new method of energy storage rate determination has been designed. The method allows obtaining distribution of this quantity along the gauge length of the specimen. The obtained results are related to the texture evolution.

2. Methods

The energy storage rate can be calculated by a determination of plastic work increment Δw_p and corresponding increment of energy dissipated as heat Δq_d, see equation 3.

The method is based on the experimental procedure for simultaneous measurements of temperature, and strain distributions on the surface of tested specimen during tensile deformation. This procedure involves two complementary imaging techniques: CCD camera and infrared thermography (IRT). In order to determine the strain distribution, markers in form of graphite dots were plotted on one surface of the specimen. In this way, the surface was divided into sections, whose sizes are specified by the distance between the dots (Figure 1).

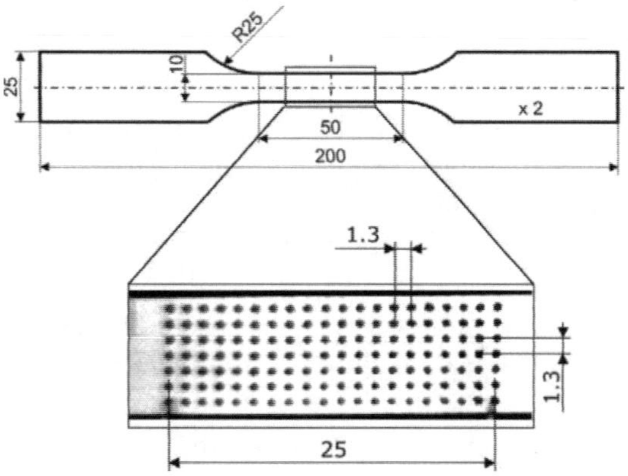

Figure 1. The graphite dots on the gauge part of the specimen and its shape and dimensions.

Displacements of the dots were recorded by means of CCD camera during deformation process. Taking into account the reference distance l_0 between the centers of the dots and recording distance between these centers as a function of deformation time $l(t, y_n)$, the true local strain $\varepsilon_y(t, y_n)$ in the direction of tension was calculated:

$$\varepsilon_y(t, y_n) = ln\left(\frac{l_y(t, y_n)}{l_0}\right), \tag{4}$$

where y_n is the vertical coordinate of the centre of section n.

The true stress σ_y was determined dividing the load by the current cross-sectional area of the specimen corresponding to the given section. Assuming that a dependence of Young's modulus E on strain is negligibly small, local plastic strain $\varepsilon_y^p(t, y_n)$ was obtained from the following formula:

$$\varepsilon_y^p(t, y_n) = \varepsilon_y(t, y_n) - \frac{\sigma_y(t, y_n)}{E}. \tag{5}$$

The strain and stress distributions along the axis of the tested specimen were used to calculate surface distribution of specific plastic work.

$$w_p(t, y_n) = \frac{1}{\rho(t, y_n)} \int_0^{\varepsilon_y^p(t, y_n)} \sigma_y(t, y_n) \cdot d\varepsilon_y^p, \tag{6}$$

where $\rho(t, y_n) = \dfrac{m_s}{V(t, y_n)}$ is the current mass density of a particular section,

$m_s = const = \rho_0 \cdot V_0 = \rho(t, y_n) \cdot V(t, y_n)$ is mass of a section and $V(t, y_n)$ is the current volume of particular section.

The energy dissipated in the form of heat causes an increase in the temperature of the specimen. This increase depends on the strain rate. In the area of the strain localization, the distribution of strain rate is not uniform. This heterogeneity is manifested by a non-uniform temperature field on the specimen surface. The temperature field during deformation process was recorded by means of the IR thermography system and the average temperature $T(t, y_n)$ as a function of time for each n-th section along the axis of the specimen was determined. Such a selection of positions of sections allows avoiding the influence of the specimen edges on the measured temperature values. The initial size of each, considered section is identified by the distance between the centres of the dots and the thickness of the specimen. Thus, the initial volume V_0 of each section was equal to $(1.3 \times 1.3 \times 2)\,mm^3$ and the initial cross-section S_0 was $(1.3 \times 2)\,mm^2$. Dimensions of the sections were changing during the tensile process. These changes were determined by tracking dot centres in the visible range by means of the CCD camera. All these data have been used to calculate the heat $\Delta q_d(t, y_n)$ generated during time interval Δt in $n-th$ section. The interval Δt is the reciprocal of the taking frequency.

In order to avoid the influence of the heat convection phenomenon, the displacement rate 2000 *mm/min* was chosen. Nevertheless, despite the relatively high displacement rate the deformation process was non-adiabatic mainly due to the heat transport to the grips of the testing machine. Thus, taking in account the phenomena considered above and the heat expended for compensation of drop in temperature caused by the thermoelastic effect, the heat $\Delta q_d(t, y_n)$ generated during time Δt in $n-th$ section has been calculated according to the following formula:

$$\begin{aligned}\Delta q_d(t, y_n) = {} & c_w \cdot \Delta T(t, y_n) + \frac{\alpha T(t, y_n)\Delta\sigma_y(t, y_n)}{\rho(t, y_n)} + \\ & + \frac{S_{n/n+1}(t, y_n) \cdot \lambda \cdot \Delta t}{\rho_0 \cdot V_0} \cdot \left[\frac{T(t, y_n) - T(t, y_{n+1})}{y_{n+1} - y_n}\right] + \\ & + \frac{S_{n/n-1}(t, y_n) \cdot \lambda \cdot \Delta t}{\rho_0 \cdot V_0} \cdot \left[\frac{T(t, y_n) - T(t, y_{n-1})}{y_n - y_{n-1}}\right],\end{aligned} \tag{7}$$

where α is the coefficient of linear thermal expansion, $T(t, y_n)$ is the mean absolute temperature of the considered section, $\Delta\sigma_y(t, y_n)$ is the increase in the stress during subsequent time intervals Δt, λ is the coefficient of thermal conductivity, c_w is the specific heat and V_0 is a volume of considered section. $S_{n/n+1}(t, y_n)$ and $S_{n/n-1}(t, y_n)$ are cross-section areas between neighbouring sections $n/n+1$ and $n/n-1$.

The first component of equation 7 expresses the heat needed to rise the temperature of a unit mass of the tested material by ΔT. The second describes the heat expended for compensation of the temperature drop due to the thermoelastic effect. The third and fourth components, according to the Fourier's law, take into account the heat transfer between given section n and its neighbours $n/n+1$ and $n/n-1$, respectively. It is worth to notice that the Fourier's law refers to a heat flux. Thus, in order to determine the heat related to a mass unit, the heat flux should be divided by the mass of the section and multiplied by Δt and by $S_{n/n+1}(t, y_n)$ or $S_{n/n-1}(t, y_n)$, respectively.

Adding up the heat increments $\Delta q_d(t, y_n)$ in the successive time intervals, the time dependence of the energy dissipated as heat $q_d(t, y_n)$ for each tested section has been obtained. The energy storage rate for a particular section is defined as

$$Z = 1 - \frac{\Delta q_d(t, y_n)}{\Delta w_p(t, y_n)}. \tag{8}$$

During non-uniform deformation in the equal time intervals Δt, the increments of plastic work Δw_p are not equal. Thus, Δq_d was determined for equal increments of Δw_p. In further analysis $\Delta w_p = 2\, J/g$ was taken. Such value was selected in order to reduce the dispersion of increments' ratio $\frac{\Delta q_d}{\Delta w_p}$. Having obtained $w_p(t, y_n)$ and $q_d(t, y_n)$ for the considered sections and using the procedure mentioned above the distribution of the energy storage rate was determined.

3. Experiments and results

Experiments were performed on austenitic stainless steel (304L). From the sheets of annealed material the specimens for tensile testing were cut out using electro-spark machining. In order to remove the technological surface layer, the specimens were electropolished. Optical metallographic and TEM observations of tested material indicated complete recrystallization and average grain size about 50 μm. The dislocation density in the specimens before deformation was low; dislocations were randomly distributed in the matrix and in the grain boundaries, and no regular dislocation arrangements were observed. Grains were randomly oriented (Figure 2).

Figure 2. Distribution of grain orientation in the initial state obtained by EBSD technique.

All specimens were strained using an MTS 858 testing machine. During the tension the displacement rate was controlled and equal to 2000 *mm/min*. For such displacement rate and given geometry of the specimen (see Figure 1) the corresponding mean value of strain rate was equal to $6.6 \cdot 10^{-1} s^{-1}$.

In the course of deformation process the infrared and visible range image sequences were recorded simultaneously. The experimental setup is presented in Figure 3.

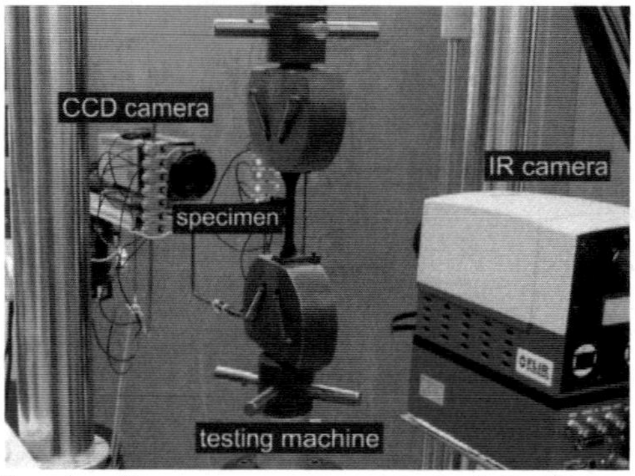

Figure 3. Experimental setup for simultaneous measurement of displacement and temperature fields.

The taking frequency of both cameras was equal to 538 *Hz*. Such frequency was selected in order to take into account the heat transfer between neighbouring sections. Simultaneously, the straining force as a function of time t was measured. The analysis of the displacement fields (based on markers tracking) and the assignment of them to the appropriate temperature distributions were performed using MATLAB software.

From the experimental data, tensile curves for selected local sections: A, B, C, D, E, lying on the axis of specimen, were calculated (Figure 4).

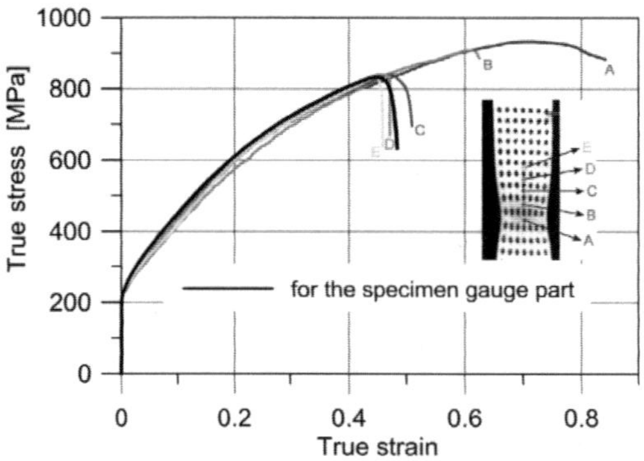

Figure 4. Stress-strain curves for selected sections of the specimen.

It is seen that with an increase in strain, the individual sections of the specimen stop to deform, whereas others are still deforming. It is result of the localization of plastic deformation. On the basis of the stress-strain dependences, the plastic work as a function of time for the selected sections lying on the specimen's axis were calculated (Equation 6).

From the surface temperature field, the distribution of energy dissipated as heat along the specimen axis was determined using a local form of heat equation (Equation 7). Then, using Equation 3 the energy storage rate Z for each of the considered sections was obtained. The results, as a function of true strain, are presented in Figure 5. Results obtained for different sections are marked by different symbols with different gray intensity, corresponding to stress-strain curves shown in Figure 5.

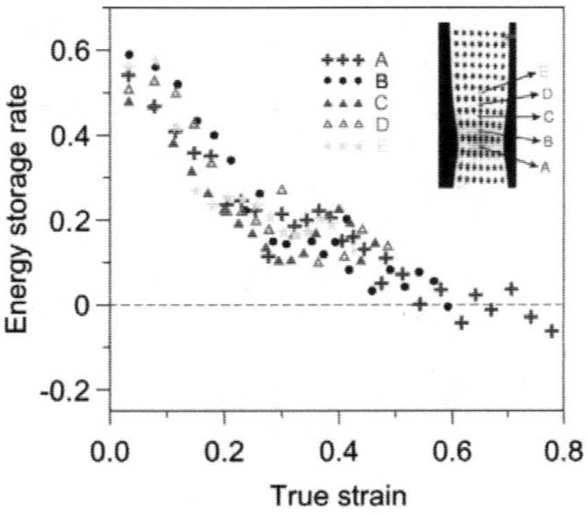

Figure 5. Energy storage rate as a function of true strain for selected sections of the specimen.

It is seen that the energy storage rate decreases with strain for all considered sections. During the evolution of plastic strain localization some sections cease to deform, which causes a stop in the energy storage process in this region, whereas the energy storage rate in the other sections (where deformation still proceeds) drops to zero. It has been shown [15] that for austenitic steel the zero value of the energy storage rate corresponds to the point of Considèr stability criterion. Thus, $\frac{\Delta e_s}{\Delta w_p} = 0$ can be treated as the plastic instability criterion based on energy conversion. A confirmation of obtained negative values of Z needs further studies. Negative values of this quantity could be explained by a release of the energy stored in previous deformation stages.

In order to find microscopic interpretation of obtained results, the study of evolution of the crystallographic orientations during the deformation process were performed using Electron Back Scattering Diffraction (EBSD) technique.

To obtain the crystallographic orientation maps for different levels of plastic deformation, three specimens were strained to the different values of strain: 0.3, 0.4 and 0.8. These values correspond to macroscopically uniform deformation, the onset of strain localization and state just before fracture, respectively. Then, the orientation maps were obtained for the selected areas of deformed specimens. The maps are shown in Figures 6-8. One can see that from the onset of plastic strain localization (Figure 7) the amount of mechanical twins increases. It means that the contribution of twinning in deformation process is growing.

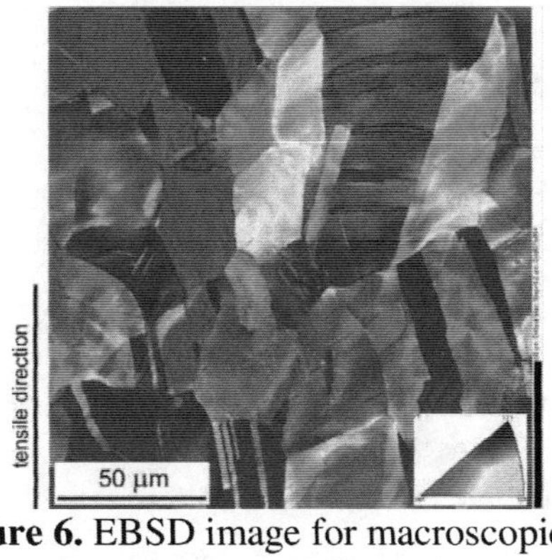

Figure 6. EBSD image for macroscopically homogeneous deformation.

Figure 7. EBSD image for the onset of plastic strain localization.

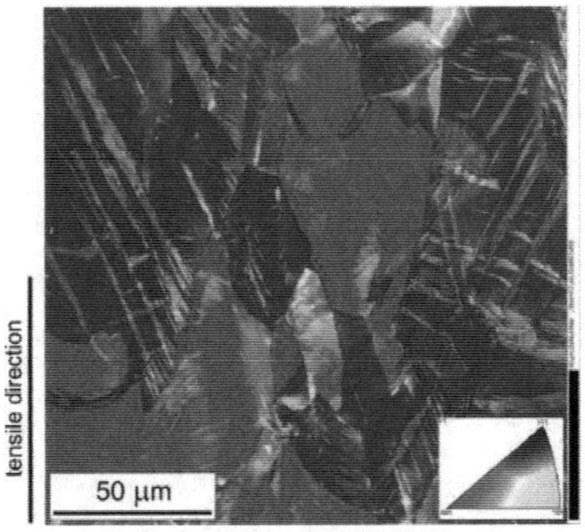

Figure 8. EBSD image for advanced localization of plastic strain (just before fracture).

In Figures 6-8 a noticeable trend in the evolution of the orientation of individual grains in direction of two dominant texture components is observed. It is seen that the development of strain localization is accompanied by further rotation of individual grains in the direction of two texture components. Rotation of grains proceeds in such a way, that the {111} type planes become parallel to the planes of maximum shear stress. The angle between the trace of these planes and the tension direction is about 40^0. The above analysis shows that in the necking zone, just before the fracture (see Figure 8), the macroscopic shear of the localized area is possible without change of a crystallographic plane. In other words, the conditions for the crystallographic shear banding are created [16]. This corresponds to the loss of stability of the plastic deformation. The presented preliminary results have shown that the zero value of the energy storage rate may be treated as a criterion of plastic instability. An interpretation of such criterion in terms of microstructure evolution requires further studies.

4. Conclusions

Experimental method of determination of energy storage rate distribution in the area of heterogeneous deformation has been proposed.

It has been shown, that in the plastic strain localization area material reaches the state, where energy storage rate reaches the 0 value and then becomes negative. This means that the material loses an ability to store the energy, though energy supply the internal energy of the specimen decreases. Thus, the 0 value of the energy storage rate could be regarded as the plastic instability criterion based on the energy conversion.

On the basis of microstructure observation it is believed that 0 value of energy storage rate corresponds to the state in which only two dominant components of the texture appear, creating conditions to crystallographic shear-banding.

References

[1] Taylor G and Quinney H 193) The latent energy remaining in a metal after cold working *Proc R Soc Lond* **143** 307-326

[2] Bever M B, Holt D L and Titchener A L 1973 The stored energy of cold work *Prog Mater Sci*, Pergamon Press **17** 23-58

[3] Oliferuk W, Gadaj S P and Grabski M W 1984 Energy storage during the tensile deformation of Armco iron and austenitic steel *Mater Sci Eng* **70** 131-141

[4] Chrysochoos A and Maisonneuve O 1985 Energy balance for elastic plastic deformation at finite strain *C R Acad Sci* **300** 985-990

[5] Oliferuk W, Świątnicki W and Grabski MW 1993 Rate of energy storage and microstructure evolution during the tensile deformation of austenitic steel *Mater Sci Eng A* **161** 55-63

[6] Chrysochoos A, Maisonneuve O, Martin G, Caumon H and Chezeaux JC 1989 Plastic and dissipated work and stored energy *Nucl Ebg Des* **114** 323-333

[7] Oliferuk W, Świątnicki W and Grabski MW 1995 Effect of grain size on the rate of energy storage during the tensile deformation of austenitic steel *Mater Sci Eng A* **197** 49-58

[8] Oliferuk W, Korbel A and Grabski MW 1997 Slip behaviour and energy storage process during uniaxial tensile deformation of austenitic steel *Mater Sci Eng A* **224-226** 1122-1125
[9] Rittel D 1999 On the conversion of plastic work to heat during high strain rate deformation of glassy polymers *Mech Mat* **31** 131-139
[10] Chrysochoos A and Louche H 2000 An infrared image processing to analyse the calorific effects accompanying strain localization *Int J Eng Sci* **38** 1759-1788
[11] Hodowany J, Ravichandran G, Rosakis A J and Rosakis P 2000 Partition of plastic work into heat and stored energy in metals *Exp Mech* **40** 113-123
[12] Oliferuk W, Korbel A and Bochniak W 2001 Energy balance and macroscopic strain localization during plastic deformation of polycrystalline metals *Mater Sci Eng A* **319-321** 250-253
[13] Oliferuk W, Maj M and Raniecki B 2004 Experimental analysis of energy storage rate components during tensile deformation of polycrystals *Mater Sci Eng A* **374** 77-81
[14] Oliferuk W and Maj M 2004 Effect of pre-strain direction on energy storage process during tensile deformation of polycrystal *Mater Sci Eng A* **387-389** 218-221
[15] Oliferuk W and Maj M 2007 Plastic instability criterion based on energy conversion *Mater Sci Eng A* **462** 363-366
[16] Paul H and Driver JH 200) Deformation structure and texture transformations in twinned fcc metals: Critical role of micro- and macro- scale shear bands *Mater Sci Forum* **550** 521-526

IOP Conf. Series: Materials Science and Engineering **71** (2015) 012056 doi:10.1088/1757-899X/71/1/012056

The applicability of acoustic emission method to modeling the endurance of metallic construction elements

S V Ponomarev[1], S Rikkonen[1], A Azin[1], A Karavatskiy[1], N Maritskiy[1] and S A Ponomarev[1]
[1]National Research Tomsk State University, Research Institute of Applied Mathematics and Mechanics,Tomsk, 634050, Russia

E-mail: antonazin@mail.ru

Abstract. Acoustic emission method is the most effective nondestructive inspection technique of construction elements. This paper considers the expanded applicability of acoustic emission method to modeling the damage and the remaining operational life of building structures, including the high-ductile metals. The modeling of damage accumulation was carried out to predict endurance using acoustic emission method.

1. Introduction

The current trend in the building industry establishes certain requirements for the materials used. There is a need to build more, faster and cheaper. For these purpose special materials providing better strength and endurance properties are used. Acoustic emission technology is the most practical way to determine such properties of the materials. The development of a reliable method for predicting the operational life and remaining life of the resource is the most important task in the modern construction industry.

Buildings and structures often fail when operating in severe weather conditions. Moreover, the actual operational life is often significantly shorter than the designed one. Materials used for manufacturing structures may contain defects further leading to serious breakdowns. Structural integrity can be broken suddenly - an item can break down. As a result there might be accidents and fatalities. These problems can be solved with effective quality control of components both during production and during installation of equipment and in-service, which improves not only competitiveness, but also minimizes the risk for consumers.

2. Methods

The presence of internal and external defects of various kinds causes difficulty in selecting effective nondestructive inspection technique (NDIT) of construction elements. When performing work the following NDIT are used: optical methods of nondestructive testing (including visual inspection); plastic methods of structural analysis; rebound method; method of separation with chipping and rib spalling design; ultrasonic method of determining the strength; acoustic emission method; shock pulse monitoring.

Traditionally, ultrasonic flaw detection method is used because it allows mainly detecting large defects such as discontinuities, exfoliation, debonding. It is used at the input control of semi manufacture and product. At the same time, defects are quite often generated during operation and have an extremely small size, so urgent is the development of new or adaptation of existing nondestructive inspection techniques that can detect defects of this kind (and size). A promising method for the detection of voids formed during the operation is the method of acoustic emission (AE). AE signals emitted by the material in the process of loading (destruction) may carry information about the size of cracks, their growth rate, allowing solving the problem of early detection and prediction of the destruction and the remaining useful life.

To describe the behavior of materials under load it is necessary to take into consideration the evolution of the real defects. In this regard, the development of a method of quantitative evaluation of structural changes in the solid, and the study of the general laws of the kinetics of damage accumulation become a necessary step in solving the actual problem of identifying real defects.

Most commonly known technology of NDIT is limited in application due to its material base and the physical capacity to research designs [1].

Acoustic emission is one of NDIT based on detection of elastic waves. Any defect produces its own acoustic signal that it can be registered at a great distance. Localization of the defect is made by calculating the difference between the times of a check-wave recorded by AE sensors. The main features of the acoustic control method that determine its capabilities and application area are the following: (i) the method provides detection of defects according to the degree of danger; (ii) the method has a high sensitivity to the growing defects and allows the operating conditions to determine the increment of crack to fractions of millimeters; (iii) the method has object orientation and its position does not affect the detection of defects. Currently, this method is widely used for diagnosing and heavily large high-risk facilities, as well as structures, where there is access to the control surface.

Many models were developed in order to link AE and the parameters of material destruction. Three main approaches can be distinguished. The essence of the first approach is to diagnose the stage of destruction on the Status of singular points (local extremum and fractures) of AE flow (Figure 1a). When measuring the concentration of defects during plastic deformation of materials in process of power impact a number of acts of acoustic emission N_a and residual strain ε were recorded simultaneously. After that, considering the parameters of acoustic emission depended on the residual deformation the starting point of defects development, the destructive strain component ε_d and the concentration of defects are determined [2].

A similar approach is described in [3], where the accumulation chart of acoustic pulses is associated with the time of static deformation (deformation degree) to identify the main stages of deformation (Figure 3) and to assess the long-term strength limit.

The essence of the second approach is to diagnose stages of deformation energy and time parameters of AE act [4]. The method is based on the fact that the spectrum of AE represents the spectrum of a Poisson flow of short elastic pulses. Consequently, one can estimate the total power W in the whole of the AE band of frequencies from the spectral density measured in the restricted from f_1 to f_2 band. This makes it possible to determine the total average energy per one act of AE by dividing the total power W to the value of the flow rate of AE acts. It is known that the normalized area of the spectral density $G(f)$ per unit of the Poisson flow of AE acts throughout the duration τ_a (0 ... ∞) has a band is as follows (Figure 1c)

$$G(f) = 4\tau_a / (1 + 4\pi^2 f^2 \tau_a^2) \quad (1)$$

The average duration of the certificate AE τ_a is estimated experimentally overlaying dependence diagrams (1) to the experimental points of the spectrum of AE $G^*(f)$ (Figure 1c and d), and choosing the parameters τ_a in the formula so that the values approximating the spectrum $G(f)$ are most closely through the measured values of $G^*(f)$. Further, equating the measured power W^* and W^{**} approximating processes in the band from $f_1$to f_2, one obtain an estimated average energy of AE act E_a in the whole band. After setting the depending of the samples of the average E_a energy and the endurance of AE τ_a act degree on the deformation and damage accumulation, these parameters can identify the stage of material deformation.

The essence of the third approach is to diagnose the degradation of materials by AE invariant. Because of the random nature of the micro-flow, at some stages of deformation and fracture of solids stable (invariant) ratio can be found $I_p^* = inv$ between recorded AE parameters. Then, the appearance of deviations of recorded values I_p of these relations from their invariant values I_p^* can serve as a criterion for the transition to the next stage of deformation or fracture (Figure 1b) [5]. In addition, there is a technique [6], in which the invariant is not equal to a constant value, and a linear dependence.

Among the three approaches of using the parameters of acoustic emission for assessing the material damage the first approach was chosen. The objectivity of selection is formulated by the following factors: (i) considering the number of acoustic emission events it is easy to monitor the degree of damage to the material, using the relation $N_a = f(\varepsilon_r^{1/2})$; (ii) AE spectral composition suffers the greatest distortions in the distribution and reception of signals, so the second approach is expanded by further relations for the restoration of the spectral composition which affects the efficiency of the data obtained; (iii) the third approach allows having an opportunity to define the beginning of the destruction of the material, but it is impossible to determine the extent of damage.

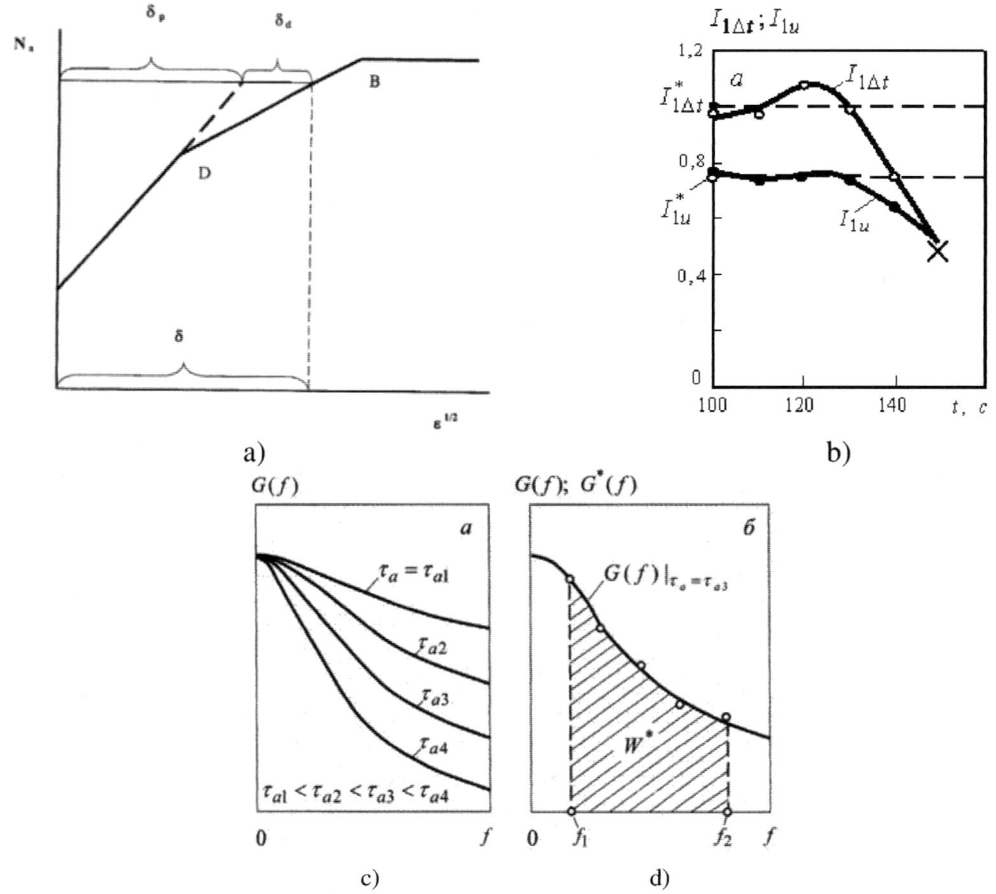

a) Peculiar relation $N_a = f(\varepsilon_r^{1/2})$ and scheme of determination of the point of destruction D and destructive components of residual strain; b) Graphs of invariant relations in the process of loading the sample steel 95X18. The cross shows the moment of destruction; c) spectrum of the Poisson process, depending on the duration of its events; d) the approximation of the spectrum of AE

Figure 1. Correlations between the parameters of acoustic emission and of materials fracture.

IOP Conf. Series: Materials Science and Engineering **71** (2015) 012056 doi:10.1088/1757-899X/71/1/012056

3. Damage modeling and lifetime prediction of the elements of metal structures

3.1. Account of damage to the constitutive relations

Currently the accounting for practical calculations of the damage accumulation in the constitutive relations (2) is implemented through the ratio (3):

$$\sigma = E\varepsilon, \tag{2}$$

$$\sigma = (1-\omega)\bar{\sigma}, \tag{3}$$

where σ – current stress tensor, *E*– matrix of material constants, ε – strain tensor, ω - a measure of the damage (varies from 0 to 1, 1 - corresponds to the destruction), $\bar{\sigma}$ – stress tensor calculated for these loading conditions in the absence of damage. The selection of this ratio is primarily dictated by the ease of determining the model parameters, and secondly, an acceptable accuracy [7].

3.2. Determining the type of damage functions

To determine the state of the material at the moment and predict the remaining life the function of material damage ω is introduced as follows (0 – no damage, 1 – the destruction of the material):

$$\omega = \frac{N_A^*}{N_A}, \qquad 0 \leq \omega < 1, \tag{4}$$

where N_A –number of acoustic pulses at the test loading (loading with a given amplitude loads or strains), N_A^* –number of acoustic pulses corresponding to the material destruction.

AE method is widely used for the diagnosis of steel and concrete structures as shown in the studies [1]. The results of this work demonstrated that the method of predicting the remaining life of the resource and using AE is applicable not only for classical structural materials (steel and concrete), but also for soft metals and their alloys (for example, tin-lead solder and aluminum alloys).

For AE method testing and determining the form of damage to the material features they carried out a series of experiments on the detection of acoustic signals by setting AE Micro2 digital AE system during the deformation of the material samples for diagrams accumulation of acoustic signals from developing the defects during deformation, determination of the accuracy of the method, parameters of the model of damage accumulation and the inception of the defects in the process of mechanical action. Based on the analysis results of the experiments they plotted the diagram of accumulation of acoustic pulses occurring during the development of defects in the sample depending on the strain of the alloy (for example, solder alloy-61) (Figure 2a). Further data are given for the retention of the acoustic signals during deformation of structural steel and aluminum alloy (Figures 2b and 3) [8].

Presented in Figures 2a and 3 experimental curve for samples of solder with a measure of in accuracy less than 3% can be approximated by the ratio (5):

$$N(x) = A1 * \left(\text{th}\left(A2(x - x0)\right) + 1\right), \quad x = \varepsilon^{1/2}, \tag{5}$$

where *N* – the number of pulses of acoustic signals ,ε – strain, *A1, A2, x0* –model parameters, th – hyperbolic tangent function.

IOP Conf. Series: Materials Science and Engineering **71** (2015) 012056 doi:10.1088/1757-899X/71/1/012056

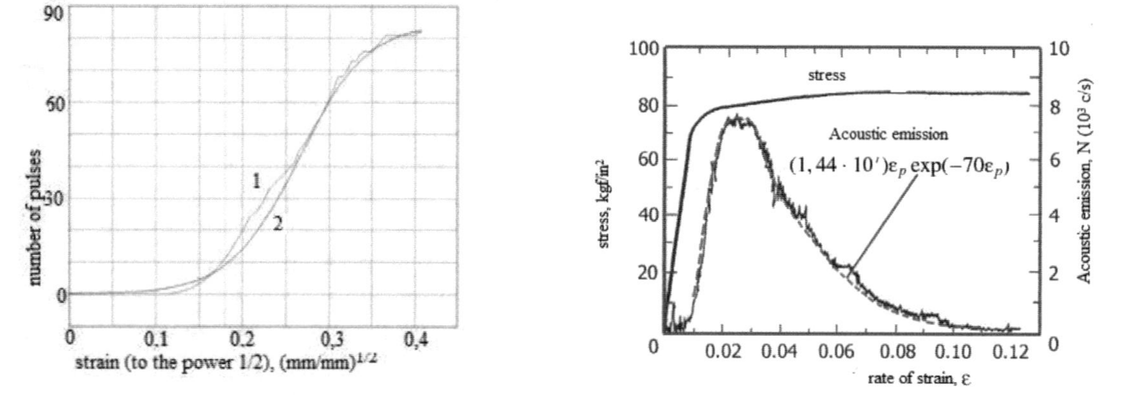

a) b)

1 - Experimental data, 2 - approximate curve.

Figure 2. a) Accumulation of AE pulses depending on the strain $\varepsilon^{1/2}$ samples of the solder; b) depending on the count rate of AE N and stressσ on the strain ε tensile specimens of aluminum alloy 7075-T6.

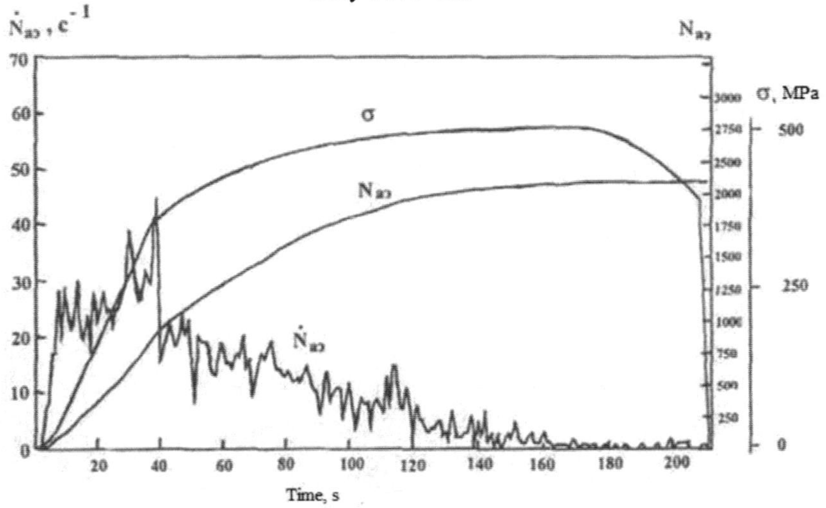

Figure 3. Flow signal intensity changes $\dot{N}_{аэ}$ and the total number of AE pulses $N_{аэ}$ depending on the time of strain (rate of strain) with static tensile specimen of steel 19G in original structural condition.

Thus, the use of acoustic emission technology allows obtaining correct results in a growth of the defect detection (such as a specific event, and the number of pulses dependent on the time) until the sample decomposition. Based on the results the damage function parameters are determined, which enables to reliably determine the remaining life of operation of the object and the relationship between stress and strain based on the accumulated damage.

The proposed method of predicting the remaining life of the operation and life of the object of the study is as follows: the finished product attached to it with AE sensors is subjected to loading test, during which the acoustic data are noted. Correlating the registered number of acoustic pulses at the test loading is determined by the point on the chart depending on the amount of acoustic pulses of strain and, consequently, to determine the current strain diagram corresponding strain of the material. A measure of accuracy in the determination of this point consists of the loading device measure of inaccuracy, a measure of inaccuracy of detection of acoustic signals and the approximation measure of inaccuracy. Thus, the total measure of inaccuracy to predict the damage degree, as a consequence, the prediction measure of inaccuracy of the remaining resource life does not exceed 9% [7].

IOP Conf. Series: Materials Science and Engineering **71** (2015) 012056 doi:10.1088/1757-899X/71/1/012056

3. Conclusion

Existing methods for assessing the endurance and the remaining operational life include insufficient mathematical modeling techniques and accurate assessment of accumulated damage during the operation and impact of this damage on the strength of quality.

Modern NDT method – the method of acoustic emission – was proposed for detailed assessment and lifetime prediction. The proposed methodology of fracture prediction using acoustic emission allows evaluating resource use of products with high accuracy, as well as to determine the place of destruction.

References

[1] Azin A, Maritsky N, Ponomarev S and Ponomarev S 2010 *Universities bulletin Physics* **12/2** 3

[2] Буйло С И 2009 *Физико-механические и статистические аспекты акустико-эмиссионной диагностики предразрушающего состояния* (Ростов-на-Дону: Автореферат на соискание ученой степени доктора физико-математических наук)

[3] Буйло С 2009 Метод акустической эмиссии в задачах ранней диагностики деструкции материалов [Электронный ресурс] (Нижний Новгород: Диагностика материалов и изделий из них) URL: http://uran.ip.rsu.ru/

[4] Berezin A and Kozinkina A Russian patent number 2298785

[5] Penkin A, Terent'ev V and Maslov L 2004 *Intercontact Science* **70**

[6] Кондранин Е 2009 *Известия Южного федерального университета. Технические науки* **14**

[7] Марицкий Н Н 2014 Моделирование механического поведения материалов элементов конструкций космических аппаратов (Томск: диссертация на соискание учебной степени кандидата физико-математических наук) 135

[8] Источники и механизмы акустической эмиссии [online source] URL:http://www.sds.ru/articles/ae_review/ar01s03.html

The effect of hydrogen embrittlement on the mechanical properties of aluminum alloy

A V Bochkaryova[1], Yu V Li[2], S A Barannikova[1,2,3] and L B Zuev[1,2]
[1]Institute of Strength Physics and Materials Science, SB RAS, Tomsk, 634055, Russia
[2]National Research Tomsk State University, Tomsk, 634050, Russia
[3]Tomsk State University of Architecture and Building, Tomsk, 634003, Russia

E-mail: avb@ispms.tsc.ru

Abstract. The effect of hydrogen embrittlement on the mechanical properties of aluminum alloy D1 was investigated. The studies were performed for the test samples of aluminum alloy subjected to electrolytic hydrogenation. It is found that the mechanical properties of aluminum alloy are affected adversely by hydrogen embrittlement. The hydrogenated counterpart of alloy has a lower degree of ductility relative to the original alloy; however, the plastic flow behavior of material remains virtually unaffected. The deformation diagrams were examined for the deformed samples of aluminum alloy. These are found to show all the plastic flow stages: the linear, parabolic and pre-failure stages would occur for the respective values of the exponent n from the Ludwik-Holomon equation. Microhardness tests were performed for as-treated aluminum alloy D1; the measurement results are presented.

1. Introduction

It is the concern of the analyst to confirm to what extent corrosion processes affect the mechanical characteristics and, consequently, the performance of construction metals and alloys [1, 2]. Among thermally hardened aluminum alloys, duralumins hold a firm place due to ready availability, satisfactory strength characteristics and high corrosion resistance owing to the protective surface inert oxide film [3]. In view of the above, these materials find wide applications in various branches of industry; to mention just a few, structural components for aircraft, aerospace, ship building and other structural applications. Due to the exposure of constructions or components to environmental factors, the protective surface oxide film would be thinned down, thus entailing corrosion process development, which would affect adversely construction reliability. By and large, aluminum alloys are thought to offer resistance to aggressive media. However, this is only true for alloys having low extent of solid solution supersaturation, whereas alloys having high concentration of alloying additions, duralumin systems included, will undergo corrosion embrittlement under exposure to corrosive media. Hydrogen is also liable to cause metal embrittlement [4-8].

To explain the effects of hydrogen, many microscopic models have been developed [9, 10] based on the dislocation theory. However, such approaches seem to be insufficient, since they do not take into consideration the fact that plastic deformation has a local character during the entire process of flow [11, 12].

In this regards, the effect of electrolytic hydrogenation on the mechanical characteristics of aluminum alloys subjected to thermal strengthening has to be elucidated.

IOP Conf. Series: Materials Science and Engineering **71** (2015) 012057 doi:10.1088/1757-899X/71/1/012057

2. Experimental procedure

The investigation was performed for the test samples of duralumin D1, which had been subjected to thermal strengthening. The original duralumin D1 had the following chemical composition: 95% Al; 3.5% Cu; 0.2% Mg; 0.5 % Mn; the remainder Si, Fe, S, P and Zn. The test samples having dog-bone shape were stamped out from hot-rolled 2-mm sheet. These were annealed for 1 h at temperatures 300; 320; 340; 350 and 360°C with subsequent cooling down in the furnace to remove internal stresses.

The mechanical tests were carried on at the rate $6.67 \cdot 10^{-5} s^{-1}$ at room temperature in a universal testing machine Walter+Bai (Switzerland). Then the plastic flow curves plotted for the samples tested in tension were analyzed to single out individual flow stages.

The electrolytic hydrogenation of alloy samples was carried on for 24 h and 96 h under a controlled cathode potential. The sample was placed in an electrochemical cell containing 1N sulfur acid solution to which 20 mg/l thiocarbonic acid diamide had been added to enhance the process at 323 K [13]. The electrochemical cell was equipped with a graphite anode; a chorine silver reference electrode was connected in the circuit to maintain a constant potential U= – 430 mV. To control the three-electrode cell and run the electrochemical reaction, a Potentiostat/Galvanostat IPC-Compact unit was employed.

Using a standard technique, microhardness measurements were made for the original alloy D1 and for the counterpart subjected to hydrogenation. The indentation testing was performed with the aid of microhardness tester PMT-3 equipped with the Wickers tip; the load on the indenter P=0.15 N. Prior to testing, the sample work part was polished. Mathematical data treatment was carried on using routine procedures; the data obtained are reported in [14, 15].

3. Experimental results

The loading curves illustrated in Figure 1 were obtained for the test samples of alloy D1, which had been annealed at different temperatures. The curves obtained for the samples tested by uniaxial tension have a characteristic serrated shape, which is indicative of plastic flow instability. The plastic flow would exhibit an unsteady-state behavior from the yield limit up to material failure.

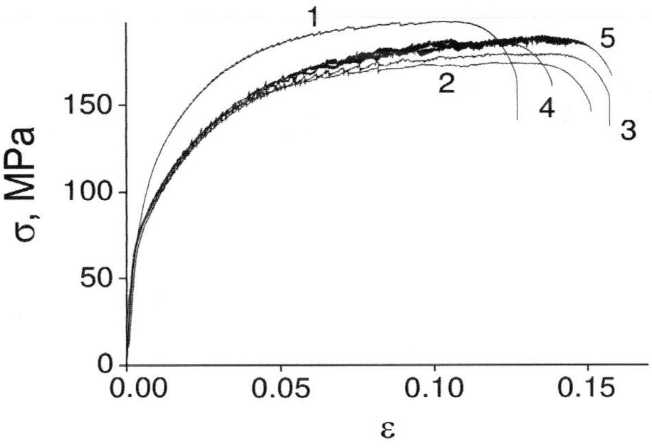

Figure 1. Loading curves of alloy D1 subjected to annealing at 300°C(1); 320°C(2); 340°C(3); 350°C(4) and 360°C(5) with subsequent cooling in furnace for t=1 h.

Provided their serrated shape is ignored, these curves might be grouped with diagrams of general type [16], which are described by the parabolic function

$$\sigma = \sigma_0 + K\varepsilon^n , \quad (1)$$

where K is the coefficient of deformation hardening and n is the exponent of deformation hardening.

Using the method reported in [17], the loading curve can be represented in the system of functional logarithmic coordinates as $\ln(s - s_0) - f(\ln e)$ (here s is the true stress which takes no account of reduction in the work cross-section and e is the true deformation). This enables individual segments to be singled out on the curve for a constant value of the exponent n, which varies discretely on going from one segment to the next. All the loading diagrams obtained in this study were analyzed by the above method. A lot of diagrams had serrated shape, which made it impossible to single out an individual stage. We succeeded in singling out the linear, parabolic and prefracture stages in the diagrams of the samples annealed at ≤340°C.

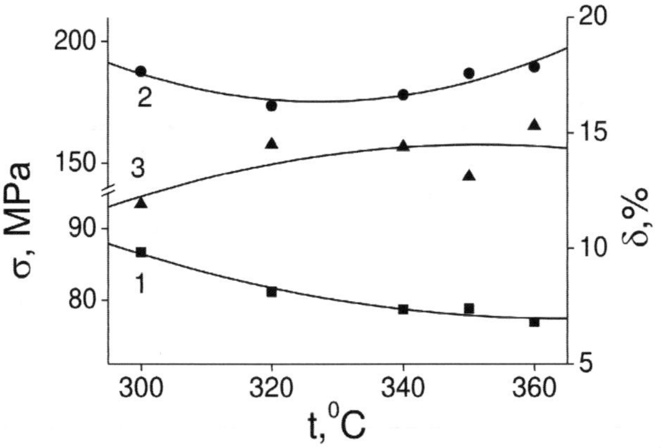

Figure 2. The values $\sigma_{0.2}$, σ_B and δ as a function of annealing temperature obtained for alloy D1 (curves 1, 2 and 3, respectively)

Evidently, an increase in the temperature of annealing up to 350°C and 360°C causes a drastic rise in the tensile stress, which will plot as a saw-tooth curve that alludes analyzing. Thus annealing schedules have been determined that are optimal in terms of expected mechanical characteristics of as-treated alloy, i.e. the thermal treatment of material is to be performed for 1 h at 340°C with subsequent cooling in furnace. Figure 2 illustrates the variation in the yield stress and ultimate strength of alloy D1 as a function of annealing temperature. It can be seen from figure 2 that the annealing carried on at T< 330°C would cause a decrease in the yield stress and ultimate strength of alloy D1 and an enhancement in its plasticity.

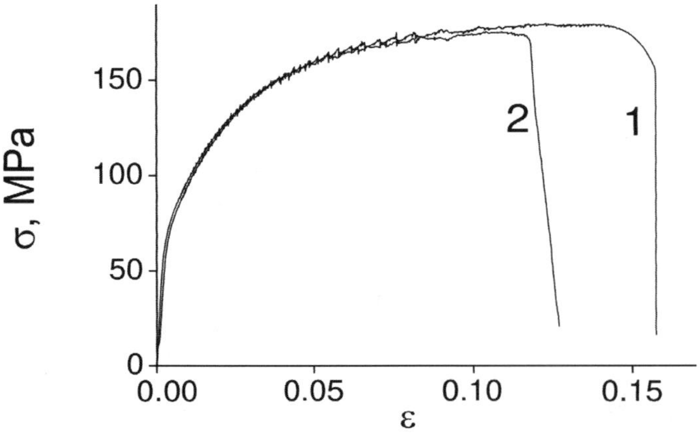

Figure 3. Loading curves obtained for the original alloy D1 and the counterpart of the same subjected to hydrogenation for t=96 h (curves 1 and 2, respectively).

IOP Conf. Series: Materials Science and Engineering **71** (2015) 012057 doi:10.1088/1757-899X/71/1/012057

The effect of hydrogenation on the mechanical characteristics of aluminum alloy D1 was studied. The test sample was subjected to tensile loading as soon as it was taken out of the electrochemical cell. The results of mechanical tests are presented in Figure 3.

Evidently, the loading curves, which have been plotted for the original alloy D1 and the counterpart subjected to electrolytic hydrogenation for 24 h, are practically identical.

Two sets of loading diagrams were matched (see Figure 3). It is found that a 15% decrease in the elongation to rupture is observed for the counterpart subjected to hydrogenation for 96 h relative to the original alloy; hence, the plasticity of alloy D1 has been impaired significantly by the hydrogenation treatment, although its strength characteristics remain unaffected.

The quantitative data on mechanical characteristics of alloys are listed in the Table 1. In the deforming sample of the original alloy, the onset of necking takes place, which is a forerunner of viscous fracture. The deformation curves plotted for the original alloy were represented in functional logarithmic coordinates. Three distinct rectilinear segments are distinguished on the flow curve for strains ε in the range 1.3-2.3%; 3.6-6.1% and 6.9-13.8% for a constant value n. These segments correspond to the linear and parabolic work hardening stages and to the prefracture stage, which occur for $n = 1$; $n \approx ½$ and $n \approx 0.3$, respectively. The data agree with the results of previous studies [18].

Table 1. Mechanical characteristics of the original duralimin D1 (1) and the counterpart of the same alloy subjected to electrolytic hydrogenation for 96 h (2).

	$\sigma_{0.2}$, MPa	σ_B, MPa	δ ,%	Linear stage, $n\approx1$		Parabolic stage, $n\approx½$		Prefracture stage, $n\leq0.3$	
				ε_i, %	ε_f, %	ε_i, %	ε_f, %	ε_i, %	ε_f, %
1	76.3	180.0	15.0	1.3	2.3	3.6	6.1	6.9	13.8
2	77.6	175.6	12.6	1.1	2.4	3.4	5.5	5.9	11.1

Note: $\sigma_{0.2}$ – proof stress; σ_B – ultimate stress; δ – relative elongation to rupture; ε_i – initial strain range; ε_f – final strain range.

4. Discussion

An analysis was made of the loading curve obtained for the test samples subjected to electrolytic hydrogenation. We distinguished three work hardening stages on the flow curve, i.e. linear, parabolic and prefracture ones. The latter two flow stages occurring in the original alloy and the hydrogenated counterpart practically coincide (see Table 1). In the case of counterpart hydrogenated for 96 h fracture occurs in the absence of necking.

Microhardness tests were performed by a standard procedure using 20 measurements. Analysis of data was made by the statistic method for physical experimental data treatment using a double t-criterion [14]. The microhardness tests were performed for the original alloy D1 and the counterpart, which had been hydrogenated for 96 h; the values obtained are $\overline{H}_\mu = 251.0 \pm 5.5$ MPa and $\overline{H}_\mu = 271.9 \pm 4.9$ MPa, respectively. Thus, using Student's t-criterion, we obtain $t = \frac{271.9 - 251}{\sqrt{136+108}}\sqrt{20} = 5.99$. For the confidence level $\alpha = 0.95$ and the number of degrees of freedom $f = 38$, Student's coefficient $t_{\alpha,f} = 2.02$ [15].

The resultant quantity $|t| \geq t_{\alpha,f}$ suggests that the average microhardness values obtained for the original alloy and the hydrogenated counterpart differ significantly, although on a percentage basis, the physical values differ by about 8%. It can be thus concluded that the hydrogenation treatment would cause an enhancement in the microhardness of aluminum alloy.

Thus, the effect of electrolytic hydrogenation on the mechanical properties of duralumin D1 has been studied. It is found that material strength and plasticity are virtually unaffected by the preliminary hydrogenation for 24 h, while the same properties would be affected adversely by the hydrogenation performed for 96 h. In the latter case, the linear and parabolic work hardening stages as well as the prefracture stage are found to occur on the flow diagram plotted for the deforming sample. However, no necking would take place at the prefailure stage; hence, the test samples undergo brittle fracture. It can thus be concluded that the plasticity of duralumin D1 is affected adversely by the electrolytic hydrogenation carried on for 96 h. Hence, the probability of hydrides forming in as-treated material is high [4-8].

For the case of cracking of aluminum alloys according to the hydrogen embrittlement mechanism there are not theoretically based values of the depth of penetration of hydrogen and its concentration in the area in front of the crack tip necessary for its further advance. The diffusion rate of the hydrogen must agree in order of magnitude with the rate of crack advance v_{cr} if the latter develops in two stages, embrittlement in the fist and in the second a jump in the crack by the distance of embrittlement at the speed of sound.

In principle, acceleration of transport of hydrogen is possible at a crack tip if the gradient in the hydrostatic component of stresses is sufficiently high [19]. However, according to improved calculations [20] the distribution of stresses is somewhat different so that the only method of acceleration of transport is transfer of hydrogen by dislocations. The rate of transfer of hydrogen by dislocations is determined by the critical rate v_c of breaking away of them from segregations (Cottrell atmospheres) [9]:

$$v_C = \frac{4kTD_H}{\beta} \quad (2)$$

Keeping in mind that dislocations effectively capture hydrogen only in the presence in the metal of excess vacancies [21] and also the formation of vacancy-hydrogen complexes it must be taken into consideration that in essence the dislocations transport these complexes. Their specific atomic volume is approximately the same as for a vacancy. In the given situation the parameter of elastic interaction β is equal to $1.6\cdot10^{-30} J/m$. In addition, in place of the coefficient D_H in equation (2) we may substitute the diffusion coefficient of vacancies, the best known determination of which at room temperature is $10^{-12} m^2/s$ [10]. As a result, we obtain $v_C \approx 10^{-3} m/s$. If the average value for all alloys of $D_H = 10^{-13} m^2/s$ calculated on the basis of data on its values in pure aluminum and in 7075-T6 alloy [8] is used as the diffusion coefficient of hydrogen, then $v_C \approx 10^{-4} m/s$.

5. Conclusion

The corrosion properties of semi-products manufactured from the Al-Cu-Mg alloy system are determined by the composition and distribution of precipitates at the grain boundaries and within the grain bulk and by the extent of copper depletion from the near-boundary regions, which in turn depends on thermal treatment conditions.

The analysis of experimental data [3-7] suggests that the duralumin systems Al-Zn-Mg and Al-Cu-Mg would undergo corrosion cracking, which is largely due to hydrogen evolving during electrochemical reactions; however, the mechanism involved in embrittlement has not been defined unambiguously as yet. There is good reason to believe that the failure in hydrogenated material occurs via a mechanism, which is essentially different from that involved in corrosion cracking. In the former case, material plasticity is affected adversely, which is liable to cause intergranular fracture due to hydride formation. At first hydrogen would diffuse at a low rate towards the grain boundaries; therefore, it takes time for hydrides to form in the material. Corrosion cracking will occur simultaneously with hydrogen embrittlement via electrochemical mechanism, which involves periodic disintegration of passive film due to the local plastic strains at the grain boundaries and to the dissolution occurring locally in the same regions.

Acknowledgement
The work was performed in the frame of Fundamental Research Program of State Academy of Sciences for the period 2013-2020 yrs.

References

[1] Holroyd N J H, Vasudevan A K and Christodolou L 1989 *Treatise on Materials Science and Technology Ch. 16: Stress Corrosion of High-Strength Aluminum Alloys* (San Diego: Academic Press) p 463-483
[2] Polyanskii V M 1985 Role of hydrogen embrittlement in the corrosion cracking of aluminum alloys *Sov Mater Sci **21*** 301-309
[3] Larignon C, Alexis J, Andrieu E et al. 2013 The contribution of hydrogen to the corrosion of 2024 aluminium alloy exposed to thermal and environmental cycling in chloride media *Corros Sci* **69** 211-220
[4] Lunarska E and Chernyaeva O 2004 Effect of precipitates on hydrogen transport and hydrogen embrittlement of aluminum alloys *Mater Sci* **40** 399–407
[5] Kannan M and Raja V 2006 Hydrogen embrittlement susceptibility of over aged 7010 Al-alloy *J Mat Sci* **41** 5495–5499
[6] Kim S, Han M and Jang S 2004 Electrochemical characteristics of Al-Mg alloy in seawater for leisure ship: Stress corrosion cracking and hydrogen embrittlement *Korean J Chem Eng* **26** 250–257
[7] Kumar S and Namboodhiri T 2011 Precipitation hardening and hydrogen embrittlement of aluminum alloy *AA7020 B Mater Sci* **34** 311–321
[8] Nykyforchyn H, Ostash O, Tsyrul'nyk O, Andreiko I and Holovatyuk Yu 2008 Electrochemical evaluation of the in-service degradation of an aircraft aluminum alloy *Mater Sci* **44** 254–259
[9] Hirth P and Lothe J 1968 *Theory of Dislocations* (New York: McGraw-Hill) 363
[10] Friedel J 1964 *Dislocations* (Oxford: Pergamon press) p 371
[11] Barannikova S A, Nadezhkin M V, Mel'nichuk V A et al. 2011 Tensile plastic strain localization in single crystal of austenite steel electrolitically saturated with hydrogen *Tech Phys Lett* **37** 793-796
[12] Barannikova S A, Nadezhkin M V, Lunev A G et al. 2014 Regularities in localization of plastic flow upon electrolytic hydrogenation of an iron bcc-alloy *Tech Phys Lett* **40** 2113-214
[13] Yagodzinskyy Y, Todoshchenko O, Papula S and Hänninen H 2011 Hydrogen Solubility and Diffusion in Austenitic Stainless Steels Studied with Thermal Desorption Spectroscopy *Steel Res Int* **82** 20–25
[14] Zazhigayev L, Kishyan A and Romanikov Yu 1987 *Methods for planning a physical experiment and data processing* (Moscow: Atomizdat) 232
[15] Kondrashov A and Shestopalov E 1977 *The Basis of the Physical Experiment and Mathematical Treatment of Measurement Results* (Moscow: Atomizdat) 200
[16] Pelleg J 2013 *Mechanical Properties of Materials* (Springer: Dordrecht) 634
[17] Trefilov V, Moiseyev V, Pechkovsky E et al. 1989 *Deformation Strengthening and Fracture in polycrystalline materials* (Kiev: Naukova Dumka) 256
[18] Danilov V, Bochkaryova A and Zuev L 2009 Macrolocalization of deformation in material having unstable plastic flow behavior *Metal Physics and Metal Science* **107** 660–667
[19] McMeeking R 1977 Finite deformation analysis of crack-tip opening in elastic-plastic materials and implications for fracture *J Mech Phys Solids* **25** 357-381
[20] Gortemaker P C M, de Pater C and Spierlng R M E 1981 *Near-crack tip finite strain analysis* in: Advanced Fracture Research. Preprint of the 5th International Conference on Fracture (ICF5), Cannes, Franc, Vol. I, Pergamon, Oxford--New York (1981) 151-160
[21] Foster L M, Jack T H, Hill W W and Jeitschko W 1970 Structure revealed in zone refined aluminumby trltlum decoration *Metallurgical Transactions* **1** 3117-3124

IOP Conf. Series: Materials Science and Engineering **71** (2015) 012058 doi:10.1088/1757-899X/71/1/012058

X-ray diffraction analysis of reorientation regions formed in [111]-single crystal of aluminum under compression

I V Bespalova[1] and L A Teplyakova[2]
[1]Kazakh National Technical University named after K.I. Satpaev, Almaty, 050013, Kazakhstan
[2]Tomsk State University of Architecture and Building, Tomsk, 634003, Russia

E-mail: besiv@mail.ru

Abstract. The paper presents results of the crystalogeometric analysis of the shear pattern formed on lateral faces of a single crystal of aluminum under compression along the direction $[\bar{1}11]$. Main features of shear strain in $[\bar{1}11]$-single crystals of aluminum have been revealed. Typical elements of strain relief have been identified and places of their location have been established. Occurrence of a bending in the crystal lattice at the macro level has been established using the Laue pattern method. The results of the X-ray analysis have been compared with the shear pattern.

1. Introduction

Currently, systematic studies on spatial organization of a shear strain at the macro level are carried out on symmetrically oriented single crystals of aluminum [1-5]. The fact of the course of macrofragmentation of the shear strain from the very beginning of the plastic flow has been experimentally established in these studies. It has been revealed that the initial macrofragmentation of the shear is caused by the heterogeneity of the slip on equally-loaded octahedral planes. At the macrolevel the heterogeneity of the slip is the reason for occurrence of bends in the crystal, as a rule, in joints of neighboring macrofragments with an asymmetric slip [2, 4]. As stated in [6, 7], the subsequent strain in regions of bending can lead to formation of discrete misorientation boundaries.

The paper presents the results of the X-ray analysis of the strained $[\bar{1}11]$-single crystal.

$[\bar{1}11]$-single crystals of aluminum with orientation of lateral faces $(\bar{1}10)$ and $(\bar{1}1\bar{2})$ have been investigated. With this orientation of the single crystal, three octahedral slip planes are equally-loaded (Figure 1, *a*). In addition, easy shear volumes (ESV) are allocated for each of them in the volume of the single crystal (Figure 1, *b, c, d*). The samples had the shape of a parallelepiped with linear dimensions of (3x3x6) mm^3. Strain was carried out under compression up to small degrees of strain ($\varepsilon<0,1$) at room temperature with a velocity of $6\times10^{-4}s^{-1}$. An optical microscope MIM- 10 and a scanning electron microscope Tesla BS – 301 were used in investigation of the strain relief pattern. The pattern of the strain relief on all of its free faces has been investigated in recreation of the spatial organization of the shear in the volume of the single crystal (macrolevel). The internal structure of single crystals after their strain under compression has been investigated using the X-ray method (shooting of Laue patterns).

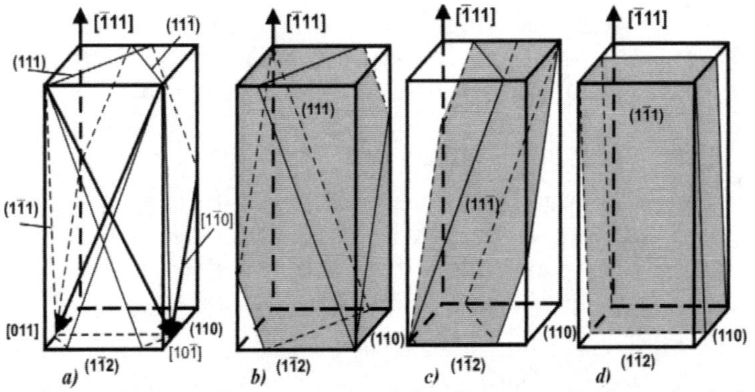

Figure 1. Crystallogeometric of the diagram of orientation of equally-loaded octahedral planes and directions in $[\bar{1}11]$ – single crystal (*a*); location of ESV for families of planes (111), $(11\bar{1})$, and $(1\bar{1}1)$ - (*b-d*).

2. Strain relief

Figure 2 shows the scheme of strain relief patterns formed under compression of the single crystal on all of its free faces after strain up to ε=0,06 (Figure 2, I-IV). Analysis of the cumulative strain relief has showed that macrofragmentation and macrolocalization of strain takes place in $[\bar{1}11]$-single crystals of aluminum from the beginning of the plastic strain. This is evidenced by formation of systems of strain macrobands of various types in the volume of the sample (Figure 2 *a-e*).

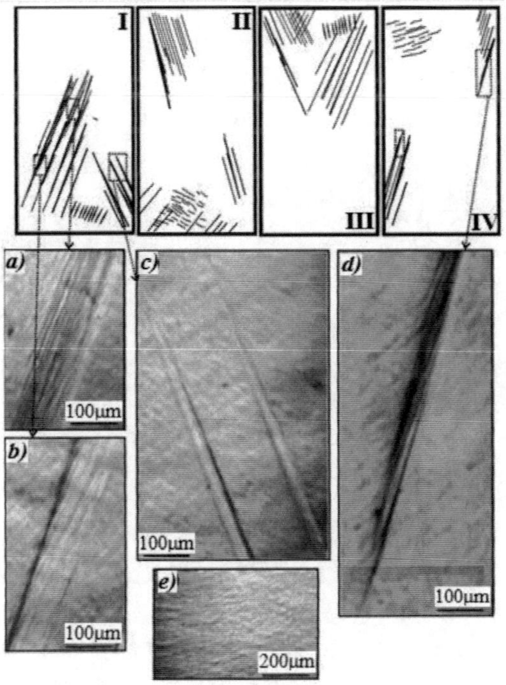

Figure 2. The scheme of macrorelief on lateral faces of the $[\bar{1}11]$-single crystal: images of different types of macrobands – (*a-d*) and (e) – the region of the face with small strain folds.

At ε=0,06 the boundaries of strain bands, as a rule, are deviated from the outcrop of the nearest equally-loaded octahedral plane at angles of 5-6°. On sections of largest shear bands observed on faces of the single crystal there are shear traces and strain folds perpendicular to bands (Figure 2, *a, b*). Systems of perpendicular folds also occur outside of bands (Figure 2, II at the bottom part of the sample and Fig. 2, IV at the top part). Apparently, they are formed by the mechanism of kink bands. The peculiarity of the strain relief pattern in $[\bar{1}11]$-single crystals is in the fact that the majority (70-80%) of face surfaces is occupied not by systems of traces of the octahedral shear, which is typical for aluminum single crystals of other orientations [1-5], but by systems of small deformation folds resembling hillocks (Figure 2, *e*). Furthermore, one bending region is allocated on each face $(\bar{1}1\bar{2})$ (Figure 2, I and III) near the vertex of the sample. The internal structure of bending regions (on faces) represents a system of shear traces and strain folds, approximately parallel to the outcrop of the plane $(11\bar{1})$.

3. X-ray diffraction analysis of the strained $[\bar{1}11]$-single crystal of aluminum

Shooting of Laue patterns of regions of the $[\bar{1}11]$-single crystal strained up to ε=0,06 has been carried out in the work. Shooting was carried out in three mutually perpendicular directions (single crystal was X-rayed through). Figure 3 shows the obtained Laue diffraction patterns and regions of the sample on which the X-ray beam was aimed. Asterism of spots in Laue patterns is clearly visible, indicating the occurrence of continuous misorientations in the strained single crystal. The results of interpretation and analysis of Laue patterns obtained in three mutually perpendicular directions of the beam to the surface of the single crystal faces should be examined.

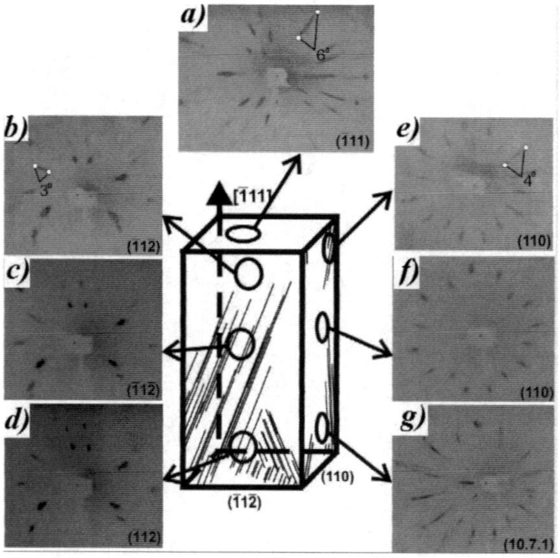

Figure 3. Laue patterns obtained during irradiation of the $[\bar{1}11]$-single crystal (ε=0,06) perpendicular to: end (*a*); face $(\bar{1}1\bar{2})$ (*b-d*), and face (110) (*e-g*).

1. As a result of interpretation of the Laue diffraction pattern taken during irradiation along the compression axis of the single crystal it has been established that two families of planes, namely $(\bar{1}11)$ and $(\bar{9}.9.10)$, make a contribution to the cumulative picture. The angle between the normals to these planes is 3 degrees. Measurement of the radial misorientation by blurring of spots on the investigated Laue pattern (Figure 3, a) yields the result from 2 up to 6^0.

2. Laue diffraction patterns obtained during shooting in three regions of the sample along the direction $[\bar{1}1\bar{2}]$ (Figure 3, *b - d*) were similar. Their interpretation showed that only the family of planes $(\bar{1}1\bar{2})$ is perpendicular to the X-ray beam. Most spots of the discussed Laue patterns (Figure 3, *b*) are indicative of radial misorientations by 2-3 degrees.

3. Finally, Laue diffraction patterns obtained from different regions of the face (110) during successive shooting along the direction $[110]$ are shown in Figure 3, *e - g*. It is seen that Laue diffraction patterns taken from the upper and the middle part of the sample are similar, and in interpretation give planes (110) (Figure 3, *e, f*). Interpretation of the Laue diffraction pattern taken near the end (Figure 3, *g*) showed that the main contribution to the pattern of spots on the Laue pattern is made by planes $(10.7.1)$. The angle between the normal to this plane and to the plane of the face (110) equals to 11^0.

4. Conclusion

The obtained results confirm the conclusion, which was made above based on the study of the strain relief, on formation of the bending region of the crystal-lattice in the $[\bar{1}11]$-single crystal under strain. Formation of bending regions from the beginning of the plastic strain may be associated both with the heterogeneity of the stress state caused by a rectangular shape of the single-crystal sample [8] and by the presence of volumes of the constrained shear in $[\bar{1}11]$-single crystals for families of equally-loaded octahedral planes.

References

[1] Teplyakova L A, Lychagin D V and Kozlov E V 2002 Localization of shear under strain of aluminum single crystals with orientation of the compression axis *Phys Mesomech* **5** 49-55

[2] Teplyakova L A, Lychagin D V and Bespalova I V 2004 Regularities of strain macrolocalization in aluminum single crystals with orientation of the compression axis *Phys Mesomech* **7** 63-78

[3] Teplyakova L A, Lychagin D V and Bespalova I V 2006 Peculiarities of spatial organization of the shear at the macrolevel in [111]-aluminum single crystals *Phys Mesomech* **9** 63-71

[4] Teplyakova L A, Bespalova I V and Lychagin D V 2006 Regularities of organization of the shear strain in [001]-aluminum single crystals with lateral faces {100} under compression *Phys Mesomech* **9** 77-84

[5] Teplyakova L A and Kozlov E V 2005 Formation of scale-structural levels of plastic strain localization in metallic single crystals. I. Macrolevel *Phys Mesomech* **8** № 6 57-66

[6] Беспалова И В, Теплякова Л А и Куницына Т С 2009 Рентгеноструктурное исследование областей переориентации деформированных $[1\bar{1}2]$-монокристаллов алюминия *Вестник. Серия физическая* **1** 128-131

[7] Gubkin S I 1961 *Plastic strain of metals* (Moscow: Metallurgy) **1** 376

IOP Conf. Series: Materials Science and Engineering **71** (2015) 012059 doi:10.1088/1757-899X/71/1/012059

Obtaining high thermally conductive materials by pressing from the granulate

A Ditts[1], I Revva[1], Y Pautova[1], V Pogrebenkov[1], Y Nepochatov[2], E Galashov[3] and R Tarnovskiy[1]
[1]National Research Tomsk Polytechnic University, Department of Silicate and Nanomaterials Technology, Tomsk, 634050, Russia
[2]Holding JSC "NEVZ-Soyuz", Novosibirsk, 630049, Russia
[3]Novosibirsk State University, Department of Applied Physics, 630090, Novosibirsk

E-mail: ditts@tpu.ru

Abstract. This work contains results of investigation of obtaining high thermally conductive ceramics from commercial powders of aluminum nitride and yttrium oxide by the method of monoaxial compaction of granulate. The principal scheme of preparation is proposed and technological properties of granulate are defined. Compaction conditions for simple items to use as heat removal in microelectronics and power electrical engineering have been established. Investigations of thermophysical properties of obtained ceramics and its structure by the XRD and SEM methods have been carried out. Ceramics with thermal conductivity from 172 to 174 W/m·K has been obtained as result of this work.

1. Introduction

Growing demand for power semiconducting modules of high power, high reliability and affordable cost is caused by permanently developing power converter installations market: drives, energy management systems, uninterrupted power supplies, switched power supplies in electrical transport means. Ceramic substrate plays role of an electrically insulating and thermally conducting layer between semiconducting crystals and basis in power modules both brazed one with insulated basis and in modules of pressure construction (IGBT, FRD and so on). Increase of power increases heat generation which leads to increase of module temperature and deterioration of performance of semiconductive crystals and element of power electric engineering in whole, decreases its reliability and service life. Ceramics on the basis of Al_2O_3, that is used at the present, cannot ensure efficient heat removal (thermal conductivity of Al_2O_3 is 24 W/m·K), use of BeO is limited by its high cost, danger of production and toxicity for human (thermal conductivity of BeO is between 180 and 250 W/m·K). Therefore problem of heat removal in power converter modules is very urgent and its solution is possible only using high thermal conducting ceramic materials. At the present there are no acting productions of high thermal conducting ceramic materials on the basis of aluminum nitride in Russia, all consumable volume of aluminum nitride items is import production.

2. Experimental procedure

Aluminum nitride industrial powders of the firm H.C. Starck grade B were used in this work. Yttrium oxide of firm H .C. Starck grade C was inserted to decrease sintering temperature and to purify aluminum nitride from oxygen admixture.

According to data of X-ray phase analysis performed with the Shimadzu XRD-7000 diffractometer, aluminum nitride powder is presented with single phase (000-25-1133). X-ray pattern of yttrium oxide powder contains only yttrium oxide reflexes (000-41-1105).

Results of laser granulometry of initial powders of aluminum nitride and yttrium oxide obtained on the SALD 7101 laser diffraction particle size analyzer are presented in Table 1.

Table 1. Powder granulometry data.

Initial powder	Particle size, μm		
	D_{10}	D_{50}	D_{90}
AlN	0.200	1.200	5.100
Y_2O_3	0.641	1.455	2.926

Obtained results of laser granulometry are confirmed by data of Scanning Electron Microscopy performed on the JSM 7500FA microscope for powders of aluminum nitride and yttrium oxide under investigation.

To calculate quantity of sintering addition (Y_2O_3) the oxygen content in aluminum nitride powder using the ONH836 device of the LEGO firm was defined; oxygen content was 1.153 wt. %.

To make compacts mixture of aluminum nitride and yttrium oxide powders was granulated because initial powders and their mixture had unsatisfactory properties (bulk density: 0.23 g/cm^3 and 0.63 g/cm^3, accordingly). Flowability of initial powders could not be measured. Granulation of powder mixture was performed with the "break of a briquette".

According to proposed scheme powders of aluminum nitride and yttrium oxide were mixed in a ball mill during 24 hours. Obtained powder mixture was dried in a vacuum drying cabinet. Bond solution (PVB solution) was added to prepared powder mixture and a briquette was compacted from plasticized mass with specific pressure 1.0 t/cm^2. Obtained briquette was crushed and sieved on a vibration sieve to obtain granulate of specified fraction from 0.063 to 0.315 μm.

To remove excessive moisture obtained granulate was dried during day at room temperature. The samples were compacted in the form of discs with diameter 30 mm and height up to 5 mm, specific compaction pressure was changed from 1.0 up to 2.0 t/cm^2 with step 0.5. Compacted samples were sintered in the high temperature graphite furnace in nitrogen medium at sintering temperature of 1800 °C.

3. Results and discussion

At production of items with monoaxial compacting it is necessary to compact fine powder materials to improve important engineering properties, such as bulk density, flowability, and natural angle of slope. Use of fine powders in initial state is connected with many difficulties: spraying powders, increase of friction force between particles that leads to decrease of compacting force. All this does not allow obtaining compacts of good quality. The properties have been defined by standard methods on initial powders, on powder mixture and obtained granulate. Obtained data are given in Table 2.

Granulation allows to obtain granulate with bulk density being more than twice as one of initial powders and flowability of 25.0 g/s.

Table 2. Engineering properties of powders and granulate.

Engineering properties	Material			
	Initial powders		Powder mixture	Granulate
	AlN	Y_2O_3		
Bulk density, g/cm^3	0.63	0.23	0.43	0.88
Flowability, g/s	0	0	0	25.0

Granulation allows to obtain granulate with bulk density being more than twice as one of initial powders and flowability of 25.0 g/s.

Important moment at production of items is obtaining required compact form. It depends on number of factors: quantity of temporary organic bond, dispersity of granulate, specific pressure, and mold construction. Granulate was prepared varying the main factors influencing compact properties: content of bond, specific pressing value, fraction composition, and granulate moisture.

All prepared granulates had similar granulometric composition, determined by sieving method, that is connected with the same method of granulate obtaining. Granulometry data of prepared granulates are presented in Figure 1.

All granulates have similar form of granules as volume particles of fragmentary form of various size, that is connected with the same method of making granulate, microphotographs of granules are presented in Figure 2.

Influence of various factors was estimated according to properties of compacts obtaining by monoaxial compaction in metallic molds, and according to relative density of compacts. Apparent density of compacts was determined by geometric way.

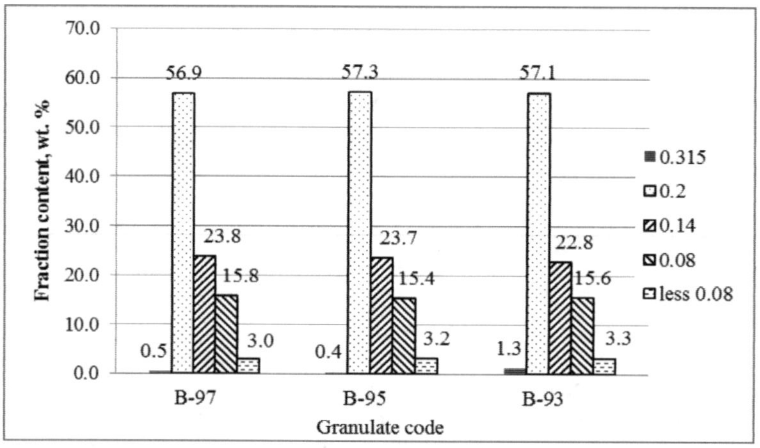

Figure 1. Granulometric composition of prepared granulates.

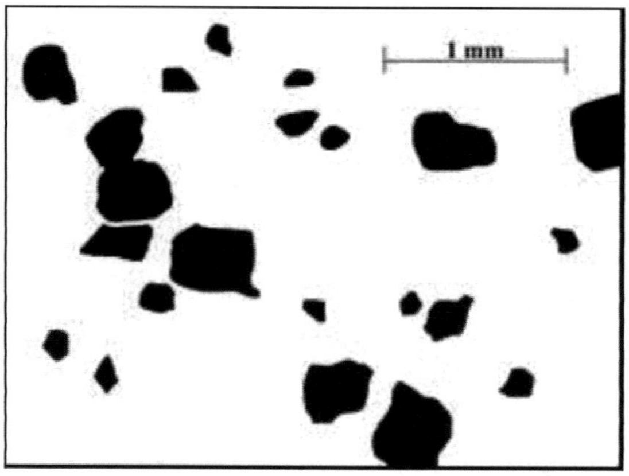

Figure 2. Microphotographs of granules B-95×25.

Solution of polyvinyl butyral was used as temporary bond, its content in granulates changed from 3 to 7 mass % (Table 3).

Results of compaction of obtained granulates are presented on the plot in Figure 3.

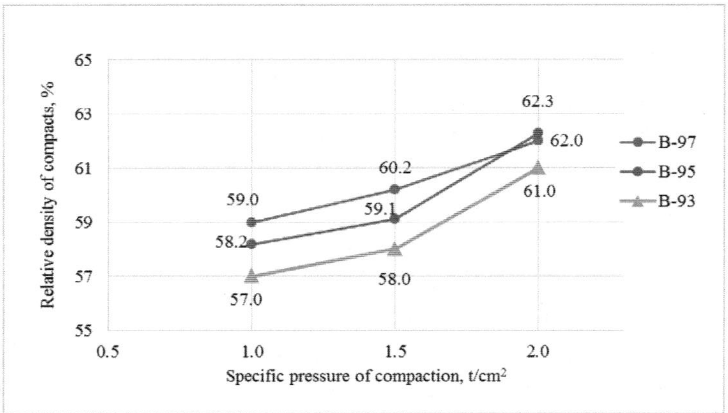

Figure 3. Dependence of relative density of compacts on content of temporary bond.

It is established, that relative density of compacts growths with specific pressure at all contents of bond. Minimal content of bond being necessary to compact samples of specified form has been determined.

Table 3. Compositions of prepared granulates.

Granulate component	Granulate Code		
	B-97	B-95	B-93
PVB, wt.%	3.0	5.0	7.0
Powder mixture, wt.%	97.0	95.0	93.0

Necessary content is 5 wt.% , at smaller content one cannot obtain good compacted samples of the specified form (edges were spalled), and at greater content of bond relative density of compacts was less at all specific pressures. Compositions of granulate with combined bond (bond + plastificator) in quantity of 5 wt. % were studied on the next stage of works, prepared compositions are presented in Table 4. Addition of plasticizer to bond decreases strength of granules and at compaction its destruction will begin at less specific pressures. Diagram of dependence of relative density on plasticizer content is presented in Figure 4.

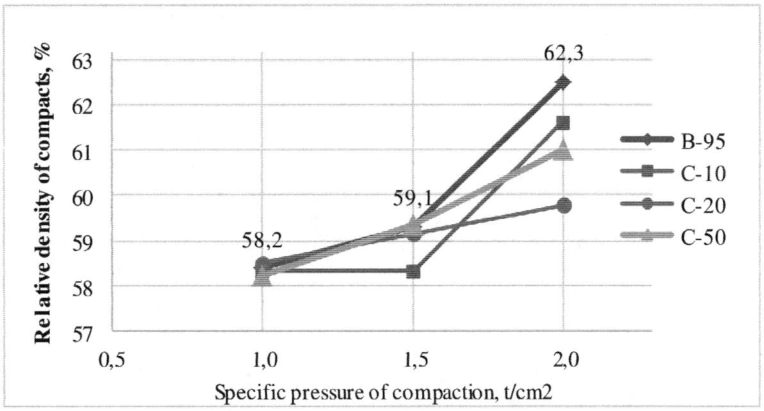

Figure 4. Dependence of relative density of compacts on bond - plasticizer ratio.

Relative density of compacts decreases with increasing of plasticizer content, in as much as density of granules is decreased and their destruction occurs on the initial stages of compaction process. In connection with this granulometric composition of granulate is sharply changed and density of its package is decreased. Force of friction between particles is increased and efficiency of compaction is decreased.

Table 4. Compositions of prepared granulates with combined bond.

Granulate component	Granulate Code C-50	C-20	C-10	B-95
PVB, wt.%	2.5	4.0	4.5	5.0
DBP, wt.%	2.5	1.0	0.5	0
Powder mixture, wt.%	95.0	95.0	95.0	95.0

Influence of granulometric composition on relative density of compacts was estimated on obtained granulates with continuous granulometric composition and two fraction composition. Granulate B-95 has continuous granulometric composition and granulate A-95 contained two fractions: large one from 0.200 to 0.315 μm and small one less than 0.063, taken in ratio 70/30 in vol. %. Results of compaction are presented in Figure 5.

Relative density of compacts for model 2 fraction granulation is more than one for granulate with continuous granulometric composition inasmuch as package density for 2 fraction granulate is more.

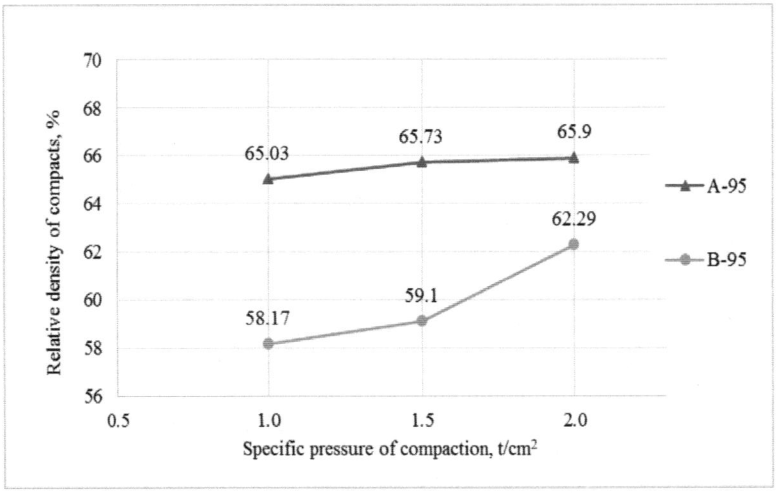

Figure 5. Influence of granulometric composition on relative density of compacts.

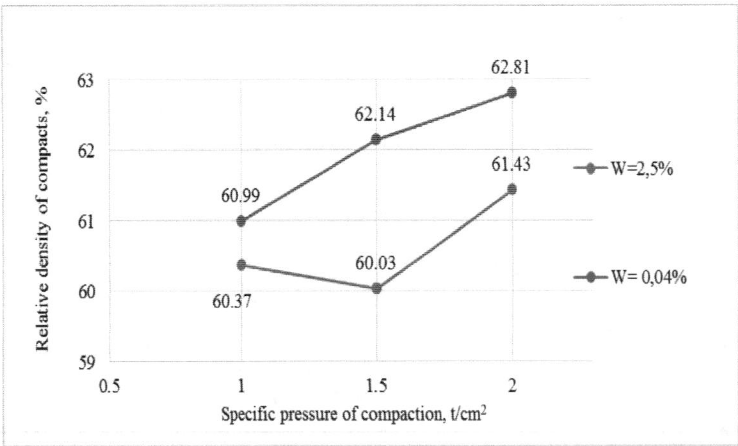

Figure 6. Influence of granulate moisture on relative density of compacts.

IOP Conf. Series: Materials Science and Engineering **71** (2015) 012059 doi:10.1088/1757-899X/71/1/012059

Obtained dependence of relative density of compacts on granulate moisture is presented in Figure 6. If moisture increases relative density is decreased at all compaction pressures. It is connected with the fact that if moisture increases, pore volume after its evaporation is increased. Air permeability of material is decreased and more quantity of air is solved in moisture. Moisture quantity on the level of 0.04 wt. % (natural) in combination of bond in quantity of 5 wt. % is enough to decrease force of friction between particles, ensure slide and rotation of particles each against other on initial stages of compaction. It was impossible to obtain compactions at granulate moisture more than 2.5 wt. %.

Data of thermal analysis of made granulate is presented on Figure 7. Four effects are determined on thermogramm: the first one at temperature of 49 °C is endothermic and connected with removal of solvent; the second one at temperature of 175 °C is exothermic and connected with oxidation of organic bond; the third peak at temperature of 236 °C is connected with oxidation of surfactant and the last exothermic effect at temperature of 376 °C is connected with oxidation of carbon residue. Temperature chart of removal of bond has been proposed on the basis of results of thermal analysis. Removal of bond occurs on air at temperature 500 °C, and heating rate 5 °/min with 30 min. exposure at temperatures 170, 230, and 500 °C.

Sintering of aluminum nitride was performed in the high temperature furnace with graphite heaters in nitrogen medium at temperature of 1800°C and isothermic exposure 4 hours.

To obtain high thermal conductivity it is required to ensure purification of aluminum nitride from oxygen admixture, direct contact between aluminum nitride grains, and continuity – absence of pores in sintered material. To ensure this requirement it is necessary to insert sintering addition – yttrium oxide - in aluminum nitride powder.

On the microphotographs presented in Figure 7 one can see that sintering addition is evenly distributed in sintered ceramics. At sintering formation of eutectic compound of yttrium aluminum garnet takes place as result of interaction of yttrium oxide and aluminum oxide, which is in the aluminum nitride powder.

Appearance of melt in the time of sintering ensures transport reaction, sintering by liquid phase mechanism that considerably accelerates process and ensures purification of aluminum nitride from oxygen admixture. After that the melt is forced in triple point ensuring direct contact between aluminum nitride grains where it crystallizes in yttrium aluminum garnet. According to data of electronic microscopy, size of aluminum nitride grain is from 2 to 7 μm, size of yttrium aluminum garnet is about 2 μm.

Apparent density, relative density, porosity and water absorption of sintered samples were determined by means of hydrostatic weighing. According to data of determination of relative density one can say that all samples are sintered, relative density growths with increasing of specific pressure of sample compaction for all granulates, the highest relative density is achieved for granulates C-10 and B-95 and is 99.96 % and 99.92 %, accordingly. Open porosity for majority of samples is close to zero value excepting B-95 with specific pressure 1.0 t/cm^2. Such high value of open porosity and low value of relative density is connected with low relative density of compact. Granules had high strength and at low pressure they are not destructed at compaction, and dense package of particles is not ensured.

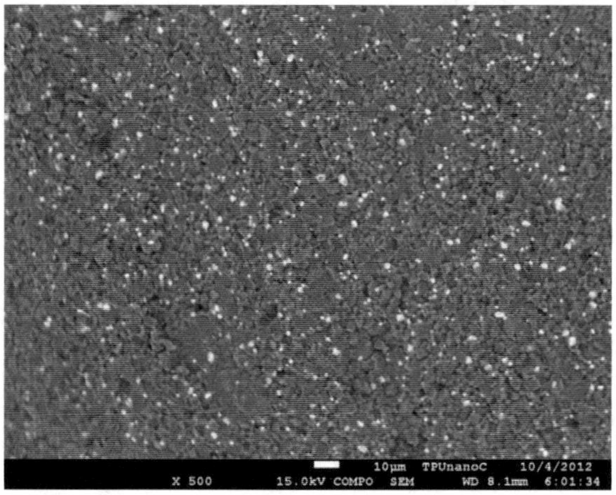

Figure 7. Ceramics microphotographs ×500.

Shrinkage of specimens at sintering is different in different planes and is in interval from 14.6 to 16.9%, when specific pressure increases, it decreases.

4. Conclusion

Principal scheme to obtain granulate from aluminum nitride and yttrium oxide is proposed in the work. Study of influence of various factors on properties of granulate, compacts and sintered ceramics has been carried out. Granulation improves engineering properties (bulk volume, flowability) and increases quality of compacts. Relative density of compacts is influenced in more degree by specific pressure of compaction and granulometric composition of granulates. The highest relative density of compacts being 62.0 % has been obtained at specific pressure 2.0 t/cm^2 for granulate with content of 5 wt. % of bond and consisting from two fractions. Sintered ceramics has high physical properties, relative density of all granulates exceeds 99 %, open porosity is close to zero. Important characteristic of ceramics is thermal conductivity that determines ability to use obtained materials as heat removal. Thermal conductivity of obtained ceramics is from 172 to 174 W/m·K, answers to requests of consumers, and corresponds to world level.

References

[1] G V Samsonov 1969 *Nitrides* (Kiev: Naukova Dumka)
[2] Gromov A A, Strokova Yu I and Ditts A A 2010 Passivation films on particles of electroexplosion aluminum nanopowders: a review *Russ. J. Phys. Chem.+* **4** 156-169
[3] Chaplina Y, Gromov A A, Ditts A A and Mitina N A 2011 Baking of oksinitridny ceramics from the fusion mixture received the self-extendingby high-temperature synthesis *News of higher education institutions. Physics* **54** 283-287

The effect of thermal cycling and stress-assistant ageing two-way shape memory effect in [$\bar{1}$23]-oriented $Co_{40}Ni_{33}Al_{27}$ single crystals

A S Eftifeeva[1], E Yu Panchenko[2] and Yu I Chumlyakov[1]
[1]National Research Tomsk State University, Department of Physics of Metals at Tomsk, 634050, Russia
[2]National Research Tomsk State University, Laboratory at Siberian Physical Technical Institute, Tomsk, 634050, Russia

E-mail: anna_eftifeeva@rambler.ru

Abstract. The effect of thermal cycling through an interval of B2-L1_0 martensitic transformation (MT) under action of external stress and tensile stress-assistant ageing on the two-way shape memory effect in [$\bar{1}$23]-oriented $Co_{40}Ni_{33}Al_{27}$ (at.%) single crystals are investigated. For the first time it is experimentally established that tensile stress-assistant 100 MPa ageing at 573 K for 1 h along [$\bar{1}$23]-direction of $Co_{40}Ni_{33}Al_{27}$ single crystals creates the necessary conditions for two-way shape memory effect (TWSME) with the reversible strain up to ε=2.4 (±0.3)% at cooling/heating. The TWSME in quenched [$\bar{1}$23]-oriented $Co_{40}Ni_{33}Al_{27}$ single crystals can be induced by thermal cycling through an interval of B2-L1_0 MT under action of constant external stress 50 MPa with the reversible strain less than 1%.

1. Introduction

In recent years CoNiAl crystals undergoing thermoelastic B2-L1_0 martensitic transformation (MT) have emerged as new ferromagnetic shape memory alloys. The shape and size of these ferromagnetic materials can be reversed not only by changes in temperature and mechanical stress, but also by applying an external magnetic field. The single crystals of CoNiAl alloys have high-temperature superelasticity at T>373 K, shape memory effect (SME) and a magneto-induced strain of 3.3%, while simultaneously showing action of external stress and magnetic field [1-3]. Fulfilled on aged poly and single crystals of CoNiAl alloys, studies show that precipitation of nanoscale particles leads to a high-strength state and high cyclic stability of functional properties [2-4]. This is very important for operation of devices in cyclic loadings. As shown in [5] the two-way shape memory effect (TWSME) in stress-assistant [$\bar{1}$23]-oriented $Co_{35}Ni_{35}Al_{30}$ single crystals aged at 673 K for 0.5 h can be attributed to the internal stress fields up to $|\langle\sigma_G\rangle|$~75 MPa by the oriented arrangement of non-equiaxed particles of ε-Co with hcp-lattice of size 20-30 μm [6,7]. In this case the material has a spontaneous recoverable shape change induced during forward and reverse MT at cooling/heating due to the variant selection of cooling-induced martensite by oriented internal stress fields. It is well known that TWSME can be induced by stress-assisted ageing, special thermomechanical treatment and/or training through a temperature interval of MT under constant stress, which generate oriented internal stress fields [8]. The control of internal stress fields and appearance of TWSME allows expansion of the scope of practical application of CoNiAl single crystals. TWSME can greatly simplify the design of devices, and reduce the

IOP Conf. Series: Materials Science and Engineering **71** (2015) 012060 doi:10.1088/1757-899X/71/1/012060

number of component parts of equipment to wear or break. Thus, the main purpose of this work is clarification of the necessary conditions for TWSME in [$\bar{1}$23]-oriented single crystals of $Co_{40}Ni_{33}Al_{27}$ (at.%) alloy quenched and aged at 573 K for 1 h.

The high-temperature B2-phase in the quenched $Co_{40}Ni_{33}Al_{27}$ single crystals involves a large volume fraction (f >10%) of the γ-phase (disordered face-centred-cubic structure) compared with $Co_{35}Ni_{35}Al_{30}$ (volume fraction (f< 2%) of the γ-phase) [9]. It permits the increasing of ductility of the CoNiAl single crystals due to the presence of the ductile γ-phase [9] and clarifies the role of γ-phase in shaping the conditions for the manifestation of TWSME during thermal cycling through an interval of MT under action of stress.

2. Experimental procedures

Single crystals of the $Co_{40}Ni_{33}Al_{27}$ alloy were grown by the Bridgeman method in an inert gas atmosphere. In order to get clear of the multi-phase state the initial specimens were annealed for 8.5 h at 1613 K and quenched in water at room temperature. Ageing of these single crystals at 573 K for 1 h was performed in a vacuum chamber of the machine for mechanical testing. For stress-assistant ageing the dog-bone shaped flat tensile specimens with the working part dimensions (2.6×1.2×16.0) mm^3 in the gauge section was placed in the grips of the machine, and constant load was applied to the sample at T=473 K. A low temperature of ageing at T=573 K (as compared to T=673 K in [5]) was selected for increase of temperature MT and, respectively, working temperatures of TWSME [2]. Prior to testing, the specimens were ground and polished in 200 ml of an electrolytic solution H_3PO_4+25 ml Cr_2O_3 at 293 K, U=20 V. Mechanical tests were carried out using a specially designed apparatus for measuring SME during cooling/heating under constant stress with output strain-temperature response ε(T) on the computer. Metallographic observations were carried out with an optical microscope EPITIP-2.

3. Experimental results and discussion

In quenched single crystals one-phase state was not obtained [6,7] and the high-temperature B2-phase contained particles of the γ-phase of size more than 100 μm and variants of residual martensite at T=295 K (Figure 1). The average volume fraction of the γ-phase in specimens is f~8%.

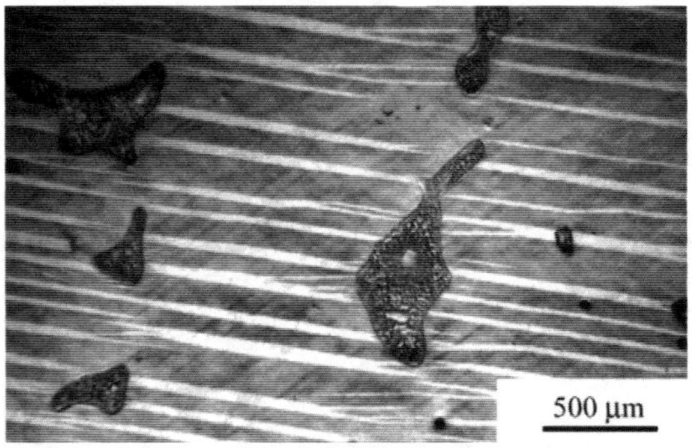

Figure 1. Optical metallography quenched [$\bar{1}$23]-oriented $Co_{40}Ni_{33}Al_{27}$ single crystals.

Figure 2 shows strain-temperature response ε(T) during cooling/heating of quenched [$\bar{1}$23]-oriented $Co_{40}Ni_{33}Al_{27}$ single crystals under a constant tensile stress σ. The sample size and shape is not changed during cooling/heating cycles under the minimum external tensile stress, necessary to secure the specimen in the grips of the machine |σ|=1.6÷3 MPa (change of size at 3 MPa less than error ±0.3%). In this case, the self-accommodation structure of cooling martensite is formed and the TWSME is not observed. During thermal cycling through the temperature interval of MT under action of external tensile stress from 10 to 50 MPa the sample size is changed on cooling and completely restored on heating; the SME is then observed. The reversible strain ε_{SME} increases with the growth of $|\sigma|$.

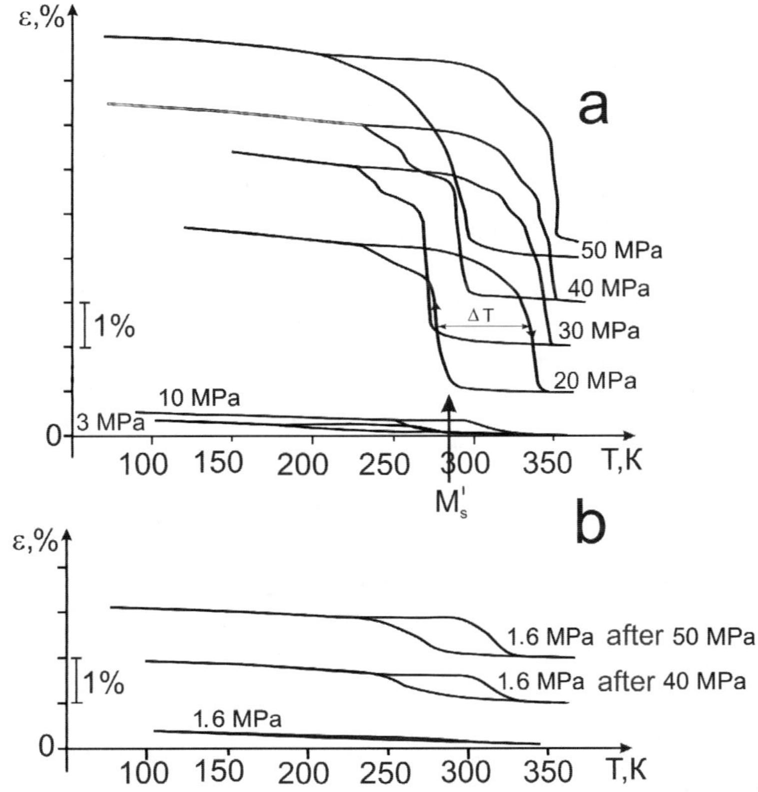

Figure 2. Strain-temperature response at cooling/heating for quenched $[\bar{1}23]$-oriented $Co_{40}Ni_{33}Al_{27}$ single crystals: (a) with applied tensile stress at $|\sigma|$=1.6 from 50 MPa; (b) with applied minimum tensile stresses of $|\sigma|$=1.6 MPa (TWSME).

The maximum value of the reversible strain is ε_{SME}=4.3 (±0.3)% at 50 MPa, which is close to the theoretical lattice strain of ε_0=4.7% in $[\bar{1}23]$ orientation at B2-$L1_0$ MT. In these crystals the linear growth of the martensite start temperature M_s' with the increase in external stress $|\sigma|$ >10 MPa is observed (Figure 3 (a)).

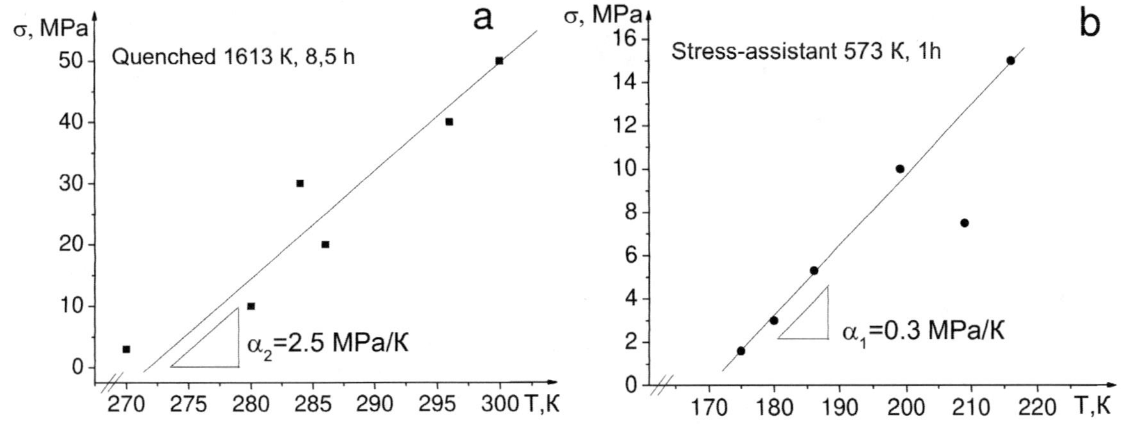

Figure 3. The dependence of martensite start temperature M_s' on external tensile stress for quenched (a) and stress-assistant ageing (b) $[\bar{1}23]$-oriented $Co_{40}Ni_{33}Al_{27}$ single crystals.

The σ(T) response can be described by the Clausius-Clapeyron relationship [8]:

$$\frac{d\sigma}{dT} = -\frac{\Delta S^{A-M}}{\mathrm{V_m} \cdot \varepsilon_{tr}^{A-M}}, \qquad (1)$$

where ΔS^{A-M} is the change of entropy associated with the forward transformation, ε_{tr} is the transformation strain and V_m is the molar volume. The coefficient of the σ(T) curve slope is α_1=d|σ|/dT=2.5 MPa/K in quenched [$\bar{1}$23]-oriented $Co_{40}Ni_{33}Al_{27}$ single crystals. TWSME is observed with a small value of reversible strain ε_{TWSME}=0.6 (±0.3)% after thermomechanical training at 40 MPa. The value of TWSME at minimum tensile stresses of 1.6 MPa increases to ε_{TWSME}=0.9 (±0.3)% after thermomechanical training at 50 MPa. In-situ observations of reversible motion of the interface in the loading-unloading cycle show the physical reason for TWSME in quenched crystals. Figure 4 shows that around particles of the γ-phase at room temperature T=293 K crystals of unoriented $L1_0$-martensite are formed. Growth of crystals of martensite oriented in accordance with external stresses with the increase of external tensile stresses is observed. After one loading-unloading cycle at T=293 K in specimens in free state, residual martensite around particles of the γ-phase become oriented in accordance with the applied-stress external stresses in cycle. At the next cooling in free state (without load) the growth of oriented martensite can be generated by oriented residual martensite, creating the necessary conditions for TWSME.

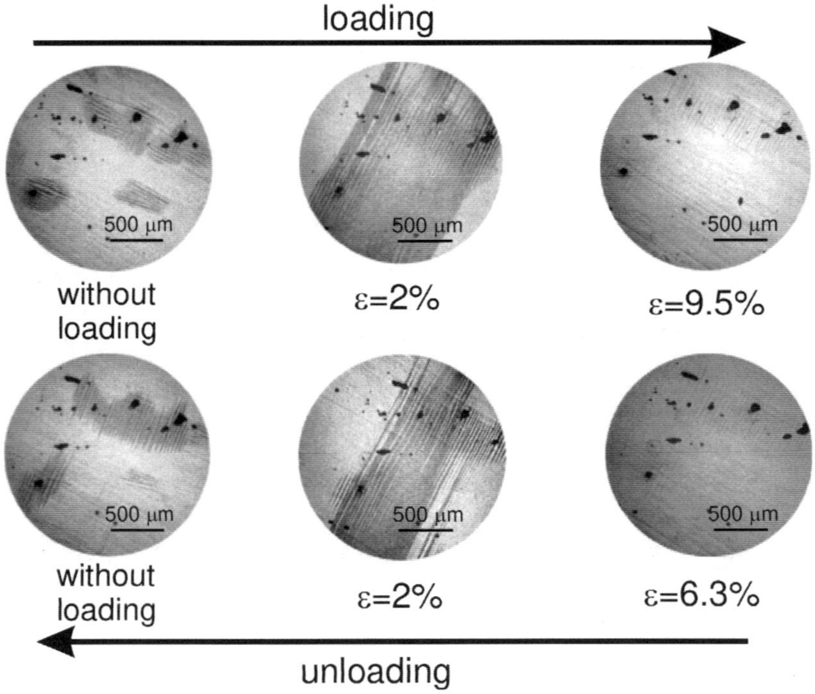

Figure 4. Optical metallography of quenched CoNiAl single crystals.

Thus, TWSME in quenched single crystals can be induced by cycle training an interval of MT at 50 MPa with reversible strain of no more 1%.

Figure 2 shows strain-temperature response ε(T) during cooling/heating of stress-assisted [$\bar{1}$23]-oriented $Co_{40}Ni_{33}Al_{27}$ single crystals aged at 100 MPa at 573 K, 1 h, under a constant tensile stress. In stress-assisted aged $Co_{40}Ni_{33}Al_{27}$ single crystals TWSME can be observed without thermomechanical training. The value of the reversible strain at TWSME is ε_{TWSME}=2.1 (±0.3)% at minimum stresses of |σ|=1.6 MPa (Figure 5 (a, b)). Growth of the value of the reversible strain with the increase of external tensile stress from 1.6 to 15 MPa is observed and temperature M_s' slowly increases with coefficient α_2=0.3 MPa/K.

The maximum reversible strain of ε_{SME}=3.7 (±0.3)% observed at stress 15 MPa is less than for quenched crystals (Figure 5 (a)). In stress-assisted aged [$\bar{1}$23]-oriented $Co_{40}Ni_{33}Al_{27}$ single crystals the maximum value of the reversible strain during the development of stress-induced MT is less than for quenched crystals. In heterophase crystals, when calculating the theoretical values of the transformation strain it is necessary to take into account that the particles of second phases do not undergo MT, and, consequently, decrease the volume fraction of the material in which the transformation occurs. This leads to a decrease of the reversible strain compared with the theoretical values of transformation strain and quenched crystals. So, in aged crystals the maximum theoretical values of the transformation strain at B2-L10 MT during the development of stress should be $\varepsilon_0'=\varepsilon_0(1-f)$=3.8% (f=20% is the average volume fraction of the dispersed particles precipitated during ageing at 573 K, 1 h). The experimental values of the maximum reversible strain in stress-assisted aged crystals are close to the theoretical values of transformation strain at MT.

Training through a temperature interval MT under the action of $|\sigma|$=10 MPa and $|\sigma|$=15 MPa of stress-assisted aged crystals has no material effect on the value of TWSME ε_{TWSME} ≈2.1-2.4 (±0.3)%, but the temperature M_s and temperature interval of direct transformation increase (Figure 5 (b)). The value of the reversible strain does not increase after training; therefore the physical reason for TWSME is that these crystals have been attributed to the internal stress fields observed in stress-assistant ageing. This thermomechanical treatment is effective for TWSME.

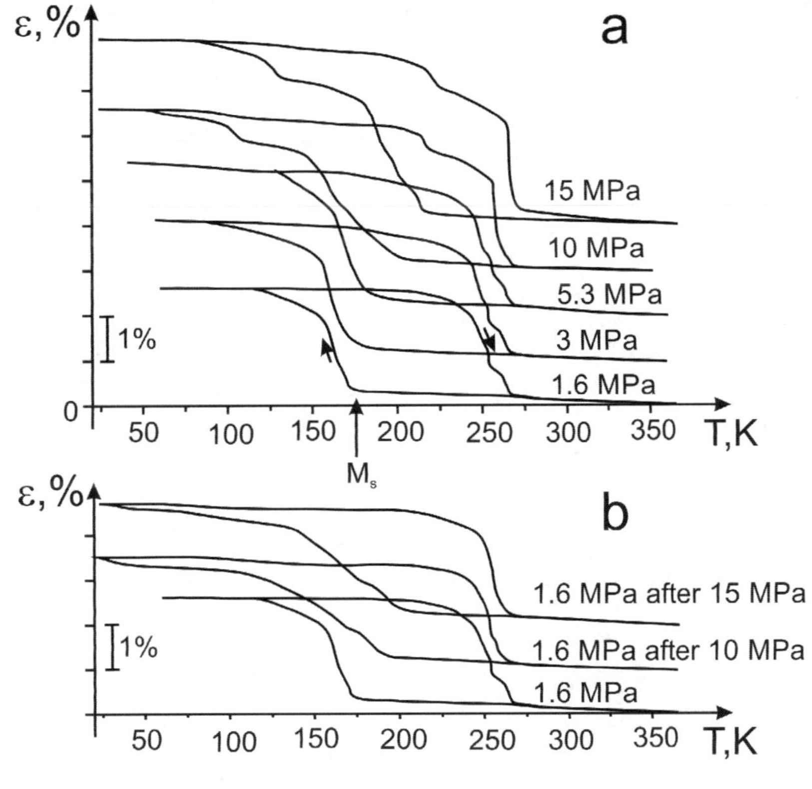

Figure 5. Strain-temperature response at cooling/heating for stress-assistant ageing [$\bar{1}$23]-oriented $Co_{40}Ni_{33}Al_{27}$ single crystals: (a) with applied tensile stress at $|\sigma|$=1.6 from 15 MPa; (b) with applied minimum tensile stresses of $|\sigma|$=1.6 MPa (TWSME).

4. Conclusions

Thus, the TWSME in quenched [$\bar{1}$23]-oriented $Co_{40}Ni_{33}Al_{27}$ single crystals can be induced by thermal-cycle training an interval of B2-L1_0 MT under action of external stress of 50 MPa with reversible strain of no more 1%. The physical reason for TWSME is that reorientation crystals of residual L1_0-martensite formed around particles of the γ-phase during the stress-induced MT. For the first time it has been experimentally shown that tensile stress-assistant 100 MPa ageing at 573 K for 1 h along [$\bar{1}$23]-direction of $Co_{40}Ni_{33}Al_{27}$ single crystals creates conditions for TWSME with values of reversible strain up to ε=2.4 (±0.3)%. The physical reason for TWSME in tensile stress-assistant ageing crystals can be attributed to the internal stress fields according to the oriented arrangement of dispersed particles due to summation of the local stress field, arising from the differences in the lattice parameters of the particle and the matrix.

The results obtained within the framework of the government contract of the Russian Ministry of Education and Science No. 16.1346.

References

[1] Morito H, Oikawa K, Fujita A et al. 2010 A large magnetic-field-induced in Co–Ni–Al single-variant ferromagnetic shape memory alloy *Scripta Mater* **63** 379-382
[2] Panchenko E Yu, Chumlyakov Yu I, Maier H, Kirillov V A and Kanafieva A S 2011 Characteristic features of development of thermoelastic transformations in aged single crystals of a CoNiAl ferromagnetic alloy *Russ Phys J* **54** 721-728
[3] Panchenko E Yu, Chumlyakov Yu I, Maier H, Kanafieva A S and Kirillov V A 2013 The effect of ageing of the [011]-oriented $Co_{35}Ni_{35}Al_{30}$ single crystals in free state and under loading on their functional properties *Russ Phys J* **55** 1123-1131
[4] Dadda J, Maier H J, Karaman I and Chumlyakov Y I 2009 Cyclic deformation and austenite stabilization in $Co_{35}N_{35}Al_{30}$ single crystalline high-temperature shape memory alloys *Acta Mater* **57** 6123-6134
[5] Panchenko E Y, Chumlyakov Y I, Maier H J and Eftifeeva A S 2014 Two-way shape memory effect in ferromagnetic $Co_{35}Ni_{35}Al_{30}$single crystals aged under stress *Scripta Mater.* **90-91**10-13
[6] Bartova B, Wiese N, Schryvers D, Chapman J N and Ignacova S 2008 Microstructure of precipitates and magnetic domain structure in an annealed $Co_{38}Ni_{33}Al_{29}$ shape memory alloy *Acta Mater* **56** 4470-4476
[7] Bartova B, Schryvers D, Yang Z, Ignacovab S and Sittnerb P 2007 Microstructure and precipitates in as-cast $Co_{38}Ni_{33}Al_{29}$ shape memory alloy *Scripta Mate.* **57** 37-40
[8] Otsuka K, Ren X 2005 Physical metallurgy of Ti–Ni-based shape memory alloys *Progress in Materials Science* **50** 511-678
[9] Chumlyakov Y I, Panchenko E Y, Ovsyannikov A V, Chusov S A, Kirillov V A, Karaman I and Maier H J 2009 High-temperature superelasticity and the shape-memory effect in [001] Co-Ni-Al single crystals *Phys Met Metallov* **107** 194-205

Influence of particles on the functional properties of single crystals of high-strength ferromagnetic alloys

I V Kretinina[1], O A Kuts[1], D A Kuksgauzen[2], V V Poklonov[1], Z V Pobedennaya[2] and Yu N Platonova[3]
[1]National Research Tomsk State University, Department of Metal Physics, Tomsk, 634050, Russia
[2]National Research Tomsk State University, Siberian Physical-Technical Institute, Tomsk, 634050, Russia
[3]National Research Tomsk Polytechnic University Department of General Physics, Physical-Technical Institute of the, Tomsk, 634050, Russia

E-mail: irbas@sibmail.com

Abstract. Single crystals of the ordered ferromagnetic $Co_{49}Ni_{21}Ga_{30}$ (at.%) alloy with B2-$L1_0$ martensitic transformation and of the disordered iron-based $Fe_{41}Ni_{28}Co_{17}Al_{11.5}X_{2.5}$ (X=Ta, Ti) (at.%) alloys, which undergo thermoelastic γ-α' martensitic transformations, were studied in terms of the influence of chemical composition, size and volume fraction of the dispersed γ'-phase particles on functional properties – shape memory effect and superelasticity. Single crystals of $Co_{49}Ni_{21}Ga_{30}$ alloy showed that the precipitation of nanometric γ'-phase particles changes the martensitic transformation characteristic temperature, reduces the value of the shape memory effect and superelasticity, increases thermal and stress hysteresis, and leads to hardening of the high temperature phase, which promotes superelasticity at a wide temperature range and at high temperatures at T>373K compared with the crystals without particles. Single crystals of FeNiCoAlX (X=Ta, Ti) alloys have experimentally demonstrated that the precipitation of ordered γ'-phase particles at a size of d=5–10 nm during aging at T=973 K, 3 h, leads to the occurrence of the shape memory effect and superelasticity.

1. Introduction

There is currently a search for new ferromagnetic materials with high strength and functional properties to be used as sensors, detectors, and actuators. One of the most promising ferromagnetic alloys with high strength and good ductility are the ordered CoNiGa alloys with B2-$L1_0$ martensitic transformation (MT) (B2 is the ordered phase based on the volume-centred cubic lattice, $L1_0$ is tetragonal martensite based on the face-centred tetragonal lattice) and the disordered Fe-based alloys with γ-α' MT (γ is a face-centred cubic lattice (fcc), α' is a volume-centred tetragonal lattice (bct)) [1–10]. It is known that, due to the precipitation of dispersed γ'-phase particles in alloys, not only the strength properties of high-temperature phase, but temperatures of MT, the temperature interval of superelasticity (SE), the type of MT, thermal and stress hysteresis can be controlled [1–10]. In contrast to the ordered CoNiGa alloy in iron-based alloys in a state without particles the MT is non-thermoelastic and shape memory effect (SME) and SE are not observed. Precipitation of dispersed γ'-phase particles with an atomic ordered $L1_2$-type structure leads to the development of the thermoelastic γ-α' MT, coherently related to high-temperature phase in iron-based alloys [6, 9–11]. Analysis of the data presented in [4, 6–9, 11] suggests that achieving the conditions for a thermoelastic

γ-α' MT in iron-based alloys plays an important role not only for the ordered γ'-phase precipitation, but also for its chemical composition, which determines its stability and strength at γ-α' MT.

Currently there have been no systematic investigations of the influence of chemical composition, size, and volume fraction of dispersed γ'-phase particles on the functional properties of single crystal CoNiGa ordered alloys and on Fe-based disordered alloys. Therefore, in this paper we present the results of studies of the effect of γ'-phase particles on the SME and SE in single crystals of ferromagnetic alloys CoNiGa and FeNiCoAlX (X=Ta, Ti). It is assumed that the regularities of the influence of dispersed particles on the development of MT, SME and SE should be general and should not depend on the specific type (B2-$L1_0$ or γ-α ') of MT.

2. Materials and discussion of the results

2.1. The shape memory effect and superelasticity in aged [001]-oriented single crystal CoNiGa ordered ferromagnetic alloys at thermoelastic B2-$L1_0$ martensitic transformation

For studies of SME and SE in the $Co_{49}Ni_{21}Ga_{30}$ (at.%) alloy crystals were selected oriented along the [001] direction, in which under compression, firstly, there is no contribution from the detwinning strain of $L1_0$-martensite at the value of transformation deformation ε_0, and development of reversible stress-induced MT has a maximum value of transformation strain ε_0=4.5%. Secondly, the [001]-oriented crystals are characterized by the high level of deforming stresses of B2-phase owing to Schmid factors equalling zero for operating a <100>{110} slip system, in contrast to other orientations of the crystals, in which the Schmid factors of these systems' slip are high [1–4]. Therefore the crystals were oriented along the [001] direction and had to demonstrate broad temperature interval of stress-induced MT, wide temperature range of SE, maximum values of SME and SE, and narrow thermal and stress hysteresis compared to the other orientations.

SME and SE studies were carried out in two states: single-phase and after aging with nanometric particles. The low-temperature heat treatment at 623 K for 15 minutes was selected for precipitation of nanometric γ'- phase particles. Electron microscopy found that after aging at T=623 K, 15 minutes, γ'-phase particles form a spherical shape 5 nm in size and a volume fraction f=10–12%.

Figure 1 shows the results of studies of critical stresses σ_{cr} for stress-induced $L1_0$-martensite within a broad temperature range of T=100K÷573K for the [001]-oriented single-phase crystals and crystals with γ'-phase particles. It can be seen that in the single-phase state, $\sigma_{cr}(T)$ is experiencing one linear stage in which σ_{cr} increases with increasing test temperature with $\alpha=d\sigma_{cr}/dT$=1.88 MPa/K. Such dependence on the $\sigma_{cr}(T)$ stage is associated with the development of stress-induced MT, which is described by the Clausius–Clapeyron relation [12]:

$$\frac{d\sigma_{cr}}{dT} = -\frac{\Delta H}{\varepsilon_0 T_0} \quad (1)$$

where ΔH and ΔS are the enthalpy and entropy changes, respectively, ε_0 is transformation strain for B2-$L1_0$ MT, which depends on crystal orientation, and T_0 is the chemical phase-equilibrium temperature. It can be seen that precipitation of the particles leads to the emergence of stages on $\sigma_{cr}(T)$ with α_1=1.10 MPa/K in the interval σ_{cr}=10÷200 MPa and α_2=1.58 MPa/K at σ_{cr}=200÷700 MPa (Figure 1), to a decrease in martensitic start (M_s) transformation temperature on 173K (±2 K) and to a decrease in the value of α compared to the single-phase state. For an explanation of stages on $\sigma_{cr}(T)$ and changes in the value of α in the aged crystals, SME was studied under stress.

Study of SME at cooling/heating at different levels of externally applied stresses σ_{ext} from 10 MPa to 300 MPa showed that for both single-phase (Figure 2a) and aged crystals (Figure 2b), SME was realized at minimum stresses of σ_{ext}=10 MPa. Consequently, these stresses are sufficient to destroy the self-accommodation structure of $L1_0$-martensite, and the growth of oriented $L1_0$-martensite occurs at the development of MT under σ_{ext}=10 MPa. In this case, the value of SME in single-phase crystals ε_{SME}=4.1% and this value is close to the theoretically calculated value of the lattice deformation ε_0 for the [001]-oriented crystal at B2-$L1_0$ MT. In crystals with nanometric γ'-phase particles the value of ε_{SME}=2% and it is less in 1.5–2 times as compared with the single-phase state. An estimate of the SME by the relation [2]: $\varepsilon_{SME}=\varepsilon_0(1-f)$ shows that the value of ε_{SME} should be equal to 3.96% and this value is greater than that experimentally observed. Therefore, the decrease in the value of ε_{SME} at aging is associated not only with a decrease in the volume of material undergoing MT by particles' precipitation as the particles themselves are not experiencing MT, but also to the influence of the particles on the nucleation and growth of martensite under σ_{ext} [5]. This trend was observed earlier in [$\bar{1}$23] $Co_{49}Ni_{21}Ga_{30}$ single crystals [1, 2].

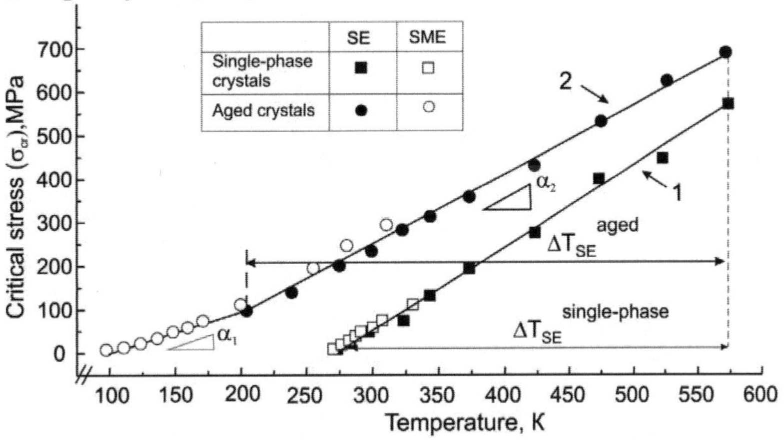

Figure 1. Temperature dependence of critical stresses σ_{cr} for [001]-oriented single crystals $Co_{49}Ni_{21}Ga_{30}$ alloy in compression: 1 – single-phase crystals and 2 – crystals with particles.

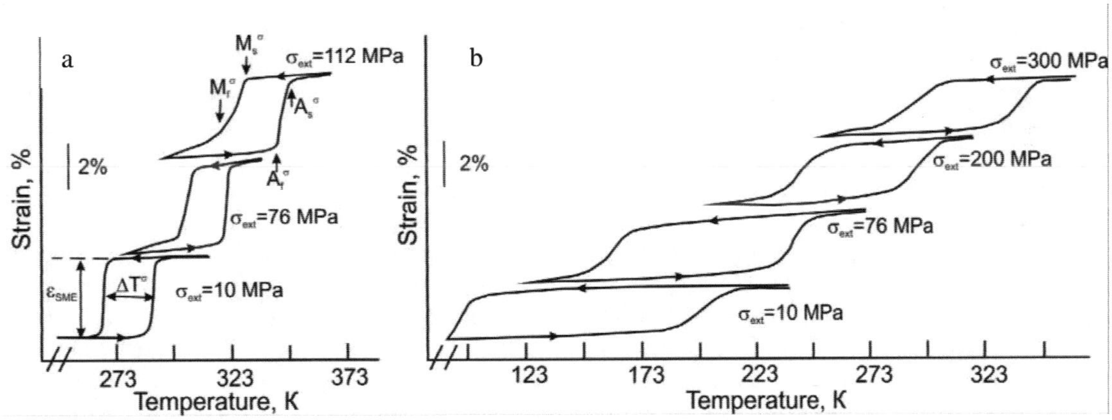

Figure 2. The curves of the dependence of the shape memory effect ε_{SME} at the level of applied external stress σ_{ext} for [001]-oriented single crystal $Co_{49}Ni_{21}Ga_{30}$ (at.%) alloy at compression deformation: a – single-phase crystals and b – crystals with particles.

In [1, 2] it is shown that the stresses σ_{cr} depend not only on the transformation strain, but also on the energy dissipation which is characterized by the thermal hysteresis $\Delta T^{\sigma}=A_f^{\sigma}-M_s^{\sigma}$. Figure 2 shows that in single-phase crystals the value of SME ε_{SME} and the thermal hysteresis ΔT^{σ}=22K with increasing σ_{ext} do not change, and curve $\sigma_{cr}(T)$ has one stage. In the aged crystals with increasing σ_{ext} the value of ΔT^{σ} decreases, ranging σ_{cr}=10÷200 MPa, and a sharp decrease is seen in the ΔT^{σ}, while σ_{cr}>200MPa the value of ΔT^{σ} varies only slightly. This may be one of the causes of the two stages on $\sigma_{cr}(T)$ depending on different values of α in the aged [001]-oriented crystals of the B2-$L1_0$ MT [2].

It is shown that the precipitation of γ'-phase particles leads to expansion of the temperature range of SE on 80K compared to the single-phase crystals, ΔT_{SE}=370K, and for the latter a perfect loop SE was observed at T≥573 K. This is due to the influence on particles of the stress level of the high-temperature phase and M_s temperature. Figure 3 shows the curves for single-phase and aged crystals.

A comparison of SE curves shows that the precipitation of particles leads to a decrease in the value of SE ε_{SE} 1.5 times, to increase the magnitude of stress hysteresis $\Delta\sigma$ 3.8 times, and the development of MT under $\sigma>\sigma_{cr}$ with a high strain hardening coefficient $\theta=d\sigma/d\varepsilon$ compared with the single-phase state.

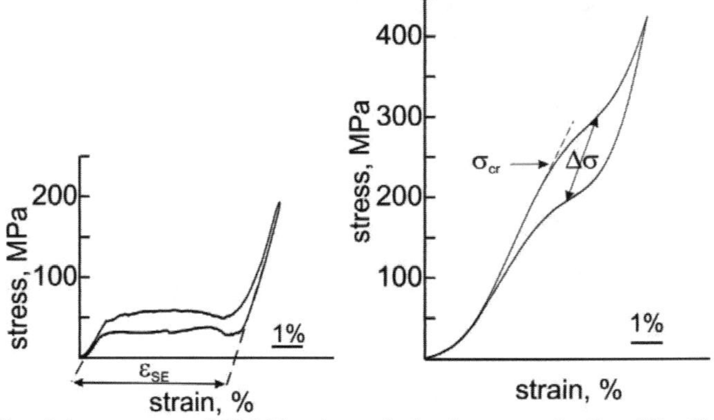

Figure 3. Superelasticity curves of [001]-oriented single crystals $Co_{49}Ni_{21}Ga_{30}$ (at.%) alloy at T=298K in: a) single phase and b) aged crystals.

2.2. Shape memory effect and superelasticity in aged [001]-oriented single crystals of ferromagnetic disordered iron-based alloys

For the study of SME and SE in disordered iron-based alloys $Fe_{41}Ni_{28}Co_{17}Al_{11.5}X_{2.5}$ (X=Ta, Ti) (at. %) at tensile deformation crystals were selected that were oriented along the [001] direction in which, during tensile deformation, the value of the lattice strain ε_0=8.7 % for γ-α' MT has a maximum value. This would provide a maximum resource of reversible deformation in the study of SME and SE [7]. For the precipitate of the ordered γ'-phase aging at T=973 K, 3 h was chosen, to clarify the effect of particles of similar sizes, d=5–10 nm, and different chemical compositions, as a result of variation of the fifth element, while maintaining the atomic concentration of the alloy on the nature of stress-induced MT and values of SME and SE.

Figure 4 shows the temperature dependence of the critical stresses σ_{cr} at tension for [001]-oriented single crystals of $Fe_{41}Ni_{28}Co_{17}Al_{11.5}X_{2.5}$ (X=Ta, Ti) alloys at a wide temperature range of 77–550 K. It can be seen that on depending of $\sigma_{cr}(T)$ two stages are observed that are characteristic for alloys undergoing stress-induced MT. At the first stage 77 K $<T<M_d$ (M_d is the temperature at which the critical stress level of the high-temperature phase is equal to the critical stress level for martensite under stress), σ_{cr} increases with increasing test temperature: it is close to a linear dependence and is described by the relation Clausius-Clapeyron (1). At $T>M_d$ is the second stage, which is connected with the deformation of the high-temperature phase and is characterized by a normal temperature dependence of the critical stress of typical FCC materials [13].

Figure 4 shows that the variation of the chemical composition of the γ'-phase particles leads to a change in the M_s temperature, in stress in temperature $T=M_d$, and therefore in the temperature range of stress-induced MT ΔT_{SIM} (first stage). The stress at $T=M_d$ in [001] single crystals of FeNiCoAlTa alloy equal 1043 MPa, and FeNiCoAlTi alloy $\sigma_{cr}(M_d)$=900 MPa. The temperature at point M_d in crystals of FeNiCoAlTa also has a maximum value and is equal to 423 K, that less on 117 K is compared to in crystals FeNiCoAlTi alloy, where $T(M_d)$=306 K. Such a difference in the temperature and stresses at the point of M_d leads to a different temperature range of stress-induced MT ΔT_{SIM}: in crystals of FeNiCoAlTa ΔT_{SIM}=346 K, and in crystals of FeNiCoAlTi, ΔT_{SIM}=229 K. The value of $\alpha=d\sigma_{cr}/dT$ after aging at 973 K for 3 h also depends on the chemical composition of the γ'- phase particles and equals 3.3 MPa/K in crystals of FeNiCoAlTa, and in crystals of FeNiCoAlTi $\alpha=d\sigma_{cr}/dT$=4.8 MPa/K.

It has been experimentally established that the precipitation of γ'-phase particles at T=973 K for 3 h in [001]-oriented crystals of FeNiCoAlX (X=Ta, Ti) alloys leads to the manifestation of SME and SE. The value of SME ε_{SME} and SE ε_{SE}, temperature range of SE ΔT_{SE} and value of thermal ΔT^{σ} and stress $\Delta\sigma$ hysteresis depends on the chemical composition of the alloy (Figs. 4, 5, 6).

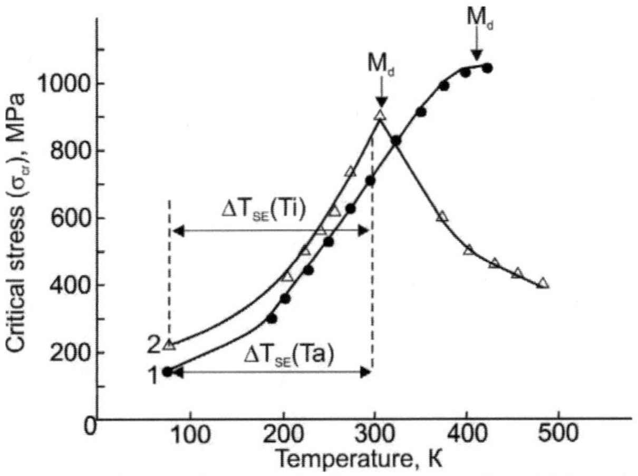

Figure 4. Temperature dependence of the critical stresses σ_{cr} for [001]-oriented crystals aged at T=973 K for 3 h at tensile deformation: curve 1 - FeNiCoAlTa alloy, 2 – FeNiCoAlTi.

Figure 5 shows the results of the research of the value of SME in the experiment during cooling/heating under constant tensile stress. From the figure it can be seen that in [001]-oriented single crystals of FeNiCoAlX (X=Ta, Ti) alloys during aging at T=973 K for 3 h at a constant external stress the reversible one-stage γ-α' MT is implemented. The value of thermal hysteresis ΔT^{σ} under maximum applied stress defined on middle-loop stress hysteresis in FeNiCoAlTa alloys is equal to 50 K. This value ΔT^{σ} at more than 1.8 times in the FeNiCoAlTi alloy (Figure 5). The maximum value of the SME in FeNiCoAlTa crystals is equal ε_{SME}=4.2 %, and in FeNiCoAlTi crystals ε_{SME}=3 %. In such a relationship the values correlated with equation (1), at that the ratio of the value α(Ti)/α(Ta)=1.4 coincides with the value ε_{SME}(Ti)/ε_{SME} (Ta)=1.4. The experimentally observed value of SME ε_{SME} in [001]-oriented crystals FeNiCoAlX (X=Ta, Ti) alloys is less than the theoretically calculated values of the lattice strain for the [001] ε_0=8.7% at the γ-α ' MT [6]. The reason for the decrease in the value of SME is to reduce the volume of material undergoing MT, the presence in the alloys of TaC, TiC particles, which are formed in the melting process of the alloy, and single crystal growth.

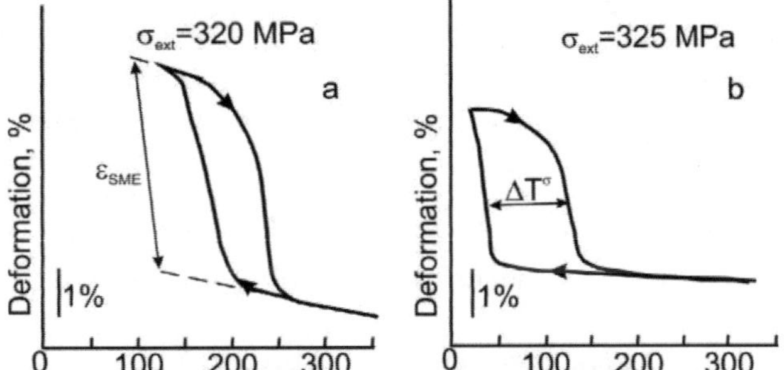

Figure 5. The value of the shape memory effect in the [001]-oriented single crystals aged at T 973 K 3 h at constant tensile stresses: a - FeNiCoAlTa alloy, b – FeNiCoAlTi.

The temperature range of SE in [001]-oriented single crystals of FeNiCoAlTi and FeNiCoAlTa alloys is close and equals ΔT_{SE}=129 K (Figure 4). A maximum value of SE ε_{SE}, defined in experiments on cycling, equals 4.1% (Figure 6) and, as with the value of SME ε_{SME} is less than the theoretical value ε_0 [6]. The value of stress hysteresis Δσ at a one-test temperature T=77 K depends on the composition of the alloy (Figure 6). In [001]-oriented single crystals FeNiCoAlTa alloy Δσ=52 MPa and in [001]-oriented single crystals FeNiCoAlTi alloy Δσ=75 MPa. Changing stress hysteresis Δσ in alloys is associated with the different critical stress levels in the high-temperature phase in these alloys.

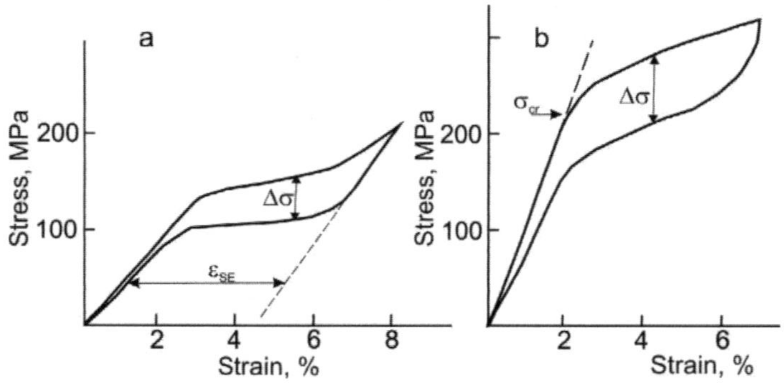

Figure 6. Stress-strain (σ-ε) curves of superelasticity in the [001]-oriented single crystals aged at T=973 K for 3 h at tensile deformation. Test temperature T=77 K: a - FeNiCoAlTa alloy, b - FeNiCoAlTi

3. Conclusion

These studies on the single crystals of ferromagnetic $Co_{49}Ni_{31}Ga_{30}$ and $Fe_{41}Ni_{28}Co_{17}Al_{11.5}X_{2.5}$ (X=Ta, Ti) alloys show that precipitation of nanometric particles of different chemical composition, size, and volume fraction is a promising approach towards the creation of a new class of high-strength materials with SME and SE. Dispersed particles do not experience MT, but enable by variations in their chemical composition, size and volume fraction control of both the mechanical and functional properties and thus facilitate acquisition of materials with the required performance properties. Precipitation of the particles in the ordered $Co_{49}Ni_{31}Ga_{30}$ alloy leads to the manifestation of SE not only with a wide interval, but also at high temperatures up to 573 K, and the precipitation of ordered γ'-phase particles in size d=5÷10 nm in disordered $Fe_{41}Ni_{28}Co_{17}Al_{11.5}X_{2.5}$ (X=Ta, Ti) alloys contributes towards a change in the kinetics of MT from non-thermoelastic to thermoelastic and observation SME and SE.

This work was supported by RSF grant № 14-29-00012, and under the guidance of PhD Professor Chumlyakov Y.I. and PhD, leading researcher Kireeva I.V.

References

[1] Kireeva I V, Pons J, Picornell C, Chumlyakov Yu I, Cesari E and Kretinina I V 2013 *Intermetallics* **35** 60–66
[2] Kireeva I V, Picornell C, Pons J, Kretinina I V, Chumlyakov Yu I and Cesari E 2014 *Acta Mat* **68** 127–139
[3] Dadda J, Maier H J, Niklasch D, Karaman I, Karaca H E and Chumlyakov Y I 2008 *J. Physical Metall and Mat Science* **39** 2026–2039
[4] Chumlyakov Y I, Kireeva I V, Panchenko E Y, Timofeeva E E, Pobedennaya Z V, et al. 2008 *Russ Phys J* **51** 1016
[5] Chumlyakov Yu, Kireeva I and Panchenko E 2011 *Russ Phys J* **54** 96–108
[6] Chumlyakov Yu, Kireeva I, Panchenko E, Karaman I, Maier H J and Timofeeva E 2013 *J of Alloys and Compounds* **577** S393–S398
[7] Tanaka Y, Himuro Y, Kainuma R, et al. 2010 *J Science* **327** 1488–1490
[8] Chumlyakov Yu I, Kireeva I V, Panchenko E Y, Zakharova E G, Kirillov VA, Efimenko S P and Sehitoglu H 2004 *Doklady Physics* **49** 47–50
[9] Omori T, Abe S, Tanaka Y, Lee G Y, Ishida K and Kainuma R 2013 *J. Scripta Mat* **69** 812–815
[10] Chumlyakov Yu I, Kireeva I V, Kretinina I V, Keynih K S, Kuts O A, Kirillov V A, Karaman I and Maier H 2013 *Russ Phys J* **56** 66–74
[11] Kokorin V V and Gun'ko L P 1995 *Metal Physics and Advanced Technologies* **17** 30–35
[12] Otsuka K and Wayman C M 1998 Shape memory materials *Cambridge University Press* 284
[13] Chumlyakov Yu I, Kireeva I V, Korotaev A D, Litvinova E I, Zuev Y L 1996 *Russ Phys J* **3** 189–210

Electron-beam modification of coating - aluminum substrate systems

Yu F Ivanov[1,2], O V Krysina[1,2], A D Teresov[1,2] and M E Gracheva[3]
[1]Institute of High Current Electronics SB RAS, Tomsk, 634055, Russia
[2]National Research Tomsk State University, Tomsk, 634050, Russia
[3]National Research Tomsk Polytechnic University, Tomsk, 634050, Russia

E-mail: krysina_82@mail.ru

Abstract. Multiphase surface alloys with improved strength and tribological characteristics have been synthesized by exposing coating/A7 substrate systems to a pulsed electron beam. Optimum modes of electron-ion-plasma treatment of commercially pure aluminum have been found at which the wear resistance and hardness of the surface layer were observed to increase by a factor of about 7.5 and up to 18, respectively.

1. Introduction

Aluminum and aluminum-based alloys have reasonable costs and are accessible for various applications. They have such advantages as high thermal conductivity, machinability, low cost, etc. In order to extend the application of aluminum products further, to endow them with high wear resistance, increased hardness, etc., a reasonable increase in costs is required. One of the modern and promising methods of modification of aluminum-based alloys is a complex electron-ion-plasma treatment of products, which combines deposition of thin one-component or multicomponent coatings followed by exposure to an electron beam to provide mixing of the coating/substrate system or to melt the hard coating in the low-melting substrate [1, 2]. In this case, in addition to the strengthening thin layer, whose properties are determined by the coating deposition conditions, an extended near-surface modified layer is formed which can consist of sublayers possessing different structure-phase characteristics and unique properties [3, 4].

Coating/substrate systems are exposed to an intense pulsed electron beam of submillisecond (up to 200 μs) duration, which provides ultrahigh rates of heating (up to 10^6 deg/s) of a surface layer of thickness tens micrometers to temperatures at which a melt can exists. This makes it possible to attain very high temperature gradients (10^7–10^8 deg/m) in the surface layer that ensure cooling of the surface layer due to heat removal to the cold bulk material at a rate of 10^4–10^6 deg/s [5-6]. Fast cooling of the surface layer conserves its non-equilibrium structure-phase state which is characterized by submicrometer and nanometer sizes of crystallites, high concentration gradients of alloying elements, and multiply increased values of physic-mechanical, tribological, electrical, and other parameters [7-8].

The goal of the study presented in this paper was to produce strengthened layers in coating/substrate systems by electron-beam treatment, to optimize the modes of complex electron-ion-plasma modification, and to examine the strengthened layers by up- to-date methods of material science.

2. Materials and procedures

The substrate material was commercially pure A7 aluminum alloy (99.7% Al, max 0.16% Fe, max 0.16% Si, max 0.02% Ti, max 0.01% Cu, max 0.01% Zn). At the first stage of electron-ion-plasma treatment thin (0.5–1 μm) Ti and Ti–Cu coatings were produced by plasma-assisted vacuum-arc deposition with the use of the QUINTA automated facility [9]. Commercially pure VT1-0 titanium and Ti–Cu (12 at. %) sintered powder [10] were used as materials for the evaporable cathode. The gas-metal plasma, which promoted the synthesis of coatings, was generated with the use of a DI-100 vacuum arc evaporator and the PINK plasma source with a hybrid (filament and hollow) cathode [11, 12].

The conditions for preliminary cleaning of the substrates with argon ions were chosen to minimize the aluminum surface heating and etching. The ion current density at the substrate was ≤1.5 mA/cm^2 and the temperature during the ion bombardment was not above 75°C. The coating deposition modes were optimized so that the coating growth rate would be high but the maximum sizes of the macroparticles present in the plasma flow would be not greater than the thickness of the coating.

The metal coatings were produced in argon used as the working gas medium with the working gas pressure $p \sim 0.1$ Pa and the evaporator arc current $I_d = 60–100$ A. The substrates were at a floating potential ($U_b \approx 15$ V). Under these conditions, the coating growth rate was 1.5–2 μm/h. The duration of the deposition process was chosen so that the coating thickness would be 0.5 or 1 μm. The temperature of the aluminum substrates at the stage of coating deposition was not over 150°C.

After deposition of a titanium-containing coating on an aluminum substrate, the coating/substrate system was irradiated with an electron beam in the SOLO automated electron-beam facility [13]. The irradiation parameters were the following: pulse repetition rate $f = 0.3$ Hz, electron beam energy density $E_s = 10–20$ J/cm^2, pulse duration $\tau = 50–200$ μs, and number of pulses $N = 3–50$.

Examination of the modified samples was carried out using optical microscopy (μVizo-MET-221 metallographic microvisor), scanning electron microscopy combined with x-ray spectral analysis (Philips SEM-515 scanning electron microscope equipped with an EDAX Genesis XM 260 SEM microanalyzer), x-ray diffraction analysis (Shimadzu XRD 6000 x-ray diffractometer), micro- and nanoindentation (PMT-3 microhardness tester and NHT-S-AX-000X Nano Hardness Tester), and tribological examinations (THT-S-AX0000 PC-Operated High Temperature Tribometer).

3. Results and discussion

3.1. The Ti /A7 coating/substrate system

Experimentally it has been found that the coating/substrate system was not melted at $W = E_s \cdot N/\tau \leq 1 \cdot 10^6$ W/cm^2 per pulse (E_s – electron beam energy density; N – number of pulses; τ – pulse duration).

In this case, the coating survived, but pronounced cracking was observed, resulting in fragmentation of the coating (Figure 1 a, b). The material separating the coating fragments was aluminum. When W was in the range $(1–5) \cdot 10^6$ W/cm^2 per pulse, an island-type structure was observed to form (the coating islands were separated by regions where cellular crystallization of the coating/substrate system had occurred) (Figure 1 c, d). At greater values of W, the coating/substrate system was inconvertibly melted (Figure 1 e–h).

The structurally non-uniform surface layer contained, conventionally speaking, three types of regions different in substructure: (i) polycrystalline regions with the grain size ranging between 1.0 and 1.5 μm (Figure 1 h) in which the concentration of aluminum and titanium was 95 and 5 wt. %, respectively; (ii) regions of preferentially rounded shape in the bulk of which crystallization cells 150–700 nm in size were observed (Figure 1 g) and the concentration of aluminum and titanium was 90 and 10 wt. %, respectively, and (iii) irregularly shaped regions formed as a result of cellular crystallization which contained numerous micropores whose size ranged from 0.2 to 1.8 μm (Figure 1 h); according to x-ray spectral analysis data, the concentration of aluminum and titanium in these regions was 80 and 20 wt. %, respectively.

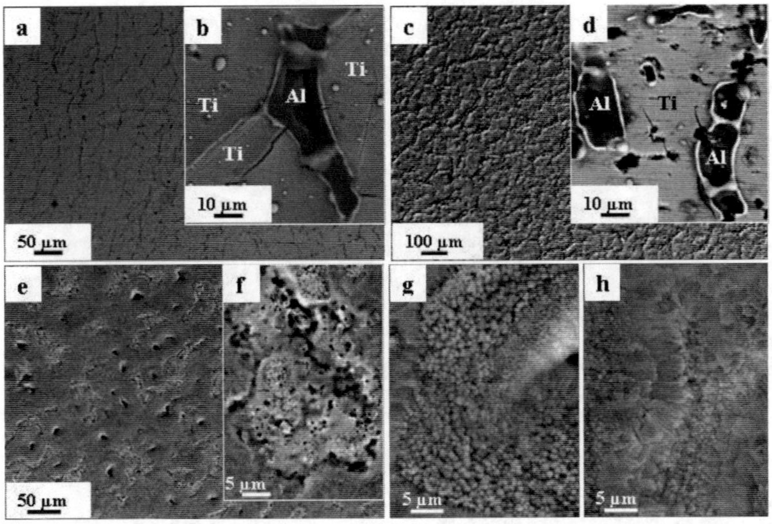

Figure 1. Surface structure of the Ti/A7 coating/substrate system treated with an intense pulsed electron beam under the following conditions: $E_s = 10$ J/cm^2, $N = 5$, $\tau = 50$ μs (a, b); $E_s = 15$ J/cm^2, $N = 3$, $\tau = 50$ μs (c, d), and $E_s = 15$ J/cm^2, $N = 30$, $\tau = 50$ μs (e–h).

X-ray diffraction analysis has revealed that the surface layer of thickness ~10 μm in the Ti/A7 coating/substrate system is multiphase and its composition depends both on the system elemental composition and on the mode of electron-beam treatment. The following phases have been detected: aluminum-based solid solutions, titanium-based solid solutions, and titanium aluminide Al_3Ti, whose volume fraction was a maximum (~20%) when the irradiation parameters were $E_s = 10$ J/cm^2, $\tau = 50$ μs, $N = 10$ pulses, and $f = 0.3$ Hz.

The formation of surface alloys is accompanied by a manifold increase in hardness and wear resistance of commercially pure aluminum. The maximum increase in hardness has been attained for the electron-beam treatment modes with $E_s = 15$ J/cm^2, $\tau = 50$ μs, $N = 3$ and 30 pulses (Figures 2).

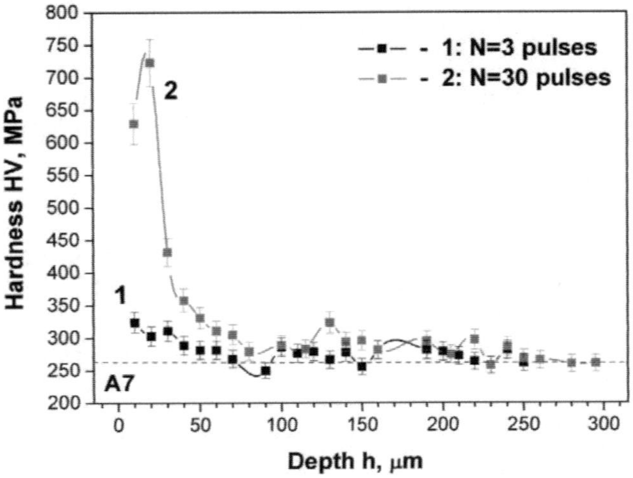

Figure 2. Microhardness depth distribution for the Ti/A7 coating/substrate system after electron-beam treatment: $E_s = 15$ J/cm^2, $\tau = 50$ μs, $N = 3$ (curve *1*) and $N = 30$ (curve *2*).

After three irradiation pulses, the maximum hardness, which was greater than the hardness of the original aluminum A7 by a factor of 40, was achieved at the irradiated surface. The hardness of the coating/substrate system exposed to 30 pulses increased in going from the surface and, at a depth of about 10 μm, it became three times greater than the hardness of the original aluminum A7 (Figure 2).

In wear resistance tests, the best result (increase in wear resistance of the original aluminum by a factor of about 7.5) has been attained for the treatment mode with E_s = 15 J/cm^2, τ = 50 μs, N = 10 pulses, and f = 0.3 Hz. In this case, the friction coefficient decreased by a factor of 1.4.

3.2. The Ti–Cu/A7 coating/substrate system

The Ti–Cu/A7 coating/substrate system, like the Ti/A7 coating/substrate system, was not melted at W ≤ $1\cdot10^6$ W/cm^2 per pulse. When W was in the range (1–6)·10^6 W/cm^2 per pulse, an island-type structure was observed to form (Figure 3 a, b). At W = 15·10^6 W/cm^2 per pulse, the coating/substrate system was inconvertibly melted (Figure 3 c, d). The surface layer became polycrystalline with the grain size ranging from 5 to 40 μm (Figure 3 d). The structure formed in the grain bulk as a result of cellular crystallization consisted of cells whose size ranged from 0.4 to 2.0 μm (inset in Figure 3 d).

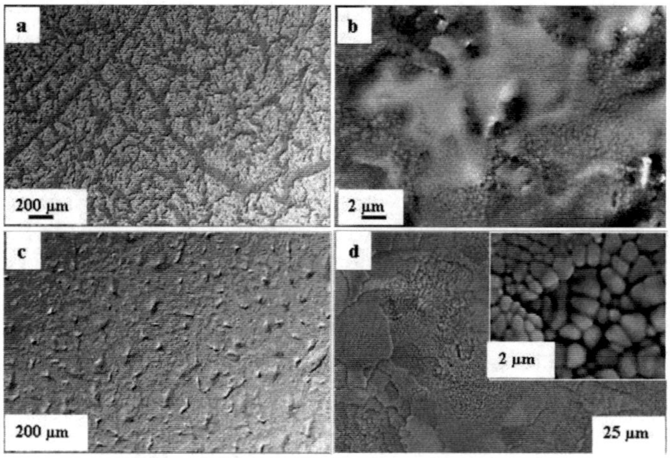

Figure 3. Structure of the surface of the Ti–Cu/A7 coating/substrate exposed to an electron beam with the parameters: E_s = 15 J/cm^2, τ = 50 μs, N = 3 pulses (a, b) and E_s = 15 J/cm^2, τ = 20 μs, N = 20 pulses (c, d).

X-ray spectral analysis has detected Al–Ti–Cu surface alloys with the element concentration strongly depending on the electron-beam irradiation parameters and ranging over wide limits (from 52 to 2.5 wt. %) (Figure 4). Thus, irradiation of the coating/substrate system with an electron beam makes it possible to vary the elemental and, hence, the phase composition of the surface layer of the material over wide limits.

X-ray diffraction analysis performed in Bragg–Brentano geometry has revealed a multiphase structure of the surface layer (~10 μm) whose qualitative and quantitative composition depended on the mode of electron-beam treatment. It has been found that the total volume fraction of the second phases (Al_3Ti, Ti_3Al, and Ti_3Cu) was a maximum (~40%) for the irradiation mode with E_s = 15 J/cm^2, τ = 100 μs, N = 20 pulses, f = 0.3 Hz in which a surface alloy of composition 79Al–16Ti–5Cu (wt. %) was formed (Figure 4).

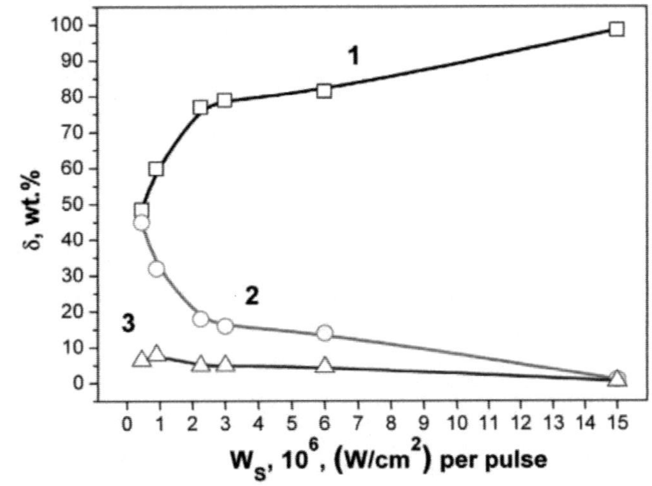

Figure 4. Concentration of aluminum (curve *1*), titanium (curve *2*) and copper (curve *3*) in the surface layer of the Ti–Cu/A7 coating/substrate system versus irradiation parameters.

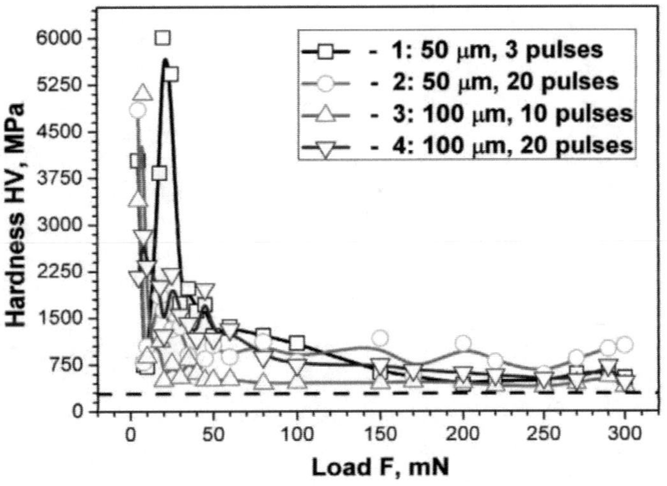

Figure 5. Hardness of the surface alloys versus load: $E_s = 15$ J/cm^2; the pulse duration and the number of pulses are specified in the respective figures.

Modification of the phase and elemental composition and of the defect substructure of the aluminum surface layer should affect the physicomechanical and tribological properties of the material. To qualify the surface alloy, the hardness, Young modulus, and wear resistance of the coating/substrate system have been examined before and after exposure to an electron beam. The results of analyses on the surface layer hardness are given in Figures 5.

It can clearly be seen that the maximum hardness (5–6 GPa) and the maximum Young modulus (200–250 GPa) are achieved at the exposed surface. Taking into account that the Young modulus of aluminum oxide Al_2O_3 ranges between 200 and 400 GPa and the Young modulus of aluminum is not above 70 GPa, it can be said that a thin oxide film was formed on the system surface at the stage of cooling. For the indenter loads ranging between 20 and 70 mN, a sublayer of hardness ~2 GPa (~7 times that of the original material) and Young modulus ~110 GPa (~1.8 times that of the original material) was detected. As the indenter load was further increased, both the hardness and the Young modulus decreased, approaching the characteristics of the original material. It should however be noted that in the coating/substrate system exposed to an electron beam with $E_s = 15$ J/cm^2, $f = 0.3$ Hz, $\tau = 50$ μs, and $N = 20$ pulses, an extended layer was formed whose hardness was more than three times and the Young modulus was 1.5 times greater than the respective values for the original aluminum.

The improvement of the strength properties of the surface layer was accompanied by an increase in wear resistance by a factor of ~1.2, which was found for the coating/substrate system exposed to an electron beam with $E_s = 15$ J/cm^2, $f = 0.3$ Hz, $\tau = 100$ μs, and $N = 20$ pulses.

4. Conclusion

For the Ti/A7 and Ti–Cu/A7 coating/substrate systems irradiated with an intense electron beam, multiphase alloys have been produced on the surface of the commercially pure aluminum. The concentrations of alloying elements in these alloys ranged up to 50 wt. %. Structural examinations have shown that the improvement of the physicomechanical characteristics of the aluminum was due to the multiphase submicrostructure and nanostructure state resulting from melting and fast crystallization of the coating/substrate systems.

Acknowledgments

The reported study was partly supported by RFBR, research Project No. 13-08-00416-a and in the framework of basic research performed by the RAS Siberian Division in cooperation with institutions of the RAS Ural Division, Project No. 107.

References

[1] Koval N *et al* 2012 *Russian Physics Journal* **54** 1024
[2] Rotshtein V *et al* 2011 *Technical Physics Letters* **37** 226
[3] Иванов Ю Ф *и др.* 2013 *Известия ВУЗов. Физика.* **56** 98
[4] Ласковнев А.П. *и др.* 2013 *Модификация структуры и свойств эвтектического силумина электронно-ионно-плазменной обработкой.*(Беларусь, Минск: Белорусская наука)
[5] Koval N and Ivanov Yu 2008 *Russian Physics Journal* **51** 505
[6] Глезер А М *и др.* 2012 *Наноматериалы: структура, свойства, применение* (Россия, Новокузнецк: Интер-Кузбасс)
[7] Rotshtein V *et al* 2006 Charter 6 in Book *Materials surface processing by directed energy techniques* (USA, New York: Elsevier)
[8] Иванов Ю.Ф. *и др.* 2011 *Усталостная долговечность стали мартенситного класса, модифицированной высокоинтенсивными электронными пучками* (Россия, Новокузнецк: Интер-Кузбасс)
[9] Shugurov V *et al* 2012 *Izv. Vyssh. Ucheb. Zaved. Fizika* 118
[10] Gurskih A *et al* 2006 *Izv. Vyssh. Ucheb. Zaved. Fizika. Supplement* №8 400
[11] Koval N *et al* 2012 *Proceedings of the 25th International Symposium on Discharges and Electrical Insulation in Vacuum* (Russia, Tomsk) **2** 537
[12] Vintizenko L *et al* 2001 *Russian Physics Journal* **44** 927
[13] Koval N *et al* 2009 *IEEE Trans. Plasma Sci.* **37** 1890

IOP Conf. Series: Materials Science and Engineering **71** (2015) 012063 doi:10.1088/1757-899X/71/1/012063

Some aspects of sherardizing implementation during anti-corrosion protection of heat-treated metal parts

V M Konstantinov[1] and I A Buloichyk[1]
[1]Belarusian National Technical University, Material Science in Mechanical Engineering Department, Minsk, 220013, Belarus

E-mail: v_m_konst@mail.ru, ilya.by@gmail.com

Abstract. The article reveals the peculiarities of obtaining Zn anticorrosive diffusion layers on heat-treated metal parts using sherardizing method. It is shown that sherardizing processes can be successfully used as a final heat treatment operation for some metal parts after hardening in a temperature gap between 350-700 °C. 380-450 °C are the most efficient temperatures of sherardizing process to reveal hardening stresses in details. At the same time, applying sherardizing method for anticorrosive protection of details subjected to final heat treatment may cause a weakening of metal accomplished by a reduction of working properties. This fact can be described by structural changes in metal caused by additional heat impact of sherardizing temperatures.

1. Introduction

Due to a range of specific coating properties, sherardizing is an actual alternative to traditional methods of zinc anti-corrosion protection. It is known, that sherardizing is widely used for anti-corrosion protection of hardware parts and, in contrast to other methods of zinc anti-corrosion protection, provides a formation of intermetallic diffusion layers. Mechanical and anticorrosive properties of zinc diffusion layers are far more exceeded than those of other zinc coatings without any diffusion interaction with metal substrate [1, 2].

Traditionally zinc anti-corrosion coatings are obtained after a final heat treatment of a detail. If it is necessary to use annealing operation as a final heat treatment of a detail, the most economically efficient way is to change traditional thermal heating by a sherardizing process. Such a replacement gives the opportunity to reduce costs on additional anticorrosive treatment of a detail as well as cuts down the amount of manufacturing equipment in the aggregate of efficient use of working area.

It is necessary to notice that a temperature interval of sherardizing processes lies between 350 and 700 °C depending on some technological characteristics of the process [3]. Such a wide temperature interval gives the opportunity to choose the necessary sherardizing conditions taking into account a prior heat treatment of a detail. The possible variants of sherardizing processes incorporation into the overall heat treatment cycle for some details are characterized in Figure 1.

Combining overstrain aging with sherardizing is efficient for steel part subjected to final thermal treatment using temperatures no less than 300 °C. The activity of sherardizing processes is extremely dependant on zinc vapor pressure inside a closed vessel as well as on the amount of oxygen inside it. The presence of oxygen inside a closed vessel may lead to formation of a thin oxide layer on metal surface that creates a diffusion barrier. To overcome these obstacles and ensure sherardizing process at minimum zinc vapor pressure it is necessary to use vacuumization or inert gas flushing with the aim to decrease the amount of oxygen inside a vessel.

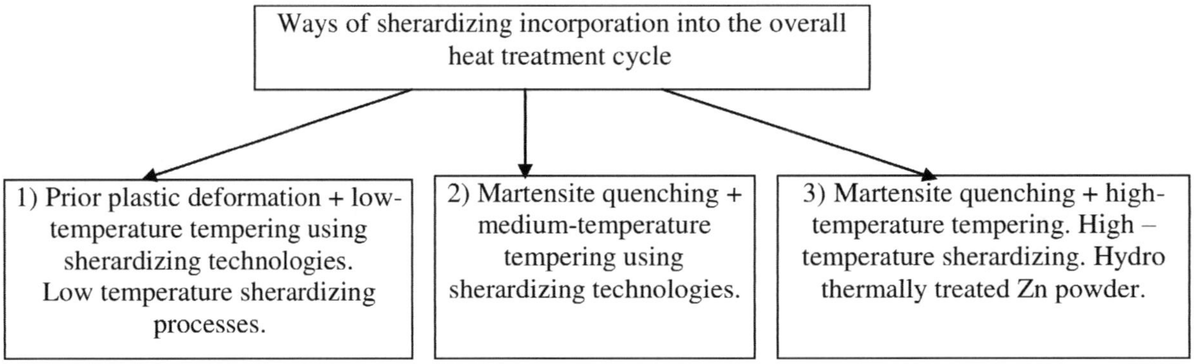

Figure 1. Incorporation of sherardizing processes into the overall heat treatment cycle for some details.

Combining martensite quenching and high temperature sherardizing processes is possible by using additional hydrothermal treatment of a zinc powder. As a result of this treatment a thin porous ZnO layer forms round the surface of each zinc globule. The ZnO layer prevents zinc powder from fritting at high temperatures and does not block zinc evaporation.

2. Experimental

The most efficient way to realize sherardizing processes is medium temperature tempering after prior quenching, which is the most widespread type of heat treatment for a range of spring elements. The implementation of sherardizing instead of tempering after prior quenching is possible using standard sherardizing temperatures (380-450 °C) without any additional treatment of zinc powder or saturating atmosphere control. Figure 2a exhibits an example of spring element with 20 μm zinc diffusion layer (Figure 2b) formed using integrated sherardizing technology.

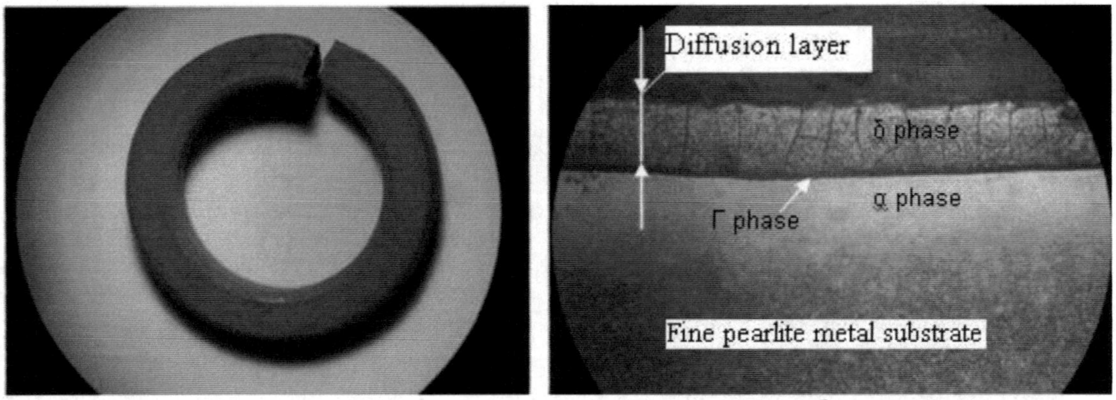

a b

Figure 2. a) Spring washer, after sherardizing X10.
b) Microstructure of zinc diffusion layer, X400.

The final microstructure of the spring washer is formed during the sherardizing process. Sherardizing temperatures provide a breakdown of martensite and its transformation into a fine pearlite (Figure 3 a, b).

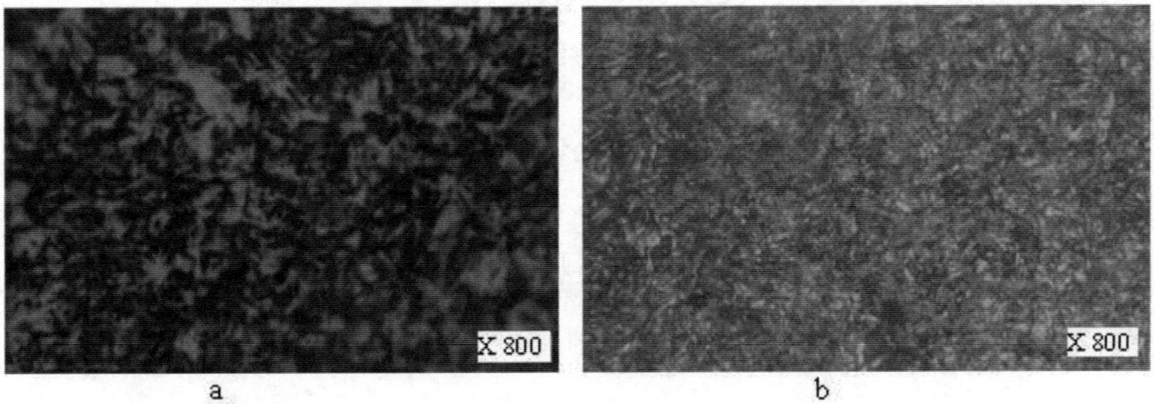

Figure 3. Microstructure of spring washer after quenching (a) and sherardizing (b).

After sherardizing hardness of spring washer structure was 43-47 HRC that corresponds to the required hardness parameters, regulated by GOST 6402-70 [5].

It is possible to use any way of obtaining zinc anticorrosive coatings on details without heat treatment or special requirements towards strength rating. The choice of zinc anti-corrosion method depends on the types of items to be treated, their dimensions and shape complexity. However, using sherardizing for anti-corrosion protection of details after the final heat treatment with special requirements towards strength class may cause the strength reduction due to thermic influence of sherardizing on metal structure. Consequently, it is the necessity to investigate the influence of sherardizing on steel structure and mechanical properties of metal parts with regulated strength rating after their final heat treatment.

With the aim to determine a possible strength reduction tensile test of bolts with different strength rating were conducted. For the bolts with each strength rating there were determined a breaking point in state of delivery, in state of delivery and subsequent sherardizing as well as after subsequent tempering according to the regimes equal to the sherardizing temperatures. The number of tests was three for each strength rating and corresponding type of treatment.

The results of tensile tests exhibit the decrease of breaking point and conventional yield strength data for bolts after sherardizing and additional tempering. This fact can be explained by activation of diffusion processes in steel due to additional thermic influence of sherardizing and additional tempering that resulted in steel microstructure change. For the high-test fasteners with a fine pearlite structure and strength rating 9.8-10.9 an additional heat over 650 °C with further soaking leads to cementite plates globulation and formation of an equiaxed grains. It was determined that for the samples with strength ratio lower that 8.8 the decrease of breaking point and conventional yield strength didn't exceed the bounds data listed in special standard GOST R 52643-2006 [6]. It's significant that the softening effect showed up after subsequent sherardizing is lower comparing with one that after subsequent tempering. This fact can be described by a decarburization effect during the process of subsequent tempering as well as by the fact that after sherardizing an intermetallic diffusion layer forms round the outer surface of coated article providing a compressing strain effect (Figure 4).

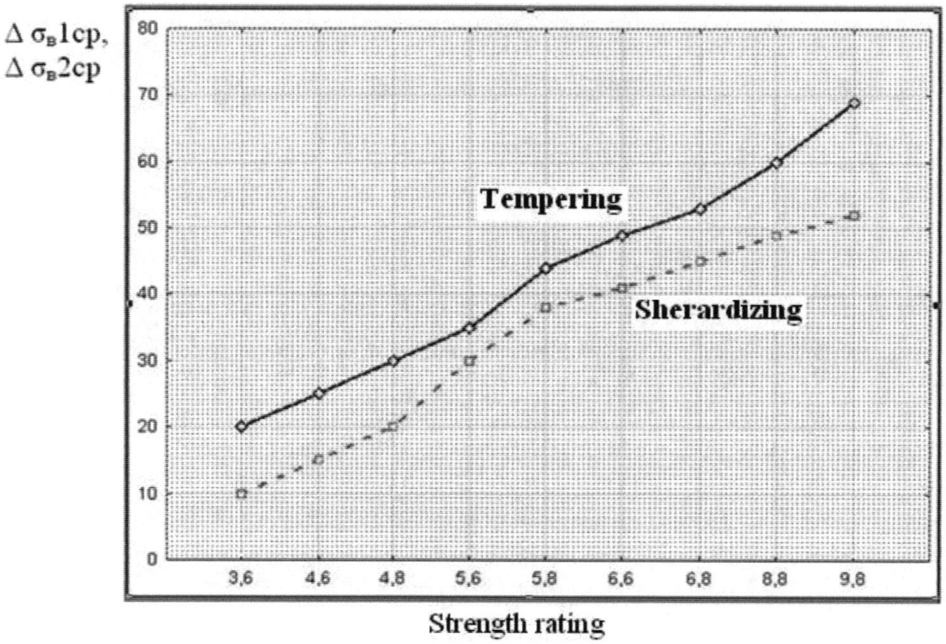

Figure 4. Variation dependence of breaking point data (average value) from strength rating for sherardizing and tempering at equal regimes (450^{0}C, 1h.).

3. Conclusions

Unlike the more widespread methods of obtaining Zn anticorrosive coating on metal articles known as hot-dip galvanizing or electrogalvanizing using sherardizing method may lead to a softening effect for some high-strength details. However, a rational choice of thermal treatment regimes and strength rating of a detail to be treated gives an opportunity to use advantages of sherardizing process for anti-corrosion protection. For some metal parts like spring elements, the most efficient decision is to combine the final stage of tempering with sherardizing. Such kind of complex heat-anticorrosive treatment gives an economy gain comparing to conventional ways of thermic and anticorrosion improvement of details. Using sherardizing to form an anticorrosive diffusion layer on details with a regulated strength rating over 8.8 is necessary to take into account the decrease of mechanical characteristics of the treated articles.

References

[1] Proskurkin E and Gorbunov N 1972 *Zn diffusion coatings* 148
[2] Konstantinov V, Ivanitsky N and Astreiko L 2013 Anticorrosive Zn coatings on metal parts: perspectives of thermodiffusion coatings *Foundry and metallyrgy* **4** 107-111
[3] Ворошнин Л, Менделеева О и Сметкин В 2010 *Теория и технология химико-термической обработки* (Москва: Новое знание) 304 с
[4] Galin R 1996 New technology of powder mixture production for thermodiffusion galvanizing *Metal science and heat treatment* **1** 21-25
[5] ГОСТ 6402-70 2006 *Шайбы пружинные. Технические условия* (Москва: Стандартинформ)
[6] ГОСТ Р 52643-2006 2007 *Болты и гайки высокопрочные и шайбы для металлических конструкций* (Москва: Стандартинформ)

IOP Conf. Series: Materials Science and Engineering **71** (2015) 012064 doi:10.1088/1757-899X/71/1/012064

Influence of crystals orientation on parameters of Ni_3Fe dislocation structure

L A Teplyakova[1], T C Kunitsyna[2] and N A Koneva[1]
[1]Tomsk State University of Architecture and Building, Physics Department, Tomsk, 634003, Russia
[2]Tomsk State University of Architecture and Building, Higher Mathematic Department, Tomsk, 634003, Russia

E-mail: lat168@mail.ru

Abstract. At present the basic evolution laws of the dislocation structure forming in the differently oriented Ni_3Fe monocrystals under pressing were established. While the data about the dislocation structure quantitative parameters and their behaviour during deformation at different crystal orientations are missing. In the paper the correlation of the sequence of change of substructures forming in the Ni_3Fe monorystals with the sequence of change of deformation stages has been revealed. The monocrystals with the deformation axis orientations [001], [$\bar{1}$11], [011] and [1.8.12] were studied. For the all mentioned deformation axis orientations the quantitative characteristics of dislocation structure were determined and their connection with the deformation curve parameters was revealed.

1. Introduction

It is known that at one axis loading the laws of formation of the dislocation structure and work hardening of the metal monocrystals and solid solutions depend significantly on their orientations [1-3]. In this case the Ni_3Fe alloy with a short range order is the most investigated one [4]. In the monocrystals with the deformation axis orientation corresponding the apexes (orientations [001], [$\bar{1}$11], [011], [1.8.12]) and the internal part (orientations [1.8.12]) of the stereographic triangular the inhomogeneous net substructure with the alternate layers with the decreased and increased dislocation density is formed. In [1.8.12] monocrystals one system of layers is observed [4, 5]. In [001], [$\bar{1}$11] and [011] monocrystals one can see two, three and four systems of layers according to numbers of equally loaded slip planes in each of them [4]. In spite of the fact that in the monocrystals with the deformation axis orientation [$\bar{1}$11] and [001] the number of equally loaded slip planes (acting within the whole volume) is higher than in [011] monocrystals, the number of locally acting planes, as a rule, is less than two [6]. Thus, during the plastic deformation the [011] monocrystal is separated into areas where the dislocation substructure evaluates as in the crystals oriented for a single slip, [001] monocrystal is devided into the areas where it develops as in [011] monocrystals and inside them, in its turn, as in the monocrystals with "mild" orientation. In the latter one sees the change of substructure types in the following succession: plane and quasi plane pile-ups and multipoles → plane and quasi plane nets → three dimensional dislocation net structure [4, 5]. The present paper is devoted to measurement of the quantitative characteristics of evaluating net substructure, construction of their distributions and determination of the connection with the hardening curve parameters for the all investigated monocrystal orientations.

2. Materials and experimental procedure

In the paper the Ni_3Fe monocrystals with the short range order oriented for single (orientation [1.8.12]) and multiple slip (orientations [001], [011], [$\bar{1}$11]) were investigated. The monocrystal orientations were determined by the Laue method and epigrams. Deformation was carried out by pressing on the samples with the size of 3 x 3 x 6 mm^2. The deformation rate was equal to 1,5 $\cdot 10^{-2}c^{-1}$.

3. Results and Discussion

3.1. Dependencies "τ – ε"

The deformation curves for different investigated orientations of the Ni_3Fe monocrystals in τ – ε coordinates (τ is the shift stress in the primary plane of octahedron slipping) are given in Figure 1. Here and next the following notations of the monocrystal deformation axis orientation are used: 1 – [001], 2 – [$\bar{1}$11], 3 – [011], and 4 – [1.8.12]. The Figure 1 shows that the monocrystal orientation essentially influences the character of hardening. The monocrystal deformation curves were analyzed by differencing the "$\tau - \varepsilon$" dependences ($d\tau/d\varepsilon = \theta$ – the deformation hardening coefficient).

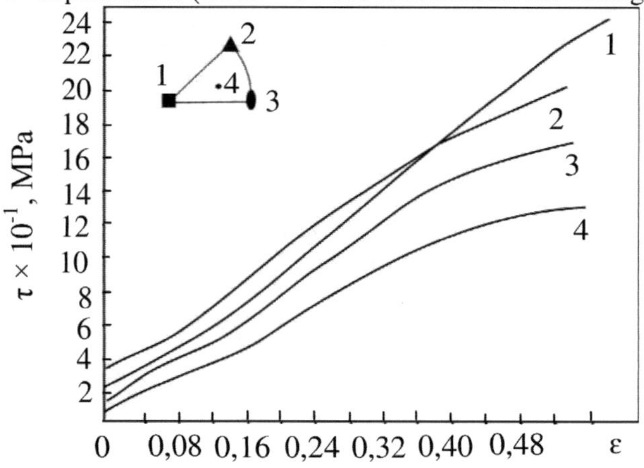

Figure 1. Strain curves of single crystals of the alloy Ni_3Fe. Orientation axis deformation of crystals.

Figure 2 shows the "d τ/d $\varepsilon - \varepsilon$" dependencies for the monocrystals of the investigated orientations. As it follows from Figure 1 and Figure 2 the staging is the most general property of all presented deformation curves. In the investigated monocrystals rather complex staging spectrum is observed. The analysis of the dependences given in Figure 2 testifies to the fact that in the monocrystals with the axis deformation orientations [001] and [111] as well as in the polycrystals [7] the deformation curve has four stages (the transitional stage with the increasing θ value, the stage II with the largest value of θ, the stage III where the θ value is decreasing and the stage IV with low and almost constant the θ value). In this case the orientation [001] appears after quite deep deformations.

The monocrystals [1.8.12] and [011] have slightly different type of staging. In the beginning of plastic deformation one observes the stage characterized by the constant value of θ. This stage cannot be classified as the true stage I because here the θ value is rather large (100÷200 MPa) [8]. It should be better named as the stage II_1. The stage II_1 is changed by the stage π which is transitional to the stage II_2 characterized by greater hardening coefficient as compared with that at the stage II_1. Finally in monocrystals [1.8.12] and [011] the stage II_2 turns gradually into the stage III. A clear idea about the θ behavior at different stages of the monocrystal deformation with the different deformation axis orientation may get from Figure 2. Thus the monocrystal deformation curves of the investigated alloys are of a staging type as well as in the monocrystals of FCC pure metals [8]. However as it has been established in the present work in the alloy the concrete type of staging depends significantly on the monocrystal orientation.

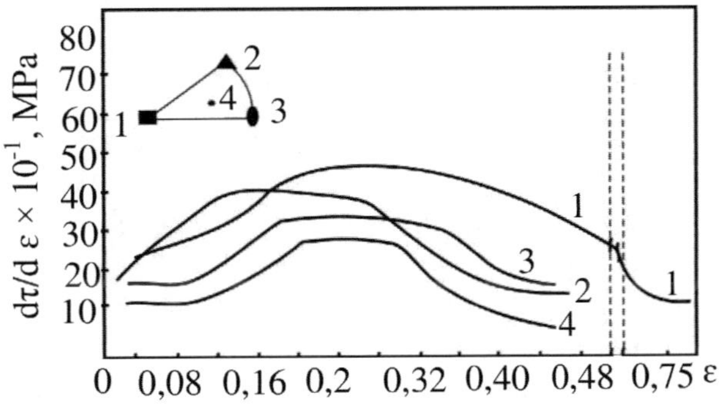

Figure 2. Behavior hardening coefficient with deformation for different orientations of the deformation axis alloy single crystals Ni_3Fe.

3.2. Dislocation substructure diagrams

The dislocation substructure diagrams represent the dependencies of the volume fraction (Pv) of every observed substructure on the deformation degree. Figure 3 shows the corresponding diagrams for the monocrystals with the orientations [1.8.12] and [011]. On the figures one sees that as the deformation develops at each its stage in the crystal volume one substructure type is actively formed and its volume fraction increases while the fraction of the other decreases. Formation of every new substructure type is connected with the appearance of new stage of hardening on the yield curve.

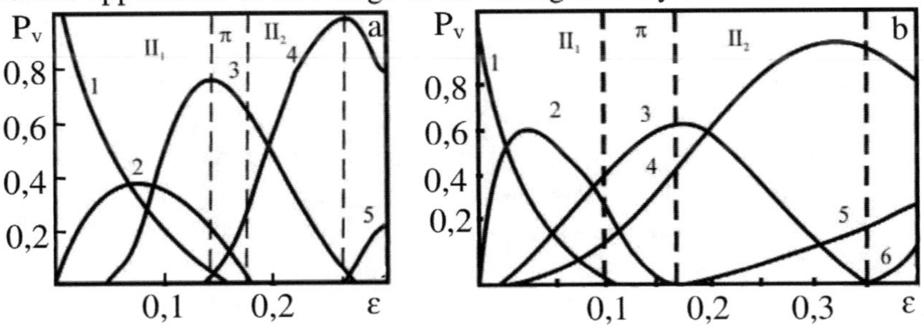

Figure 3. Depending on the volume fraction of the substructures (Pv) on the degree of deformation for [1.8.12] - (a) and [011] - (b) single crystals studied alloy. Legend substructures: curve 1 - initial chaos; 2 – pile-ups and multipoles; 3 - flat and quasiplanar grid; 4 - three-dimensional network; 5 - three-dimensional grid with misorientation; 6 - net - cellular substructure pa. Dashed lines are deformation stage, ε - the degree of deformation.

So in [1.8.12] monocrystals the following transformation chain is implemented: initial chaos → pile-ups and multipoles → plane and quasi plane net → three-dimensional net without disorientations →three-dimensional net with misorientations corresponding to the succession of the deformation stage change: $II_1 \rightarrow \pi \rightarrow II_2 \rightarrow III$. Every part in the substructure transformation chain, i.e. every following substructure needs for its formation new acting slipping system which had not taken part in the formation of the previous substructure. Inclusion into deformation of a new slipping system and the appearance of new dislocation interactions connected therewith changes the behaviour of the deformation hardening coefficient: a new stage of work hardening originates. Succession of inclusion into deformation of the slipping systems: primary → conjugated → critical → cross corresponds to the above said sequence of deformation stage changes in the monocrystals [1.8.12]. New substructure generation occurs when the developing substructure volume part reaches its maximum and the substructure being prior to it disappears (Figure 3a).

It should be pointed out that the substructure diagram for the monocrystals with orientation [011] is more complex (Figure 3b) than that for the orientation [1.8.12]. This complexity lies in the fact that the three-dimension dislocation net (curve 3) develops at all stages of work hardening.

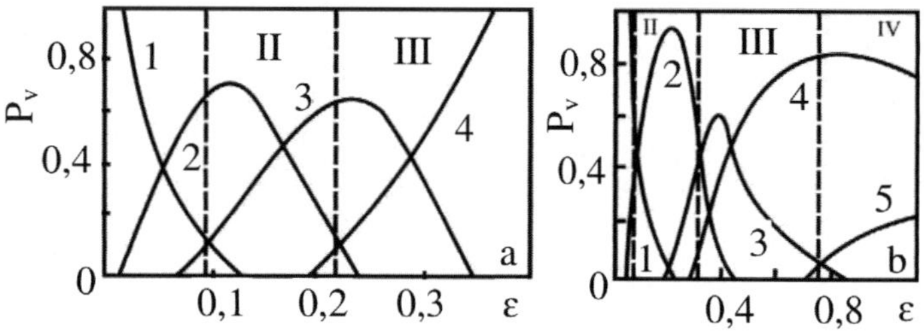

Figure 4. Diagrams of the volume fractions of substructures for monocrystals [$\bar{1}$11] (a) and [001] (b): 1 - initial chaos, 2 – pile-ups; 3 - net; 4 – microband, 5 – strips. Dotted lines are a hundred- deformation stage, ε - the degree of deformation.

This depends on the formation of two systems layers with higher dislocation density at the beginning of deformation. Presence of the volume net at the stage II_1 corresponds to the cross sections of the dislocation layers. Aside from that the laws of substructure volume parts changes determined for the orientation [1.8.12] are valid in this case. Correlation of the substructure change succession with the deformation stage change succession also follows from the diagrams of the Ni_3Fe monocrystal substructure oriented for multiple slipping (Figure 4).Such correlation was also observed in Ni_3Fe and Cu-Al polycrystals [7]. It was established that the completion of every deformation stage is connected with occupation of the most part of the material by the same substructure type.

The results obtained in the given work confirm the established law and also let us to conclude that the completion of every stage may be connected not only with filling the most part of material by the same substructure type but also with escape either of the prior or appearance of a new substructure, or with all three events together.

3.3. Distribution of dislocations

Figure 5 shows the histograms of the distances between the nearest dislocations (ℓ) in the slipping planes for all investigated monocrystals. The histograms were plotted for different monocrystal deformation degrees. One sees that at all monocrystal orientations the distributions ℓ being single-modal from the beginning of the plastic deformation remain the same approximately up to 20% of deformation. With the increase of the deformation degree the distributions ℓ become more narrow that in general testifies to the increase of the net substructure homogeneity during the plastic deformation of the alloy Ni_3Fe with short-range atomic order since more and more dislocations lie at the distances close to the average along the crystal. Meanwhile from the beginning of the plastic deformation the net structure in the crystal-edge distance is inhomogeneous at all investigated axis compression orientations. It can be described as "composite" consisting from the dense plane dislocation layers with rectangular net built in the three-dimension net substructure.

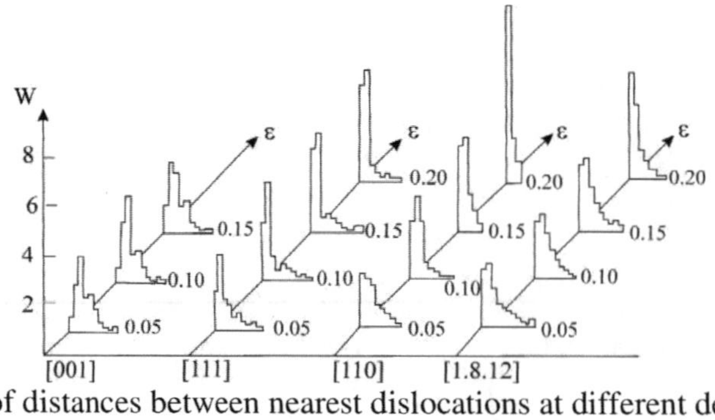

Figure 5. Distribution of distances between nearest dislocations at different degrees of deformation.

In this connection it should be interesting to investigate the ℓ value behavior separately in the dense and loose layers. Such investigation was done on monocrystals. The results are given in Figure 6. One sees that during the deformation the form of dislocation distribution in the plane net layers – dense layers – is not changed, the average net section size is only changed (Figure 6a). Consequently, for the net structure with "rectangular" net section form in the deformation degree interval where it is formed has the principle of similarity. In the three-dimension dislocation net at small deformation degrees the form of the distribution ℓ (Figure 6b) differs from that in the layers with the plane net. If in the plane net layers it is exponential within the whole investigated deformation degree interval, at the stage II in the volume net the distribution ℓ is close to the logarithmically normal and with the deformation evolves to the exponential one. By the beginning of the stage III the distribution ℓ became practically similar for both net substructure types (Figure 6).

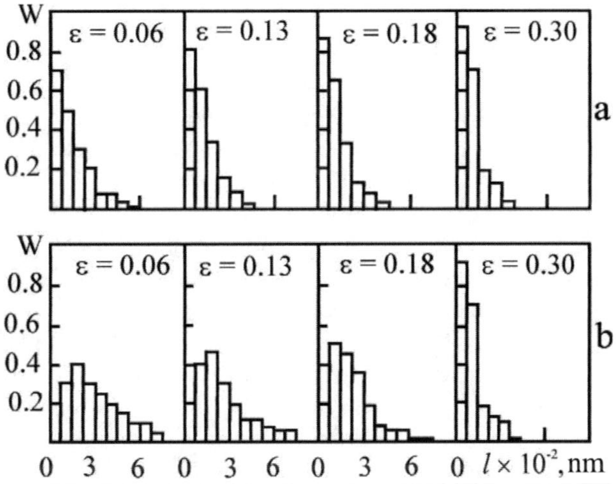

Figure 6. Distribution of dislocations over distances (l) at different degrees of deformation: a - dense; b - loose dislocation layers.

3.4. Scalar dislocation density

The average scalar dislocation density <ρ> is the most common parameter characterizing the dislocation structure. Figure 7 shows the <ρ> dependencies on the deformation degree for four investigated Ni_3Fe monocrystal orientations. It is seen that in all cases the <ρ> value is monotonically increased with the deformation according to the same law. The differences have qualitative character. As it follows from the comparison of the corresponding curves <ρ> - ε (Figure 7) and τ – ε (Figure 1), in a number of monocrystal orientations [1.8.12], [011], and [001] the following correlation is revealed: the higher monocrystal orientation symmetry the larger the dislocation accumulation rate and the value of the given shear stress. This is evidence of an important role played by the dislocation accumulation process in the formation of the flow stress.

4. Conclusion

1. The change in Ni_3Fe monocrystal orientation from [1.8.12] → [011] → [111] → [001] in the main does not influence on the types of the structures forming in the local volumes but results in displacement of their formation start sideward smaller deformation degrees. This correlates with displacement in the crystals with the orientations [011], [111] and [001] in the same direction of the start of corresponding deformation stages. Identical nature of the deformation curve stages in the

crystals with the orientations [1.8.12] and [011] is explained by the fact that in [1.8.12] the second slip plane enters the deformation early (at the stage II_1).

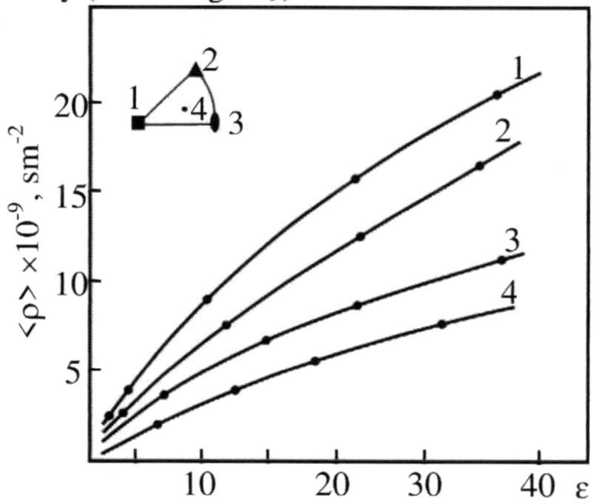

Figure 7. Dependence of the mean dislocation density <ρ> the degree of deformation (ε) for single crystals of the alloy Ni_3Fe.

2. Influence of the monocrystal deformation axis orientation on the rate of dislocation accumulation was established. The monocrystal orientation change in the direction of the increase of the number of equally loaded octahedron slipping systems (1→ 4 → 8) leads to the increase of scalar dislocation density accumulation rate. This dependence correlates with the $\tau - \varepsilon$ curve behaviour in the orientation series [1.8.12], [011] and [001].

3. In the Ni_3Fe monocrystals oriented for single slip one sees the limiting value of the scalar dislocation density in the inhomogeneous parts of the net structure which remains practically invariable within the whole investigated deformation interval. Reaching equality of the average scalar dislocation density means the transition to the stage III.

References

[1] Sastry S M L and Ramaswami 1980 *J. Mater. Sci. Eng.* **43** 2 231-234
[2] Karnthaler H P and Schugert B 1980 *Proc. 5th Int.Conf. Strength Met. And Alloys (Aachen)* **1** 205-210
[3] Starenchenko V A, Kozlov E V, Solov'eva Yu.V, Abzaev Yu. A and Koneva N A 2008 *J. Mater. Sci. and Eng.* **483** 179-183
[4] Teplyakova L A, Kunitsyna T S and Koneva N A 2014 *Adv. Mater. Res.* **1013** 54-61
[5] Teplyakova L A, Kunitsyna T S., Koneva N A and Kozlov E V 2004 *J. Bulletin of the Russian Academy of Sciences: Physics* **68** 10 1629-1634
[6] Teplyakova L A and Kozlov E V 2005 *J. Phys. Mesomech* **8** 57-66
[7] Koneva N A and Kozlov E V 1991 *J.Proc. of Higher Education. Phys.* **34** 56-70
[8] Honeycombe R W K 1972 *The Plastic Deformation of Metals* (Moscow: World) p 408

IOP Conf. Series: Materials Science and Engineering **71** (2015) 012065 doi:10.1088/1757-899X/71/1/012065

Non-destructive testing of the metal-insulator-metal using miniature eddy current transducers

S F Dmitriev[1], A V Ishkov[1], V N Malikov[1], A M Sagalakov[1] and A O Katasonov[1]
[1]Altay State University, Barnaul, 656049, Russia

E-mail: osys@me.com

Abstract. The paper puts forward a modified eddy current testing method based on the use of miniature eddy current transducers within a localized controlled area of the material surface of up to 50 μm^2. Measurement results are provided for a model composite material made up of alternating conducting and dielectric layers of system AI- PAPER -A1. Structural flaws are specified as changes in the number and position of layers in the sequence. Prospects of the proposed method and a measuring device operating on its basis for flaw detection in composite aluminum alloy materials are shown.

1. Introduction

Despite our great experience in using vortex-current transducers/transformers (VCT) in the sphere of nondestructive control, several important aspects still need to be studied nowadays.

In particular, the most vital problems are: carrying out local contactless measurements combined with controlled object scanning; price reduction in sensor engineering; making the control locality more accurate and the quick allocation of the frequency range, the most proper to scan the controlled object.

The worked out conception of the virtualized measuring device (transformer) enables a researcher to make a great number of measurements in different spheres with the help of one sensor only. To these spheres belong fault detection of conducting materials and stratified compos, thickness measuring, profilometry, tension measuring of permanent and variable magnetic fields and conductivity measuring of unferromagnetic materials and some others. In the past there were several attempts to use highly sophisticated magnetic field sensors like SQUID and Fluxgate sensors for sensitive low frequency eddy current testing to detect deep defects in metal parts [1–3]. Although very good results could be achieved such testing systems can hardly be used in real industrial applications because of their complexity, prices and insufficient robustness [4]. To make the price lower the authors came to the conclusion that one should replace the expensive hardware blocks with the software for personal computers (PC). As a result, developed device consists of a vortex-current transformer (VCT) only, which is connected to the sound card of the PC, managed with the special software. The software manages the voltage given to the generative transformer winding, it also reads out the values of the voltage from the measuring winding in conventional units. Later these units are transferred into the values of conductivity taking into account the preliminary calibration. A very promising sensor performance could be obtained by using highly sensitive inductive coils. The problem is that such sensors have to be produced by skillful specially trained operators; it results in low reproducibility and low productivity [4]. To solve these problems the VCT is often provided with an extra magnetic conductor [4].

One of its terminals has a shape of a truncated cone. This solution of the problem has only one disadvantage. Despite the increased localization of the magnetic flux, the construction of the magnetic conductor has become considerably more complicated. It makes the measuring accuracy worse, because the output signal of the VCT depends to a great extent on the interaction between two magnetic conductors, which can influence the intensification of the vortex-current field in a rather unpredictable way.

There are different well-known constructions of superimposed VCTs, whose working surface has either a plane or hemispherical form. Such kind of a surface provides a satisfactory contact of the VCT with the surface under control, but the tension quantity, sent to the VCT, greatly depends on the curvature of the controlled surface. The edge effect influences the work of the VCT considerably. This effect makes it impossible to control the details of a complicated configuration and of small sizes. If the size of the sample is small, the measurement error is unavoidable. The results show that the relative errors increase with the decrease of the size of the sample [6]. The small size of the cores (starting with 1 millimeter), used in the virtualized VCT, gives an opportunity to improve the control locality, without applying additional complication of the construction. Due to that the influence of hindrances on the VCT is considerably reduced.

The core, shaped either as a cone or pyramid, supports a high degree of localization of the magnetic field that is why the influence of the edge effects and the surface curvature is practically avoided. The existence and shape of a defect can be recognized by observing the difference in variation of the equivalent impedance. Magnetic systems with various interior deep defects are numerically analyzed by finite element method and their data are compared with those of the experimental systems. The measured data of the impedance variation are distinguishable enough to be used to estimate the existence and shape of defects in the steel structure [7]. In the laid on VCTs theory the most general objective is the study about the distribution of the electromagnetic field, created by the circular wind with the alternating current, in the multilayer media [8]. Let us consider that the turn of the radius R1 is placed above the multilayer medium in the plane, parallel with the ambit (boundary) of the section layers. Taking into account the resulting ratios, we have made hodographs, illustrating the influence of different parameters of the medium andthe sensor on the values of the applied voltage. Defects in welding joints of the LSMP may significantly reduce the material strength. Eddy Current Testing (ECT) is considered as a powerful NDT tool for the detection and sizing of defects [9]. Earlier the eddy current control method could be used to investigate only surface defects (such as cracks, cuts and other examples of metal surface discontinuity), now, due to using subminiature VCTs and special software, it is getting possible to localize the magnetic field in a small zone of the controlled object and to achieve a high degree of the (magnetic) field penetration depth into the investigated object. The only condition is to choose a proper field frequency, created by the actuating winding.

2. Materials and methods

In general the scheme of the sensor is presented in Figure 1. In this scheme there are actuating (AW) and measuring (MW) windings, placed in the VCT.

A digital signal comes from the virtual generator to the input of the digital analog transformer (DAT) of the sound card and then it is transformed into the analog one. After passing though the power-amplifier (PA) the analog signal is sent to the actuating winding (AW) of the transformer. While passing through the actuating winding of the VCT, the sinewave signal makes the electromagnetic field, that creates EMF in the measuring winding (MW) of the VCT. The tension is received by the microphone input of the sound card and, after coming though the preamplifier (PA), it goes to the input of the analog digital transformer (ADT) sound card. The analog signal is transformed into the digital one and is sent to the treatment and operation unit of the software. The treatment and control unit fixes the level of the digital signal in conventional units that correspond to the voltage values in the measuring winding.

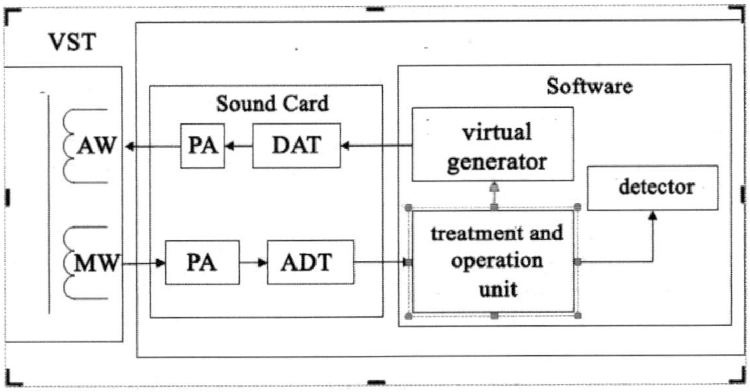

Figure 1. The scheme of sensor operation.

This level is accepted as a zero one that corresponds to the voltage level on the measuring winding without the control object. Without the control object the indicator shows zero, it corresponds to the zero value of conductivity.

Along with the considerable price reduction of the device, the virtualization of the generator and signal receiver makes it possible to mark out the frequency range, at which the process of scanning of the control object gets very effective. The effectiveness is reached due to the quick varying of the current frequency, given by the virtual generator to the VCT. Thus, researchers can define the frequency, at which changing the amplitude of the signal on the measuring winding (when it is passing through the imperfect area) is at a maximum. Besides, frequency varying enables us to find the depth of the defect hiding, because the depth of the electromagnetic field distribution of the VCT directly depends on the frequency of the field sent to the actuating winding [1].

The principle of operation of the VCT is based on the alternating magnetic field, localized in the controlled object with the help of the oxide pyramid shaped core. The shape of the core is conditioned with the necessity to localize the magnetic flux from the actuating (generator) coil. The core of this model was made of a midrange metadiscipline ferrite NM3, chosen according to the value of the maximum initial magnetic conductivity 500. The actuating winding of the subminiature transformer consists of 10 winds and its diameter is 0.12-0.13 mm. The measuring winding includes 130 winds and its diameter is 0.05-0.08 mm. The compensation winding is included into the scheme to minimize the influence of the actuating winding on the received signal. The compensation winding is connected with the measuring winding so that the researcher could subtract the tension of the actuating winding. It consists of 20 winds. A copper wire, that is 0,005 mm thick, is used to reel the winds. The windings are wound round the pyramid shaped core. The scheme of the subminiature VCT is represented on the Figure 2.

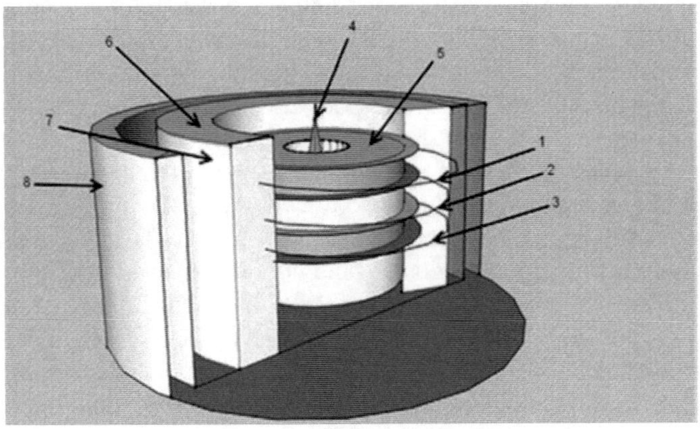

Figure 2. The scheme of the SMVCT.

The VCT is a transformer. It has 1) an actuating 2) a compensation windings 3) and a magnetic conductor 4) which is placed inside the cylindrical platform 5) with the tracks for the windings, cut on the outer side 6) which is impregnated with a compound at 200 degrees centigrade in order to prevent the windings from destruction while the ferrite screen is being put on 7) which is meant for the magnetic field localization on the controlled object. From the outside the sensor is covered with a corundum washer 8 that protects the core 4 from contacting with the controlled object.

The electromagnetic field of the generating winding gives an impulse to eddy currents in the electroconductive controlled object. Eddy current density in the object depends on the geometric and electromagnetic values of the object and on the back-to-back location of the measuring VCT and the object. The magnetic field of the eddy currents is opposed to the primary magnetic field of the generating winding, as a consequence the resulting field depends on the electromagnetic characteristics of the controlled object and on the distance between the VCT and the object, as the distribution of the eddy current density depends on these factors. There is EMF in the measuring winding of the VCT; it serves as a signal, hands over the information about the object into the measuring unit.

In this case EMF of the measuring winding decreases due to the opposite magnetic field produced by the eddy currents.

3. Experimental results

The worked out VCT enables the researchers to investigate the defects effectively. These defects may be found in the places of metal-insulator transition in the extra-small layered metal-polymeric composite objects. Such composites can contain several metal layers, separated with thin polymeric dielectric interlayers. To typical defects of such materials refer, for instance, the disturbance of layer continuity, the formation of crossbars between the layers. To investigate layered structures of the metal-dielectric-metal type we used the MNME – 5FA device (the Meter of Nonferromagnetic Material Electroconductivity), constructed earlier. A modified Fourier analyzer was used in a very special way to measure the decibel-log frequency characteristic. To demonstrate the operational capability of the suggested method we used the structure that was based on the alternation of the aluminum foil and paper, both of which were 100 micrometers thick. A hollow parallelepiped was placed between the layers as a model defect. Its sides were 300 micrometers thick.

In Figure 3 there is a spectral picture, observed when the sensor is moved above the layered medium, inside of which there is a defect. The signal level from the measuring winding characterizes the values of conductivity on the survey plot. For the fundamental frequency at 1000 Hz the level of the voltage, sent to the measuring winding, was (130±2) mW. Area 1 and area 2 on the diagram, where the voltage level falls up to 115 mW, correspond to the walls of the parallelepiped. These changes in the signal amplitude make 11 percent from the signal level, corresponding to the nondefective zone of the sample. Thus, the signal amplitude in the nondefective zone does not exceed 4mW, it makes 3 percent from the signal level, corresponding to the nondefective zone of the sample.

Judging by the diagrams in Figure 4, 5 one can notice the amplitude changes when the transformer is being tested at other frequencies. These changes, in case when the frequency is increased, are determined by the lesser penetration depth of the eddy currents into the depth of the layered structure and by the increasing influence of different small cracks on the surface of the layered structure. If the frequency decreases the eddy current field goes much deeper into the investigated object. Thus, the influence of the model defect practically cannot be traced. At the frequency of 6000 Hz (Figure5) the model defect is found as before but now the amplitude fluctuations in the nondefective zone of the sample exceed the signal level that corresponds to the nondefective zone of the sample, by 7 per cent. When the controlled object is investigated with unknown defects, such amplitude changes can be interpreted as defects by mistake. At the probing frequency at 500 Hz (Figure 6), the amplitude fluctuations in the nondefective zone are insignificant but the amplitude changes in the defective area do not exceed 3 per cent from the signal level, corresponding to the nondefective zone of the sample.

When laboratory or production measurements are carried out such amplitude fluctuations can be caused by external influence and not by defects

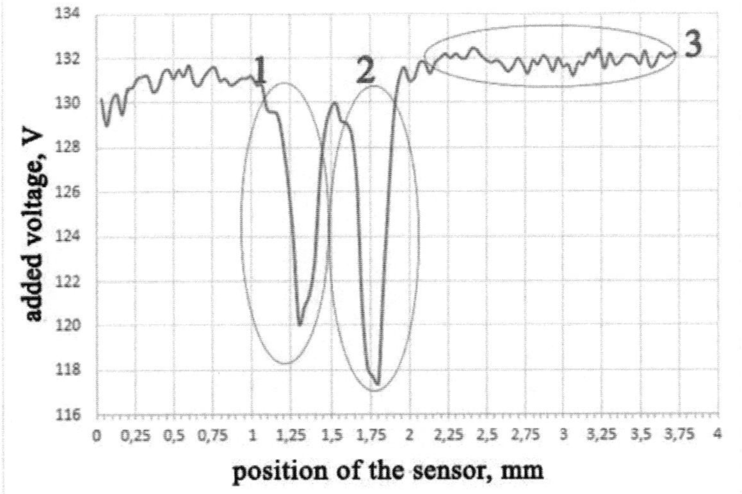

Figure 3. A spectral picture, observed when the sensor is moved along the layered medium with a defect. The transformer frequency – 1000 Hz. 1, 2 – the sides of the parallelepiped, 3 – a nondefective zone of the sample.

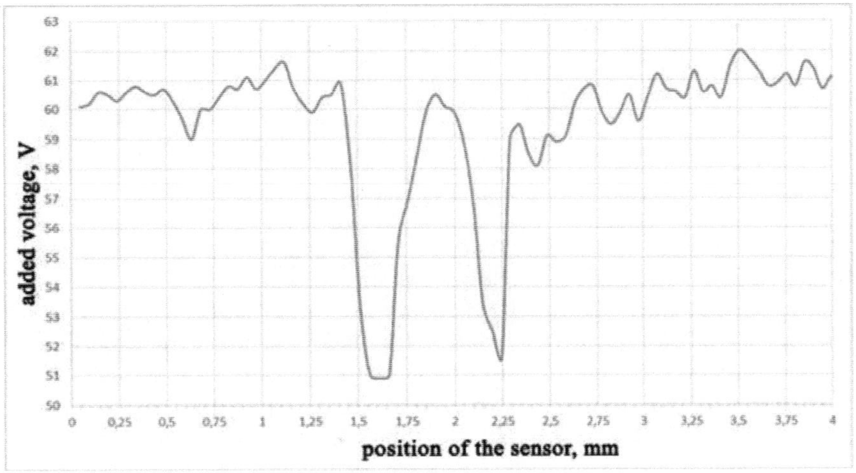

Figure 4. A spectral picture, observed when the sensor is moved along the layered medium with a defect. The transformer frequency – 6000 Hz.

When the operating frequency of the device exceeds the pointed limits the results of the measurements will be skewed by the amplitude fluctuations, caused by microcracks on the surface of the sample or by decreasing the localization of the field that is inside the layered structure. The amplitude changes, caused, in this case, by the microcracks on the sample surface, are much higher than the amplitude changes, caused by the defect directly.

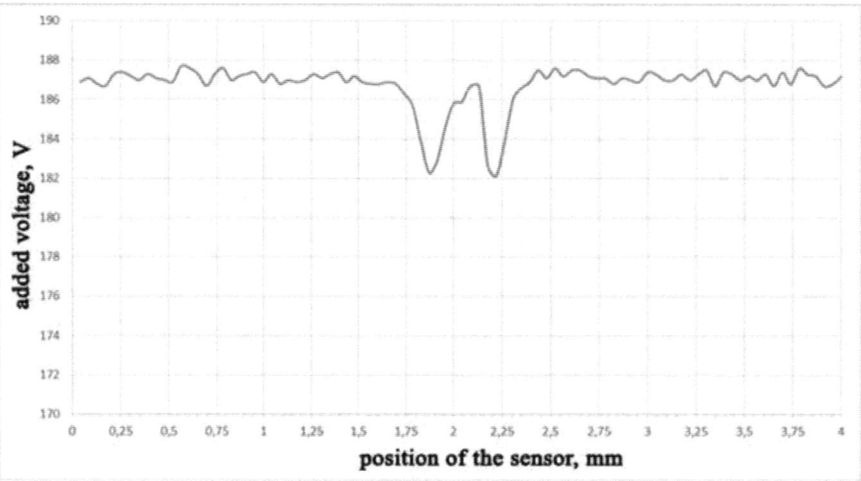

Figure 5. A spectral picture, observed when the sensor is moved along the layered medium with a defect. The transformer frequency – 500 Hz.

The defect was 600 micrometers from the sensor in the depth of the layered structure. Right up to the depth of the defect location that was 1400 micrometers we could trace a direct dependence between the response of the transformer and its position above the defect. Stating the amplitude changes of the transformer response, determined by the defect, it is possible to change the current frequency in the actuating winding so that eddy currents could concentrate in the composite layers above the defect. The inverse problem solution enables us to define the occurrence depth of the defect. After the typical defect calibration of Fourier analyzer we can use the MNME – 5FA device to diagnose multi-layer composite materials which are from 1 up to 1400 micrometers thick.

4. Conclusion

Therefore, the results of the experiment demonstrate the wide range of possibilities provided by the eddy-current method in the study of defects concealed in metal-insulator-metal type structures. Earlier the eddy current control method could be used to investigate only surface defects (such as cracks, cuts and other examples of metal surface discontinuity), now, due to using subminiature VCTs and special software, it is getting possible to localize the magnetic field in a small zone of the controlled object ant to achieve a high degree of the (magnetic) field penetration depth into the investigated object. The only condition is to choose a proper field frequency, created by the actuating winding.

References

[1] Hohmann R 1999 *SQUID-System mit Joule-Thompson-KühlungzurWirbelstromprüfung von Flugzeugfelgen* Dissertationsschrift Justus-Liebig-UniversitätGießen

[2] Gabor Vertesy, Antal Gasparics Fluxset Sensor Analysis *Journal of Electrical Engineering* **53** 2002, ISSN 1335-3632

[3] Kreutzbruck M, Allweins K and Heiden C 2000 Wirbelstromprüfsystemmitintegriertem Fluxgate-Magnetometer. *Proc. of DACH-JahrestagungDGZfP, ÖGfZP, SGZP, Innsbruck*, 29.-31.5.2000 BB 73.2 pp. 871-881

[4] Hesse O, Pankratyev S 2005 Usage of magnetic field sensors for low frequency eddy current testing *Measurement science review* **5**

[5] Lemeshko M A, Adigamov K A and Osackiy S S Patent RU 2200299 "Eddy current probes"

[6] Wang J, Fu Y, Sun M, Liu Z, Fan Z and Shi K 2012 Thickness measurement of ferromagnetic material based on pulsed eddy current testing technology *Journal of Beijing Jiaotong University* **36** 133-136

[7] Lee K H, Baek M K and Park I H 2012 Estimation of deep defect in ferromagnetic material by low frequency eddy current method *IEEE Transactions on Magnetics* **48** 3965-3968
[8] Kluev V V 2008 *Non-destructive testing* (Moscow) Vol **1**
[9] Zhang D, Li H, Chen Z and Wang Z 2012Applications of multi-frequency inversion algorithm to quantitative NDE of cracks in welding joints of a metallic lattice sandwich plate, *Int J Appl Electrom* **39** 137-143
[10] Polyakov V V, Dmitriev S F, Ishkov AV, Kolubaev EA and Malikov VN 2014 Non-destructivetesting of aluminium alloys by using miniature eddy-current flaw transducers *Advanced Materials Research* **880** 105-108

IOP Conf. Series: Materials Science and Engineering **71** (2015) 012066 doi:10.1088/1757-899X/71/1/012066

Structure-phase states and wear resistance of deposition surface formed on steel by electric arc method

S V Raikov[1], E V Kapralov[1], E S Vashchuk[1], E A Budovskikh[1], V E Gromov[1], Yu F Ivanov[2,3] and K V Sosnin[1]
[1] Siberian State Industrial University, Novokuznetsk, 654007, Russia
[2] Institute of High Current Electronics SB RAS, Tomsk, 634055, Russia
[3] National Research Tomsk Polytechnic University, Tomsk, 634050, Russia

E-mail: gromov@physics.sibsiu.ru, yufi55@mail.ru

Abstract. Investigations of elementary and phase structure, state of defect structure and tribological characteristics of deposition, formed on a low carbon low-alloy steel by a welding method were carried out. It was revealed that deposition, formed on a steel surface is accompanied by the multilayer formation, and increase of wear resistance of the layer deposition as determined.

1. Introduction

Electric arc deposition takes strong positions in the improvement of structure of machine parts and mechanisms in different branches of industry. As a result of deposition and in manufacture of machine parts it is possible to obtain a working surface possessing a necessary complex of properties: wear resistance, heat stability, heat resistance, corrosion resistance, etc. Using a deposition, it is possible to reduce the consumption of the expensive nonferrous metals and alloys and high carbon steels. Besides, repeated restoration of the worn parts decreases to a large extent the consumption of metal for the manufacture of spare parts of the equipment. The increase of service life of equipment parts is especially important, if the performance of high efficient equipment depends on their reliability and durability. A high economical and technical efficiency of deposition in metallurgical machine building is specified by it.

The efficiency of structure improvement by deposition is determined by the right choice of deposited metal composition based on the conditions of work of a part and main type of wear. The choice of the deposited material is done with regard to exploitation conditions of the part being restored, type of protection, design features of a part and the equipment available. At present a large number of electrode materials has been developed on the base of iron for arc deposition.

2. Materials and research method

Steel Hardox 400 was used as the test material; its element composition is shown in Table 1. Thick deposited layers (3 - 5 mm in thickness) were formed on steel surface by welding method. Element composition of welding wire used for formation of a layer is shown in Table 1.

Investigation of element and phase composition, state of defect substructure of deposited layer was done by methods of optic (metallographic microvisor mVizo-MET-221), scanning (scanning electron microscope Philips SEM-515 with microanalyzer EDAX ECON IV) and transmission (transmission electron microscope EM-125) electron microscopy, X-ray structural analysis (X-ray diffractometer DRON-7). Tribological characteristics of coating were revealed by wear resistance and friction coefficient determination (tribotester Tribotechnic). Tribological tests were done under the following conditions: a ball with diameter of 3 mm from a hard alloy VK-8 was used as a counterbody. A counterbody was shifted on the sample surface along the circle of 4 mm diameter at linear velocity 2 cm/sec under normal load 5 N. The total number of revolutions of counterbody was 5000.

Table 1. Chemical composition of test materials (the remainder Fe, weight%).

Chemical element	C	Si	Mn	P	N	B	S	Mo	Cr	Nb
Steel Hardox 400	0,18	0,70	1,60	0,01	0,004	0,025	0,01	0,25	-----	
Welding wire SKA 70-G	2,6	0,6	1,7	-----	------	2,2	-----	-----	14,8	4,7

3. Results of investigations and their discussion

Tribological test has shown that the formation of deposition on the surface of steel Hardox 400 results in the increase of wear resistance of material surface layer by a factor of ~ 2.25 and friction coefficient – by a factor of – 1.05.

To establish the physical mechanisms of wear resistance increase of steel with deposited material the studies of structure, phase and element composition of deposition/steel system were carried out. Structure and element composition analyses were done in two sections: in section parallel to the surface of deposition and in section perpendicular to the deposited layer.

Structure of deposition is shown in Figure 1. A large number of inclusions are of edged shape. Inclusion sizes vary within 1 mcm to 5 mcm. The second morphological element of the deposited layer is a structure of dendritic (Figures 1, b, c) and cellular (Figures 1, c,d) crystallization. The cell sizes vary within 0.3 mcm to 0.8 mcm. The cells are separated by interlayers of 50 to 100 nm thickness.

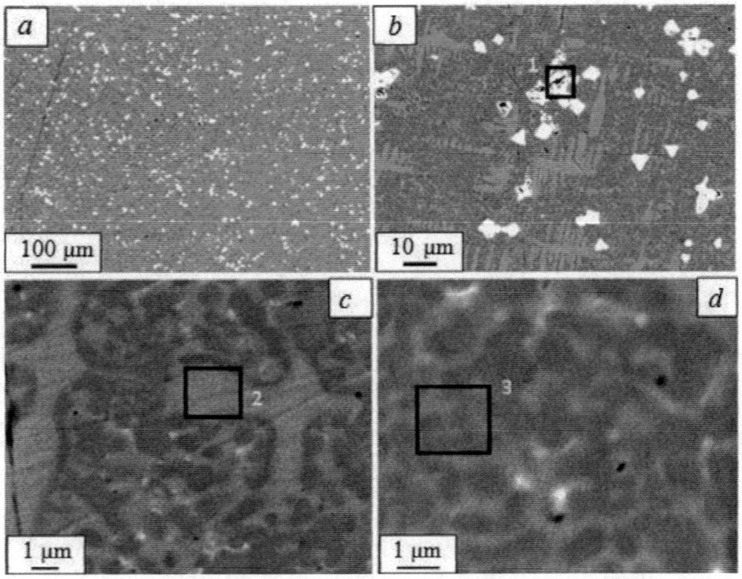

Figure 1. Structure of deposition being formed on the surface of steel. Section is parallel to the surface of deposition. Frames show the sites of X-ray spectrum analysis of the material.

X-ray microspectrum analysis of deposition sites, designated by frame in Figure 1, has shown that particles of the edged shape are enriched by atoms of niobium (Figure 1, b, zone 1); zones of dendritic crystallization (Figure 1, c, zone 2) are enriched by atoms of iron, chromium and carbon. A characteristic feature of the cellular crystallization structure being formed largely by atoms of iron is the presence of carbon and chromium atoms of large concentration (Figure 1, c, zone 3). The given results are shown in Table 2 in a quantitative ratio.

Table 2. Results of X-ray microspectrum analysis of deposition structure formed on the surface of steel.

Spectrum	Alloying elements , weight %							Result
	C	Si	Ti	Cr	Mn	Fe	Nb	
Field 1	0,0	0,0	1,02	6,50	0,0	7,07	85,42	100.00
Field 2	3,74	1,40	0,0	7,89	1,44	85,53	0,0	100.00
Field 3	4,11	0,71	0,0	13,07	1,71	80,40	0,0	100.00

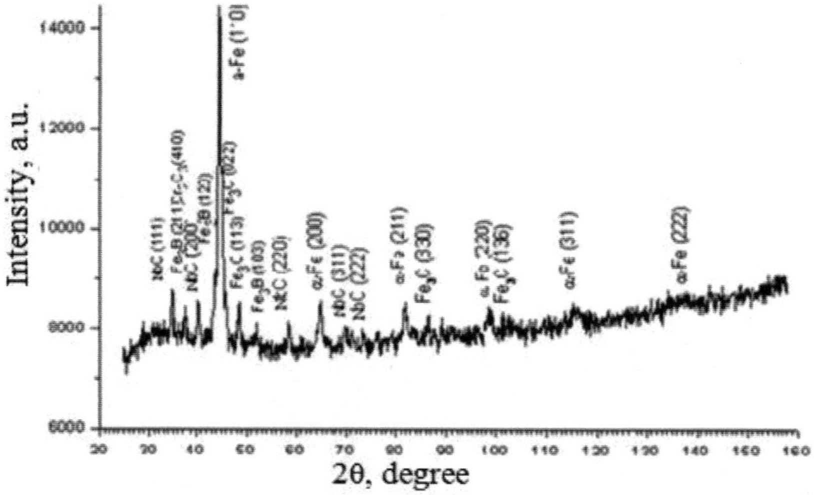

Figure 2. Part of of X-ray photograph obtained from the surface layer of deposition.

4. Conclusions

The carried out studies of structure, element and phase composition of deposition formed on the low carbon low-alloyed steel Hardox 400 make it possible to make the following conclusions:

1. Formation of deposition on the surface of steel is accompanied by the formation of multi-layer structure, the layer of which differs in the morphology of element substructure.

2. Deposition is a multi-phase material and is presented by solid solution grains on the basis of α-iron, by particles of iron carbide of composition Fe_3C (cementite), the volume fraction of which is 10%, of niobium and chromium carbides NbC and Cr_3C_2 (sum volume fraction of carbides is 20%) and of boride of iron Fe_3B, the volume fraction of which is 10%.

3. The revealed multiple (more than twice) increase of wear resistance of deposited layer (para to volume of steel) is specified by formation of submicro- and nanosize structure of α-phase crystallization and precipitation of large volume (~40%) of particles of high strength of carbide and boride phases.

Electric arc deposition with powder wire is used for strengthening the excavator's buckets and bodies of large-size automobiles.

References

[1] Babichev A P, Babushkin N S and Bratkovskii A M 1991 Physical magnitudes: Reference book (Moscow: Energoatomizdat) 1232

[2] Kurdyumov V G, Utevskii L M and Entin R I 1977 Transformation in iron and steel (Moscow: Nauka) 236

[3] Fasiska E J and Wagenblat H 1967 Dilatation of alpha-iron by carbon *Trans.Met.Soc.* **239** 11 1818–20

Modelling of creep curves of Ni_3Ge single crystals

V A Starenchenko[1], S V Starenchenko[1], O D Pantyukhova[1] and Yu V Solov'eva[1]
[1]Tomsk State University of Architecture and Building, Tomsk, 634003, Russia

E-mail: sve-starenchenko@yandex.ru

Abstract. In this paper the creep model of alloys with $L1_2$ superstructure is presented. The creep model is based on the idea of the mechanisms superposition connected with the different elementary deformation processes. Some of them are incident to the ordered structure $L1_2$ (anomalous mechanisms), others are typical to pure metals with the fcc structure (normal mechanisms): the accumulation of thermal APBs by means of the intersection of moving dislocations; the formation of APB tubes; the multiplication of superdislocations; the movement of single dislocations; the accumulation of point defects, such as vacancies and interstitial atoms; the accumulation APBs at the climb of edge dislocations. This model takes into account the experimental facts of the wetting antiphase boundaries and emergence of the disordered phase within the ordered phase. The calculations of the creep curves are performed under different conditions. This model describes different kinds of the creep curves and demonstrates the important meaning of the deformation superlocalisation leading to the inverse creep. The experimental and theoretical results coincide rather well.

1. Introduction

The creep of the Ni_3Ge alloy with $L1_2$ superstructure possesses a number of peculiarities that differ it from the creep of pure metals with fcc structure. Low rate of creep under moderate temperature, anomalous creep dependence under certain temperature intervals and the presence of creep with constant rate in the high temperature areas refer to them [1-6].

The temperature influence on dislocations movement in the alloy can explain these peculiarities. The influence has a double nature. On the one hand the temperature increase causes the superdislocations mobility rise. On the other hand a part of superdislocations are happened to be blocked into stable configurations, being higher with the temperature increase [7]. This paper is dedicated to the model where the peculiarities of superdislocations movement and the influence of temperature changes on it are considered. The calculations are performed and the possibility of some peculiarities of the Ni_3Ge alloy creep to be described in the suggested model is shown.

2. Modelling

The papers [8-10] give the formulation of the mathematical model of single crystal alloys creep with $L1_2$ superstructure, the concept of strengthening and rest being the basis of it [11-12].

Due to the concept mathematical formulation of the model of single crystal alloys creep with superstructure $L1_2$ oriented for multiplication of slipping is the following [8, 9] (1).

$$\left\{\begin{aligned}
&\frac{da}{dt} = \rho_s^* b \nu_D (\xi\rho)^{-1/2}\left(\exp\left[-\frac{U-\tau_s b^2 l_{cp}}{kT}\right]+\exp\left[-\frac{U+\tau_s b^2 l_{cp}}{kT}\right]\right),\\
&\frac{d\rho}{dt} = \dot{a}\left(C_1\frac{(\alpha G b)^2\rho}{\tau_f^0+\tau_f+\alpha G b\rho^{1/2}} + \frac{C_2 e^{-U_1/kT}+C_3 e^{-U_2/kT}}{G b \rho^{1/2}}\right) -\\
&\qquad - \min(r_a, [\theta\rho]^{-\frac{1}{2}})\frac{a_l^3 D_0}{\chi k T b}\left(\exp\left[\frac{-E_i}{kT}\right]\cdot C_i + \exp\left[\frac{-E_v}{kT}\right]\cdot C_v\right)\tau\rho^2\theta^2,\\
&\frac{dC_i}{dt} = \dot{a}\frac{1}{\kappa(\zeta)}p_j\xi B\frac{\tau}{G} - \frac{a_l^3 D_0}{\chi k T}\exp\left[\frac{-E_i}{kT}\right]C_i\tau\rho\theta - \mu_r D_0\exp\left[\frac{-E_i}{kT}\right]C_i C_v,\\
&\frac{dC_v}{dt} = \dot{a}\frac{1}{\kappa(\zeta)}p_j\xi B\frac{\tau}{G} - \frac{a_l^3 D_0}{\chi k T}\exp\left[\frac{-E_v}{kT}\right]C_v\tau\rho\theta - \mu_r D_0\exp\left[\frac{-E_i}{kT}\right]C_i C_v,\\
&\tau = const \quad (P = const).
\end{aligned}\right. \tag{1}$$

The materials ductile deformation is known to be realized by the dislocation movement thus forming shear zones. In the model for the alloys with $L1_2$ superstructure all dislocations are divided into peripheral and intrazone dislocations. The former are retarded shear-forming dislocations at the periphery of the shear zone, the latter are dislocation barriers inside the shear zone. Dislocation barriers are subdivided into two types: i) Kear-Wilsdorf barriers formed due to thermoactivating selfblock of screw dislocations and ii) the barriers formed due to thermoactivating blocking of edge superdislocations when the trapping of point defects on them. The temperature increase of deformation causes the accumulation intensity increase of dislocation barriers [3, 4] that results in the moving dislocations density decrease. Therein superdislocations mobility not included in barriers configuration increase. All mentioned above conditions are taken into account in the first equation in the system (1) that gives the creep rate occurring through thermoactivating dislocations movement (2):

$$\frac{da}{dt} = \rho_s^* b v_D (\xi\rho)^{-1/2} \times \left(\exp\left[-\frac{U-\tau_s b^2 l_{cp}}{kT}\right]+\exp\left[-\frac{U+\tau_s b^2 l_{cp}}{kT}\right]\right). \tag{2}$$

where b is the modulus of the Burgers vector, ν_D is the Debye frequency, ξ is the fraction of forest dislocations, ρ is the dislocations density, U is the activation energy of dislocations thermo fluctuation movement along a slip plane, τ_s is the thermo activating fraction of deforming slip stress, $l_{aver} = l(1-0{,}7\exp(-U_1/kT))$ is the average length of free segment, l is the average length of free segment while Kear-Wilsdorf barriers absence, U_1 is the formation energy of Kear-Wilsdorf barriers, k is the Bolzman constant, T is the deformation temperature, $\rho_s^* = \rho[1-0{,}5(e^{-U_1/kT}+e^{-U_2/kT})]$ is the density of mobile dislocations dependent on temperature, U_2 is the formation energy of diffusion barriers.

The second equation (3) of the system (1) gives dislocations accumulation in the time of deformation period. Both the accumulation of periphery as well as intrazone dislocations and their character are considered [11-14]. The intrazone dislocations (Kear-Wilsdorf barriers and diffusion barriers) are formed both in low and high temperature areas but with different rates. Kear-Wilsdorf barriers are intensively formed and accumulated under low and moderate temperature up to 500 K [12-14], diffusion barriers are under higher temperature (due to the higher mobility of deformation point defects). With temperature increase diffusion barriers displace Kear-Wilsdorf barriers. This is taken into account in the second equation (3) of the system (1) by means of factors C_2, C_3:

$$\frac{d\rho}{dt} = \dot{a}\left(C_1 \frac{(\alpha G b)^2 \rho}{\tau_f^0 + \tau_f + \alpha G b \rho^{1/2}} + \frac{C_2 e^{U_1/kT} + C_3 e^{U_2/kT}}{G b \rho^{1/2}} \right) - \min(r_a, [\theta\rho]^{-\frac{1}{2}}) \frac{a_l^3 D_0}{\chi k T b}\left(\exp\left[\frac{-E_i}{kT}\right] \cdot C_i + \exp\left[\frac{-E_v}{kT}\right] \cdot C_v \right) \tau \rho^2 \theta^2. \quad (3)$$

Here C_1,C_2 and C_3 are constants, $\alpha = \alpha_0 - \beta T$ is the parameter characterizing interdislocation interaction, α_0, β are constants derived from experimental dependence $\alpha(T)$, for a certain $L1_2$ alloy, G is the shear modulus, τ_f^0 is the friction stress related to overcoming of the Peierls barrier, $\tau_f = \gamma_1 \tau_0^{(1)} \exp(-U_1/kT) + \gamma_2 \tau_0^{(2)} \exp(-U_2/kT)$ is the friction stress due to the formation of Kear-Wilsdorf barriers on screw dislocations and the trapping of point defects on edge dislocations, γ_1, γ_2 is contribution factor (for uniform expanding rectangular dislocation loop $\gamma_1 = \gamma_2 = 1/2$), $\tau_0^{(1)}$, $\tau_0^{(2)}$, α_0, β, are temperature independent constants, v is Poisson's ratio, θ is the edge dislocation fraction, a_l is the lattice parameter, D_0 is the preexponential factor, χ is the orientation factor, E_i and E_v is the activation energy of interstitial atoms and vacancies, respectively; C_i и C_v is the concentrations of interstitial atoms and vacancies.

The first term of the equation (3) determines dislocations [11)] accumulation and dislocation barriers [12]; the second one determines their annihilation.

The third (4) and the forth (5) equations of the system calculate the accumulation of deformation point defects: interstitial point atoms and vacancies, respectively.

$$\frac{dC_i}{dt} = \dot{a}\frac{1}{\kappa(\zeta)} p_j \xi B \frac{\tau}{G} - \frac{a_l^3 D_0}{\chi k T} \exp\left[\frac{-E_i}{kT}\right] C_i \tau \rho \theta - \mu_r D_0 \exp\left[\frac{-E_i}{kT}\right] C_i C_v, \quad (4)$$

$$\frac{dC_v}{dt} = \dot{a}\frac{1}{\kappa(\zeta)} p_j \xi B \frac{\tau}{G} - \frac{a_l^3 D_0}{\chi k T} \exp\left[\frac{-E_v}{kT}\right] C_v \tau \rho \theta - \mu_r D_0 \exp\left[\frac{-E_i}{kT}\right] C_i C_v \quad (5)$$

here p_j is the fraction of forest dislocations forming jogs, $\kappa(\zeta)$ is the coefficient taking into account the fraction of jogs generating point defects, ξ is the fraction of forest dislocations, B is the constant, μ_r is the temperature-independent coefficient of own interstitial atoms and vacancies recombination [15], D_0 is the pre-exponential factor.

The equations (4), (5) presuppose the point defects to be generated (the first term) due to the non-conservative dragging of jogs by means of screw dislocations or near to it orientation [16].

Annihilation of deformation point defects in the model (1) is defined by their trapping on edge dislocations and their interdependent annihilation [17], the second and the third terms in the equations (4), (5) respectively.

The presented model in the paper gives the calculations for a number of conditions: i) tension at the constant load, ii) compression at the constant load, iii) the constant stress.

For i) and ii) conditions the different degree of plasticity deformation localization in the sample was considered. Under outer applied constant load the tension is changed because of the sample section value change during its deformation. In case of homogeneous microscopic creep the stress depends on deformation and it seems to be equal: $\tau = \tau_0(1+\varepsilon)$ is in case of tension. $\tau = \tau_0(1-\varepsilon)$ is in case of compression. Under microscopic deformation localization more intensive change of sample section happens and that results in stress change in the creep zone. In this case the stress is as follows: $\tau = \tau_0(1 \pm K_3\varepsilon)$, where K_3 being the sample fraction where creep is confined.

3. Calculation results and discussion

The relation of deformation from time $\varepsilon(t)$ in case of creep of single crystal alloys with $L1_2$ superstructure is calculated according to the model (1). The parameters values needed for solving the system of equations were taken closely spaced to parameters values for Ni_3Ge alloy. The parameters values are as follows: $C_1 = 6\cdot10^4$ N/m^2; $C_2 = C_{12} - T\cdot C_{12}/1000$; $C_{12} = 5\cdot10^{16}$ N/m^4; $C_3 = T\cdot C_{13}/1000$; $C_{13} = 3\cdot10^{22}$ N/m^4 [12]; $U_1 = 0.0094$ eV; $U_2 = 0.72$ eV; $\tau_0^{(1)} = 300$ MPa; $\tau_0^{(2)} =$ 1700 MPa; $\chi=0.48$ [14]; $\alpha_0 = 1.8$; $\beta = 1.9\cdot10^{-3}$ [11]; $\tau_f^0 = 50$ MПa; $G = 8\cdot10^{10}$ N/m^2; $b = 2.5\cdot10^{-10}$ m; $a_l = \tau_f^0\, 2.5\cdot10^{-10}$ m; $D_0 = 7.7\cdot10^{-5}$ m^2/s; $E_i = 0.15$ eV; $E_v = 1.27$ eV; $E_i^0 = 1.6\,eV$; $E_v^0 = 4.8\,eV$; $\theta = 0.5$; $\nu = 1/3$; $\Gamma = 1$; $\kappa(\zeta) = 30$; $p_j = 0.5$; $\xi = 0.5$; $B = 340$; $\mu_r = 10^{19} m^{-2}$; $k_0 = 0.5$; $\gamma_1 = \gamma_2 = 0.5$.

The creep rate (the first equation of the system (1)) is the following:

$$\frac{da}{dt} = K_1 \rho_s^* b \nu_D (\xi\rho)^{-1/2} \exp\left[-\frac{U - K_2 \tau_s b^2 l_{cp}}{kT}\right] + K_1 \rho_s^* b \nu_D (\xi\rho)^{-1/2} \exp\left[-\frac{U + K_2 \tau_s b^2 l_{cp}}{kT}\right]$$

K_1 and K_2 values should be considered as adjusted parameters. Coefficient values of K_1 and K_2 used in calculation are given in the Table 1.

Table 1. Coefficient values of K_1 and K_2

U eV	K_1	K_2
3.4	$1.6*10^{-9}$	0.1
2.0	$1.0*10^{-14}$	0.08
1.0	$1.0*10^{-14}$	0.04

The calculation results with initial conditions

$a(0) = 0$, $\rho(0) = 10^{10}$ m^{-2},

$C_i(0) = \exp(-E_i^0/kT)$,

$C_v(0) = \exp(-E_v^0/kT)$, where $E_i^0 = 1{,}6$ eV

and $E_i^0 = 4{,}8$ eV is formation energy of interstitial atoms and vacancies are respectively given in Figures 1, 3, 4.

Figure 1 gives creep curves (Figure 1 *a* and *b*) and creep rate dependence on time (Figure 1 *c* and *d*) calculated in the model. Applied stresses were considered as $\tau_0 = 166.5$ MPa ($\sigma = 370$ MPa) and $\tau_0 = 450$ MPa ($\sigma = 1000$ MPa).

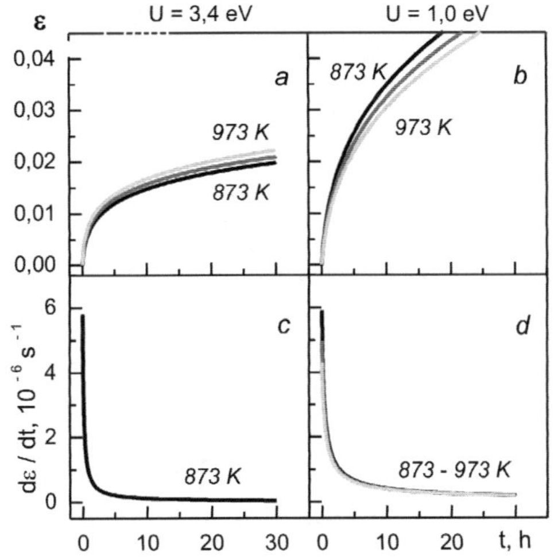

Figure 1. Calculated creep curves (*a*), (*b*) and time dependent creep rates (*c*), (*d*) for alloys with the $L1_2$ superstructure at different deformation temperatures, the activation energy of the thermoactivating superdislocation movement U, and the constant stress. Stress σ=370 MPa

Figure 2 gives the experimental creep curves (Figure 2 *a*) and the time dependence of creep rate (Figure 2 *b*) obtained for Ni_3Ge alloy single crystals oriented along direction [001] under its compression [18]. As well as in the experiments (Figure 2) calculated curves $\varepsilon\,(t)$ and $\dot{\varepsilon}\,(t)$ (Figure 1) demonstrate two stages of creep: the first stage is the stage of primary creep with continuous rate decrease, the second one being the stage of steady-state creep where the rate keeps stable.

The activation energy U in the range between 2.0 and 1.0 eV demonstrates anomalous creep rate dependence on temperature: temperature increase from 823 to 973 K results in creep rate decrease (Figure 1). Experimentally that situation is observed at the initial stage of the creep for Ni_3Ge alloy single crystals under the same initial load.

Load change to higher ones up to 1000 MPa (Figure 3) considerably changes the character of the creep curves due to the activation energy of the thermoactivating superdislocation movement U. During the test the linear creep with stable rate is observed in the range of activation energy U being between 1.0 to 2.0 eV. The creep anomaly characteristic to lower loads disappears in this case (Figure 3). Similar changes are observed in the experiments of Ni_3Ge alloy single crystals [18] (Figure 2.II).

I) II)

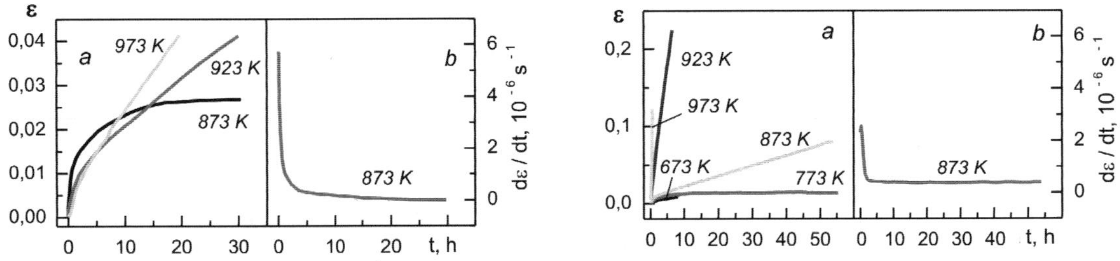

Figure 2. Experimental creep curves (*a*) and time dependent creep rates (*b*) at different deformation temperatures of the Ni_3Ge alloy single crystal [18]. The deformation axis [001]: (I) stress value σ = 370 MPa; (II) stress value σ =1000 MPa.

The consideration of deformation localization in the creep process also shows its significant influence on creep curves (Figure 4). The influence is only quantitative in the case of compression (Figure 4 *a* and 4 *b*): without changing the character of the creep stages it results to the creep rate decrease the higher the degree of localization is.

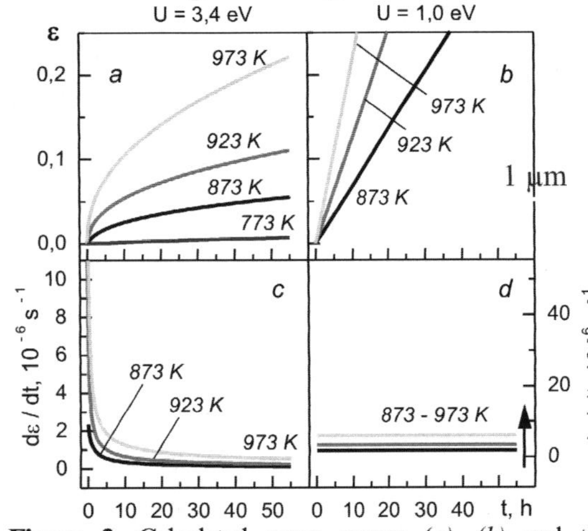

Figure 3. Calculated creep curves (*a*), (*b*) and time dependent creep rates (*c*), (*d*) for alloys with the $L1_2$ superstructure at different deformation temperatures, the activation energy of the thermoactivating superdislocation movement U, and the constant stress. Stress σ=1000 MPa

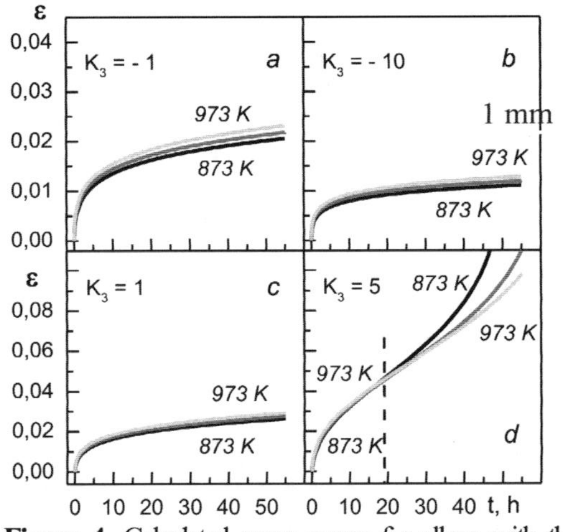

Figure 4. Calculated creep curves for alloys with the *L*12 superstructure at different deformation temperatures, the activation energy of the thermoactivating superdislocation movement U=3.4eV in case of taking into account deformation localization: (*a*), (*b*) compression, (*c*), (*d*) tension. Stress σ =1000 MPa.

4. Conclusions

Thus, calculations testify that the creep peculiarities characteristic to $L1_2$ alloys can be defined by the model where the accumulation peculiarities and self-locking are included. Competition of the self-locking and thermoactivating superdislocation mobility leads to the creep rate anomaly in examined superstructures. The high energy of the thermoactivated superdislocation movement and low density of free and mobile superdislocations determine the low creep rate in the $L1_2$ superstructure.

Deformation localization in the creep process may cause the appearance of the inverse creep stage.

The reported study was partially supported by RFBR, research project No. 14-02-92605 KO_a.

References

[1] Nicholls J R and Rawlings R D 1977 *J. Mater. Sci.* **12** 2456
[2] Hemker K J and Nix W D 1989 *Mater. Res. Soc. Symp. Proc.* **133** 481
[3] Hemker K J, Nix W D and Mills M 1991 J. *Acta Metall. Mater.* **39** 1901
[4] Hemker K J and Nix W D 1993 *Metall. Trans.* A. **24A** 335
[5] Rong T S, Jones I P and Smallman R E 1995 *Acta metal.* **43** 1385
[6] W H Zhu, D Fort, I P Jones, R E Smallman 1998 *Acta mater.* **46** 3873
[7] B H Kear, H G B Wilsdorf: Trans. Metallurg 1962 *Soc. AIME* **224** 382
[8] Yu V Solov'eva, B I Burtsev and V A Starenchenko 2005 *Izv. Vyssh. Uchebn. Zaved., Fiz.* 9 28 *(in Russian)*
[9] Starenchenko V A, Pantyukhova O D, Solov'eva Yu V, Burtsev B I and Starenchenko S V 2005 *Phys. Met. and Metallogr.* **100** 399
[10] Starenchenko V A, Pantyukhova O D and Starenchenko S V 2002 *Physics of the Solid State* **44** 994
[11] Starenchenko V A, Chernykh Yu A and Abzaev Yu A 1987 *Physics of Metal* **2** 22
[12] Starenchenko V A, Solov'eva Yu V, Abzaev Yu A 1999 *Physics of the Solid State* **41** 407
[13] Starenchenko V A, Solov'eva Yu V, Abzaev Yu A, Kozlov E V, Shpeĭzman V V, Nikolaev V I and Smirnov B I 1998 *Physics of the Solid State* **40** 618
[14] Starenchenko V A, Solov'eva Yu V, Abzaev Yu A, Nikolaev V I, Shpeĭzman V V and Smirnov B I 1996 *Physics of the Solid State* **38** 1668
[15] K P Gurov, A B Tsepelev 1991 *Nuclear Materials* **182** 240
[16] Starenchenko V A, Starenchenko S V, Kolupaeva S N and Pantyukhova O D 2000 *Russian Physics Journal* **43** 61
[17] Starenchenko V A, Pantyukhova O D, Solov'eva Yu V 2004 *The Physic of Metals and Metallography* **97** 545
[18] Solov'eva Yu V, Gettinger M V, Starenchenko S V and Starenchenko V A 2009 *Russian Physics Journal* **52** 390

IOP Conf. Series: Materials Science and Engineering **71** (2015) 012068 doi:10.1088/1757-899X/71/1/012068

Localization of deformation in [$\bar{1}23$] - monocrystals of aluminum under compression

L A Teplyakova[1], T S Kunitsyna[2], M A Poltaranin[3] and A D Kashin[4]
[1,4]Tomsk State University of Architecture and Building, Physics Department, Tomsk, 634003, Russia
[2]Tomsk State University of Architecture and Building, Mathematics Department, Tomsk, 634003, Russia
[3]Institute of Strength Physics of Materials Science SB RAS, Tomsk, 634021, Russia

E-mail: lat168@mail.ru

Abstract. The study of the deformation relief of aluminum monocrystals oriented for a single slip has been carried out using methods of optical and interference microscopy. It has been established that in the volume of a single crystal several macrofragments are formed, differing in size and direction of shear. Reasons for formation of two systems macrobands of shear are identified. It has been shown that formation of the deformation relief in the investigated crystal occurs at three scale-structural levels.

1. Introduction

Monocrystals of metals and alloys with FCC lattice, oriented for a single slip, are one of the most difficult objects in revealing the laws of plastic strain and hardening. The complexity is caused by the pronounced inhomogeneity of strain behavior in such single crystals due to the anisotropy of shear stresses in slip systems [1]. It is the main reason in the fact that single crystals with orientations in the center of a standard stereographic triangle are poorly studied at all scale-structural levels, including the macro level. In recent years, for a number of single crystals of metals and alloys it has been established that the inhomogeneity of plastic strain behavior under compression is manifested, primarily, in formation of shear macrofragments [2-6]. At the same time, the laws of microfragmentation depend on the availability in a single crystal of possibilities for shear carriers to emerge onto free faces of the crystal. The volumes of the crystal in the first case became known as volumes of the lightweight shear (VLS), in the second – the volumes of the constrained shear (VCS). Usually, in single crystals with VLS the localization of the shear strain occurs in these volumes from the beginning of the plastic flow, resulting in formation of slip macrobands in single crystals [2, 5, 7]. It is natural that in single crystals oriented for a single slip there is always a volume of the lightweight shear. It occupies a large part of the crystal, and a significant localization of the shear strain with formation of a distinct slip macroband is expected, as it is observed, for example, in [1.8.12] - single crystals of Ni_3Fe [2]. The present work is devoted to the study of laws of the plastic strain macrolocalization in aluminum single crystals oriented for a single slip.

2. Materials and methods

Aluminum single crystals of commercial purity have been investigated in the work. Single crystals had the shape of a parallelepiped and were oriented for compression along the crystallographic direction $[\bar{1}23]$, i.e. for a single octahedral slip. The side faces of the crystal were parallel to the planes $(3\bar{6}5)$ and (210).

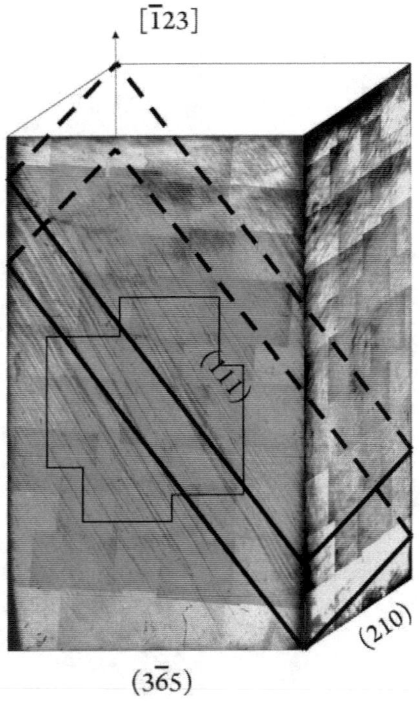

Figure 1. Strain relief of the crystal surface.

At the chosen orientation of the single crystal the primary slip system was the system (111) [110] with a Schmid factor m=0.47. The direction of the primary slip in the crystal had an orientation at which it was possible to expect an emergence of edge dislocations on the face (210), and screw dislocations on the face $(3\bar{6}5)$. In crystals with a selected crystallogeometric setting, for families of primary (111) and conjugate $(\bar{1}11)$ slip planes it is possible to allocate the volumes of the lightweight shear. For the primary plane, the volume of the lightweight shear is shown in Figure 1. Strain of the single crystal was carried out under compression at room temperature with the velocity of $1{,}5 \times 10^{-2}$ s^{-1} using the device “Instron” up to strain degree of d ε=0,06. Shooting of the strain relief was carried out using an optical microscope MIM-10 with a 250-fold magnification. With the help of special output devices the record of optical images of the strain relief was carried out directly on the computer. Along with the metallographic method, the study of the strain relief was carried out using optical (interference) profilometer Zygo New View 6000.

3. Results and Discussions

3.1. Fragmentation of shear strain at the macro-level

Figure 1 shows optical images of the strain relief formed by two mutually perpendicular lateral faces of the investigated aluminum single crystal. Although, in general, reliefs on the faces (210) and $(3\bar{6}5)$ differ considerably, the main components of the relief at the macro-level are: the system of primary slip lines and two subsystems of large-scale bands of the localized shear. Tracks of the primary slip are typical elements of the strain relief. Their average length is 320 μm. The most striking part of the strain relief is coarse shear bands. They can be observed on one of these faces in Figure 1. All bands are combined into two groups. Bands of the first group are concentrated in the lower bottom part of the face and begin at the left vertical edge, and the bands of the second group are localized in the upper right part of the face and begin from the right vertical edge. Most bands taper as they move away from the edge. The length of strips varies in the range of 0.5...5.0 mm; the width is 20...200 μm. The broadest bands begin in the volume of the lightweight shear of the primary system (Figure 1, for example, at the upper left and the lower right corner of the face). However, during the promotion of bands of both subsystems depthward the single crystal, they deviate from primary slip lines towards corresponding vertical edges of the single crystal at angles 4-8°. As a result, the bands that "deviated" to the opposite sides of these two systems become approximately parallel to each other. This is clearly seen in Figure 2, which is an enlarged part of the surface of the face $(3\bar{6}5)$ shown in Figure 1.

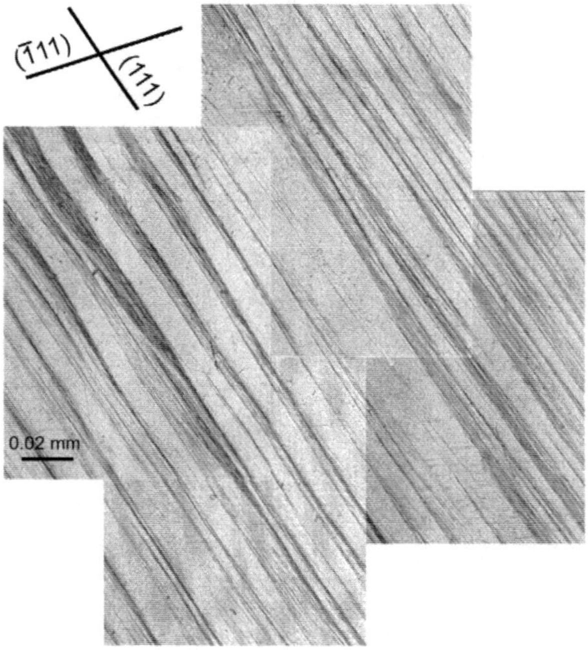

Figure 2. Enlarged images of the area highlighted in Figure 1.

As noted above, along with the optical study, the study of the strain relief for all free faces of the single crystal has been carried out in the work with the use of an interference profilometer. The corresponding photographs of two parallel faces with indicated systems of parallel bench marks are presented in Figures 3 and 4.

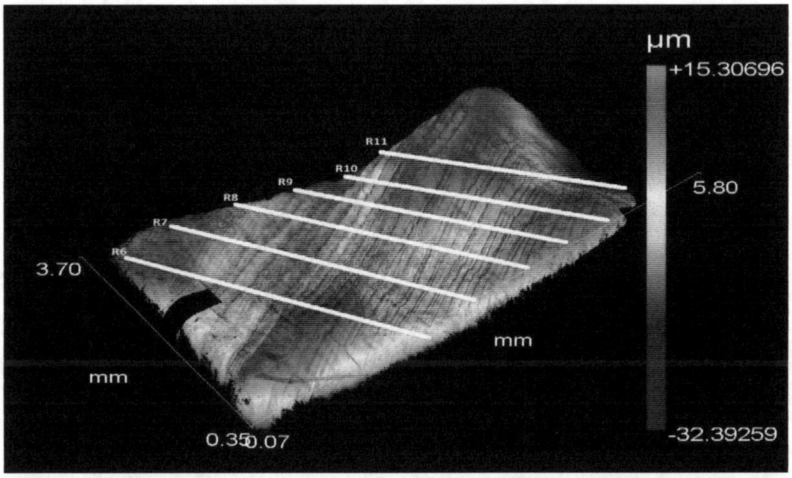

Figure 3. Surface images of faces $(3\bar{6}5)$ obtained using the interference microscopy method. The lines of bench-marks, along which the profiles of the strain relief were photographed, are indicated.

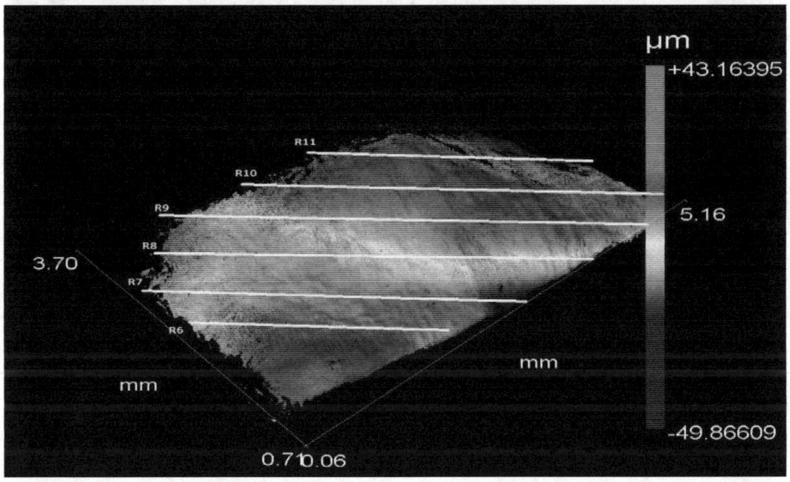

Figure 4. Surface images of faces $(\bar{3}6\bar{5})$ obtained using the interference microscopy method. The lines of bench-marks, along which the profiles of the strain relief were photographed, are indicated.

From the analysis of the relief it follows that in consequence of the plastic strain (ε = 0,06) several distinct shear macrofragments are formed in the volume of the single crystal. Their linear dimensions vary in the range of 1...6 mm. Figures 3 and 4 give an idea on the shape of macrofragments. As can be seen from the scale in Figures 3 and 4, the maximum spread of the height of the relief elements on faces $(3\bar{6}5)$ is 50...60 μm. On faces (210) – it is twice as large. On boundaries of macrofragments the difference in heights of the relief does not exceed few micrometers. At the same time, the difference in heights occurs smoothly. As a result, almost all borders of macrofragments of the shear strain at ε = 0,06 are blurred, i.e. they are boundaries with continuous (smooth) misorientation. From the comparison of Figure 1, 3 and 4 it follows that at the macro level the largest in size changes occur in VLS for primary slip planes and in areas adjacent to its vertices. At the same time, in the central part of VLS on one of the faces $(3\bar{6}5)$ a convex (red area in Figure 3) is observed, and on the opposite side – a concave (green area in Figure 4). This means that in the central part of VLS there was a shear along the primary slip system. As a result, there is a definite correlation between the shape of macrofragments and the shape of VLS for the primary plane in its central part. However, in near-end areas of VLS such correlation is absent. In general, in a single crystal the main part of shear bands is located outside VLS for the primary slip system.

3.2. Shear bands

Figure 5 and 6, as an example, illustrate profiles of the strain relief that correspond to bench-marks R6 and R9 indicated on images of the relief (Figure 3). From the analysis of profilograms it follows that the relief, formed during plastic strain, can be characterized as a periodic, three-level. The first one, the largest scale-structural level is associated with formation of shear macrofragments. The scale of this level is millimeters, which corresponds to linear dimensions of macrofragments. The height of elements of the first level is on average few micrometers. The periodicity of the relief profile at the second scale-structural level on faces $(3\bar{6}5)$ can be easily traced for all bench-marks, including those presented in Figure 5, 6. Formation of the second scale-structural level is associated with formation of macrobands of the localized strain (Figure 1 and 2) and, of course, the scale of this level is determined by their quantitative characteristics. The period of the relief at the second level is 150-250 μm, which corresponds to the thickness of slip macrobands. The height of elements of the relief at this level varies in the range of 1-5 μm.

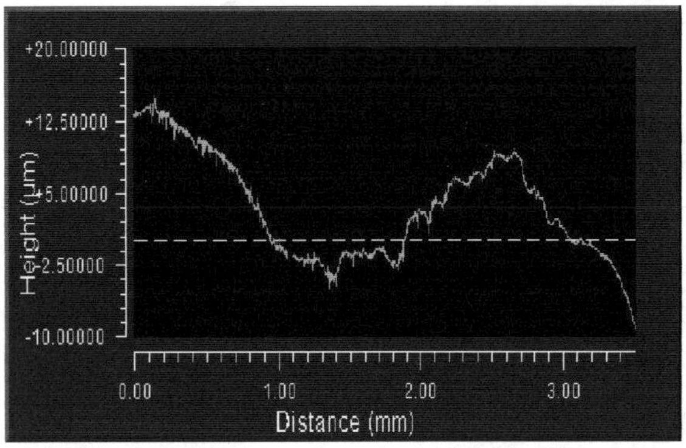

Figure 5. Profiles of strain relief, corresponding to the bench marks R6, at the surface of the face $(3\bar{6}5)$ (Figure 3*a*).

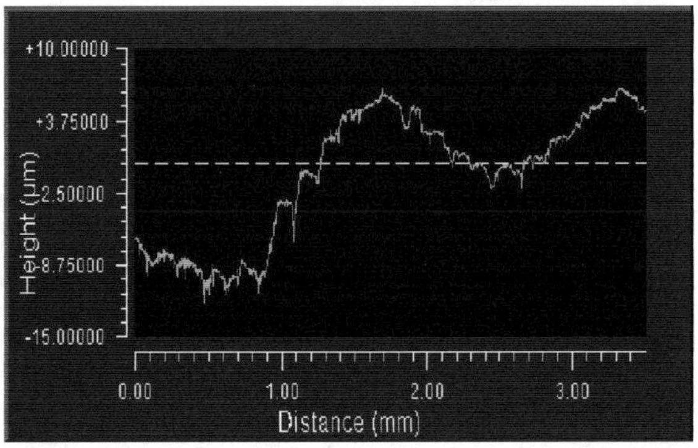

Figure 6. Profiles of strain relief, corresponding to the bench marks R9, at the surface of the face $(3\bar{6}5)$ (Figure 3*a*).

4. Conclusion

The investigated in the work $[\bar{1}23]$ - single crystal of aluminum in its crystallogeometry falls into the category of FCC single crystals with a volume of the lightweight shear. As a rule, macrolocalization of plastic strain with formation of shear macropacks in VLS takes place under uniaxial compression of such single crystals up to small degrees of strain. This mechanism of shear strain localization is characteristic of single crystals of aluminum and alloy Ni_3Fe, oriented for a symmetrical slip along close-packed planes [6, 8, 9]. A condition necessary for formation of shear macropacks is: the presence of a common edge (linear concentrator) and/or vertices (point concentrator) in the crystal and the volume of lightweight shear in it [9]. In the investigated aluminum single crystal these conditions are met. Meanwhile, even though macrolocalization of shear strain in VLS takes place the typical shear macropacks are not observed. As mentioned above, in VLS $[\bar{1}23]$ - aluminum single crystal the macrolocalization takes place, but it is achieved by a slip at the mesoscopic level rather than at the macroscopic as in the abovementioned crystals. It is to be supposed that the main reason for another mechanism of strain localization in aluminum single crystal oriented for easy slip are significant frictional force between the end faces of the sample and the testing machine punches. In crystals of the same metal but with symmetric even-numbered orientations ([110] and [001]) the mechanical stability of single-crystal samples at the macro level was achieved by two conjugate families of an active octahedral slip planes. In this case, forces of mechanical friction were compensated symmetrically. In a single crystal with an asymmetric orientation this mechanism is impossible. Due to friction at the ends, geometrically necessary dislocations gradually accumulate in the crystal (excess density of

IOP Conf. Series: Materials Science and Engineering **71** (2015) 012068 doi:10.1088/1757-899X/71/1/012068

dislocations), and long-range elastic-plastic stresses appear in the volume of the crystal. Their relaxation takes place upon reaching the critical density of such dislocations and, as a consequence, high local stresses. It happens with formation of multiple shear bands in the scale of the sample. Although formation of the most representative bands begins from the areas of lateral edges located in VLS of the crystal, most of them lie outside of VLS.

References

[1] Honeycombe R W K 1972 *The Plastic Deformation of Metals* (Moscow: World) p 408
[2] Teplyakova L A and Kozlov E V 2005 *J. Phys. Mes.* **8** 6 57-66
[3] Ki-Hyin Kim, Heung-Hyin Kim, Soo-In Oh 2006 *J. Mater. Proc. Technol.*205-213
[4] Lychagin D V 2006 *J. Phys. Mes.* **6** 3 103-113.
[5] Zuev L B, Danilov V I, Barannikova S A 2008 Physics macrolocalization of plastic (Novosibirsk: Science) p 328
[6] Teplyakova L A, Bespalova I V, Kunitsyna T S 2014 . *Adv. Mater. Res.* **1013** 54-61
[7] Harren S V, Deve H E, Asaro R J 1988 *J Acta Met.* 36 **9** 2435-2480
[8] Teplyakova L A and Kunitsyna T S 2014 *J. Proc. of H. Education. Phys.***57** 2 61-70
[9] Bespalova I V and Teplyakova L A 2014 *J. Proc. of H. Education. Phys.***57** 2 85-90

IOP Conf. Series: Materials Science and Engineering **71** (2015) 012069 doi:10.1088/1757-899X/71/1/012069

Spall destruction of coarse-grained and ultrafine-grained titanium on exposure to nanosecond heavy-current electron beam

S Afanasyeva[1], G Bakach[1], N Belov[2], E Dudarev[1], A Markov[3], A Tabachenko[1], M Khabibullin[1] and N Yugov[1]
[1]National Research Tomsk State University, Tomsk, 634050, Russia
[2]Tomsk State University of Architecture and Building, Tomsk, 634003, Russia
[3]Institute of Heavy-Current Electronics of the Siberian Branch of Russian Academy of Sciences, Tomsk, 634055, Russia

E-mail: s.a.afanasyeva@mail.ru

Abstract. This paper presents results of experimental and theoretical research of spall destruction of volume coarse-grained and ultrafine-grained titanium when using it as a generator of shock wave of nanosecond relativistic heavy-current electron beam. Computer modeling of effect of an intensive electron beam on the condensed target has been carried out taking into account destruction, phase transitions, dependence of strength characteristics of materials on internal energy. This is considered within elastic, ideally plastic Prandtl-Reiss model. The results of calculations are compared with experimental data.

1. Introduction

Nonconventional sources of thermo-mechanical impact on solid bodies: electron and ionic beams, lasers, x-ray radiations are promising in high pressure physics for research of deformation behavior of materials and constructions at intensive pulse loads [1] and to investigate the spall fracture of metallic materials along with high-speed collision of plates using shock waves generated by the impact of high-current ion and electron beams [2,3]. Experimental and theoretical research of deformation behavior and destruction at shock-wave loading of only small amount of ultrafine-grained metals and alloys have been so far conducted. It turned out that when ultrafine-grained structure is formed, spall strength of some metals and alloys increases, and that of others, on the contrary, decreases. However, the mechanisms of plastic deformation responsible for such different change of spall strength under transition from coarse-grained to ultrafine-grained structure have not been explored yet. This requires additional studies on shock-wave dynamics, deformation behavior, evolution of dislocation substructure and the mechanism of spall fracture of both grain structures.

In the present work, a similar research has been done for ultrafine-grained and coarse-grained VT1-0 titanium. As a generator of shock wave, the SINUS - 7 accelerator was used radiating a nanosecond relativistic heavy-current electron beam: electron energy 1,35 MeV, pulse duration 45 ns, power density $3{,}4 \cdot 10^{10}$ W/cm^2 [4]. The feature of this beam is volume nature of energy release that makes it possible to study spall fracture of massive targets. Targets were ultrafine-grained (the average grain size $d < 1$ μm) and coarse-grained ($d > 10$ μm) disks of diameter ~ 20 mm and thickness from 1,67 to 4,05 mm.

Calculations have been carried out within the mathematical model of continuum mechanics [5].

2. Experimental data

Fast heating of the target has volume character and leads to formation of high pressure zone, and the compression wave starts propagating deep into targets. It gradually fades because of interaction with unloading waves from free front surface of the target. After shock wave comes to free back surface of a solid, there is a reflected centered depression wave. At the same time the back surface gets speed. Spall fracture is a consequence of interaction of counter unloading waves (if the tension stress amplitude exceeds dynamic strength). Material destruction causes tension stress value to drop to zero, there arises compression wave (spalling pulse), propagating on the stretched substance from destruction plane to free back surface of solid. By the time the spalling pulse came to surface, the falling depression wave had reduced its speed. Surface speed increases again in the front of the spalling pulse, and in the course of subsequent repeated reflections it undergoes damped oscillations.

It was experimentally established that at the specified parameters of electron beam, when irradiating ultrafine-grained and coarse-grained titanium targets of thickness ranged from 1 to 6 mm, the crater was formed on the face of targets which was the consequence of material ablation. Table 1 gives dependence of thickness of the spalled layer on thickness of the irradiated sample of the ultrafine-grained and coarse-grained VT1-0 titanium. At the same thickness of a sample, the thickness of the spalled layer is approximately identical to coarse-grained and ultrafine-grained structures.

Table 1. Dependence of thickness of spalled layer on sample thickness.

	Sample thickness (mm)	2,08	3,04	4,05
Coarse-grained state	Experimental spall thickness (mm)	0,24	0,32	0,37
	Calculated spall thickness (mm)	0.26	0,32	0,38
	Sample thickness (mm)	1,67	2,30	3,40
Ultrafine-grained state	Experimental spall thickness (mm)	0,23	0,31	0,35
	Calculated spall thickness (mm)	0,25	0,29	0,33

Researches of fracture surfaces of coarse-grained and ultrafine-grained titanium plates using the scanning electronic microscope are shown in Figures 1 and 2.

On the surface of spallation fracture of coarse-grained 2,08 mm thick titanium plate (Figure 1 *a*) there are practically no cracks normal to fracture surface.

A great number of material chipping areas are observed as result of crack extension at small angles to fracture surface. The basic elements of the relief are dimples of ductile cleavage whose average size (~ 8 μm) is approximately three times more than the size of the same cleavage dimples in ultrafine-grained titanium plate of the same thickness. On cleavage crests in many fracture fragments (the average size ~ 13 μm) there are cleavage steps that indicates the presence of a brittle component in the overall picture of spallation fracture.

On the surface of spallation fracture of ultrafine-grained 1,67 mm thick titanium plate (Figure 2 *a*) the general relief of fracture surface is rather uniform, with large-scale fracture fragments from 25 to 100 μm in size. The fragments are located on different levels relative to surface and formed at origin, concentric growth and coalescence of cracks. There is a number of short cracks, which are almost normal to fracture surface, and material exfoliation. The surface of large-scale fragments consists of cleavage dimples of various depth, with average size of dimples ~ 2,5 μm. A part of them is almost flat that prove a small plastic cleavage deformation in the area. Along with cleavage dimples there are larger fragments with cleavage steps on the surface. It indicates the presence of a brittle component when forming such fraction fragments. On the surface of some flat cleavage dimples there are very small (up to 0,5 μm and less) deep dimples or pores.

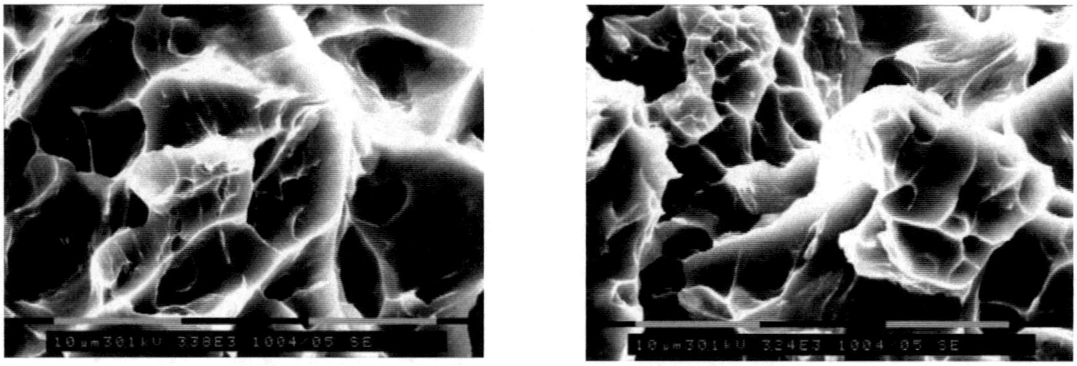

a *b*

Figure 1. The surface of spallation fracture of coarse-grained VT1-0 titanium formed on exposure to nanosecond relativistic heavy-current electron beam on the plate of thickness: *a* – 2,08 mm; *b* – 3,04 mm.

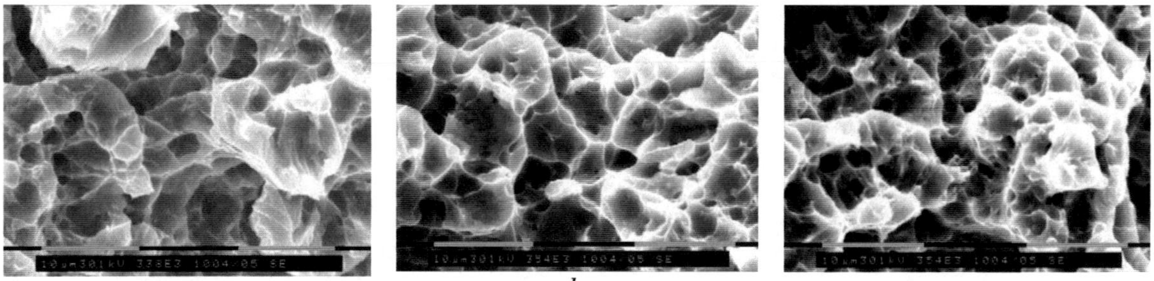

a *b* *c*

Figure 2. The surface of spallation fracture of ultrafine-grained VT1-0 titanium formed on exposure to nanosecond relativistic heavy-current electron beam on the plate of thickness: *a* – 1,67 mm; *b* -2,3 mm; *c* – 3,4 mm.

A characteristic difference in spallation fracture of the coarse-grained and ultrafine-grained titanium is different size of cleavage cells. Dimple diameter of ultrafine-grained titanium is less than that of the coarse-grained one but their quantity is more.

3. Mathematical modeling

The results of mathematical modeling of irradiation of VT1-0 titanium samples at electron energy 1,35 MeV, pulse duration 45 ns, and power density $3{,}4 \cdot 10^{10}$ W/cm^2 are given in Figures 3 - 5. Parameters of the mathematical model [1] titanium VT1-0 are given in Table 2, where μ_m - the shear modulus; σ_{sm} - yield strength; ρ_0 - initial density; c_0 - the speed of sound; γ_0 - thermodynamic Gruneisen coefficient; R_μ - the universal gas constant, A - atomic weight; T_{n0} - the melting point of the initial isochor; $T_0 = 293$ °K; α_0 - initial porosity; α_{00} - residual porosity; α_* - critical porosity; - ψ_1 ψ_1 ψ_3 ψ_4 a_s - constants of the material.

Table 2. Constants titanium VT1-0, included in the mathematical model.

μ_m, GPa	σ_{sm}, GPa	ρ_0, g/sm^3	c_0, m/s	ψ_1	ψ_2	ψ_3	ψ_4	γ_0
48,0	0,6	4,51	5110	3,465	1,44	0,479	3,52	1,33
$3R_\mu T_0/A$, KJ/kg	T_{n0}, °K	α_0	α_{00}	α_*	a_s, GPa			
153,0	2651	1,0001	1,0001	1,43	0,42			

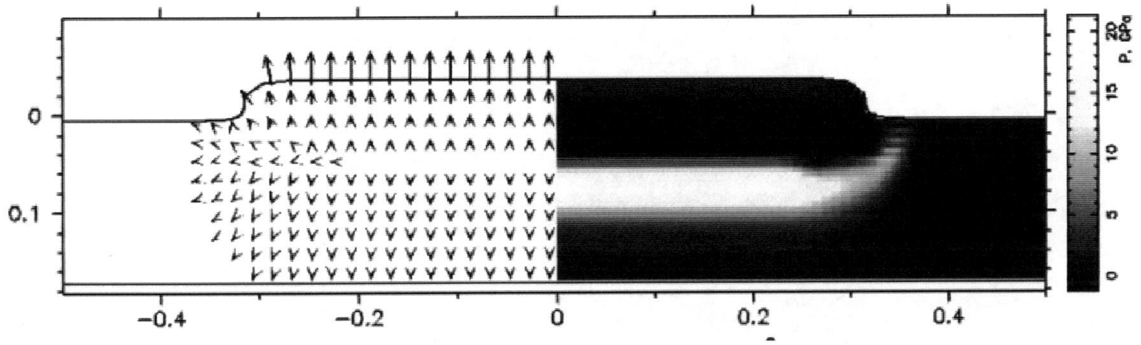

t = 0,12 μs, $U_{\max}$ = 4730 m/s

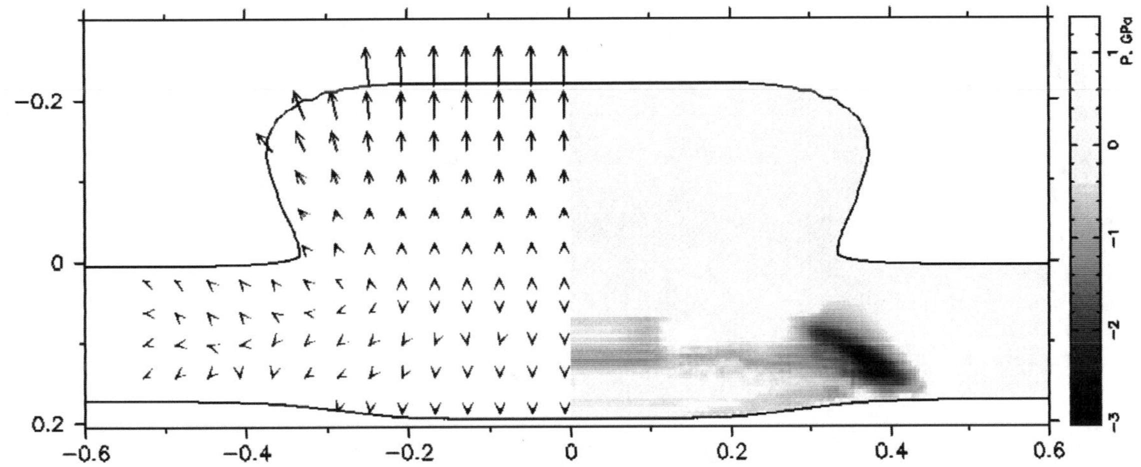

t = 0,42 μs, $U_{\max}$ = 7179 m/s

Figure 3. Flow field on irradiation of ultrafine-grained 1,67 mm thick titanium.

The presented calculations involve the following: energy-absorbing zone radius – 3 mm, linear coefficient of absorption - 46 cm^{-1}, $U_{\max}$ – modulus of the longest vector. In the right half-plane of figures the pressure field (GPa) is shown. A zone of projecting particles of the melted and evaporated material is formed near the high bound of the target. At the projection of substance the recoil momentum propagates in the solid part of the material. Its interaction with the free back bound of the sample leads to formation of a zone of spall back fractures.

The stress profile σ_{11} from time t is given for the sample section which is originally remote from the front surface at the distance of 0,0125 cm. In this section the maximum stress ($\sigma_{11} \approx 35$ GPa) is realized. The stress distributions σ_{11} (GPa) and relative void volume ξ (%) along the sample axis characterize the occurrence of spall fracture. Tension stresses emerge in the spall zone. The porosity grows at the front surface of the target in the projection zone of liquid particles and in place of spall formation.

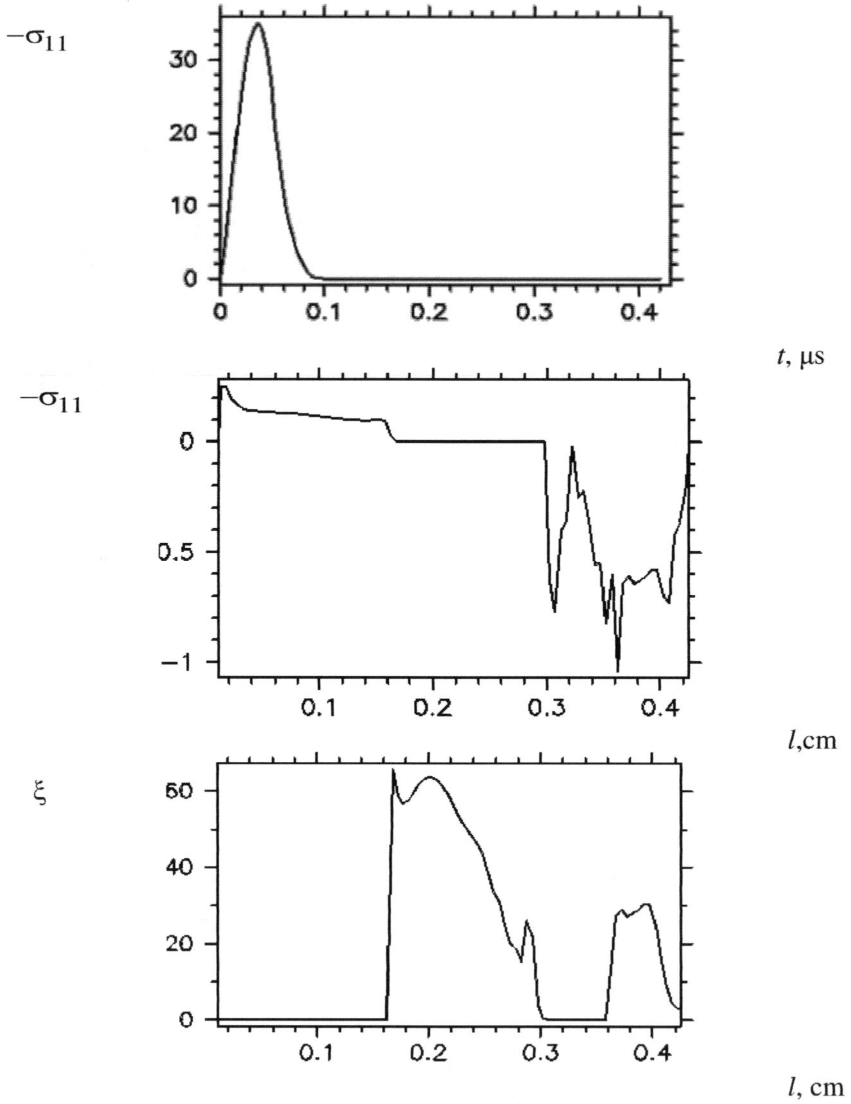

Figure 4. Dependence of stress $-\sigma_{11}$ on time, stress distribution $-\sigma_{11}$ and relative void volume ξ along the sample axis of ultrafine-grained VT1-0 titanium at 0,42 µs.

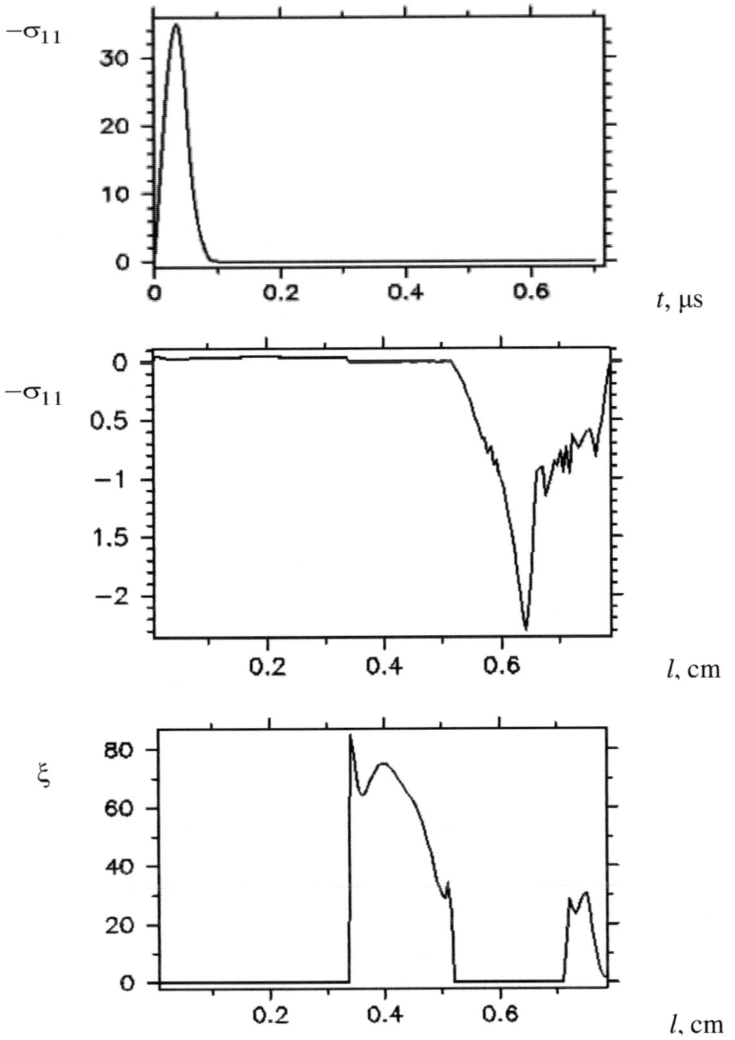

Figure 5. Dependence of stress $-\sigma_{11}$ on time, stress distribution $-\sigma_{11}$ and relative void volume ξ along the sample axis of coarse-grained 3,04 mm thick titanium at 0,7 μs.

Acknowledgement
The present work is performed in the framework of the Programme of raising competitive capacity of National Research Tomsk State University.

References

[1] Kanel G, Razorenov S, Utkin A and Fortov V 1996 *Shock-wave phenomena in the condensed media* (Moscow: Yanus-K) 408
[2] Antoun T, Seaman L and Currant D *et al.* 2003 *Spall fracture* (N. Y.: Spinger Verlag) 404
[3] Baumung K, Bluhm H, Kanel G *et.al.* 2001 *Intern. J. of Impact Engineering* **25** 631-639
[4] Dudarev E, Markov A, Tabachenko A *et al.* 2007 Spall fracture of coarse-grained and ultrafine-grained aluminum on exposure to nanosecond relativistic heavy-current electron beam *Izv. vuzov. Physics* **12** 32-38
[5] Afanaseva S, Belov N, Dudarev E *et al.* 2013 Target destruction on exposure to nanosecond relativistic heavy-current electron beam *Izv. vuzov. Physics* **12** 80-88

IOP Conf. Series: Materials Science and Engineering **71** (2015) 012070 doi:10.1088/1757-899X/71/1/012070

Stress-strain state simulation of large-sized cable-stayed shell structures

S Ponomarev[1], A Zhukov[1], A Belkov[1], V Ponomarev[1], S Belov[1] and M Pavlov[1]
[1]National Research Tomsk State University, Research Institute of Applied Mathematics and Mechanics, Tomsk, 634050, Russia

E-mail: belovsv@niipmm.tsu.ru

Abstract. This paper studies the opportunities for applying framed cable-stayed shell structures to generate innovative structures in civil engineering. Numerical solution methods for stress-strain state problems of these kinds of geometrically nonlinear structures were developed. Developed methods efficiency is presented by a range of large-dimensional space antenna reflectors.

1. Introduction

At the moment, civil engineering is influenced by distinct tendencies in generating ultra-lightweight transformable structures of large dimensions, having the potential to sustain heavy loads. The application “tensegrity” is evidence for popularization of framed cable shelled structures (FCSS).

Tensegrity is a structural principle based on the use of prestressed and compressed elements. The term was coined by scientist and architect Richard Buckminster Fuller [1]. Structures of this kind became widely used in architecture, art, urban construction, the aerospace industry, architectural hardware, and biomechanics [2]. In some cases, these structures are of hundreds of meters in characteristic dimension. The experimental development of large-sized structures requires expenditures in material and time. For this reason, computer - generated simulation takes center stage in the design and development of these structures. It allows tracing the process of structure generation, its settings, etc. and predicting its behavior under different conditions.

In order to provide the most efficient performance of every acting element of the structure, it is necessary to carry out a careful review of the stress - strain state of the whole structure. This creates the need to generate the proper structural model, not merely carry out a limited examination of representative elements. At another point, the analysis of structures of large dimensions composed of cable rods, and thin-walled shells is obstructed by the geometrical nonlinearity of these kinds of systems, which creates the need to develop an approach permitting the numerical analysis of the above described structures.

2. Application of cable-stayed shell structures in the building industry

2.1. Examples of tensegrity structures

Figures 1 - 4 show tensegrity structures used to create space antennas and civil engineering facilities [3, 4]. The advantage of the structures shown is their small volume in folded position, relatively small design and material costs, and eco-friendliness. The advantage of space antennas design is the ability to withstand heavy loads at launch.

Figure 1. Deployable space antenna AstroMesh.

Figure 2. The Millenium dome in London.

Figure 3. Roof of the stadium "La Plata Stadium" in Argentina.

Figure 4. Frame cable-stayed bridge in Australia.

2.2. Shell structures

Shell structures represent a surface stretched by elastic cords or special power frame made of rigid bars and beams (Figures 1,2,3).

The level of prestressing in the surface of the shell should be sufficient within the operational lifetime to maintain the level of tension and the shape of the object at one hand, and to allow the shell material to remain in the elastic deformation zone on the other. Under the influence of loads such as wind or snow, the tension on the surface may be increased by 6-8, even 10 times. For these reasons, the shell material must have a prestress equal to 1/20 of breaking strength.

The strength of the material may change due to changes in temperature, humidity, and creep. Therefore, the production requirements of the materials' mechanical properties are constantly verified using special tests. In general, research is necessary to find material which is not expensive, fire-resistant, and easy to use and transport [5].

When designing space antennas, apart from the effect of prestressing and the temperature on the reflecting surface, it is very important to control such a geometric characteristic as a reflecting surface root-mean-square deviation (RMSD). RMSD should be within 2 - 3% of the operating wavelength. Such restrictions impose more stringent requirements on antenna design and material properties of the reflecting surface than on some civil engineering projects.

Materials for membranes can be made of canvas, fiber glass, and organic synthetic fibers such as polyester. At the moment there is a large selection of materials on the market, but the most commonly used are polyvinyl chloride (PVC), polytetrafluoroethylene (PTFE) , and glass silicone coating. For example, the Millennium Dome, located on the Greenwich Peninsula in south-east London (Figure 2), was originally planned to be made of PVC. However, to extend its operational life, it was made of PTFE [5].

For the manufacturing of the reflective surface of space antenna AstroMesh, a thin conductive mesh made of molybdenum wire with diameter 0.03mm, coated with gold is used. [6]

3. Physical-mathematical model of stress-strain state of cable-stayed shell structures

There is considerable number of research on physical and mathematical models of FCSS. For example, Correa [7] deals with the formulation and solution of the problem of finding the FCSS state of tensegrity structures affected by external forces, using the virtual work. Murakami [8, 9] describes static and dynamic analyses of these kinds of structures, as well as the quasi-static approach to their analysis.

The authors consider modeling designs from a perspective of nonlinear elasticity theory. Shell elements are modeled by a thin membranous screen. FCSS extended structural elements, such as ropes, bars and other elements, were modeled by the core elements with effective characteristics, which give these model elements the same stiffness properties of the model as in real structural elements.

Stress-strain state of FCSS can be described by a stationary nonlinear system of equations of elasticity theory

$$\frac{\partial}{\partial x_k}\left(\sigma_{kj}\left(\delta_{ij}+\frac{\partial u_i}{\partial x_j}\right)\right)=0 \tag{1}$$

$$\varepsilon_{ij}=\frac{1}{2}\left(\frac{\partial u_i}{\partial x_j}+\frac{\partial u_j}{\partial x_i}+\frac{\partial u_k}{\partial x_j}\frac{\partial u_k}{\partial x_i}\right) \tag{2}$$

$$\sigma_{ij}=\frac{E_m}{2(1+\nu_m)}\left(\varepsilon_{ij}+\frac{\nu_m}{1-2\nu_m}\delta_{ij}\varepsilon_{kk}\right)+\sigma_{ij}^0\ , \tag{3}$$

where δ_{ij} – Kronecker index; u_i, σ_{ij}, σ^0_{ij}, ε_{ij} – second Piola-Kirchhoff stress, prestressing tensor, strain tensor; E_m, ν_m – modulus of elasticity and Poisson ratio of m-material.
Boundary conditions are as follows:

$$u_i(\bar{x})=u_{i0},\quad \bar{x}\in S_1 \tag{4}$$

$$n_k\sigma_{kj}\left(\delta_{ij}+\frac{\partial u_i}{\partial x_j}\right)=p_i^n,\quad \bar{x}\in S_2\ , \tag{5}$$

where $S=S_1+S_2$ – domain boundary Ω FCSS; p_i^n – boundary stress S_2, characterized by the normal vector $\bar{n}$, u_{i0} - boundary displacement S_1 [10].

4. Numerical solution to the problem of the stress-strain state of FCSS

Problem (1) - (5) was solved with the finite element method using the software package ANSYS [10]. In the numerical solution of equations (1) - (5), a major role is played by the choice of the initial approximation. Since the convergence domain is small, it is difficult to determine a good initial approximation which makes it possible to obtain a solution to the stationary problem. Therefore, we make a sequence of solutions where each new solution uses the previous one as the initial approximation (Figure 5).

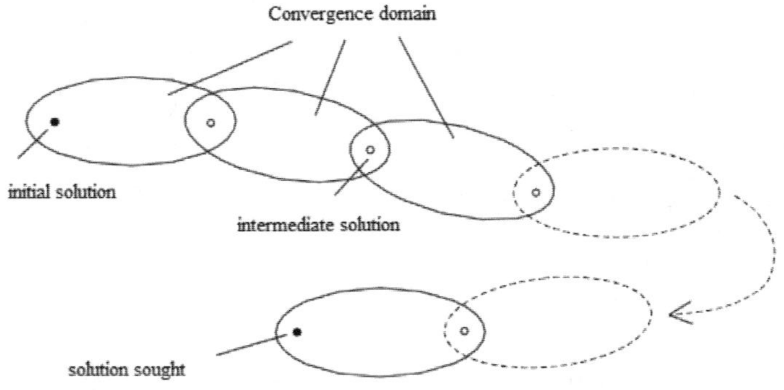

Figure 5. The sequence of solutions used to find the sought.

Moreover, the solution sought would be the last in this sequence. Solutions correspond to different boundary conditions. For the initial solution, an additional boundary condition is applied:

$$u_i(\bar{x}) = 0, \quad \bar{x} \in \Omega \tag{6}$$

Condition (6) corresponds to the full consolidation of FCSS. The process of solving will continue until the solution with the boundary conditions (4) and (5) is worked out. This procedure was developed by S. V. Ponomarev and V. A. Solonenko and used for numerical simulation of a single umbrella reflector of a large dimension. The procedure for finding the initial state does not have a clear algorithm yet, so changing the design of FCSS can lead to a substantial change in the sequence of removal of restraints [11].

5. Numerical models of cable-stayed shell structures on the example of large-sized space reflectors

When modeling large-sized space reflector the assumptions for application of finite element method are introduced:

• the dimension of the problem decreases; for example, for a reflective surface with a thickness on the order of a millimeter we can assume that the variables of the problem are constant throughout the thickness, and thus solve the two-dimensional problem;

• it is presumed that cable-stayed antenna design elements make no resistance to compressive forces, which introduces significant nonlinearity to the behavior of the structure [12].

Figures 6-13 show the finite-element model of large-sized space reflectors and their corresponding numerical model as developed in the software package ANSYS. Figure 7 shows a front-tension network reflector due to its tension cords. Figure 9 presents the strain of the reflector as a result of pressure on its surface. Figure 11 shows the displacement of the reflector assemblies reflecting the surface tension.

The calculation of the deviation of the reflecting surface of a paraboloid, given by the equation $z = (x^2 + y^2) / 4F$, where F - focal length, was carried out for the reflector in Figure 12. More information about these models can be found in works [13, 14].

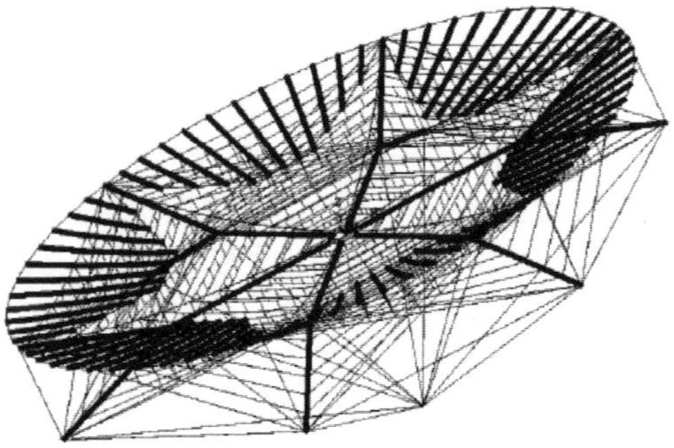

Figure 6. Umbrella type reflector with aperture of 50 m.

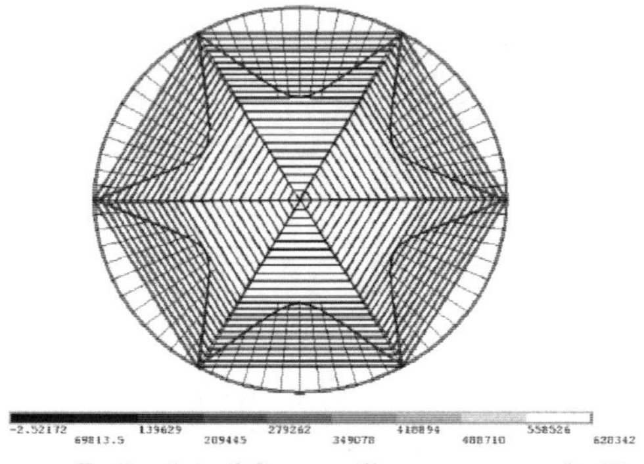

Figure 7. Strain of front reflector network, Pa.

Figure 8. Inflatable reflector with aperture of 50 m.

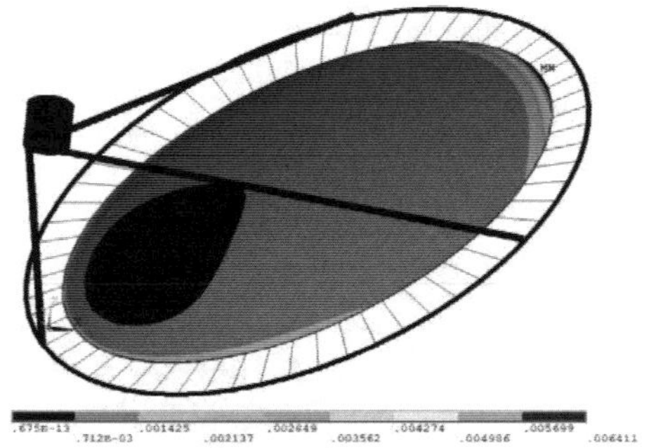

Figure 9. Reflector strain.

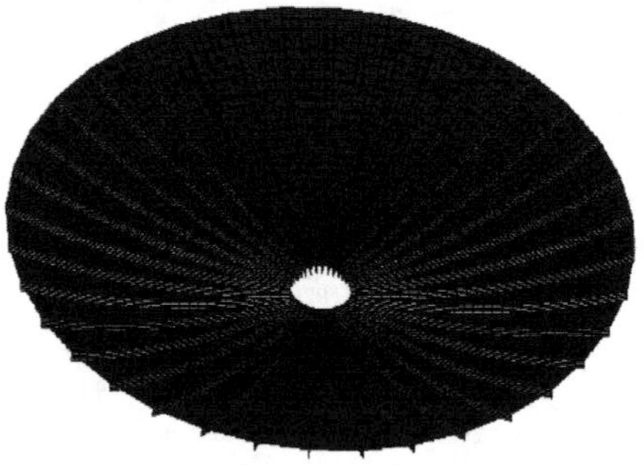

Figure 10. Reflector with flexible spokes with aperture of 50 m.

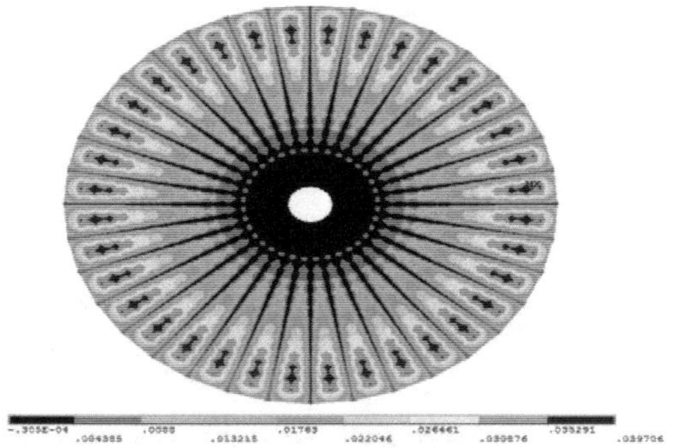

Figure 11. Displacement of reflector assemblies, m.

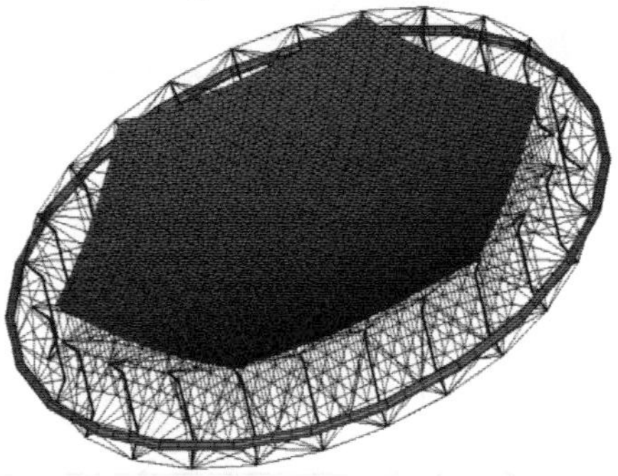

Figure 12. Reflector with tensegritied rim with aperture of 48 m.

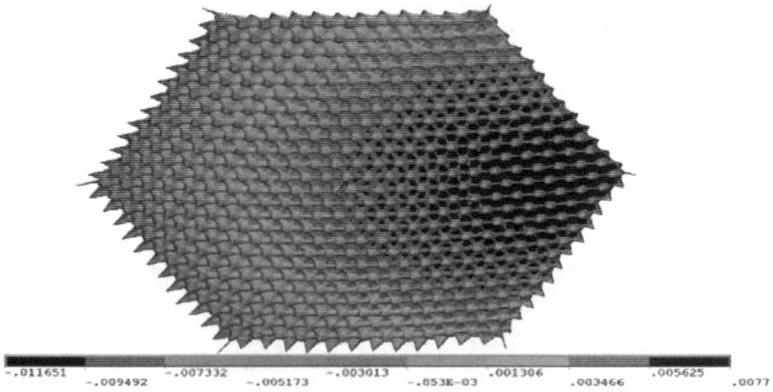

Figure 13. Distribution of deviations from the reflecting surface of paraboloid where RMSD is $4{,}05 \cdot 10^{-3}$ m.

All the numerical models are developed by using the approach proposed in Section 4. For all prescribed designs and initial solutions were predetermined zero boundary conditions on the displacement of assemblies. Further, to obtain the intermediate solutions, these boundary conditions for part of the assemblies are not predetermined, and the process continued until the solution sought was found. The main reason for its use is the large geometric nonlinearity arising in cable-stayed shell structures.

6. Conclusion

The problem of cable-stayed shell structures applicability was considered to generate innovative designs in building industry. The technique of numerical solution of the problem of stress-strain state of structures with high geometric nonlinearity was developed. The effectiveness of the developed technique is shown in a number of designs of large-sized space antenna reflectors.

References

[1] Fuller and Buckminster R 1982 (MacMillan Publishing Co. Inc 876)
[2] Levin S 2006 *Textbook of Musculoskeletal Medicine* (Oxford University Press)
[3] Gomez-Jauregui 2010 296
[4] Tibert A 2002 *Doctor Thesis Royal Institute of technology* SE-100 44
[5] Lewis W *2003Tension Structures. Form and Behaviour* (Thomas Telford: London) 201
[6] Morterolle S, Maurin B, Quirant J and Dupuy C 2012 Numerical form-finding of geotensoid tension truss for mesh reflector *Systèmes Acta Astronautica* **76** 154
[7] Correa J 2001 *Static Analysis of Tensegrity Structures* (University of Florida) 218
[8] Murakami H 2001 *Int J Solids Struct* **38** 3599
[9] Murakami H 2001 *Int J Solids Struct* **38** 3615
[10] Zhukov A, Ponomarev V and Ponomarev S 2011 *Universities bulletin. Physics* **54** 10/2154
[11] Zhukov A and Ponomarev S 2012 *Universities bulletin. Physics* **55** 7/2 72
[12] Yashchuk A 2005 Thesis... Cand. Sc. Tomsk State University
[13] Belov S, Bel'kov A, Evdokimov S, Zhukov A, Pavlov M, Ponomarev S, Ponomarev V, Solonenko V and Yashchuk A 2012 *Universities bulletin. Physics* **55** 9/13
[14] Velichko A, Belov S and Ponomarev S 2013 *Universities bulletin. Physics* **56** 7/3143

Investigation of the features of polycrystals complex loading using a two-level crystal plasticity theory

P S Volegov[1], P V Trusov[1] and D S Gribov[1]
[1]Perm National Research Polytechnic University, Department of Mathematical Modeling of Systems and Processes, Perm, 614990, Russian Federation

E-mail: crocinc@mail.ru

Abstract. The article considers a two-level mathematical model of inelastic deformation of metal polycrystals taking into account evolution of the structure. The structure of the model was considered, some special features of its application to describe the intensity of inelastic deformations were marked. The need for a careful physical analysis of the hardening laws construction was highlighted. To evaluate the applicability of multi-level models to describe the known experimental effects of cyclic deformation a number of results of field experiments on the complex proportional and disproportional cyclic deformation was considered, some specific effects that appear in these processes were identified: stresses amplitude output at the stationary value; additional cyclic hardening at a disproportionate loading, which magnitude depends on the so-called degree of disproportionality. Numerical experiments on the disproportionate cyclic loading were carried out, the possibility of modified hardening laws to describe access to the stationary values of the stress intensity was noted, and also the possibility of a qualitative description of the effect of additional cyclic hardening was demonstrated. The A.A. Ilyushin hypothesis by isotropy and the principle of vector properties delay at the turn of the deformation path were validated.

1. Introduction

Severe plastic deformation processes (SPD) are of great concern in the modern material processing technologies, especially in production of textured sub-microcrystalline and nanocrystalline materials with high performance properties. Development of process conditions of SPD requires setting up and solving of corresponding boundary value problems of solid mechanics. The most important element in the process of setting up is formulation (or selection from existing ones) of constitutive relations (CR). From the positions of classical (macrophenomenological) solid mechanics, the materials used in SPD processes are mediums with memory [1-2], which reaction on external action is determined by their previous history of loading and its complexity. In order to execute quantitative estimation of the later the plasticity theory widely uses concepts introduced by A.A. Ilyushin (stress and strain vectors, deformation path, loading process images).

On the other side, the recent decades have shown an intensive development of approaches to building material models at the “junction” of nonlinear solid mechanics and solid-state physics. Models of this class, usually called crystal plasticity theories, are based on introduction of inner variables, i.e. parameters describing evolving meso- and microstructure of polycrystalline materials [3].

2. Materials and methods

In this paper it is considered only a two-level model of polycrystalline material. Crystallites (grains, subgrains) are the elements of lower level (mesolevel), polycrystalline aggregate (aggregation of several hundreds of crystallites) is the element of an upper level. Congeneric parameters of macrolevel and mesolevel are marked by similar letters, but congeneric parameters of macrolevel are marked by capital letters and of mesolevel – by lower-case letters.

Constitutive model for description of behavior of the representative volume of macrolevel is represented by the following combination of relations:

$$\begin{cases} \Sigma^{\mathrm{r}} \equiv \dot{\Sigma} + \Omega^{\mathrm{T}} \cdot \Sigma + \Sigma \cdot \Omega = \Pi : D^{e} = \Pi : (D - D^{in}), \\ \Omega = \Omega\left(\omega_{(i)}, \pi_{(i)}, \sigma_{(i)}\right), i = 1, ..., N, \\ \Pi = \Pi(\pi_{(i)}, o_{(i)}), i = 1, ..., N, \\ D^{in} = D^{in}(d^{in}_{(i)}, \pi_{(i)}, \omega_{(i)}), i = 1, ..., N, \end{cases} \tag{1}$$

where Σ – Cauchy stress tensor, Π – elasticity moduli tensor, D, D^{e}, D^{in} – strain rate tensor, its elastic and inelastic components, index “r” means corotation derivative which is independent from coordinate system, Ω – macro level spin; $\pi_{(i)}, \sigma_{(i)}, d^{in}_{(i)}, \omega_{(i)}, o_{(i)}$ – tensors of elastic moduli, stress tensor, inelastic component of strain rate and orientation of lattice of i crystallite on mesolevel, N – number of crystallites that create the representative volume.

The following system of relations (crystallite number omitted) is used on mesolevel of two-level model:

$$\begin{cases} \sigma^{\mathrm{r}} \equiv \dot{\sigma} - \omega \cdot \sigma + \sigma \cdot \omega = \pi : d^{e} = \pi : (d - d^{in}), \\ d^{in} = \sum\limits_{i=1}^{K} \dot{\gamma}^{(i)} m^{(i)}_{(S)}, \\ \dot{\gamma}^{(i)} = \dot{\gamma}_0 \left| \dfrac{\tau^{(i)}}{\tau^{(i)}_c} \right|^{1/n} H(\tau^{(i)} - \tau^{(i)}_c), \ i = 1, ..., K, \\ \dot{\tau}^{(i)}_c = f(\gamma^{(j)}, \dot{\gamma}^{(j)}), \ i, j = 1, ..., K, \\ \hat{\nabla} v = \hat{\nabla} V, \end{cases} \tag{2}$$

where σ – Cauchy stress tensor, π – crystallite elastic properties fourth-rank tensor, d, d^{e}, d^{in} – strain rate tensor, its elastic and inelastic components on mesolevel, $\gamma^{(i)}, \tau^{(i)}_c$ – accumulated shear and critical shearing stress by k slip system, $\dot{\gamma}_0, n$ – material constants, $\tau^{(i)}$ – tangential strain acting in k slipping system, $H(\cdot)$ – Heaviside function, K – number of crystallographic systems for considered type of lattice.

In [4] the problem of matching of the defining relations of different scale levels in two-level models of inelastic deformation were considered:

$$\Pi = <\pi>, \quad \Sigma = <\sigma>, \quad D = \langle d \rangle . \tag{3}$$

It was shown that the macro spin tensor Ω and inelastic component of the strain rate tensor D^{in} should be determined by the relations:

$$\Omega = <\omega>, \tag{4}$$

$$D^{in} = <d^{in}> + \Pi^{-1} : <\pi' : d^{in\prime}> - \Pi^{-1} : (<\omega' \cdot \sigma'> - <\sigma' \cdot \omega'>), \tag{5}$$

where primes denote the deviation of the corresponding quantities from their average values for representative macrovolume.

As the basic hardening law [5-6] in $(2)_4$ the relation of the form [7] was used:

$$\dot{\tau}_c^{(k)}\left(\gamma^{(i)},\dot{\gamma}^{(i)}\right)=G\left(\sum_{i=1}^{24} a_i^{(k)}\frac{\left(\gamma^{(i)}\right)^{\psi}\dot{\gamma}^{(i)}}{\left(\sum_{i=1}^{24}\gamma^{(i)}\right)^{\delta}}\right),\ k=\overline{1,24},\ \psi>0,\ \gamma^{(i)}\geq 0, \quad (6)$$

where $a_i^{(k)}$ – matrix of dimensionless hardening moduli; G – shear modulus.

3. Results

In this paper a series of experiments on the disproportionate cyclic deformation is considered, given in [8]. In [8] the disproportionate loading of the sample was considered, the type of the control signal is given below:

$$l=l_m\sin(\omega t),\ \gamma=\gamma_m\sin(\omega t-\varphi),\ \rho=\gamma_m\sqrt{3l_m}, \quad (7)$$

where l – current sample elongation, γ – current sample twist angle, parameter φ – phase difference between the torsion and tension (in [8] φ called the degree of loading disproportionality), ρ – factor which relates the amplitudes of the two components of deformation. Under disproportionate cyclic loading increase of shear stress, depending on the degree of disproportionality of deformation is observed. The effect was named an additional hardening, and for the convenience of this additional hardening was correlated with yield strength during deformation in compression - tension. The value of additional hardening could be up to 100% of the hardening that occurs with a proportional cyclic loading.

We marked out on the stationary values of the stress intensity at the optimal choice of the parameters of the law hardening in numerical experiments (Figure 1).

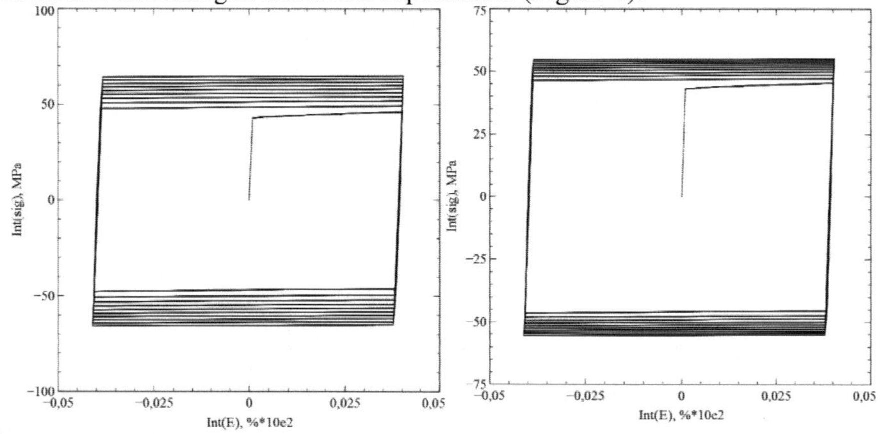

Figure 1. Dependence of the stress intensity on the strain intensity under cyclic tension - compression, 10 cycles, the law (6) is used, the parameter $\psi=1{,}3$ (on the left), $\psi=1{,}5$ (on the right).

Disproportionate cyclic loading leads to a substantial change in the structure of the material at the micro level, there is a sliding edge dislocations across multiple evolving slip systems. According to the data obtained in [8], depending on the angle disproportion φ (7) additional hardening appearing at disproportionate loading changes significantly.

Figure 2(a) represents two dependences of stress intensity on strain intensity using only the main term in the hardening law (6). It should be mentioned that further stress intensity will have a strictly positive sign, i.e. one cycle of loading corresponds to even and odd cycles in the diagram according to the intensity of the stresses and strains.

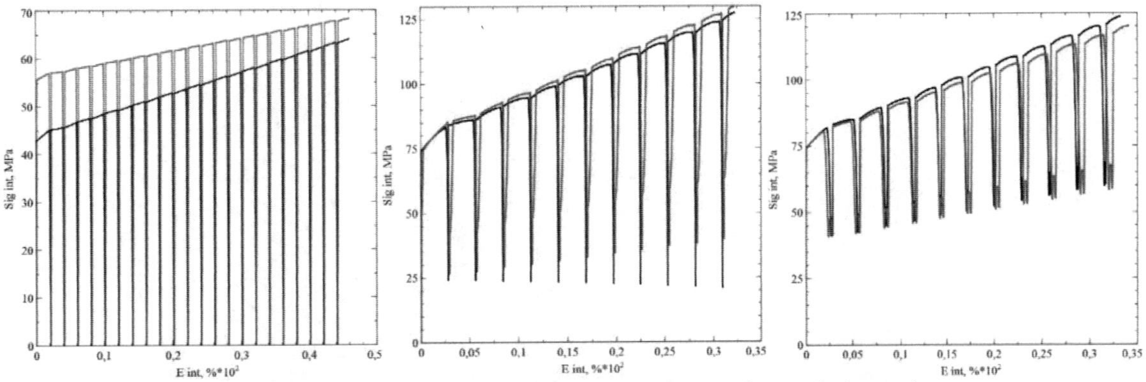

Figure 2. Dependence of the stress intensity on the strain intensity:
a) under cyclic tension - compression (gray) and shear (black), 11 cycles; b) under disproportionate cyclic loading, at values of disproportionality angle φ of 0 (black) and 15 (red) degrees; c) under disproportionate cyclic loading, at values of disproportionality angle φ of 0 (black) and 15 (red) degrees, 6 cycles.

When considering the loading diagram in Figure 2(a) it was found out that value of the stress intensity is different due to the fact that the shear is implemented different picture of edge dislocations than in tension-compression. Thus, in the tension test slip systems are oriented in such a way that the critical stresses in them have come at lower values of the stress intensity.

Figure 2(b-c) shows the results of the disproportionate loading at various disproportionate angles φ (at 0, 15, 30, 45 degrees). On the diagrams, one can see a nonlinear variation of stresses in the zone of plastic deformation. This is related to the more complex activities of slip systems in this type of loading due to the fact that the slip systems are activated depending on the value of the tangential stresses in each. It should be noted, that disproportionate loading becomes a significant change of shear stresses, and the slip system, as activation, make a comprehensive contribution disproportionate to the growth of the critical stress.

Based on the numerical experimental data table was constructed according to the stationary values of the stress intensity on the angle of disproportionality (Table 1).

Table 1. The dependence of the steady-state values of the stress intensity on the disproportionate angle.

№	φ, degrees	σ_u, MPa
1	0	79.2
2	15	79.3
3	30	79.7
4	40	80.5

Hence, it is interesting to verify fulfillment of the postulate of isotropy for the case of complex loading of representative macrolevel ("macro specimen") [9]. The following two paths are taken as two paths that differ only by their orientation in deformation space and have the same geometry:

$$1)\ \textit{stretching along axis } Э_1 \rightarrow \textit{deformation along axis } Э_3, \quad (8)$$

$$2)\ \textit{deformation along axis } Э_3 \rightarrow \textit{stretching along axis } Э_1. \quad (9)$$

Figure 3 represents loading process images for different deformation breaking paths (8) and (9), stress intensities marked by numbers on vector ends.

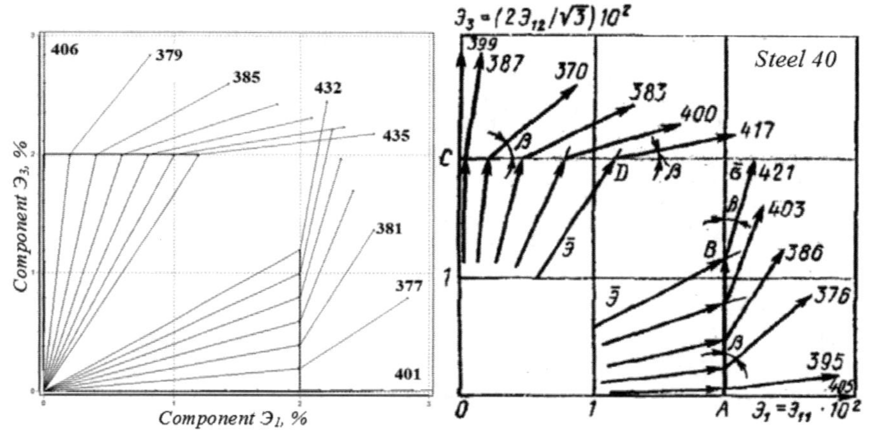

Figure 3. Process image for different deformation breaking paths (8) and (9); on the left – numerical experiment [10], on the right – natural experiment [11].

Projections of loading process images on plane $Э_1$–$Э_3$ are obtained for paths under consideration, by this projections we can judge on fulfillment of isotropic postulate, namely: before path breaking angles between stress and strain rate are the same and close to 0°, thereat design and experimentally measured moduli of stress vector differ not more than by 1%, after path breaking angles between tangent line to deformation path and strain vectors and strain vector moduli differ not more than by 2%. Comparison of strain vector lengths shows us lowering (just about by 7-8%) of intensity of flow stress (the effect of "stress dive"). On the basis of considered information we can make a conclusion on good qualitative and quantitative conformity of vector and scalar properties established experimentally and obtained in numerical experiments [12-13].

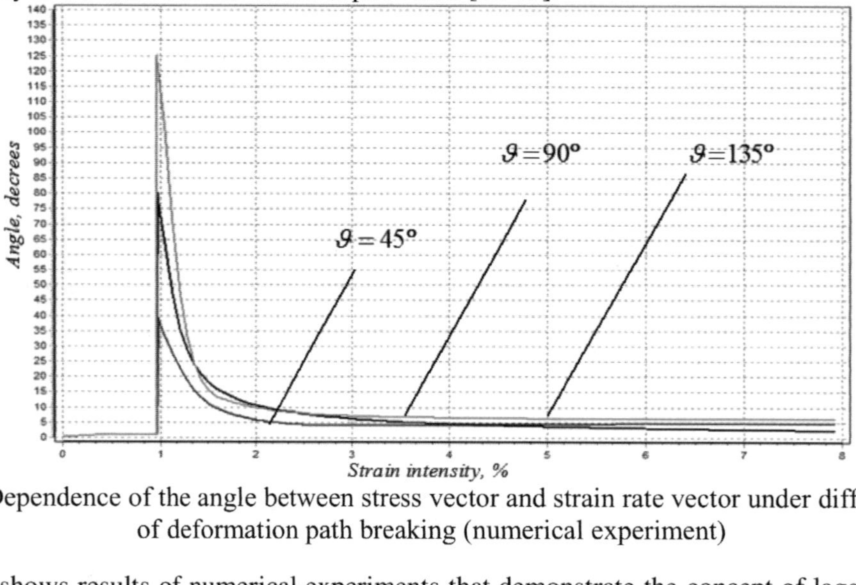

Figure 4. Dependence of the angle between stress vector and strain rate vector under different angles of deformation path breaking (numerical experiment)

Figure 4 shows results of numerical experiments that demonstrate the concept of lagging of scalar properties. Diagrams represent dependence of angle between stress vector and strain rate vector on full cumulative deformation intensity. Let us note that the data of natural experiments [12] leads to certain questions associated with the value of angle between stress vector and tangential line to path just after breaking, which is in some cases higher than the value of path breaking angle. Comparative analysis of dependences helps us to make a conclusion on good qualitative and quantitative correspondence of results, but velocity of approaching of angle to 0° in numerical experiments slightly differs from those that are registered in natural experiments. It is worth mentioning, that both in numerical and natural experiments the velocity of approaching the angle θ to zero almost does not depend on the value of path breaking angle: after ending of 1-2% of deformations after breaking, diagrams of dependence of angle on deformation arc length are very close.

4. Conclusions

Thus, in the article the results of a series of field experiments on the cyclic deformation were considered, some characteristics of these processes were identified. The applicability of the hardening law was validated; critical orientation in which there is linear plastic deformation was marked. The hardening law was construct that allows describing the phenomena observed in the experiments: access to the stationary value stress-strain curve. It was considered a set of numerical experiments on cyclic loading disproportionate, we noted the possibility of modified hardening law to describe access to the stationary values of the stress intensity, and also demonstrate the possibility of a qualitative description of the effect of additional hardening.

This work was supported by RFBR (grants №13-01-96006 r_ural_a, 14-01-00069-a, 14-01-96008 r_ural_a), President Grant № MK-4917.2015.1.

References

[1] A A Ilyushin 1963 Plasticity. The bases of general mathematical theory (Moscow: Science) 270 (*in Russian*)

[2] P V Trusov, A I Shveykin, E S Nechaeva and P S Volegov 2012 Multilevel models of inelastic deformation of materials and their application for description of internal structure evolution *Phys Mesomech* **15** 155-175

[3] K W Neale 1993 Use of Crystal Plasticity in Metal Forming Simulations *Int. J. Mech. Sci.* **35**(12) 1053–1063

[4] P V Trusov, P S Volegov and A Yu Yanz 2014 Two-scale models of polycrystals: Macroscale motion decomposition *Phys Mesomech* **17** 116-122

[5] G J Weng 1980 Dislocation theories of work hardening and yield surfaces of single crystals *Acta Mechanica* **37**(3–4) 217–230

[6] N Bonfoh, A. Carmasol P. Lipinski 2003 Modeling of intra–crystalline hardening of materials with particles *Int. J. Plasticity* **19** 1167–1193

[7] P V Trusov and P S Volegov 2011 Crystal plasticity: theory and applications to the description of inelastic deformation of materials. Part 3: Hardening theories, gradient theories *PNRPU Mechanics Bulletin* **3** 146-197 (*in Russian*)

[8] A Benallal and D Marquis 1988 Constitutive Equations for Nonproportional Cyclic Elasto-Viscoplasticity *J. Eng. Mater. Technol.* **109** 326-336

[9] P V Trusov, P S Volegov and A.Yu. Yanz 2014Two-Scale Models of Polycrystals: Independence of the loading process image of a representative macrovolume *Phys Mesomech*, **17** 190-198

[10] P V Trusov, P S Volegov, A Yu Yants Two-scale models of polycrystals: application to the analysis of the effects of complex loading *Phys Mesomech* **16** 43-50 (2013) (*in Russian*)

[11] V I Gultyaev 2007 Experimental investigation of the complex deformation of materials on multilink trajectories *Problems of Strength and Plasticity* **67** 95-98 (*in Russian*).

[12] V G Zubchaninov 1990 *Fundamentals of the theory of elasticity and plasticity* (Moscow: High school) 368 (*in Russian*)

[13] V G Zubchaninov 2010 Mechanic processes of plastic media (Moscow: High school) 352 (*in Russian*)

Numerical simulation of deformation and fracture in a coated material using curvilinear regular meshes

A Zinoviev[1], R Balokhonov[1,2], S Martynov[1], V Romanova[1,2] and O Zinovieva[1,3]
[1]Institute of Strength Physics and Materials Science SB RAS, Laboratory of Mechanics of Heterogeneous Media, Tomsk, 634021, Russia
[2]National Research Tomsk Polytechnic University, Tomsk, 634050, Russia
[3]National Research Tomsk State University, Physical-Technical Faculty, Tomsk, 634050, Russia

E-mail: zav@ispms.tsc.ru

Abstract. This paper presents a computational analysis of the deformation and fracture mechanisms of a material with a porous polysilazane coating under tension and compression. A dynamic boundary-value problem in the plane strain statement is solved numerically by the finite-difference method. The coating-substrate interface and porous coating microstructure corresponded to configurations found experimentally are accounted for explicitly in the calculations. For this purpose an algorithm for curvilinear finite-difference meshing based on the solution of the elasticity theory has been developed. The algorithm implemented offers several benefits over the rectilinear meshing. Local regions experiencing bulk tension are shown to form along pore surfaces that control the fracture mechanisms at the mesoscale level.

1. Introduction

In recent years, advanced metal-ceramic composites have been widely used instead of pure metallic compounds and identified as important materials in high-precision manufacturing. Polysilazanes are extensively used to manufacture polymer-derived ceramics (PDCs), which can be employed to produce fibers, coatings and fiber-reinforced ceramic matrix composites. The resulting PDC-based coatings exhibit optimum adhesion ability, good corrosion and heat resistance [1]. However, due to the shrinkage occurring during the polymer-to-ceramic conversion, PDC coatings generally suffer from unwanted porosity, cracks and defects [2]. There is an increasing number of experimental and theoretical works devoted to investigation of pore characteristics and to analysis of methods to estimate the porosity and to produce porous coatings (see, e.g., [3]).

In this study we analyze numerically the mechanisms of deformation and fracture involved in material with a porous polysilazane-based ceramic coating under uniaxial tensile and compressive loading. A role of the geometry of the coating-substrate interface and of the porous coating microstructure is investigated. The simulated microstructure of the coated material corresponds to the experimental data on the composite (Figure 1a) and is accounted for explicitly in the calculations. In order to describe the mechanical behavior of the substrate and coating materials, we use an elasto-plastic model accounting for isotropic strain hardening and of a modified fracture criterion of the Huber type [4, 5].

There are two stages of the work. In stage 1, an algorithm for 2D regular curvilinear meshing based on the approach proposed in [6] is implemented, and a pertinent finite-difference solution is compared to a solution obtained using a uniform rectilinear mesh (see, e.g., [4, 5]). The curvilinear mesh is shown to play an important role in the quality of the description of real pore shapes and geometry of the curvilinear coating-substrate interface. Stage 2 involves a numerical simulation study of the deformation and fracture of a mesovolume of the coated material subjected to compression and tension (Figure 1b), with the interfaces being taken into account explicitly.

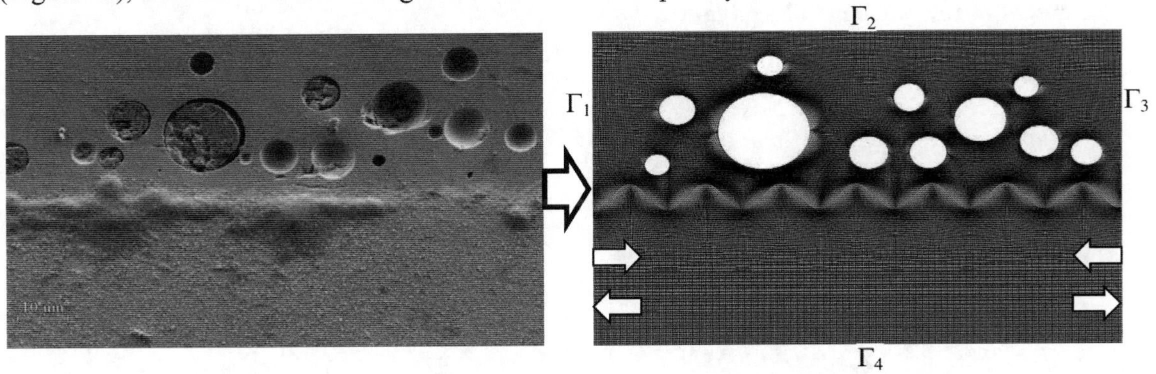

a *b*

Figure 1. Scanning electron microscope image of a material with a porous ceramic coating [1] (*a*) and regular curvilinear mesh (*b*).

2. Algorithm of curvilinear mesh generation

In our previous work on numerical study of the deformation behavior of coated materials [4, 5], curvilinear interfaces are represented as step-wise lines of the rectilinear rectangular mesh (Figure 2a). The main idea of an approach proposed in [6] to generate 2D regular curvilinear meshes is to deform a rectangular mesh by solving the boundary-value problem in the plane stress formulation. The geometry of computational domain under study being meshed is considered as an elastic solid. The following conforming model is used to calculate the deformation of the initially regular rectangular mesh:

$$\begin{aligned} \frac{\partial^2 u_x}{\partial x^2} + \frac{(1-\vartheta)}{2}\frac{\partial^2 u_x}{\partial y^2} + \frac{(1+\vartheta)}{2}\frac{\partial^2 u_y}{\partial x \partial y} = 0 \\ \frac{\partial^2 u_y}{\partial y^2} + \frac{(1-\vartheta)}{2}\frac{\partial^2 u_y}{\partial x^2} + \frac{(1+\vartheta)}{2}\frac{\partial^2 u_x}{\partial x \partial y} = 0 \end{aligned}, \qquad (1)$$

Here u_x and u_y are the displacement vector components, ϑ is the computational parameter varying between 0 and 1. By specifying the boundary displacements $u_{1,j}$, $u_{n,j}$ and $u_{i,1}$, $u_{i,n}$, where i and j vary from 1 to n, we define a curved geometry of boundary. The boundary displacements can be found analytically. The final configuration, i.e. the displacements of internal nodes ($i = 2..n - 1, j = 1..n - 1$), is defined by solving Equation (1) numerically with the use of the finite-difference Seidel method. The following finite-difference approximation is considered:

$$(u_x)_{i,j} = \frac{1}{A_{i,j}}\left\{\frac{2}{\Delta x_i}\left(\frac{(u_x)_{i+1,j}}{\Delta x_{i+1/2}} + \frac{(u_x)_{i-1,j}}{\Delta x_{i-1/2}}\right) + \frac{(1-\vartheta)}{\Delta y_i}\left(\frac{(u_x)_{i,j+1}}{\Delta y_{j+1/2}} + \frac{(u_x)_{i,j-1}}{\Delta y_{j-1/2}}\right) + \frac{(1+\vartheta)\left[(u_y)_{i+1,j+1} - (u_y)_{i+1,j-1} - (u_y)_{i-1,j+1} + (u_y)_{i-1,j-1}\right]}{2\Delta x_i \Delta y_j}\right\},$$

$$(u_y)_{i,j} = \frac{1}{B_{i,j}}\left\{\frac{2}{\Delta y_i}\left(\frac{(u_y)_{i,j+1}}{\Delta y_{j+1/2}} + \frac{(u_y)_{i,j-1}}{\Delta y_{j-1/2}}\right) + \frac{(1-\vartheta)}{\Delta x_i}\left(\frac{(u_y)_{i+1,j}}{\Delta x_{i+1/2}} + \frac{(u_y)_{i-1,j}}{\Delta x_{i-1/2}}\right) + \frac{(1+\vartheta)\left[(u_x)_{i+1,j+1} - (u_x)_{i+1,j-1} - (u_x)_{i-1,j+1} + (u_x)_{i-1,j-1}\right]}{2\Delta x_i \Delta y_j}\right\}. \qquad (2)$$

Here $\Delta x_{i+1/2} = x_{i+1} - x_i$, $\Delta y_{j+1/2} = y_{j+1} - y_j$, $\Delta x_i = x_{i+1} - x_{i-1}$, $\Delta y_j = y_{j+1} - y_{j-1}$, and

$$A_{i,j} = \frac{2}{\Delta x_i}\left(\frac{1}{\Delta x_{i+1/2}} + \frac{1}{\Delta x_{i-1/2}}\right) + \frac{(1-\vartheta)}{\Delta y_i}\left(\frac{1}{\Delta y_{j+1/2}} + \frac{1}{\Delta y_{j-1/2}}\right), \quad B_{i,j} = \frac{2}{\Delta y_i}\left(\frac{1}{\Delta y_{j+1/2}} + \frac{1}{\Delta y_{j-1/2}}\right) + \frac{(1-\vartheta)}{\Delta x_i}\left(\frac{1}{\Delta x_{i+1/2}} + \frac{1}{\Delta x_{i-1/2}}\right) \qquad (3)$$

Figure 2b shows the algorithm output for a rectangular region of the coating material with a single perfectly round pore at the centre. A curvilinear mesh corresponding to the experimental microstructure of the coated material containing multiple pores in the coating and a curvilinear coating-substrate interface is presented in Figure 1b.

3. Statement of the mechanical problem

The coated material deformation was simulated in the plane strain formulation. A total system of equations of mechanics includes equations of motion and continuity, expressions for components of the strain rate tensor and constitutive relations [4, 5]. Numerical solutions were performed in terms of Lagrangian variables using the finite-difference method of the second-order of accuracy [7].

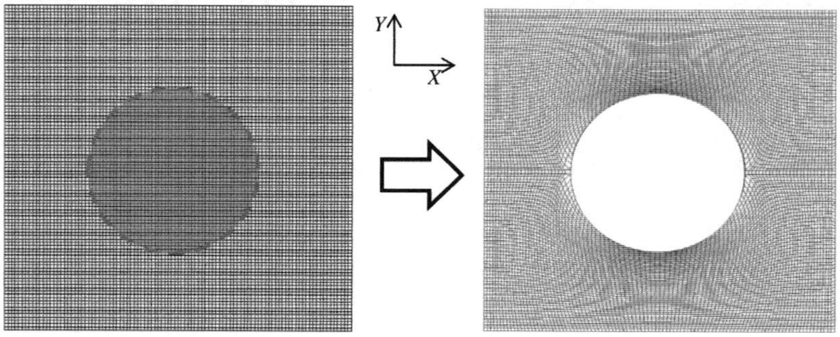

Figure 2. Rectilinear (a) and curvilinear (b) meshes for a region with a single pore in the center.

For plane strain the following non-zero components of the strain rate tensor are given by

$$\dot{\varepsilon}_{xx} = \frac{\partial \dot{u}_x}{\partial x},\ \dot{\varepsilon}_{yy} = \frac{\partial \dot{u}_y}{\partial y},\ \dot{\varepsilon}_{xy} = \frac{1}{2}\left(\frac{\partial \dot{u}_x}{\partial y} + \frac{\partial \dot{u}_y}{\partial x}\right). \tag{4}$$

The upper dot stands for the time derivative. The mass conservation law and equations of motion in the absence of body forces take the forms:

$$\dot{V}/V = \dot{\varepsilon}_{xx} + \dot{\varepsilon}_{yy}, \tag{5}$$

$$\frac{\partial \sigma_{xx}}{\partial x} + \frac{\partial \sigma_{yx}}{\partial y} = \rho \ddot{u}_x,\ \frac{\partial \sigma_{xy}}{\partial x} + \frac{\partial \sigma_{yy}}{\partial y} = \rho \ddot{u}_y, \tag{6}$$

where V is the specific volume, σ_{ij} are the stress tensor components, ρ is the mass density. Taking into account the stress tensor decomposition in the spherical and deviatoric part $\sigma_{ij} = -P\delta_{ij} + S_{ij}$, the strain tensor decomposition in the elastic and plastic part $\dot{\varepsilon}_{ij} = \dot{\varepsilon}^e_{ij} + \dot{\varepsilon}^p_{ij}$, and the hypothesis of plastic incompressibility $\dot{\varepsilon}^p_{kk} = 0$, we get the pressure rate and deviatoric stress rate components

$$\dot{S}^*_{ij} = 2\mu\left(\dot{\varepsilon}_{ij} - \frac{1}{3}\dot{\varepsilon}_{kk} - \dot{\varepsilon}^p_{ij}\right),\ \dot{S}^*_{ij} = \dot{S}_{ij} - S_{ik}\omega_{jk} - S_{jk}\omega_{ik}, \tag{7}$$

$$\dot{P} = -K\dot{\varepsilon}_{kk}, \tag{8}$$

where K and μ are the bulk and shear moduli, $\dot{\varepsilon}^p_{ij}$ is the plastic strain rate tensor, ω_{ij} is the material spin tensor. Summation is implied over the repeated index.

The coating is elastic-brittle ceramics. The substrate material is assumed to be isotropic and follow the von Mises yield criterion with isotropic hardening. In order to describe the plasticity of the steel substrate, we use the plastic flow law $\dot{\varepsilon}^p_{ij} = \dot{\lambda}\partial f/\partial S_{ij}$ associated with the von Mises yield condition $f(S_{ij}) = \sigma_{eq} - Y_0(\varepsilon^p_{eq}) = 0$, where λ is a scalar parameter, σ_{eq} and ε^p_{eq} are the equivalent stress and equivalent plastic strain, respectively. The strain-hardening function for the high-strength austenitic steel used as the substrate material is given in [8] as

$$\sigma_y(\varepsilon_{eq}^p) = \sigma_s - (\sigma_s - \sigma_0)\exp(-\varepsilon_{eq}^p / \varepsilon_r^p), \tag{9}$$

where σ_s and σ_0 are the ultimate tensile and yield strengths, respectively, ε_r^p is the reference value of the plastic strain.

To describe the fracture of the PDC-based coating, use was made of a maximum distortion energy criterion of Huber type accounting for crack initiation and growth in local regions experiencing bulk tension. According to this criterion, fracture occurs in any local coating region if the equivalent stress σ_{eq} reaches its critical value: C_{ten} or C_{com} in local tension or compression cases, respectively. Local tension means that the cubic strain ε_{kk} takes on a positive value. Here, C_{ten} and C_{com} are the constants that characterize the tensile and compression strength of porous ceramic coating, respectively.

The boundary conditions on surfaces Γ_1 and Γ_3 of the computational domain simulate uniaxial tension/compression along the X-axis. The bottom surface Γ_4 is the symmetry line and the top surface $\mathbf{\Gamma_2}$ is free from loads (Figure 1b).

4. Simulation results

4.1. Convergence analysis

For convergence test we have performed two sets of numerical simulations for rectilinear and curvilinear meshes (see Figure 2) that approximate a rectangular region with a single perfectly round pore under tensile loading. Material inside the pore region possessed very low elastic properties.

The calculated equivalent stress patterns for the models with rectilinear and curvilinear meshes are shown in Figure 3a-b. Comparison of the stress measured along *A-A'* and *B-B'* lines is illustrated in Figure 3c-d to confirm the robustness of the implemented approach. Convergence of the numerical solution was found for both cases (Figure 4). The total number of nodes per pore diameter along the *X*-axis was varied from 10 to 300. Due to the fact that the geometry of the pore material is accounted for more accurately in the calculations, the use of the curvilinear mesh allows accelerating the approach of a maximum stress concentration to a saturation physical value.

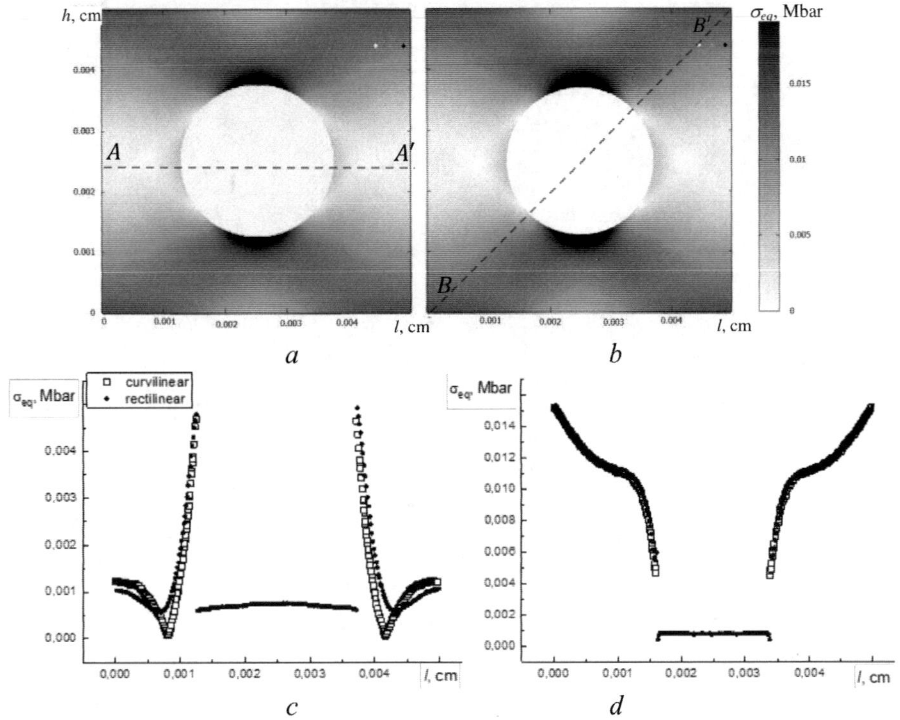

Figure 3. Equivalent stress patterns obtained from the rectilinear (*a*) and curvilinear models (*b*) and

equivalent stress values taken along *A–A'* midlines (*c*) and *B-B'* diagonals (*d*).

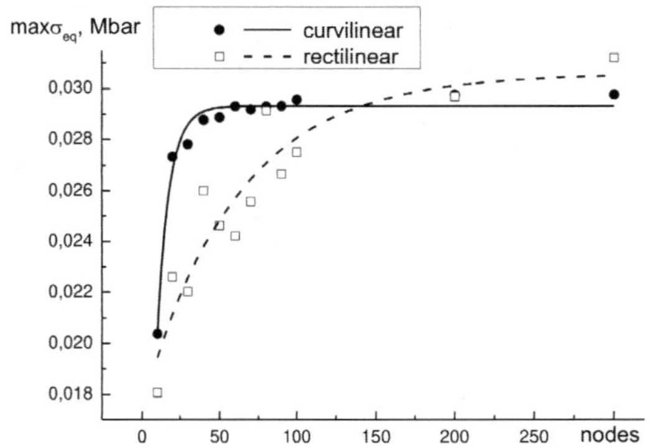

Figure 4. Maximum equivalent stress versus the number of nodes per pore diameter.

4.2. The deformation and fracture of the steel substrate with porous PDC-based coating

Let us analyse the mechanisms for stress concentration in the porous coating using the coated material (Figure 1a) approximated by the curvilinear mesh (Figure 1b). At earlier stages of the deformation, both the tensile and compressive stresses are found to occur along pore surfaces in both cases of external tension or compression. In both cases, cracks originate in regions of local tension. Under tension and compression tensile regions occur at different points. On further loading tensile cracks originate and propagate in different ways, i.e., along the loading axis in the case of external compression (Figure 5b, d) and in the perpendicular direction in the case of external tension (Figure 5a, c).

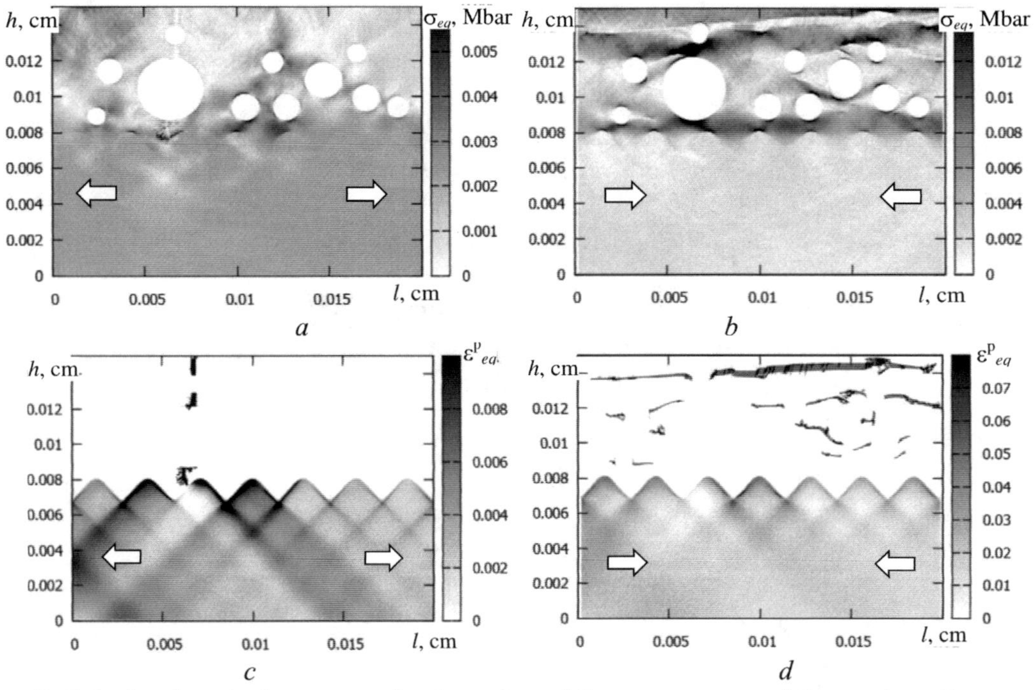

Figure 5. Calculated equivalent stress, plastic strain and fracture patterns of the coating of a composite material for the case of tension (*a, c*) and compression (*b, d*).

5. Conclusions

An algorithm for 2D curvilinear finite-difference meshing based on the solution of the elasticity theory has been implemented that allows describing the pore shape and geometry of the curvilinear coating-substrate interface more precisely. The convergence of the numerical solution has been verified for both rectilinear and curvilinear model, while the use of the curvilinear mesh is shown to accelerate an approach of the maximum stress concentration to a saturation physical value dictated by the geometry of the pore material. Using the curvilinear model, we have performed numerical simulations of tension and compression of coated steel specimens by solving a dynamic boundary-value problem. The pores in coating are shown to be the sources of a geometrical stress concentration on mesoscale level, where thereafter cracks originate. Both under external tension and compression, cracks in the coating tend to originate in the regions experiencing tensile loading. The fracture patterns depend heavily on the type of the applied load, that coming from different locations of local tensile regions under different types of the external loading.

6. Acknowledgements

The work was supported by Russian Science Foundation (grant No. 14-19-00766) and a program to increase the competence of National Research Tomsk Polytechnic University.

References

[1] Günthner M, Wang K, Bordia R K and Motz G 2012 *J. Eur. Ceram. Soc.* **32** 1883–1892
[2] Torrey J D and Bordia R K 2008 *J. Eur. Ceram. Soc.* **91** 41-45
[3] Kohn D H 2011 *Porous Coatings in Orthopedics* (Comprehensive Biomaterials) ed P Ducheyne *et al.* (Netherlands: Elsevier) **6** 65-77
[4] Balokhonov R R and Romanova V A 2009 *Int. J. Plast.* **25 11** 2225-2248
[5] Balokhonov R R, Romanova V A, Schmauder S, Martynov S A and Kovalevskaya Zh G 2014 *Composites: Part B* **66** 276–286
[6] Игнатьев А А 2000 *Математическое моделирование* **12 2** 101-105
[7] Wilkins M 1999 *Computer simulation of dynamic phenomena* (Berlin Heidelberg: Springer-Verlag)
[8] Koval A V and Panin S V 2000 *Theoret. Appl. Fract. Mech.* **34** 117-121

IOP Conf. Series: Materials Science and Engineering **71** (2015) 012073 doi:10.1088/1757-899X/71/1/012073

Two dimensional cellular automata simulation of grain growth during solidification and recrystallization

O Zinovieva[1, 2], A Zinoviev[2], V Ploshikhin[3], V Romanova[2,4] and R Balokhonov[2,4]
[1]Physical-Technical Faculty, National Research Tomsk State University, Tomsk, 634050, Russia
[2]Laboratory of Mechanics of Heterogeneous Media, Institute of Strength Physics and Materials Science, Tomsk, 634021, Russia
[3]Airbus Endowed Chair for Integrative Simulation and Engineering of Materials and Processes, University of Bremen, Bremen, 28359, Germany
[4]National Research Tomsk Polytechnic University, Tomsk, 634050, Russia

E-mail: emelyanova@ispms.tsc.ru

Abstract. A two dimensional cellular automata model has been developed to simulate and to study grain growth behavior during solidification and recrystallization processes, including predicting mechanical properties of metallic materials or investigating mechanical defects. Using the algorithm implemented, grain growth in model materials (nickel-based and titanium alloys) has been calculated in different thermal fields.

1. Introduction

Microstructure is known to play a key role in prediction of material thermo-mechanical behavior by a number of manufacturing processes such as casting, welding or additive manufacturing. In order to examine solidification and recrystallization microstructures, a variety of experimental methods can be successfully used but the dynamic evolution of microstructures in these processes is hard to be observed. In order to study grain growth behavior and predict material properties, numerical simulations of evolving microstructures have been performed quite a while and various models based on phenomenological, analytical or statistical approaches which take into account mechanisms of grain nucleation and growth have been developed in order to do it.

The paper [1] gives an overview of advances in solidification science and technology including numerical techniques. Cellular automata (CA) technique is one of the most successful methods allowing to investigate numerically microstructure features from about 1 μm to several centimeters, due to its adaptivity, low computational costs, etc. (see, e.g., [1-8]). In 1993 Rappaz and Gandin [2] proposed an approach to the simulation of grain structure formation during solidification process. Their model includes the mechanisms of heterogeneous nucleation and grain growth. Using this model they simulated two dimensional (2D) [4] and then, in 1997 [5], three dimensional (3D) grain structures by coupling CA model with finite element method for heat flow calculations. Using the combined CA finite difference (FD) model Lee et al. [6-8] simulated evolving microstructures of binary alloys during solidification. Pavlyk and Dilthey [4] proposed an original algorithm to determine interface curvature and orientation of growing grains as well as transition rules for cells.

Grain growth during recrystallization is also successfully simulated using CA models. In describing this process it is important to pay careful attention to grain boundary mobility. Humphreys and

Hatherly have written a monograph [9] on the subject of recrystallization emphasizing the theoretical description of the various phenomena taking place during recrystallization. Raabe [10] reviewed the fundamentals and some CA applications, for example, recrystallization simulation by use of a probabilistic CA model or by coupling CA with crystal plasticity FE model. Raghavan and Sahay [11, 12] developed 2D CA model for the grain growth in order to study grain growth kinetics. Gruber [13] simulated grain growth in 2 and 3 dimensions using Monte-Carlo method.

In this work the 2D grain growth model is developed using the CA method. The implemented algorithm of grain growth is based on the approach of Rappaz and Gandin [2]. Using this algorithm, grain growth in model materials has been calculated in different thermal fields. In this work we have chosen nickel-based and titanium alloys as model materials.

2. Model description

The implemented algorithm includes the following procedures: (i) calculation of the temperature field, (ii) calculation of the number of grains nucleated at each time step, (iii) nucleation of grains with randomly chosen crystallographic orientations, satisfying certain conditions and (iv) calculation of the solid fraction increment using grain growth algorithm developed by Rappaz and Gandin [2]. In order to calculate the temperature field, the heat transfer equation with Robin boundary conditions (BCs) is solved using the finite difference method [14, 15]. Let us dwell briefly on the main stages.

2.1. Grain nucleation

Grain nucleation is based on the approach of Rappaz and Gandin [2], Dong and Lee [8]. According to the approach of Oldfield [16] a continuous nucleation distribution $dn/d(\Delta T)$ can be used in order to describe the grain density increase dn induced by an increase in the effective undercooling $d(\Delta T)$. In order to describe the grain nucleation at the mould wall and in the bulk of the liquid Rappaz and Gandin [2], Dong and Lee [8] have used Gaussian functions which are characterized by mean undercooling ΔT_N and standard deviation ΔT_σ. In this work, a continuous Gaussian nucleation distribution [8] is used to describe the grain nucleation in the bulk of the melt

$$n(\Delta T) = \frac{N_{\max}}{\Delta T_\sigma \sqrt{2\pi}} \int_0^{\Delta T} \exp\left[-\frac{(\Delta T - \Delta T_N)^2}{2(\Delta T_\sigma)^2}\right] d(\Delta T), \tag{1}$$

where N_{max} is the maximum nucleation density. The undercooling within the volume is determined by the following formula $\Delta T = T_L + m_L \left|\langle C_L\rangle^L - C_0\right| - \overline{T_b}$ [17]. Here T_L is liquidus temperature, the m_L is the liquidus slope, $\langle C_L\rangle^L$ is the intrinsic volume-averaged liquid concentration, C_0 is the initial melt concentration, and $\overline{T_b}$ is the bulk temperature defined as the average temperature in the volume.

The calculated number of nuclei $n(\Delta T)$ is randomly distributed among cells. A grain nucleates in the i-th cell only if i) the cell has one of the randomly distributed nuclei within it; ii) the total undercooling in the cell $\Delta T_{ij} > \Delta T_N$, where $\Delta T_{ij} = T_L - T_{ij}$. After nucleation, the cell state changes into "solid", and the neighboring cells are considered as "growing" cells.

2.2. Calculation of the solid fraction increment

The concept of CA method is that a simulation object is considered as a set of interacting elements which dynamics is defined by rules of their relationships. In this work the simulated area is discretized by a regular rectangular mesh. At certain time each cell can have one of three states: liquid, solid and growing (i.e. a mixture of solid and liquid).

Grain orientation is defined by θ randomly where θ is the orientation angle of a principal crystallographic direction. Angle θ changes from 0° to 45°. Every nucleus has four growth directions perpendicular to each other and four vectors are associated with these directions. The first growth direction aligns with the preferential <10> direction. At each time step the ends of the vectors are connected by straight line and the centers of adjacent cells are checked to be within the area of the growing grain. Schematic representation of 2D CA algorithm of crystallite growth is shown in Figure 1. New growing cells develop by the similar algorithm. When the nucleus has taken all the neighboring cells, it stops growing. Growth rate for all grains is the same.

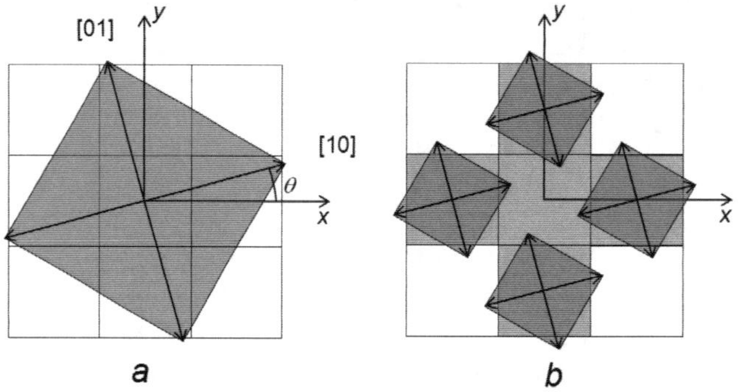

Figure 1. Schematic representation of 2D CA growth algorithm.

2.3. Simulation of recrystallization

Grain boundaries move driven by temperature gradient at a velocity established by the formulae $v = \frac{M(\theta_{ij},T)}{M_{\max}}\frac{\gamma(\theta_{ij})}{\gamma_{\max}}$ [11, 13]. Constants $M_{\max}$ and $\gamma_{\max}$ are the grain boundary mobility and grain boundary energy, respectively, when the boundary becomes a high angle grain boundary. In the calculations these constants are chosen to be equal to 1 [14]. The grain boundary energy can be calculated [18] by

$$\gamma = 3\gamma_{bound}\Delta D / D_1 D_2, \tag{2}$$

Where $\Delta D = D_1 - D_2$, D_1 and D_2 are the diameters of neighbouring grains, and $\gamma_{bound} = \frac{\theta}{\theta_{\max}}\left[1-\ln\left(\frac{\theta}{\theta_{\max}}\right)\right]$ [13]. Within the numerical implementation the grain diameter is calculated as the diameter of a circle of the same area:

$$D = \sqrt{\frac{4S_c}{\pi}} \approx h\sqrt{\frac{4n_c}{\pi}}, \tag{3}$$

where S_c is the area of a circle and n_c is the number of cells within the grain.

The grain boundary mobility is a function of misorientation between the two grains, which is given as [11, 13]:

$$M(\theta_{ij},T) = M_m \frac{\theta}{\theta_{\max}}\left[1-\ln\left(\frac{\theta}{\theta_{\max}}\right)\right], \tag{4}$$

$$M_m = M_0 \exp\left(-\frac{Q}{RT}\right), \tag{5}$$

where R is the universal gas constant, the constant M_0 and the activation energy for migration Q are chosen according to the data [9] for titanium alloy: $M_0 = 0.45$ m/s, $Q = 25$ kJ/mol.

3. Results

In order to demonstrate capabilities of the implemented algorithm based on the model mentioned above the simulations of grain growth of a nickel-based superalloy [6, 19, 20] and a titanium alloy [21, 22] during solidification process under different temperature conditions are performed.

3.1. Evolution of growing polycrystalline structure in a nickel-based superalloy

The computational domain is divided into 2000 × 500 regular square cells with a cell size of 50 μm. The time step in all calculations is 1 ms. The initial bulk temperature is 1609 K. Cooling is set at the boundaries of the computational domain, the temperatures are given in Table 1. Material properties and model parameters used in the simulations are listed in Table 1 [6, 19, 20]. Figure 2 shows the calculated temperature field at the beginning (5 ms) and at the end (35 ms) of solidification process. The corresponding solidification process is demonstrated in Figure 3, where different colours represent different crystallographic orientations of grains.

Table 1. Material properties and model parameters used in the simulations [7, 19, 20].

Property	Variable	Value
Density	ρ	7950 kg/m^3
Heat capacity	c_p	668 J/(kg·K)
Thermal conductivity	λ	30.3 W/(m·K)
Convection heat transfer coefficient	κ_i	1000 W/(m^2·K)
Environment temperature at the left boundary	T^{e_L}	1550 K
Environment temperature at the right, top and bottom boundaries	$T^{e_R} = T^{e_T} = T^{e_B}$	1605 K
Liquidus temperature	T_L	1609 K
Liquidus slope	m	-10.9 K/wt%
Initial concentration	C_0	4.85 wt%

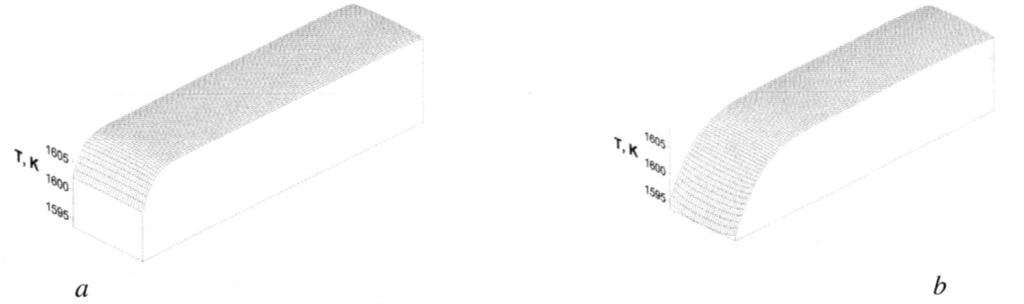

Figure 2. Temperature field evolution: $\tau = 5$ (*a*) and 35 ms (*b*).

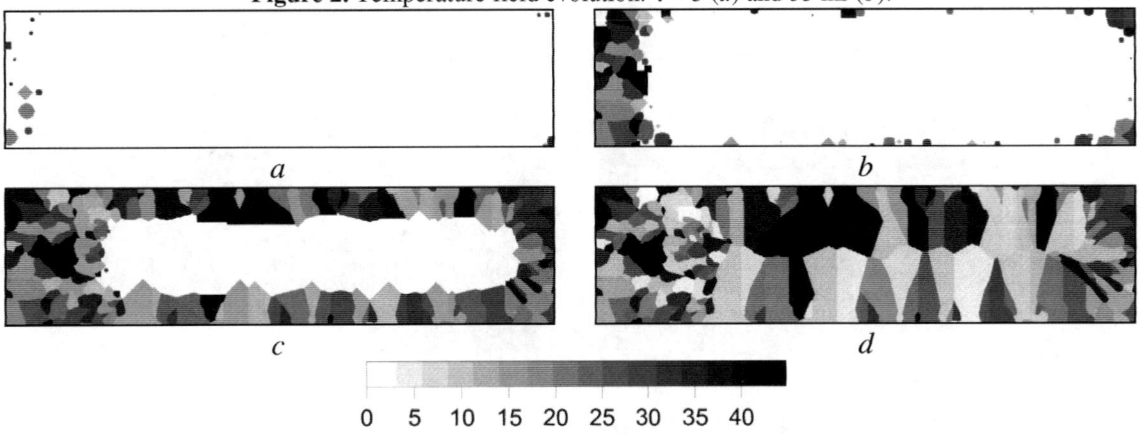

Figure 3. Numerical simulation for evolution of nickel-based superalloy microstructure: $\tau = 5$ (*a*), 10 (*b*), 20 (*c*) and 35 ms (*d*).

IOP Conf. Series: Materials Science and Engineering **71** (2015) 012073 doi:10.1088/1757-899X/71/1/012073

Figure 3a illustrates that the grains initially nucleate and grow at the left boundary in the area of the most intensive cooling and then at the other boundaries where the temperature is higher than at the left boundary but lower than in the center. Due to a high temperature area in the middle of the specimen, the probability of grain nucleation decreases towards the center of the ingot and growth of near-boundary grains primarily occurs (Figure 3d). Figure 3e illustrates the final microstructure. The shape and size of grains are irregular. While an equiaxed grain structure is observed at the left size, the most part of the ingot has a columnar microstructure.

3.2. Recrystallization of a titanium alloy under temperature gradient

Using the implemented algorithm we generated the initial microstructure. A titanium alloy was chosen as a model material. A detailed description of the thermo-mechanical properties of titanium specimens is given in [21, 22].

The computational domain is divided into 1200 × 400 regular square cells with a cell size of 50 μm. The time step is 1 ms. The 2D polycrystal is subjected to the temperature gradient so that T_{max} = 800 K on the right boundary, T_{min} = 400 K on the left boundary (Figure 4).

Figure 5a demonstrates the initial microstructure. Typical example of the microstructural evolution with time is shown in Figure 5. At the right side the growth of larger grains at the expense of smaller grains can be observed, which is a typical recrystallization feature [9, 11].

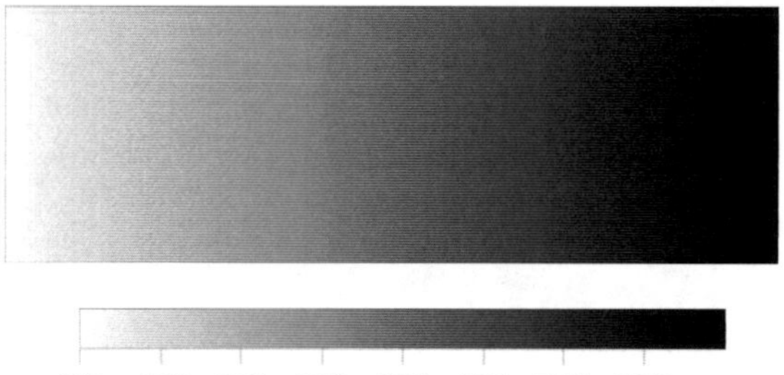

Figure 4. Temperature gradient.

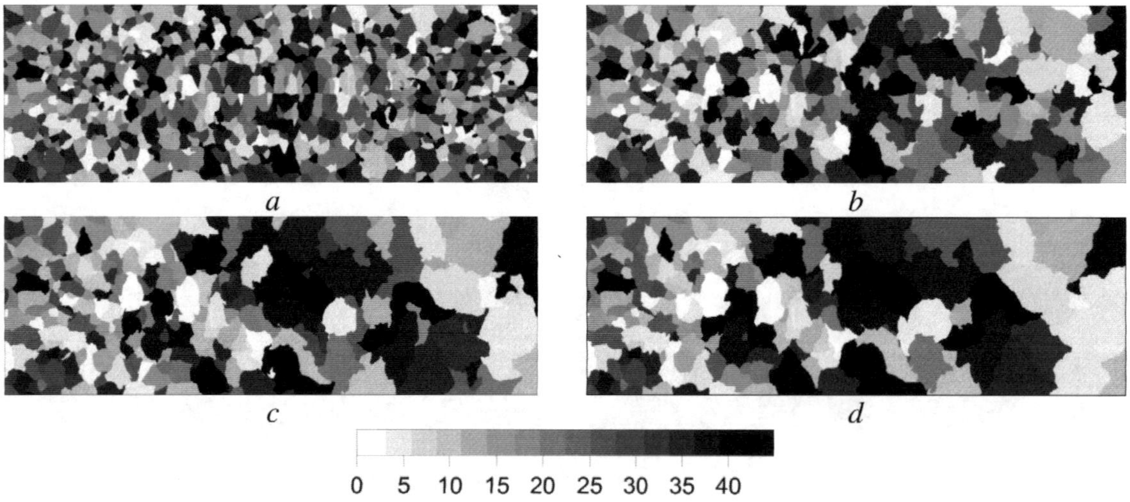

Figure 5. Initial calculated polycrystalline microstructure (a) and its evolution during recrystallization: 11 (*b*), 21 (*c*), and 33 ms (*d*).

IOP Conf. Series: Materials Science and Engineering **71** (2015) 012073 doi:10.1088/1757-899X/71/1/012073

4. Conclusions

The CA model realized enables us to describe the microstructure evolution during solidification and recrystallization processes. The grain growth model implemented offers various opportunities as, for example, to predict mechanical properties of metallic materials or to investigate mechanical defects arising during solidification and/or recrystallization such as e.g. hot cracks.

It should be mentioned that the results obtained have not been compared with concrete experimental data (grain size, shape, grain distribution within the mesovolume etc) which should be done to improve simulation results. Future work will be focused on the development of a 3D finite element and CA multiscale coupled model.

5. Acknowledgements

Results were obtained in the framework of international cooperation with the University of Bremen with partial funding from the National Research Tomsk State University (TSU) (International Youth Mobility Program) and Russian Science Foundation (grant № 14-19-00766) and a program to increase the competence of Tomsk Polytechnic University.

References

[1] Boettinger W J, Coriell S R, Greer A L *et al* 2000 *Acta Mater.* **48** 43-70
[2] Rappaz M and Gandin Ch-A 1993 *Acta Metall. Mater.* **41** 345-360
[3] Gandin Ch-A and Rappaz M 1994 *Acta Metall. Mater.* **42** 2233-2246
[4] Pavlyk V and Dilthey U 2004 *Numerical simulation of solidification structures during fusion welding* (Continuum scale simulation of engineering materials) ed D Raabe *et al* (Wiley-VCH Verlag GmbH & Co. KGaA) 727-743
[5] Gandin Ch-A and Rappaz M 1997 *Acta Metall.* **45** 2187-2195
[6] Wang W, Lee P D and McLean M 2003 *Acta Mater.* **51** 2971-2987
[7] Atwood R C and Lee P D 2003 *Acta Mater.* **51** 5447-5466
[8] Dong H B and Lee P D 2005 *Acta Mater.* **53** 659-668
[9] Humphreys F J and Hatherly M 2004 *Recrystallization and related annealing phenomena* (Kidlington, Oxford, UK: Elsevier Ltd) 574
[10] Raabe D 2002 *Annu. Rev. Mater. Res.* **32** 36-76
[11] Raghavan S and Sahay S S 2007 *Mater. Sci. Eng. A* **445-446** 203-209
[12] Raghavan S and Sahay S S 2009 *Comp. Mater. Sci.* **46** 92-99
[13] Gruber J 2007 *Interface texture development during grain growth* (Pittsburgh, US: a thesis ... for the degree of Doctor of Philosophy in Material Science)
[14] Kuznetsov G V and Sheremet M A 2007 *Difference methods for solving heat transfer problems* (Tomsk, Russia: Publishing house of Tomsk Polytechnic University) (*in Russian*)
[15] Samarskii A A 2001 *The theory of difference schemes* (New York, US: Marcel Dekker Inc.)
[16] Oldfield W 1966 *Trans. Am. Soc. Metals* **59** 945-960
[17] Nastac L 2004 *Modeling and simulation of microstructure evolution in solidifying alloys* (Boston, US: Kluwer Academic Publishers)
[18] Gorelik S S, Dobatkin S V and Kaputkina L M 2005 *Recrystallization of metals and alloys* (Moscow, Russia: MISIS) (*in Russian*)
[19] Zieliñska M, Yavorska M, Porêba M and Sieniawski J 2010 *Archives of Materials Science* **36** 35-38
[20] Fieldhouse I B and Lang J I 1961 *Measurement of thermal properties* (No. WADD-TR-60-904, Chicago IL: Armour Research Foundation)
[21] Fan Z, Sparks T E and Liou F *et al* 2007 *Numerical simulation of the evolution of solidification microstructure in laser deposition* (Rolla, MO, USA: report: Air Force Research Laboratory, United States Air Force OMB No. 0704-0188)
[22] Fallah V 2011 *Solidification in laser powder deposition of Ti-Nb alloys* (Waterloo, Canada: a thesis ... for the degree of Doctor of Philosophy in Mechanical Engineering)

IOP Conf. Series: Materials Science and Engineering **71** (2015) 012074 doi:10.1088/1757-899X/71/1/012074

On slow wave process in rocks

L B Zuev[1,2], S A Barannikova[1,2,3] and M V Nadezhkin[1,4]
[1]Institute of Strength Physics and Materials Science, SB RAS, Tomsk, 634021, Russia
[2]National Research Tomsk State University, Tomsk, 634050, Russia
[3]Tomsk State University of Architecture and Building, Tomsk, 634003, Russia
[4]National Research Tomsk Polytechnic University, Tomsk, 634050, Russia

E-mail: bsa@ispms.tsc.ru

Abstract. A study was made of the general regularities of localized plasticity development in deforming rocks. The potential usefulness of speckle photography techniques in applications to problems of deformation and fracture in rocks was explored and circumscribed. The evolution of localized plasticity in rocks was addressed using wave ideas. The tests were performed for the compression samples of marble, silvinite and sandstone; the deformation occurred in these materials via different micromechanisms of plasticity. By the deformation, autowaves would form in the compression samples of rocks. The autowave propagation rates are in the range ~10^{-5}...10^{-4} m/s (0.3...3 km/yr), which is close to slow motion rates observed in the earth crust after an earthquake or a rockslide. A correlation has been established between the calculated and experimental data on the time and coordinates of fracture in rocks.

1. Introduction

The search for law-like regularities of plastic flow development enabled one to predict novel phenomena in deforming solids. Thus on the macro-scale level the deformation was found to develop by stages, with each stage clearly distinguishable from the rest. The success in using wave ideas has reached its peak in a new concept of plastic deformation; an autowave model has been put forward to account for the inhomogeneous behavior of plastic flow in metals and alloys [1-3]. This has prompted a search for equivalent effects in other fields, in particular, geophysics, which might provide an insight into the problem of so-called slow motions, i.e. waves propagating at low rates of a few kilometers per year in rocks after an earthquake or a rockslide [4-7]. The very existence of slow motions, to say nothing of their nature, remains an open question thus far [7].

On the base of available evidence for mechanisms involved in viscously deforming rocks [9], we examined the different ways in which rock materials respond to stress. W-+e have also attempted to give numerical evaluation of the same response. The recognition and interpretation of deformation mechanisms was performed for rocks deforming via different mechanisms, i.e. dislocation glide or twinning in silvinite (NaCl+KCl) and marble ($CaCO_3$) or grain boundary sliding in sandstone (SiO_2).

2. Materials and experimental techniques

The ultimate goal of this study was to define deformation mechanisms which can operate in rocks and give numerical evaluation of the same. Therefore, the choice of marble, sandstone and silvinite was guided by a knowledge that these materials differ from one another in deformation mechanisms (Table 1). The rock samples were tested in compression along the axis x in a testing machine 'Instron-1185'

IOP Conf. Series: Materials Science and Engineering **71** (2015) 012074 doi:10.1088/1757-899X/71/1/012074

at cross head motion rates V_{mach} = $1.67 \cdot 10^{-3}$ and $1.67 \cdot 10^{-2}$ m/s. The compression-test diagrams were plotted for the test samples of rocks. These differ from the smooth flow curves $\sigma(\varepsilon)$ of most metals and alloys [9] in that they have 'saw-tooth' segments, which evidently correspond to a jump-wise decrease in the stress due to rock cracking locally [13]. The rock samples would undergo brittle fracture at the total deformation ε_{tot}= 1.7…3.5 %.

Using digital speckle image technique and an automated complex ALMEC-TV [1,10], the compression-test diagram was recorded for the test sample of rock; simultaneously, local strain patterns were being registered for the rock sample face [15,16]. The technique employed enables one to define all the components of the plastic distortion tensor $\beta_{ik} = \nabla \boldsymbol{r}$ (here r is the displacement vector, which can be determined from the speckle patterns). In what follows, we shall restrict our consideration to the axial strain $\varepsilon_{xx} = \partial u / \partial x$ (here u is displacement along the axis x). The spatial patterns $\varepsilon_{xx}(x)$ were analyzed and the distribution of local strain zones was thus obtained. Using the local nuclei's positions, X, as a function of deformation or time, the kinetics of macrolocalization patterns was examined.

3. Experimental results

Numerical evaluation of local strains was performed for the test samples of rocks. The results obtained suggest that on the macro-scale level the deformation would exhibit a localization behavior. Non-deformed material volumes are found to alternate with local strain zones; the deformation fronts travel at varying rates along the compression sample.

A localized deformation pattern obtained for the test sample of marble is illustrated in Figure 1; the deformation fronts are designated by the dashed lines and are numbered in the order of appearance. We take it as convincing evidence that on the macro-scale level the deforming material volume would undergo fragmentation with the fragment boundaries changing with time.

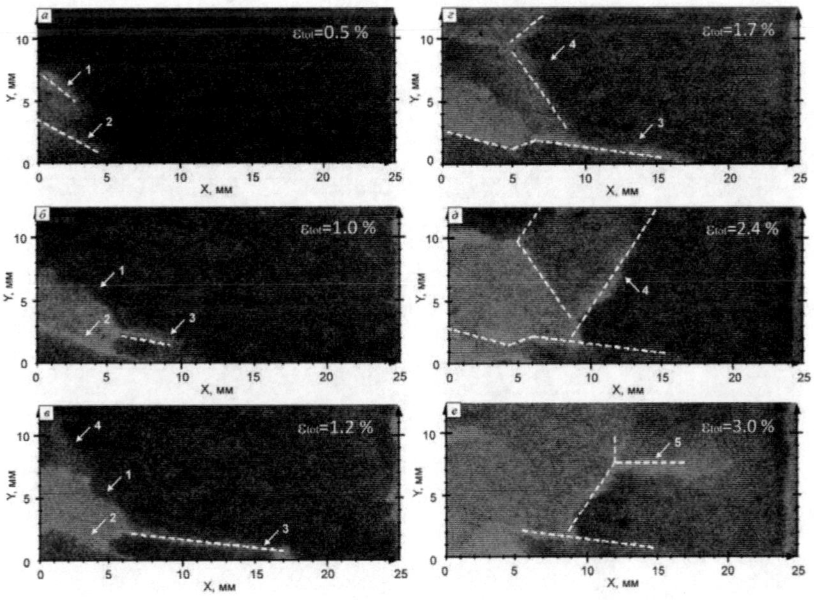

Figure 1. The localized deformation zones moving at different compressive strain levels (%) in the test sample of marble; the arrows point to the front numbers.

With growing total deformation, the emergent localized plasticity zones will start traveling over the sample surface, forming thereby a kind of autowave. The behavior of local strain zones will plot on the flow curve $\sigma(\varepsilon)$ as a linear segment. The motion rates of localized plasticity nuclei (autowaves) are listed in Table 1 for the studied rocks. It is found experimentally that the autowave motion rate V_{aw}, depends on the sample loading rate, V_{mach}, as $V_{aw} \approx 20V_{mach}$, which agrees with the experimental data obtained previously for metallic materials [1]. The motion rates of localized plasticity fronts were determined for the test samples of rocks using the complex ALMEC-TV; the values obtained are in the range (2.8…4.4)·10^{-5} m/s, i.e. 0.9…1.3 km/yr, which is close in order of magnitude to slow motion rates. One cannot but infer that slow motions are localized plasticity autowaves, which form deep in the earth crust where earthquakes or rockslides are generated. Thus, in the laboratory studies we examined the different ways in which rock materials respond to stress, in particular, the deformation processes occurring at rates close to slow motion rates.

Another regular feature of the autowave evolution is worthy of notice. It has been found previously that the evolution of localized plastic flow in metals ends in collapse of the autowave [1]. In the case of rocks, sandstone and marble the same regularity is observed. Of particular note is the point on the plot $X(t)$ at which collapse of the autowave occurs, since the point's coordinates coincide with the place and location of imminent sample fracture [10].

Table 1. The kinetics of localized deformation fronts in rock samples.

Rock	Silvinite	Marble	Sandstone
Fronts' motion rate at the linear stage, m/s	~2.8·10^{-5}	~4.2·10^{-5}	~3.0·10^{-5}
Wavelength, m	~10·10^{-3}	~(4±1) ·10^{-3}	~(4±1) ·10^{-3}
Interplanar distance, m	-	0.386·10^{-9}	0.41·10^{-9}
Transverse elastic wave rate, m/s	-	1905	1860
Coefficients in Equations (2) and (3)			
$\alpha\cdot 10^3$, c		3.08	1.7
$\alpha_0\cdot 10^6$, м/с	Failed to	-2.15	-1.9
$\xi^* = \alpha_0/\alpha$, мм	establish	-0.7	-1.1
$t^* = t_0 + 1/\alpha$, c		415	730
Matching of experimental and calculated times (4) and coordinates (5) of fracture			
t^*_{exp}/t^*_{calc}	Failed to	380/415	600/730
X^*_{exp}/X^*_{calc}	establish	6/3	5/6

The motion of local plasticity nuclei plots as a bundle of straight lines, since their rate depends linearly on the coordinate of the point of nucleation ξ. Hence, the relationship established in [1] will hold good, i.e.

$$V_{aw}(\xi) = \alpha\xi + \alpha_0, \tag{1}$$

where ξ – a coordinate counted from the stationary localization nucleus; α and α_0 – empirical constants. The values t^* and X^* are, respectively, the time of nucleus arrival and the place of sample fracture (see Table 1). These are calculated as

$$t^* = t_0 + 1/\alpha, \tag{2}$$

$$X^* = X_0 + \alpha_0/\alpha, \tag{3}$$

where t_0 – the instant of time at which the prefracture stage begins; X_0 – the position of the stationary nucleus calculated in the laboratory system of coordinates for the beginning of prefracture stage. The coordinates X^* and times t^* were calculated for marble and sandstone; the values obtained are $X^* \approx$ 3 mm and $X^* \approx$ 6 mm, respectively; $t^* \approx$ 415 s and 730 s, respectively. In point of fact, fracture occurred in the test samples of marble and sandstone within 380 s and 600 s, respectively, due to sample cracking at 6-mm and 5-mm distance, respectively, from the immovable clamp.

The values t^* and X^* were calculated and obtained experimentally. The correlation of two sets of data obtained for metals and rocks is given as

$$t_{calc} = t_0 + b \cdot t_{\exp}, \tag{4}$$

$$X_{calc} = X_0 + a \cdot X_{\exp}, \tag{5}$$

where t_0 = -18 s; b = 0.96; X_0 =-0.5 mm; a = 1.02.

There is a satisfactory agreement between the calculated and experimental sets of data. This suggests that the point where plots intersect (so-called 'pole') has coordinates which are close to the time and place of real fracture; hence, formulas (2) and (3) enable one to predict with sufficient accuracy the time and location of imminent fracture in the test sample.

On the base of the above, it is hypothesized that the place and time of fracture in rocks can be predicted using experimental linear dependencies $X(t)$, which admit of extrapolation to larger times. Similar problems were previously formulated in [11]. The response to stress of rock samples is in many ways similar to that of metals and alloys, which were used for plastic flow localization investigations; hence, one can look forward to a successful solution of this problem.

The work was performed in the frame of Fundamental Research Program of State Academy of Sciences for the period 2014-2020 yrs.

4. Conclusions

1. The development of localized plastic deformation in metals and rocks is found to have similar general features.
2. Quantitative assessment has been made of the autowave features of localized plasticity evolution in studied rocks.
3. It is shown that slow motions in rocks occur at rates close to the propagation rate of localized plasticity autowaves observed for the rock samples studied.

References

[1] Зуев Л Б, Данилов В И и Баранникова С А 2008 *Физика макролокализации пластического течения* (Новосибирск: Наука)

[2] Zuev L B, Danilov V I, Barannikova S A and Gorbatenko V V 2009 *Phys. of Wave Phen.* **17** 66-75

[3] Barannikova S A, Nadezhkin M V and Zuev L B 2009 On the localization of plastic flow under compression of NaCl and KCl crystals *PSS* **51 No 6** (Pleiades Publishing, Ltd.) pp 1142-1148

[4] Митлин В С и Николаевский В Н 1990 Нелинейная диффузия тектонических напряжений *ДАН СССР* **315** 1093-1096

[5] Быков В Г 2005 Деформационные волны Земли: концепция, наблюдения и модели. *Геология и геофизика* **46** 1176-1190

[6] Гольдин С В 2008 *Сейсмические волны в анизотропных средах* (Новосибирск: Изд-во СО РАН)

[7] Опарин В Н, Тапсиев А П, Розенбаум М А 2008 *Зональная дезинтеграция горных пород и устойчивость подземных выработок* (Новосибирск: Изд-во СО РАН)

[8] Ботвина Л Р 2008 *Разрушение. Кинетика, механизмы, общие закономерности* (Москва: Наука)

[9] Pelleg J 2013 *Mechanical properties of materials* (Dordrecht: Springer)

[10] Zuev L B, Barannikova S A, Gorbatenko V V and Nadezhkin M V 2013 On the observation of slow wave processes in deforming rock samples *Journal of Mining World Express* **2** 31-39

[11] Введенская А В 1969 *Исследование напряжений и разрывов в очагах землетрясений при помощи теории дислокаций* (Москва: Наука)

The study on the effect of the dislocation density on formation of a closed piecewise-continuous dislocation loop in aluminum

S N Kolupaeva[1], A E Petelin[2], Yu P Petelina[2] and K A Polosukhin[1]
[1]Tomsk State University of Architecture and Building, Department of Applied Mathematics, Tomsk, 634003, Russia
[2]National Research Tomsk State University, Department of Information Support of the Innovation Activity, Tomsk, 634050, Russia

E-mail: ksn58@yandex.ru, aepetelin@gmail.com

Abstract. Mathematical model allowing carrying out studies on energy, scale and time characteristics of formation of closed dislocation loop and crystallographic shear zone as a whole, and forming of a dislocation loop for FCC-materials, has been developed. Using the mathematical model study on the effect of the dislocation density on the dynamics of formation of an elementary crystallographic slip bounded by a closed piecewise-continuous dislocation loop in aluminum has been carried out. It has been shown that with decrease in the density of dislocations in metal from 10^{-12} up to $4\cdot10^{-11}$ along different orientations of the expansion of the dislocation loop the maximum value of the dislocation velocity increases by 20-30% and the maximum value of the kinetic energy and the dislocation path increases by almost 2 times. It has been shown that in aluminum the area of elementary crystallographic slip, swept out by the dislocation loop, increases with a decrease in the dislocation density in metal. But, at the same time, regardless of the value of the dislocation density in metal the radius of the dislocation loop in the final configuration along the screw orientation is practically by an order of magnitude less than along the edge orientation.

1. Introduction

Studies on strength and plasticity of metals on the basis of the dislocation theory have a long history. The number of proposed dislocation mechanisms, experimental observations, and theoretical calculations is very large. But in view of the complexity of studies on the dynamics of dislocations this line of research is still relevant.

Peculiarities of formation of a dislocation structure depend on various factors, including: the applied stress, the temperature, and the dislocation density in material, as well as many others. During the expansion the dislocation loop overcomes many obstacles of various nature and strength [1–3]. For several reasons the dislocation is usually not the only one. A series of dislocations that form the crystallographic shear zone occurs at a high velocity (significantly higher than the velocity of the macroscopic strain) as a result of the stability loss by a dislocation source. In turn, stress fields that occur during formation of the crystallographic shear zone can activate other dislocation sources, resulting in expansion of plastic strain autowaves along the strained crystal [4]. The dynamic behavior of the dislocation under conditions of the stability loss can lead to multiple effects that cannot be predicted within the framework of the conventional concept of the stationary thermally activated motion of dislocations [5, 6].

2. Mathematical model of the dislocation dynamics

To describe the dynamics of expansion of dislocation loops emitted by one dislocation source and formation of the crystallographic shear zone a mathematical model has been developed [7], which enables to study energy, scale and time characteristics of a closed piecewise-continuous dislocation loop in FCC-metals:

$$
\begin{aligned}
d\varepsilon_k^{(j)}/dt = \Bigg[& \frac{1}{4}(\pm\tau b - \tau_R b - B\mathrm{v}^{(j)})T^{(j)} - \left(8\sigma\sqrt{2\pi}\right)^{-1} e^{-\frac{\sin(\beta^{(j)})^2}{2\sigma^2}} p_j p_s \xi G b^2 \rho r^{(j)} \mp \\
& \mp \frac{Gb(i-1)\left(\cos^2(\beta^{(j)}) + \sin^2(\beta^{(j)})/(1-\nu)\right)}{2\pi(\underline{r_1^{(j)}} - r^{(j)})} \mp \frac{(\mu_0^{(j)} + \varepsilon_k^{(j)})}{2} Q^{(j)} \Bigg] \mathrm{v}^{(j)},\ j = \overline{1,m} \qquad (1) \\
dr^{(j)}/dt = \pm & \, \mathrm{v}^{(j)},\ j = \overline{1,m}.
\end{aligned}
$$

where $r^{(j)}, \mathrm{v}^{(j)}$, $\varepsilon_k^{(j)}$ are: the current radius, the velocity, and the kinetic energy along the *j*-th orientation; t is the motion time of the dislocation loop; τ is the applied stress; τ_R is the stress of the lattice strain and the impurity strain, and the dislocation resistance to the dislocation motion; G is the shear modulus; b is the Burgers vector module; B is the coefficient of the viscous deceleration of the dislocation; ρ is the density of dislocations in the material; p_j is a fraction of jog-forming dislocations of non-coplanar systems, p_s is a fraction of jogs located on near-screw segments of the dislocation loop, ξ is the Smallmen factor; ν is the Poisson's ratio; σ is the dispersion of a normal distribution of the intensity of point defect generation beyond jogs on the dislocation; $T^{(j)}$, $Q^{(j)}$ is the combination of physically determined parameters [7]; $\mu_0^{(j)}$ is the linear tension of the stationary immobile dislocation; $\beta^{(j)}$ is the angle between the Burgers vector and the normal to the *j*-th orientation; $\beta^{(m+1)} = \beta^{(1)}$; $r^{(m+1)} = r^{(1)}$, m is the number of sides of the polygonal shape of the dislocation loop.

When writing the mathematical model (1) an assumption was used that the calculation was carried out in a polar coordinate system, and the dislocation loop in the initial configuration was represented in the form of a regular polygon with infinitely small sides and the center coinciding with the center of the coordinate system. For the dislocation loop, the boundary components are normal to the polar axis, the screw components are parallel to it, and the generation of point defects, according to the experimental data [8], is carried out mainly within the limits of 15° from the screw orientation of the dislocation loop.

In the model (1) Peach-Koehler forces are taken into account, as well as the dissipative forces caused by: the lattice, the impurity and dislocation deceleration; the linear tension; the viscous deceleration; the stress from accumulation of dislocations and generation of point defects beyond jogs on the dislocation. Besides, the dependence of the linear tension and the intensity of generation of point defects on the orientation of the Burgers vector with respect to the dislocation line are taken into account.

The article presents the results of the study on the evolution of the first dislocation loop emitted by a dislocation source at various densities of dislocations in aluminum, obtained using the mathematical model (1).

3. Results of the computational experiment

For implementation of the mathematical model (1) and implementation of computational experiments a set of software (Dislocation Dynamics of Crystallographic Slip [9, 10]) has been created, which implements the most appropriate for the solution of the system (1) numerical Gere method of the variable order and the integration step. The software complex has a developed user interface that allows researchers without a broad experience in programming and knowledge in the field of numerical methods to work with a set of programs.

Computational experiments have been carried out at values of parameters of the mathematical model specific to aluminum at room temperature: the stress of the lattice and impurity deceleration of

1 MPa acting on the dislocation source with a stress of 10.7 MPa (equal to the critical stress of a dislocation source at a dislocation density of 10^{12} m^{-2}). The density of dislocations in metal in various calculations has ranged from $4{\cdot}10^{11}$ m^{-2} up to 10^{12} m^{-2}.

As the result of the carried out study it has been shown that the dislocation loop, emitted by a dislocation source, quickly gains the terminal velocity (Figure 1, *a, d*), which is consistent with the experimental results [11]. At the same time, the terminal velocity of the dislocation is nearly by an order of magnitude less than the transverse sound velocity in metal.

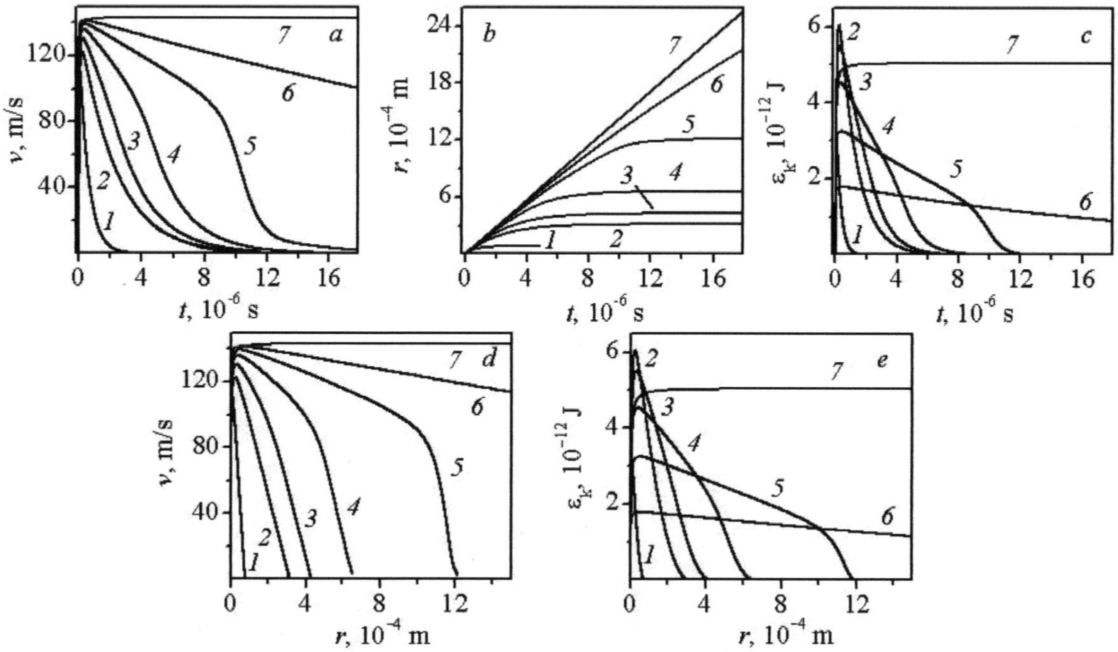

Figure 1. The dependency of the current velocity (*a*), the dislocation radius (*b*), and the kinetic energy (*c*) per unit length of the dislocation on the path time, and the dependency of the velocity (*d*) and the kinetic energy (*e*) per unit length of the dislocation on the dislocation radius in aluminum along different orientations of the dislocation loop: *2* – β = 0° (screw dislocation), *3* – β = 30°, *4* – β = 45°, *5* – β = 60°, *6* – β = 90° (edge dislocation).

The velocity of the dislocation loop decreases after reaching the limiting value at the initial path of the dislocation. A three-stage decrease in the velocity is clearly observed in the dislocation along the edge and nearby orientations: 1) the stage of slow decrease in the velocity; 2) the stage of rapid decrease in the velocity; 3) the stage of the dislocation deceleration. The closer the orientation of the dislocation loop to the orientation of the edge dislocation, the slower the velocity decreases at the stage 1, the faster the velocity decreases at the stage 2, and, hence, the stage 3 is shorter. The velocity of the dislocation along the screw and nearby orientations after reaching the terminal velocity at the initial path significantly decreases, reaching a value close to zero, and a long-term stop of the dislocation takes pace (Figure 1, *a, d*).

The radius of the dislocation increases quasi-linearly along the edge and nearby orientations of the dislocation loop. For dislocations along other orientations, after the stage of an increase in the radius, a deceleration stage is observed at which the radius remains practically unchanged (Figure 1, *b*). At the same time, the deceleration stage becomes more pronounced at a decrease in the angle between the dislocation line and the Burgers vector (angle β).

At the initial path of the dislocation loop a sharp increase in the kinetic energy per unit of the dislocation length along all orientations is observed, alternating with a smooth decrease in the kinetic energy. The smaller the angle between the dislocation line and the Burgers vector, the greater the maximum value of the kinetic energy; and for the screw dislocation it is almost 3 times greater than for the edge dislocation. Further expansion of the dislocation loop leads to a decrease in the kinetic energy along all orientations; and the greater the limiting value of the kinetic energy of the dislocation, the higher the rate of a decrease in the kinetic energy (Figure 1, *c*, *e*).

As a result of the carried out study it has been shown that the velocity and the radius of the dislocation, as well as the kinetic energy per unit length of the dislocation, increase along all orientations at a decrease in the density of dislocations in metal at the same path time and path length, (Figure 2).

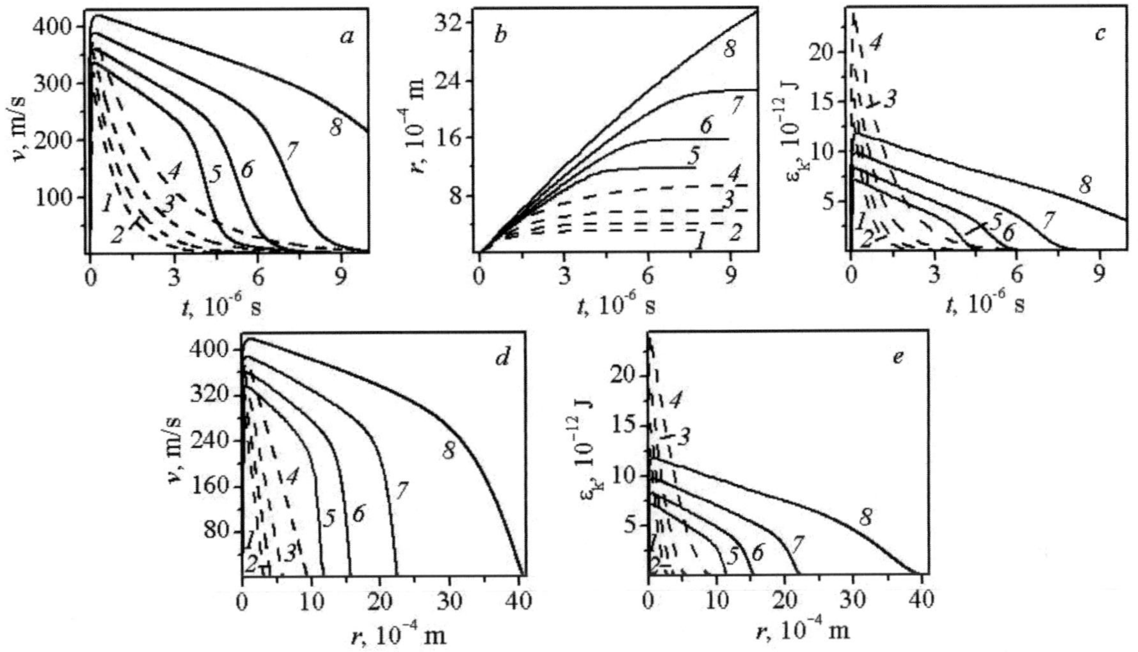

Figure 2. The dependency of the current velocity (*a*), the radius of the dislocation (*b*), and the kinetic energy (*c*) per unit length of the dislocation on the path time, and the dependency of the dislocation velocity (*d*) and the kinetic energy (*e*) per unit length of the dislocation on the dislocation radius in aluminum at various densities of dislocations in metal m^{-2}: 1, 5 – 10^{-12}, 2, 6 – $8\cdot10^{-11}$, 3, 7 – $6\cdot10^{-11}$, 4, 8 – $4\cdot10^{-11}$; and along various orientations of the dislocation loop: broken curve – $\beta = 0°$ (screw dislocation), continuous curve – $\beta = 60$.

For the screw dislocation, with a decrease in the density of dislocations in metal from 10^{-12} up to $4\cdot10^{-11}$ the maximum value of the velocity increases by nearly 30%, the kinetic energy – practically by 2 times, and the radius increases by more than 2 times. For the edge and mixed dislocations, with a decrease in the density of dislocations in metal from 10^{-12} up to $4\cdot10^{-11}$ the maximum value of the velocity, the kinetic energy, and the path length of the dislocation is slightly smaller (the difference is 15 % maximum) than for the screw dislocation. At a decrease in the dislocation density a three-stage decrease in the velocity of mixed dislocations, observed after reaching a limiting value of the velocity at the initial path of the dislocation, becomes less pronounced (Figure 2, *a*), the velocity and the kinetic energy decrease, and the radius increases more evenly (Figure 2, *b*–*e*) .

It has been shown that at the initial path the dislocation loop has the shape close to elliptical with a minor semi-axis perpendicular to the Burgers vector (Figure 3, *a*). After the initial path along the screw and nearby orientations in aluminum as well as in copper [7] a concavity area in the dislocation loop occurs (Figure 3, *b*), increasing until the dislocation stops (Figure 3, *c*, *d*). In the final configuration, the radius of the dislocation loop along the screw orientation is practically by an order of magnitude less than along the edge orientation. The width of the dislocation loop increases along all orientations at a decrease in the dislocation density in material and at the same path time.

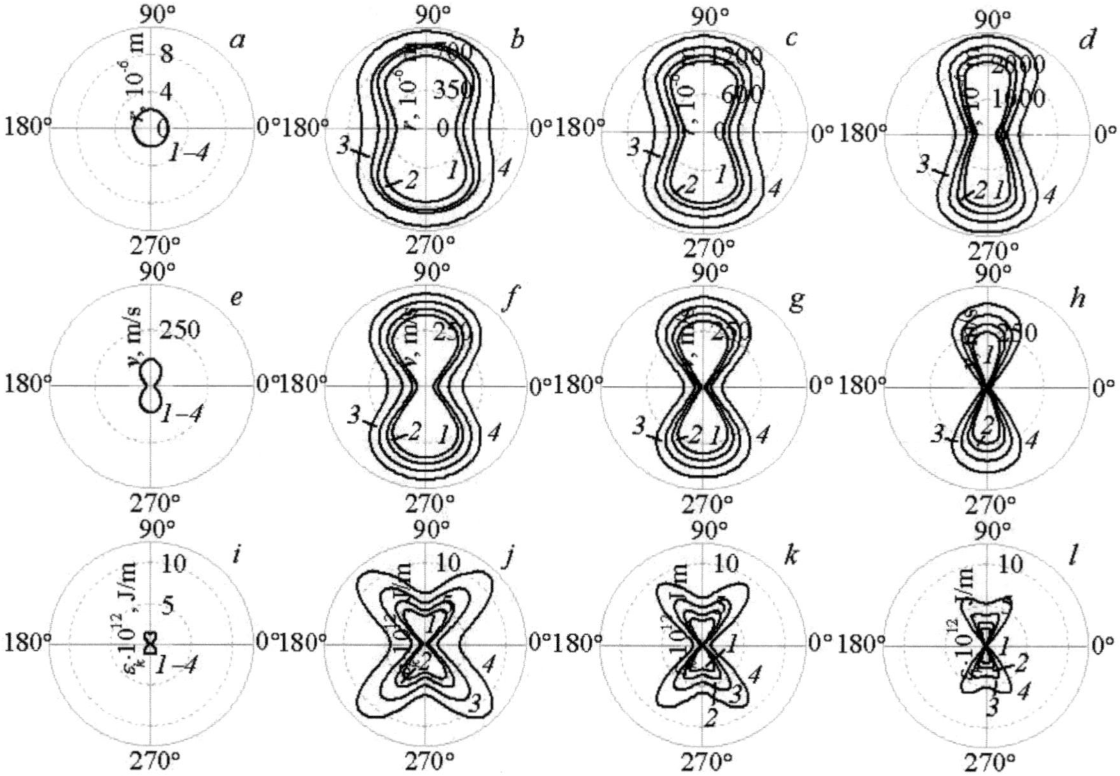

Figure 3. The shape of the dislocation loop, the velocity, and the kinetic energy per unit length of the dislocation loop along various orientations in aluminum, calculated at different time points from the beginning of the motion, s: *a, e, i* – 10^{-9}; *b, f, j* – $2{\cdot}10^{-6}$; *c, g, k* – $3.5{\cdot}10^{-6}$; *d, h, l* – $7{\cdot}10^{-6}$; at various density of dislocations, m^{-2}: 1 – 10^{-12}, 2 – $8{\cdot}10^{-11}$, 3 – $6{\cdot}10^{-11}$, 4 – $4{\cdot}10^{-11}$.

The expansion velocity of the dislocation loop is maximal in the direction of the edge and nearby orientations, and is minimal in the direction of the screw orientation (Fig. 3, *e–h*). Deceleration of the dislocation loop along the screw orientation is caused by the intensity of generation of point defects beyond jogs on the dislocation. The edge dislocation is stopped by the action of the linear tension during deceleration of the dislocation loop along other orientations. Both of these factors affect mixed dislocations in a proportion that depends on the proximity to the screw or the edge orientation. The kinetic energy per unit length of the dislocation loop, on a larger part of the path, has a maximum value in the direction of 60–70° from the screw orientation of the dislocation (Figure 3, *i–l*), reaching the value 1.5–2 times higher than the kinetic energy along the edge orientation (Figure 3, *j*). During deceleration of the dislocation loop the kinetic energy per unit length of the dislocation decreases the quickest along the screw and nearby orientations (Figure 3, *k, l*). With a decrease in the density of dislocations in aluminum the velocity and the kinetic energy of the dislocation loop increases along all orientations.

IOP Conf. Series: Materials Science and Engineering **71** (2015) 012075 doi:10.1088/1757-899X/71/1/012075

4. Conclusion

As a result of the carried out study it has been shown that a decrease in the density of dislocations in metal leads to an increase in the velocity and the radius of the dislocation, as well as the kinetic energy per unit length of the dislocation. With a decrease in the dislocation density in metal from 10^{-12} m^{-2} up to $4{\cdot}10^{-11}$ m^{-2} along different orientations of expansion of the dislocation loop the maximum velocity of the dislocation increases by 20–30% , the maximum value of the kinetic energy and the dislocation path increase by almost 2 times. It has been shown that in aluminum, as well as in copper, at the initial path the shape of the dislocation loop, initially presented in the form of a regular polygon, is close to elliptical with a minor semi-axis perpendicular to the Burgers vector. After the initial path a concavity area of the dislocation loop occurs, increasing until the dislocation stops. The area of the elementary crystallographic slip, swept out by the dislocation loop, increases with a decrease in the density of dislocations in metal. The radius of the dislocation loop in the final configuration along the screw orientation is almost by an order of magnitude less than along the edge orientation, regardless of the value of the dislocation density.

References

[1] Veselov V I, Nichugovskii G I and Predvoditelev A A 1981 *Sov Phys J* **24** 857–861

[2] Starenchenko V A, Zgolich M V and Kurinnaya R I 2009 Formation of extended compound dislocations and barriers in the course of interdislocation reactions in FCC crystals *Russ Phys J* **52** 245–251

[3] Starenchenko V, Cherepanov D, Kurinnaya R, Zgolich M and Selivanikova O 2014 The influence of dislocation junctions on accumulation of dislocations in strained FCC – single crystals. *Advanced Materials Research* **1013** 272–279

[4] Panin V E, Zuev L B, Danilov V I and Mnikh N M 1989 *Akademiia Nauk* SSSR **308** 1375–1379

[5] Slobodskoi M I and Matyushchenko A V 1997 *Russ Phys J* **40** 696–700

[6] Popov L E, Slobodskoi M I and Kolupaeva S N 2006 *Russ Phys J* 49 62-73

[7] Kolupaeva S N and Petelin A E 2014 *Russ Phys J* **57** 152–158

[8] Pfeeffer K H, Schiller P and Seeger A 1965 *Phys. Status Solidi* **8** 517–532

[9] Kolupaeva S N and Petelin A E 2011 *Vestnik of TSUAB* **3** 159–163

[10] Kolupaeva S N, Petelin A E, Petelina Y P and Polosukhin K A 2014 *Advanced Materials Research* **1013**. 280–286

[11] Gilman John J. John 2009 *Chemistry and Physics of Mechanical Hardness* (New Jersey: Wiley & Sons, Inc.) p 214

IOP Conf. Series: Materials Science and Engineering **71** (2015) 012076 doi:10.1088/1757-899X/71/1/012076

Mathematical modeling of the deformation behavior of heterophase alloys at various characteristics of the strengthening phase

O I Daneyko[1], T A Kovalevskaya[1], S N Kolupaeva[2] and N A Kulaeva[1]
[1]Tomsk State University of Architecture and Building, Department of Theoretical Mechanics, Tomsk, 634003, Russia
[2]Tomsk State University of Architecture and Building, Department of Applied Mathematics, Tomsk, 634003, Russia

E-mail: olya_dan@mail.ru

Abstract. Further development of techniques and technologies urgently requires the creation of new materials with unique properties. One of the most popular and promising areas in materials science is the creation of composite materials with desired properties. Currently, in practice peening by dispersed particles is widely used to improve the properties of metal materials. It is necessary to investigate in detail the influence of various factors on the behavior of dispersion-hardened materials in a variety of conditions to control the properties of materials. Physical experiment, in addition to large investments, often takes a long time. This paper presents a mathematical model of plastic deformation of dispersion-hardened materials with a face-centered cubic (fcc) matrix and with incoherent and coherent nanoscale particles. Application of this model can significantly reduce the time of research and necessary amount of physical experiment. The results allow optimizing of experiment, identifying ways to improve the properties of the materials.

1. Introduction

Composite materials of various types are widely used in modern industry. The application of composites based on metal matrices strengthened by second phase remains of interest in materials science to improve the mechanical properties of materials for various applications. In this work, we present a model of plastic deformation in disperse-strengthened materials. Our investigation of the effect characteristics of the dispersed phase and conditions of deformation have on the plastic behavior of materials with FCC matrices is described. Mathematical modeling makes it possible to vary the characteristics of a material and the actions on it over quite wide ranges, anticipating and possibly even modifying the experiments. The mathematical model used here consists of a system of differential equations of balance of the elements of the deformational defect medium of the material.

2. The mathematical model

The mathematical model used here consists of a system of differential equations of balance of the elements of the deformational defect medium [1-3], allowing to conduct a study of the plastic behavior of a material with nanoscale particles of incoherent and coherent types [4, 5]. Dislocations bend around incoherent particles during plastic deformation, which, in addition to the hardening effect, causes the appearance of the new elements of the dislocation structure. Coherent zones and particles in the deformation process are cut by dislocations, and the work done during the passage through the allocation of the first dislocation, further contributes to the flow stress [6-8]. Resistance to the motion of dislocations in the alloy with coherent and incoherent particles caused by stress of interaction with forest dislocations, by the friction stress τ_f, by the stress of cutting of particles, by the stress of bend around incoherent particles obtained in the form $\tau_a = \tau_f + \alpha G b\sqrt{\rho} + (1-\mu)\Gamma_s \delta^2 / 2\sqrt{6}\Lambda_p^3 + \mu G b/\Lambda_p - \delta$, where Γ_s is the surface energy of the matrix–segregation interface per unit area, Λ_p is the distance between particles of the second phase, δ is the diameter of the particles, α is the parameter of interdislocation interactions, G is the shear modulus, b is the magnitude of the Burgers vector, ρ is the density of dislocations, μ is fraction of incoherent particles in the material.

If the strengthening particles are coherent with the matrix, then in the course of the plastic deformation the emergence of the following types of defects is possible: shear-forming dislocations (we denote their density as ρ_m) and point defects: interstitial atoms (with concentration c_i), monovacancies (c_{1v}), and bivacancies (c_{2v}). If the material contains incoherent non-deformable hardening particles, the system of equations is supplemented by the balance equations of dislocation prismatic loops of interstitial (ρ_p^i) and vacancy (ρ_p^v) type, and by the dislocations in dipole configurations of vacancy (ρ_d^v) and interstitial (ρ_d^i) type. Moreover, in the equation for shear-forming dislocations the summands appear typical for materials with incoherent particles [1, 9-13]. The mathematical model in this case has the form:

$$\frac{d\rho_m}{da} = (1-\omega_s P_{as})\frac{F}{Db} - \frac{2}{\dot{a}}(1-\omega_s)\rho_m^2 b \min(r_a, \rho_m^{-1/2})(c_{2v}Q_{2v} + c_{1v}Q_{1v} + c_i Q_i) + \\ + \frac{2\alpha}{\dot{a}} b\sqrt{\rho}(\rho_p^v(c_{1v}Q_{1v} + c_{2v}Q_{2v}) + \rho_p^i c_i Q_i) + \frac{2b}{\dot{a} r_a}(\rho_d^i c_i Q_i + \rho_d^v(c_{1v}Q_{1v} + c_{2v}Q_{2v})), \tag{1}$$

$$\frac{d\rho_p^i}{da} = \frac{\mu < \chi > \delta}{2\Lambda_p^2 b} - \frac{2\alpha}{\dot{a}}\sqrt{\rho}\rho_p^i b(2c_{2v}Q_{2v} + c_i Q_i + 2c_{1v}Q_{1v}), \tag{2}$$

$$\frac{d\rho_p^v}{da} = \frac{\mu < \chi > \delta}{2\Lambda_p^2 b} - \frac{2\alpha}{\dot{a}}\sqrt{\rho}\rho_p^v b(c_{2v}Q_{2v} + 2c_i Q_i + c_{1v}Q_{1v}), \tag{3}$$

$$\frac{d\rho_d^v}{da} = \frac{\mu}{\Lambda_p b} - \frac{2b}{\dot{a} r_a}\rho_d^v(c_{2v}Q_{2v} + c_i Q_i + c_{1v}Q_{1v}), \tag{4}$$

$$\frac{d\rho_d^i}{da} = \frac{\mu}{\Lambda_p b} - \frac{2b}{\dot{a} r_a}\rho_d^i(c_{2v}Q_{2v} + c_i Q_i + c_{1v}Q_{1v}), \tag{5}$$

$$\frac{dc_i}{da} = q\frac{\tau_{dyn}}{G} - \frac{c_i}{\dot{a}}[((1-\omega_s)\rho_m + \rho_p + \rho_d)b^2 Q_i + Q_{1v}c_{1v} + Q_{2v}c_{2v} + Q_i(c_{1v} + c_{2v})], \tag{6}$$

$$\frac{dc_{1v}}{da} = \frac{q\tau_{dyn}}{6G} - \frac{1}{\dot{a}}[(((1-\omega_s)\rho_m + \rho_p + \rho_d)b^2 + c_i + c_{1v})Q_{1v}c_{1v} + Q_i c_i c_{1v} - (Q_{2v} + Q_i)c_i c_{2v}], \tag{7}$$

$$\frac{dc_{2v}}{da} = \frac{5q\tau_{dyn}}{6G} - \frac{2}{\dot{a}}[(((1-\omega_s)\rho_m + \rho_p + \rho_d)b^2 + c_i)Q_{2v}c_{2v} + Q_i c_i c_{2v} - Q_{1v}{c_{1v}}^2]. \tag{8}$$

Here a is the shear strain, D is the diameter of the shear zone, τ_{dyn} is the excess tension over the static resistance to motion of dislocations, F is a parameter defined by the shape of the dislocation loops and their distribution over the shear zone, ω_s is the fraction of screw dislocations, q is a parameter defining the intensity of generation of point defects; $<\chi>$ is the mean value of a parameter characterizing the *geometry* of the dislocations on the particles, $Q_j = Z_j \nu_D \exp(-U_j^{(m)}/kT)$ is kinetic coefficient, $U_j^{(m)}$ is the activation energy of migration of point defects of jth type, Z_j is the number of sites accessible for a jump of a defect of jth type ($j = i, v$), ν_D is the Debye frequency, k is the

Boltzmann constant, T is the deformation temperature, $\rho_p = \rho_p^i + \rho_p^v$ is the density of prismatic dislocation loops, $\rho_d = \rho_d^i + \rho_d^v$ is the density of dislocations in dipole configurations.

The model takes account of the main processes of generation, annihilation, and relaxational transformation of dislocations of various types and also of point defects. Mutual conversions are possible between the structural elements in the course of deformation and subsequent relaxation processes. Here the nature and result of interactions of elements of the dislocation structure with the particles can vary with variation of the relationships between the scale characteristics of the strengthening phase (size and shape of particles and distance between them) and the distance between dislocations [3]. The balance equations of the dislocations take into account annihilation of screw dislocations by cross-slip and of non-screw dislocations by climb on account of precipitation on them of point defects. The balance equations of the shear-forming dislocations take into account the fact that with growth of the prismatic dislocation loops and increase in size of the shoulder of the dislocation dipoles, they lose stability and in their subsequent behavior become similar to shear-forming dislocations [1, 2]. The balance equations of the point defects take into account the full set of pairwise interactions between point defects.

The model takes into account that when the critical density of dislocations ρ_c (whose magnitude is determined by the scale characteristics of the strengthening phase) is reached in the material [1], dislocation dipole structures begin to form as a result of incomplete annihilation of the dislocation branches bending the non-deformable particles as they overcome them.

The system of equations of the mathematical model also includes an equation linking the strain rate with the applied action and characteristics of the defect medium [2]:

$$\dot{a} = \frac{8}{\pi}\frac{v_D B\beta_r^{1/2}}{\xi^{1/6}F(1-\beta_r)}\frac{\tau^3(((1-\beta_r)\rho_m + \rho_p + \rho_d)(\tau-\tau_a))^{1/3}}{G^{4/3}b^{1/3}(\tau^2 - G^2b^2\xi\beta_r\rho_m)\rho_m^{1/2}}\exp[-\frac{0{,}2Gb^3 - (\tau-\tau_a)\Lambda b^2}{kT}].$$

Here B is a parameter defined by the probability of formation of the dislocation barriers bounding the shear zone, β_r is the fraction of reacting forest dislocations, ξ is the fraction of forest dislocations, Λ is the average length of a free dislocation segment and τ is the applied tension.

3. The results of research

The calculations assumed that deformation of the crystal occurs to a constant strain rate $\dot{a}$. The initial concentration of point defects is determined by the thermodynamic equilibrium concentration of point defects of the corresponding type at a given temperature, the initial density of shear-forming dislocations taken equal to 10^{12} m^{-2}. The size of hardening particles varied from 10 to 50 nm, the distance between the particles varied from 200 to 1000 nm.

Consider a material with hardened coherent and incoherent particles (Figure 1), deforming stress in the material increases during the whole process of plastic deformation. However, at temperatures above room temperature rate of increase the deforming stress is reduced. Decrease in the flow stress caused by more intense annihilation processes in dispersion-hardened material at temperatures above 293 K, compared with low temperatures (Figure 1).

Form of the curves of the flow stress depends strongly on the ratio of scale characteristics of the hardening phase (Figure 1). If the material in the deformation process has not achieved the critical density of dislocations, the stress-strain curves are smooth and monotonically increasing (Figure 1 *a, b, c, e*). If the material reaches a critical density of dislocations, the stress-strain curves have a kink due to appearance of new types of defects, it is the dislocation dipoles. With increasing temperature the length of the subcritical region increases. When the distance between the particles (Figure 1 *c, f, i*), as well as particle size increases, the critical density of dislocations is attained with a smaller amount of deformation (Figure 1 *g, h, i*), i.e. begin to form dipoles. Thus the rate of generation of dislocation dipoles is significantly higher in materials with a smaller distance between the particles. Consequently, when dipoles are begun to be formed, their density increases rapidly and becomes higher than the density of dislocation dipoles in the material with a large distance between the particles. This causes the greater strain hardening in the transcritical region in materials with a smaller distance between the strengthening particles (Figure 1 *c, f, i*).

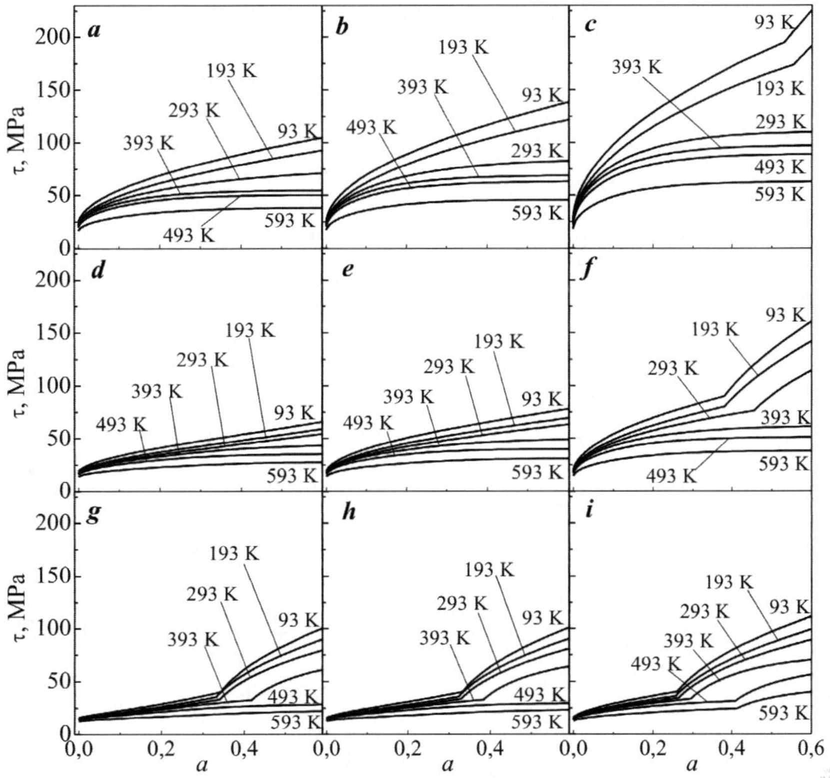

Figure 1. Stress-strain curves on the strain for an aluminum-based alloy at different temperatures. The fraction of incoherent particles is 25%. Distance between particles, nm: *a, b, c* – 200, *d, e, f* – 400, *g, h, i* – 1000; particle diameter, nm: *a, d, g* – 10, *b, e, h*– 20, *c, f, i* – 50. Strain rate 10^{-2} s^{-1}.

The greatest strain hardening at all temperatures is observed in the material having the greatest volume fraction of the strengthening phase (Figure 1 *c*). When increasing the size strengthening particles and reducing the distance between them the strain hardening increases in the subcritical region of dislocation densities. But when the critical density of dislocations is reached in the material, this dependence is not unequivocal (compare curves at temperatures of 93-393 K in figure 1 *b, e, h*).

The density of shear-forming dislocation (Figure 2 *a-c*) increases during deformation, where in the shape of the curves are different dependence on the temperature of deformation. The density of shear-forming dislocations as a function of the scale characteristics the strengthening phase is ambiguous: at low temperatures (93-193 K) the density decreases with increasing size of strengthening particles, whereas at higher temperatures (293 K or more) it increases.

The dependence of the density of shear-forming dislocations on the temperature is nonmonotonic. This is due to the fact that at moderate temperatures the contribution from the transition of the dislocations of prismatic loops and dipole configurations in the category of shear-forming dislocations is quite large, and at high temperatures it is reduced due to increased intensity of the annihilation process (Figure 2 *a, b, c*). At low temperatures the density of shear-forming dislocations is the lowest, since the contribution from the transition of prismatic loops and dipole configurations is negligible.

The dislocation density in the prismatic loops decreases with increasing strain temperature and decreasing the size of strengthening particles (Figure 2 *d-f, g-i*). At high temperatures the annihilation processes are more intensive. Dipole configurations begin to form in materials with larger reinforcing particles (Figure 2 *j-l*), at low temperatures it occurs earlier than at high. In the material with strengthening particles of diameter 50 nm (Figure 2 *l*), the contribution of dislocation dipoles in the total dislocation density becomes dominant at low and medium temperatures, but at a lower particle size dipole configurations in the material is absent.

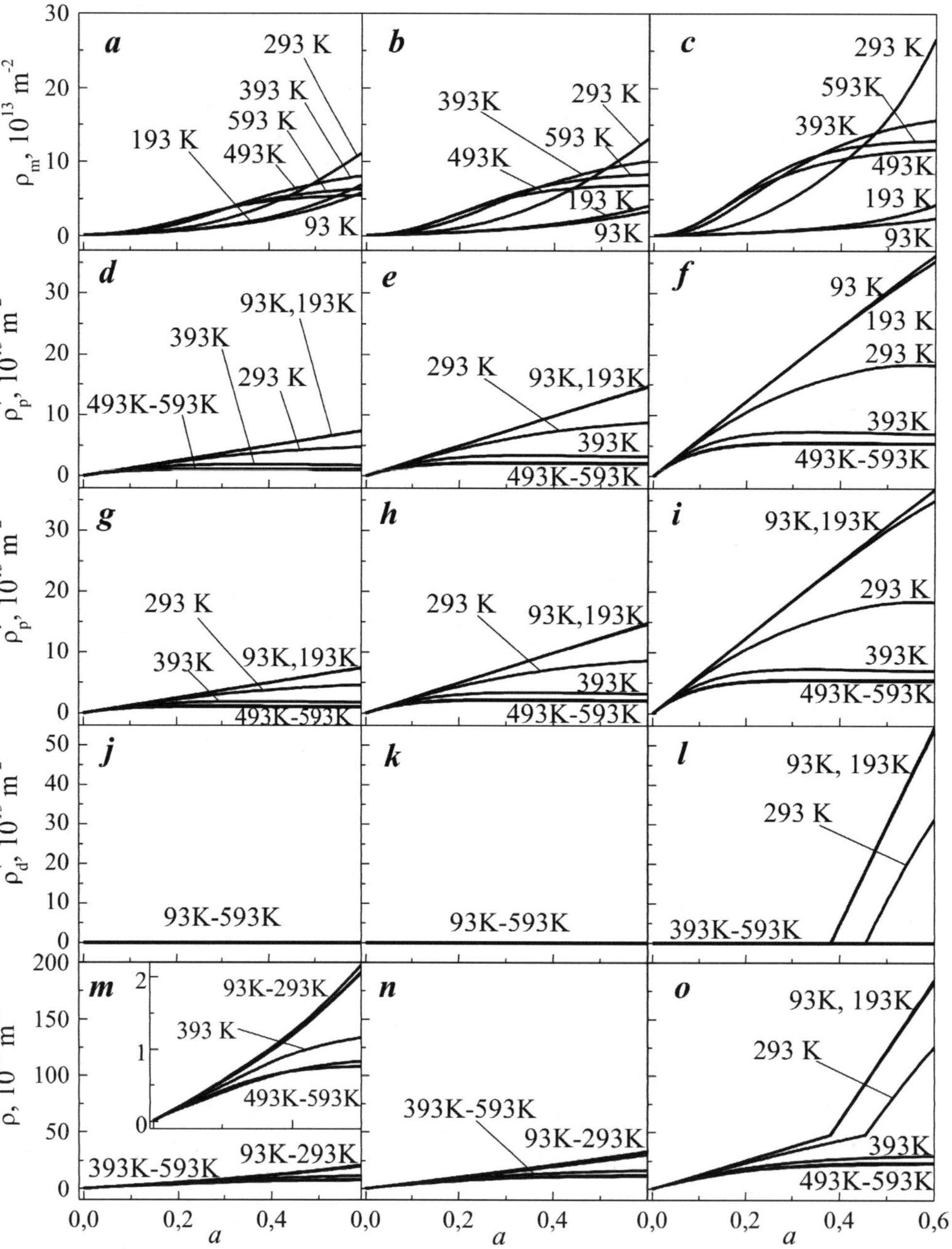

Figure 2. The dependence of the density of the shear-forming dislocations (*a*, *b*, *c*), dislocations in prismatic loops of vacancy (*d*, *e*, *f*) and interstitial (*g*, *h*, *i*) types, dislocations in dipoles (*j*, *k*, *l*), and total dislocation density (*m*, *n*, *o*) on the strain for an aluminum-based alloy at different temperatures. The fraction of incoherent particles is 25%. Particle diameter, nm: *a*, *d*, *g*, *j*, *m* – 10, *b*, *e*, *h*, *k*, *n* – 20, *c*, *f*, *i*, *o* – 50. Distance between particles is 400 nm. Strain rate 10^{-2} s^{-1}.

In a material with a particle size of 10 nm and 20 nm when the dislocation dipoles are not formed, the prismatic loops are dominant element of the dislocation structure (Figure 2). In view of the fact that the shear-forming dislocations behave in a complicated manner, at certain temperatures deformation they contribute more to the total dislocation density. Thus, at medium (293 K-393 K) and high (493 K-593 K) temperatures in the material with particle size of 10 nm and a distance between them of 400 nm contribution to the total dislocation density is greater of the shear-forming dislocation (Figure 2 *a*) than the contribution of prismatic dislocation loops (Figure 2 *d*). Such situation is observed for the material with strengthened particles of size of 20 nm at medium and high temperatures (Figure 2 *b, e*). In a material with a particles size of 50 nm is achieved critical dislocation density at temperatures of 93 K - 293 K, there are dislocations in dipole configurations, which become the dominant type of defects and make a decisive contribution to the total dislocation density. At higher temperatures the dislocation dipoles are not formed, and shear-forming dislocations become by dominant element of the dislocation structure.

4. Conclusions

Studies have shown that in materials with nanodispersed phase (δ = 10-50 nm) at a fixed volume fraction of the strengthening phase a material with smaller particle sizes reaches the greatest strain hardening in the subcritical region of dislocation densities. On reaching a critical dislocation density the dislocation dipoles begin to form in the material with the larger strengthening particles. This is reason of the enough strong strain hardening, which is ahead of the strain hardening of material with smaller particles. Such a pattern is observed at all temperatures of deformation.

References

[1] Kovalevskaya T A, Kolupaeva S N, Daneyko O I, Semenov M E and Kulaeva N A 2011 Mathematical modeling of work hardening of heterophase materials with nanosized strengthening particles *Russian Metallurgy (Metally)* **2011** 966–972

[2] Daneyko O I, Kovalevskaya T A, Kolupaeva S N, Kulaeva N A and Semenov M E 2012 Influence of the temperature and strain rate on the evolution of the dislocation structure of a dispersion-hardened material with FCC matrix *Russ Phys J* **54** 989-993

[3] Kolupaeva S N, Kovalevskaya T A, Daneyko O I, Semenov M E and Kulaeva N A 2010 Modeling of the temperature and rate dependence of the flow stress and evolution of a deformation defect medium in dispersion-hardened materials *Bulletin of the Russian Academy of Sciences: Physics* **74** 1527-1531

[4] Shewfelt R S W and Brown M L 1974 High-temperature strength of dispersion-hardened single crystals *Phil Mag* **30** 1135-1145

[5] Estrin Y, Arndt S, Heilmaier M and Brechet Y 1999 Deformation behaviour of particle-strengthened alloys: a Voronoi mesh approach *Acta mater* **47** 595-606

[6] Honeycombe R W K 1984 *The Plastic Deformation of Metals* (2nd ed Edward Arnold London)

[7] Humphreys F J and Hirsch P B 1970 The deformation of single crystals of copper and copper-zinc alloys containing aluminia particles. II. Microstructure and dislocation-particle interactions *Pros Roy Soc Lond* **318** 73-92

[8] Kelly A and Nicholson R 1966 *Precipitation hardening* (Moscow: Metallurgiya) 298

[9] Ashby M F 1966 Work hardening of dispersion-hardened crystals *Phil Mag* **14** 1157-1178

[10] Humphreys F J and Hirsch P B 1978 Work-hardening and recovery of dispersion hardened alloys *Phil Mag* **34** 373-399

[11] Yoo M H 1979 Growth kinetics of dislocation loops and voids – the role of divacancies *Phil Mag* **40** 193-211

[12] Lloyd D J and Martin J W 1978 The initial yield stress of Cu-SiO_2 single crystals between 77 K and 1200 K *Scrypta met* **12** 217-221

[13] Ebeling R and Ashby M F 1966 Dispersion hardening of copper single crystals *Phil Mag* **13** 805-834

IOP Conf. Series: Materials Science and Engineering **71** (2015) 012077 doi:10.1088/1757-899X/71/1/012077

The stored energy of plastic deformation in crystals of face-centered cubic metals

S Kolupaeva[1] and M Semenov[2]
[1]Tomsk State University of Architecture and Building, Department of Applied Mathematics, Tomsk, 634003, Russia
[2]National Research Tomsk Polytechnic University, Department of High Mathematics and Mathematical Physics, Tomsk, 634050, Russia

E-mail: ksn@tsuab.ru

Abstract. This article presents the results of investigation into changes in the stored energy during the process of plastic deformation. It has been shown that the value of stored energy increases with the degree of deformation and the higher this value the lower the deformation temperature. The stored energy decreases with increasing temperature. Intervals of strong and weak temperature dependence are associated with changes in the intensity of accumulation of deformation defects of different types. The stored energy also increases with increasing stress and the degree of deformation, which corresponds to the experimental data.

1. Introduction

Metals and alloys are in a state far from thermodynamic equilibrium in the process of plastic deformation. Excess energy that is stored in plastic deformation is an important integral characteristic of the defect structure of deformed material [1-3]. During the process of deformation and after its finish, relaxation processes occur in the deformed material: the restructuring in deformation defect structure in the direction of reducing the stored energy. After the termination of the deforming impact absorbed energy is the driving force behind relaxations' changes of the structure.

A lot of experimental data on the energy absorption in metals during plastic deformation and release of this energy during annealing post-deformation were obtained [1-3, 4].

In publication [4] were presented results of study on the thermo-mechanical responses and properties of single crystal high purity copper. The authors obtained that the stored energy of cold work depends almost linearly on the plastic strain for all different strain rates from 1000 to 8000 s^{-1}, and therefore the stored energy is on relative independence on the strain rate.

In theoretical study [5] was demonstrated that the significant part of the energy stored during deformation of tested materials is the energy of statistically stored dislocations and their mutual interactions. A method for the calculation of the stored energy was presented in publication [6]. The authors indicate that the stored energy increases with increasing hardening capacity of the metal.

Only a small part of the energy consumed by plastic deformation of the metal is absorbed by the crystal, thereby increasing its internal energy. The increase of internal energy is connected to defects arising during deformation; and it generates a driving force of recovery and recrystallization. The contribution to the change in the stored energy can produce various defects, including dislocations, point defects, microcracks, etc. Value of stored energy depends upon different factors associated with the process of deformation, and with the nature of the deformation material undergoing deformation.

The aim of the present work is to research of the stored energy of plastic deformation in crystals of face-centered cubic metals over a wide range of temperatures.

2. Mathematical model

In previous studies [7-9] the stored energy for isothermal and adiabatic deformation at a constant strain rate, annealing of the stored energy of deformation during isothermal annealing and annealing with a constant heating rate of material were calculated on based model [10]. The model [10] takes into account generating of deformation of point defects at a stationary density of thresholds in dislocations and without account the generation of dislocation dipoles.

The model [10] qualitatively correctly describes the general patterns of energy-absorbing deformation and its release during annealing. We have also shown that the value of the stored energy, taking into account the dynamic generation of point defects and dislocation dipoles leads to a better agreement with the experimental data.

In the present study the stored energy is defined by $E = E_s d\rho + E_{id}$, where E is the stored energy per unit length of the dislocation, E_s is the self-energy per unit length of dislocation [11]:

$$E_s = \frac{Gb^2}{4\pi K} \cdot \ln\frac{R}{r}, \tag{1}$$

where G is the shear modulus, b is Burgers vector module, R is the characteristic size of the region outside of which the energy of elastic distortions caused by the presence of dislocations is negligible; $K = 1$ for screw dislocations, $K = 1 - \nu$ for screwless dislocation respectively, E_{id} is the energy of the interaction of dislocations, and r is the radius of the dislocation core.

According to the publication [12] the self-energy per unit length of the dislocation can be calculated taking into account the energy of the dislocation core as follows:

$$E_s = \frac{Gb^2}{4\pi K} \cdot \ln\frac{R}{r} + E_c \tag{2}$$

The energy of the dislocation core E_c was estimated in the study [13] and it can be written as:

$$E_c = 0.1 E_s. \tag{3}$$

It is used in calculations that $r = 2b$, $R = 6\rho^{-1/2}$. The relations (1)-(3) show that the energy of the dislocation core is small compared with the first term included in (2), as a consequence the energy of the dislocation core is often neglected. In further calculations for the contribution to the stored energy due to the accumulation of dislocations with the energy of the dislocation core will use the relation:

$$\Delta E_m = 1.1\frac{Gb^2}{4\pi} K^{-1}(\psi)(\rho_m - \rho_m^0) \ln\frac{R}{r}, \tag{4}$$

where $K(\psi) = \frac{1}{2}(2-\nu)/(1-\nu)$, ρ_m^0 – the initial density of shear-forming dislocations. The energy of the dislocations in the dipole configurations is represented in the form

$$\Delta E_d = \frac{Gb^2}{4\pi(1-\nu)}\rho_d \ln\frac{\langle h\rangle}{r}, \quad \text{and } \langle h\rangle = \frac{G(2-\nu)}{8\pi\tau_f(1-\nu)} b, \tag{5}$$

where $\langle h\rangle$ is a average dipole arm, which is used to work in a uniform distribution of dipoles along the length of the arm.

The main source of strain hardening of metals is the accumulation of dislocations during plastic deformation, so the existing theoretical works [14] devoted to the stored energy usually considered only the contribution due to the accumulation of dislocations, the contribution of point defects are usually not taken into account. In this paper, the calculation takes into account the contribution of the stored energy of point defects, which is also associated with increase in stored energy. If the concentration of defects exceeds the thermodynamic equilibrium, the stored energy of the crystal is increased by an amount equal to the number of defects multiplied by the energy of formation of a single defect, assuming the formation of isolated point defects assuming the formation of isolated point defects:

$$\Delta E_v = n_v U_v^f \approx \frac{c_{1v}}{b^3} U_v^f, \qquad \Delta E_{2v} = n_{2v} U_{2v}^f \approx \frac{c_{2v}}{2b^3} U_{2v}^f, \qquad \Delta E_i = n_i U_i^f \approx \frac{c_i}{2b^3} U_i^f, \tag{6}$$

here n_v, n_{2v}, n_i – the number of vacancies, divacancies and interstitial atoms per unit volume, respectively; c_v, U_v^f, c_{2v}, U_{2v}^f, c_i, U_i^f – concentration of point defects and their formation energy, respectively. Then using relations (4)-(6) for the absorbed energy of deformable material assuming an additive contribution can be written $\Delta E = \Delta E_m + \Delta E_d + \Delta E_p + \Delta E_v + \Delta E_{2v} + \Delta E_i$.

Dependences of the stored energy from the flow stress and shear strain $\rho(\tau)$, $\rho(a)$, $c_{1v}(\tau)$, $c_{2v}(\tau)$, $c_i(\tau)$ can be determined by solving the system of equations of mathematical model of homogeneous plastic deformation in FCC materials [14]. The mathematical model [14] together with the relations (4)-(6) allows us to calculate the absorbed energy of deformable material and contributions to it energy of deformation defects of various types.

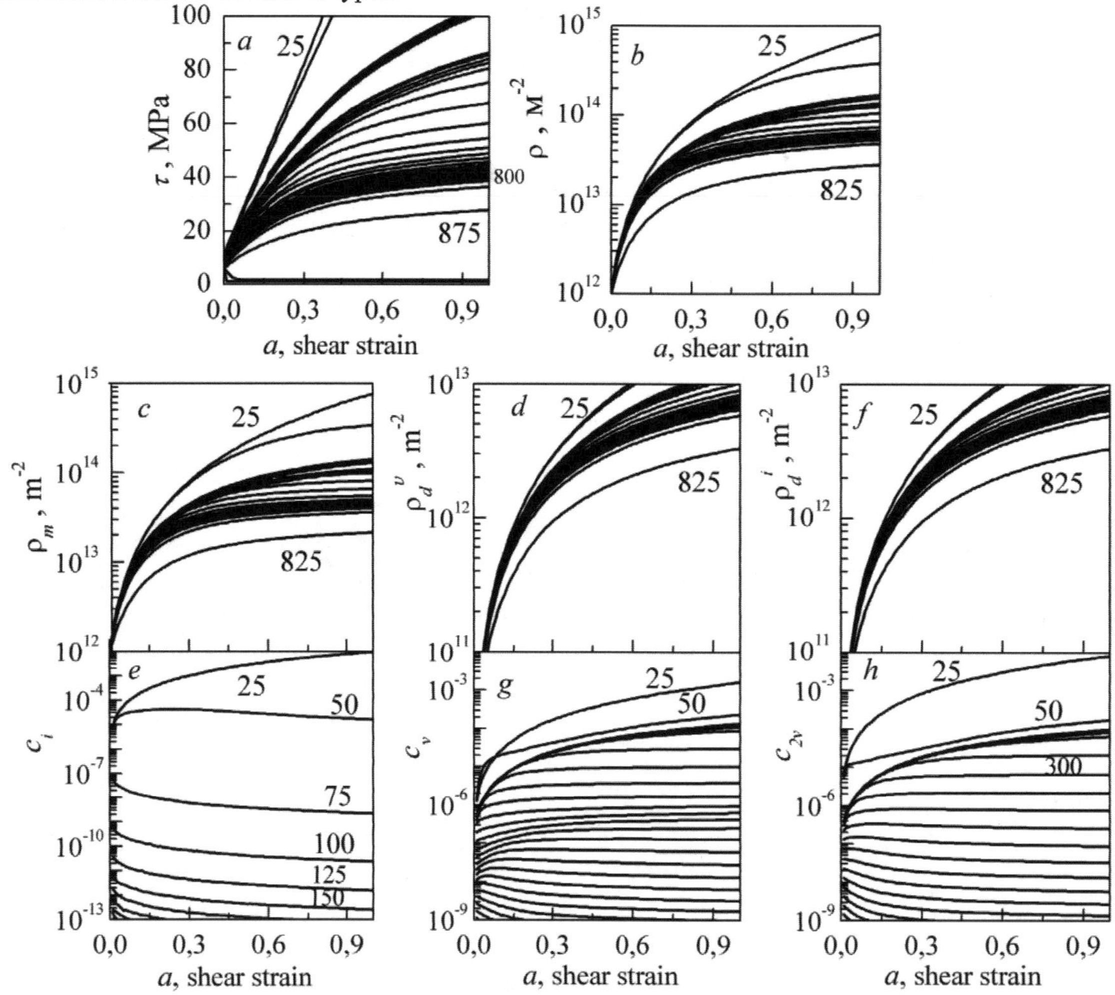

Figure 1. Flow stress (a), density of shear-forming dislocations (b), density in the dipole configuration (b), (c), concentration of interstitial atoms (e), vacancies (g), and divacancies (h) versus strain rate. Copper, temperature from 25 to 900 K in steps of 25 K.

3. Results of numerical experiments

Stress-strain curves and the corresponding dependence of the density of dislocations, shear-forming dislocations, dislocations in dipole configurations of vacancy and interstitial type and concentration of interstitial atoms, vacancies, divacancies versus strain for copper are shown in Figure 1, for deformation temperatures from T=25 K to 900 K in step of 25 K. As the temperature decreases the strain hardening coefficient and steady-state value of the flow stress are reduced (Figure 1, *a*). The temperature intervals of strong and weak temperature dependence associated with changes in the intensity of the annihilation of deformation defects of different types were found. It should be noted that unlike the model-shear deformation diffusion, in which the density of dislocation dipole is greater that the density of shear-forming dislocations, in presented model [14] the density of shear-forming dislocations is higher than the density of the dislocation dipole or comparable with it.

For each type of dislocation range of temperature dependence can be identified corresponding to the intervals of the temperature dependence of strain hardening. At temperatures above T=100 K the concentration of interstitial atoms in deformed crystal is negligible, less than c_i=10^{-9}. The concentration of vacancies even during deformation at high temperatures of at least c_v=10^{-9}.

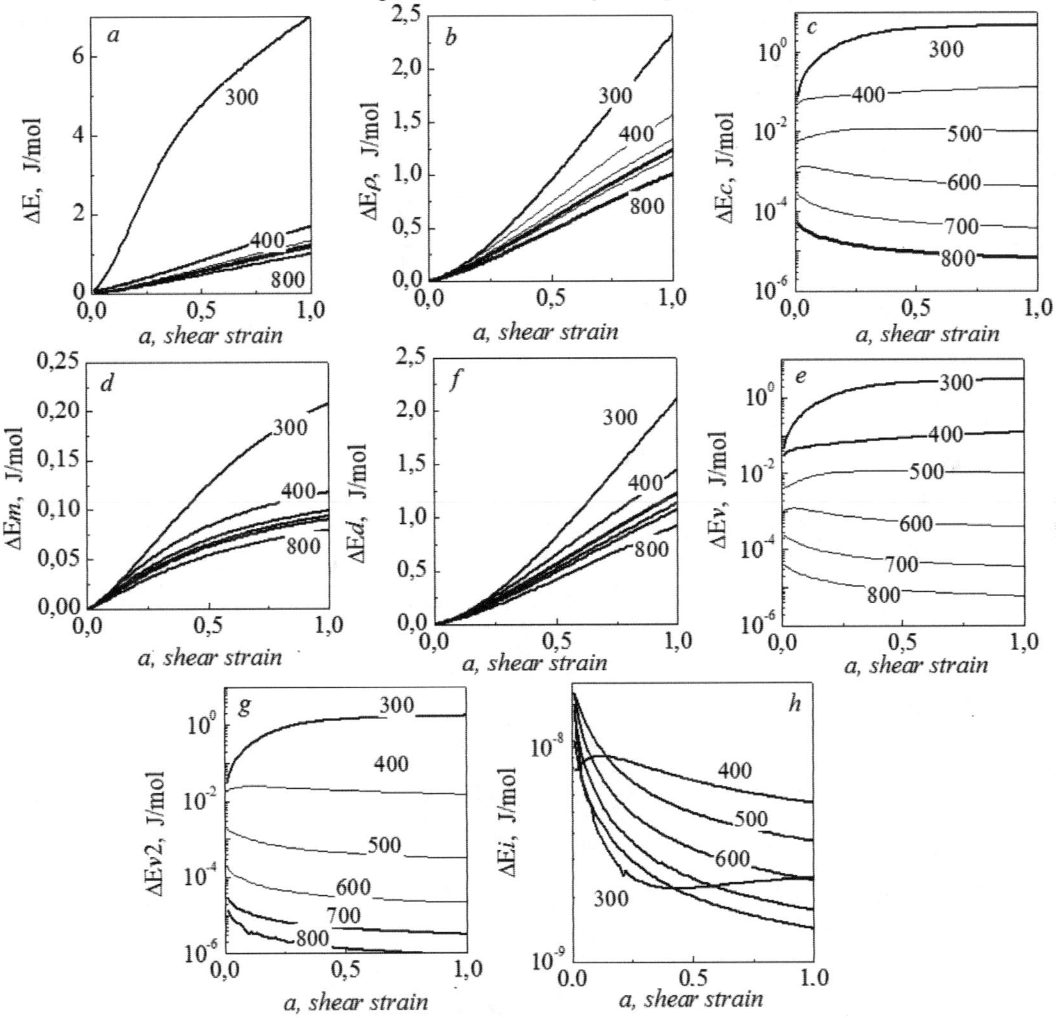

Figureure 2. Stored energy (a) and contributions to it dislocations (b), point defects (c) shear-forming dislocations (d), dislocations in the dipole configuration (e), vacancies (f), divacancies (g), and interstitial atoms (h) versus the shear strain for copper. Temperature varied from T=300 to 800 K in steps of 100 K.

Figures 2-4 show the results of our calculations of stored energy based on the developed model [14]. It has been shown that the quantity of stored energy increases with the degree of deformation and the more intense the lower the deformation temperature. With increasing temperature, the stored strain energy is reduced (Figure 2).

It was found that the temperature intervals of strong and weak temperature dependence is associated with changes in the intensity of accumulation of deformation defects of different types. The contribution of point defects in the stored energy at low temperatures may exceed the contribution of dislocations or be commensurated with it. The contribution of point defects in the stored energy is determined by the contribution of di- and vacancy, contribution of interstitial atoms is negligible. But starting with the temperature of about *T*=400 K for copper the contribution of point defects by orders of magnitude is smaller than that of dislocations.

Figure 3 shows the curves of the stored energy at low temperatures (up to *T*=300 K in steps of 50 K). By increasing the temperature to room temperature the contribution into the stored energy of point defects is significantly reduced, while the contribution from the dislocation remains virtually unchanged. Stored energy increases with increasing tension (Figure 4) and the degree of deformation, which corresponds to the experimental data.

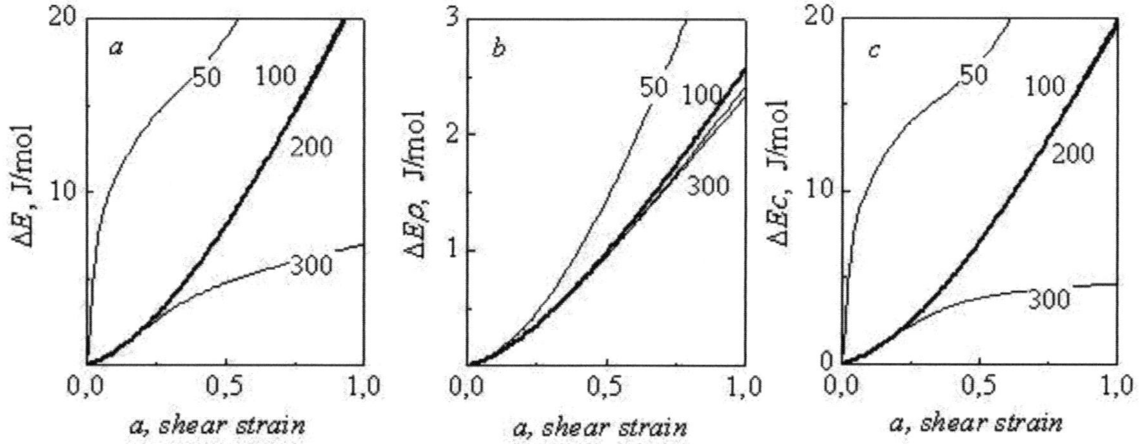

Figure 3. Stored energy (a) and deposits it dislocations (b), and point defects (c) versus the shear strain for copper. Temperature varied from *T*=50 to 300 K in steps of 100 K.

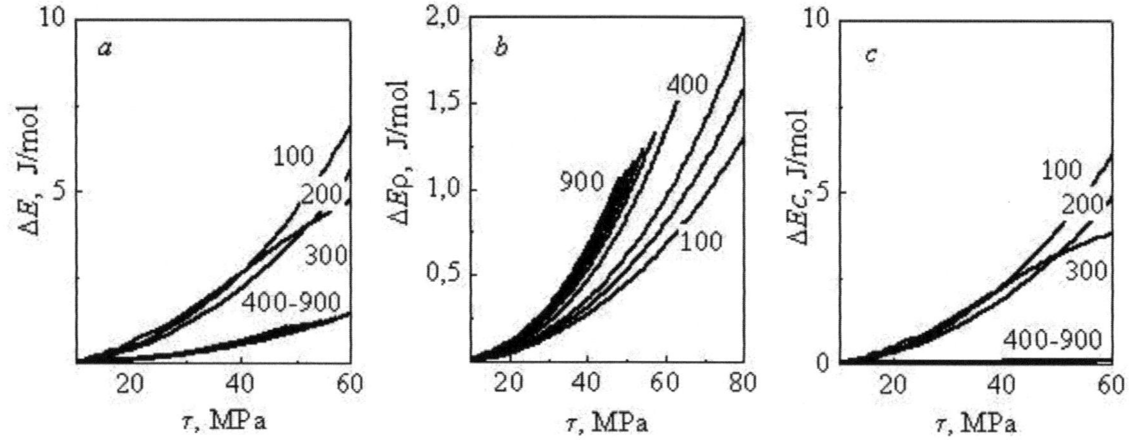

Figure 4. Stored energy (a) and contributions to it dislocations (b), point defects (c) versus the flow stress for copper.

4. Conclusion

The absolute values of the stored energy correspond to the order of data in the literature [1, 3, 11, 14, 15, 16]. The following basic laws of energy absorption of single crystals of fcc metals are noted, in agreement with the experimental data: 1) the amount of stored energy increases with the degree of deformation and the more intense the lower the temperature; 2) the quantity of stored energy higher at higher deformation strain rate, at low temperatures increasing strain rate has a little effect on the accumulation of energy. From the numerical experiments embodied in Figures 1-4 it may be concluded that the proposed mathematical model can be used in the prediction of plastic deformation processes under different external impacts.

References

[1] Большанина М А и Панин В Е 1957 *Скрытая энергия деформации* (Издательство АН СССР: Москва) 146-151
[2] Лариков Л Н и Юрченко Ю Ф 1985 *Тепловые свойства металлов и сплавов* (Наукова думка: Киев) 438
[3] Bever H B, Holt D L and Titchener A L 1973 *The stored energy of cold work* (Pergamon Press: London)
[4] Rittel D *et al* 2012 On the dynamically stored energy of cold work in pure single crystal and polycrystalline copper *Acta Materialia* **60** 3719-3728
[5] Oliferuk W and Maj M 2009 Stress-strain curve and stored energy during uniaxial deformation of polycrystals *European J. of Mechanics A/Solids* **28** 266–272
[6] Aravas N, Kim K-S and Leckie F A 1990 On the Calculations of the Stored Energy of Cold Work *J. of Engineering Materials and Technology* **112** 465-470
[7] Лазарева Л И, Колупаева С Н и Пудан Л Я 1989 *Накопление деформационных дефектов и поглощенной энергии адиабатической деформации* (Изд-во ТПИ: Томск) 71-78
[8] Колупаева С Н, Лазарева Л И и Попов Л Е 1990 *Характерные интервалы значений поглощенной энергии деформации* (Изд-во АПИ: Барнаул) С. 58-66
[9] Колупаева С Н, Лазарева Л И и Попов Л Е 1994 *Эволюция дефектных структур в металлах и сплавах* (Изд-во АГТГУ: Барнаул:) С.194-195
[10] Кобытев В С, Колупаева С Н, Пудан Л Я и Попов Л Е 1984 Эволюция дефектов структуры кристаллов. Моделирование на ЭВМ (Изд-во ЛИЯФ: Ленинград) с 106-107
[11] Seeger A, Kronmuller H 1962 Stored energy and recovery of deformed f.c.c. metals *Phil. Mag.* **7**(5) 897-913
[12] Friedel J 1967 *Dislocations* (Pergamon Press: Oxford)
[13] Yurchenko Yu F and Kononenko V L 1979 Energiya dislokatsionnyih ansambley v plasticheski deformirovannyih nikele, zheleze i titane (review) *Metallofizika* **75** 68-74
[14] Колупаева С Н, Пуспешева С И и Попов Л Е Математическое моделирование деформации скольжения *Известия РАН. Сер. физическая* **67** 1267-1276
[15] Полухин П И, Горелик С С и Воронцов В К 1982 *Физические основы пластической деформации* (Металлургия: Москва) 584 с
[16] Titchener A L and Bever M B 1958 The stored energy of cold work *Metal. Phys* **7** 247-350

IOP Conf. Series: Materials Science and Engineering **71** (2015) 012078 doi:10.1088/1757-899X/71/1/012078

Influence of stresses on structure and properties of Ti and Zr-based alloys from first-principles simulations

S A Barannikova[1,2,3], A M Zharmukhambetova[2], A Yu Nikonov[1], A V Dmitriev[1], A V Ponomareva[4] and I A Abrikosov[5]
[1] Institute of Strength Physics and Materials Science SB RAS, Tomsk, 634021, Russia
[2] National Research Tomsk State University, Tomsk, 634050, Russia
[3] Tomsk State University of Architecture and Building, Tomsk, 634003, Russia
[4] National University of Science and Technology MISIS, Moscow, 119049, Russia
[5] Department of Physics, Chemistry and Biology (IFM), Linköpings University, Linköping, 581 83, Sweden

E-mail: bsa@ispms.tsc.ru

Abstract. Computer simulations in the framework of the Density Functional Theory have become an established tool for computer simulations of materials properties. In most cases, however, information is obtained at ambient conditions, preventing design of materials for applications at extreme conditions. In this work we employ *ab initio* calculations to investigate the influence of stresses on structure and stability of Ti-Mo and Zr-Nb alloys, an important class of construction materials. Calculations reproduce known phase stability trends in these systems, and we resolve the controversy regarding the stability of body-centered cubic solid solution in Mo-rich Ti-Mo alloys against the isostructural decomposition. Calculated results are explained in terms of the electronic structure effects, as well as in terms of physically transparent thermodynamic arguments that relate phase stability to deviations of concentration dependence of atomic volume from the linear behavior.

1. Introduction

Growing economy constantly requires new materials, and about 150 years ago their discoveries accelerated substantially. The success is associated with increasing information technology contents in new materials. Novel phases are continuously synthesized, and the challenge is to identify potentially significant materials, to develop efficient technological processes, and to optimize their functionality. Emergence of new and powerful instruments such as computers initiated further development of computer modeling as a qualitatively new branch of science. The major goal of the theory in applications within the materials sciences is to assist experimentalists and engineers in a design of new materials and to discover phenomena with high potential to improve the existing products, as well as to create qualitatively new products.

One of the most impressive developments within the theoretical simulations of materials properties during the last four decades is associated with an approach of a new field, *ab initio* computer simulations based on the electronic structure theory [1]. Within this method simulations are carried out on the basis of the most fundamental lows of physics, e.g. quantum mechanics, statistical physics, theory of elasticity, and in principal without any experimental input. Accelerated by a revolutionary progress in computer technologies, the electronic structure theory has become a mature field of the

IOP Conf. Series: Materials Science and Engineering **71** (2015) 012078 doi:10.1088/1757-899X/71/1/012078

modern science. Computers rapidly became an essential and integrated part of materials design. Already at present one can witness a gradual substitution of a traditional trail-end-error approach by a knowledge-based materials design. Nowadays, computer simulations, experiment, and theory contribute on equal footing to the accumulation of new knowledge, to accessing and assembling materials parameters, and to establishing novel design concepts.

Unfortunately, most of the simulations reported in the literature are carried out at ambient conditions. Moreover, most applications of thermodynamic databases to materials design are limited to ambient pressure. Thus, the consideration of elastic contributions to thermodynamic stability is highly desirable, but not straight-forward to realise [2]. Moreover, a complexity of the problem increases substantially with increasing complexity of materials when, in addition to the pressure one varies their chemical composition. On the other hand, an ability to understand and a possibility to predict a response of a physical system upon a variation of external parameters is one of the core tasks for theoretical simulations.

The goal of this work is to investigate, by means of state-of-the-art tools for *ab initio* calculations, the influence of stresses on structure and stability of Ti and Zr-based alloys. Ti alloys are important, e.g. for aerospace and medical applications. However, hcp Ti alloys do not have high strength, though they show strong hardening during cold deformation. Their disadvantage is reduced technological plasticity. Thus, there is increasing interest in bcc Ti alloys. Mo, among few other elements, is known as bcc-Ti stabilizer. Zr-based alloys represent a material system widely used in nuclear energy applications, e.g. as fuel cladding materials. Requirements to increase fuel efficiency and to reduce the amount and activity of nuclear waste are directly related to the need to improve the properties of zirconium alloys, including their resource characteristics: durability, fatigue strength and corrosion resistance. In this work the results of our *ab initio* study of mixing enthalpies and lattice stabilities for binary Ti-Mo and Zr-Nb alloys are reported and general understanding of modifications of these properties upon application of high pressure are established.

2. Theoretical methodology

Materials properties and their performance in applications are determined by chemical composition, temperature, and pressure. Their variation can give rise to phase transitions, e.g. associated with the change of the underlying crystal structure, for instance from hexagonal closed packed (hcp) to body centred cubic (bcc) or with a modification of distributions of the alloy components on the lattice sites on the same underlying lattice, like an isostructural spinodal decomposition [3]. These two effects will be in focus for the present study.

2.1. General considerations

A phenomenological theory of phase transitions follows from equilibrium thermodynamics. However, it is important to notice that while experimentally the phase equilibria are most often studied at constant pressure conditions, the first-principles calculations are most often done at constant volume. In the former case the equilibrium is described by the Gibbs free energy:

$$G = E + PV - TS, \tag{1}$$

where E is the energy of the system, S denotes the entropy, and thermodynamic variables P, V, T represent pressure, volume, and temperature, respectively. General conditions for equilibrium are represented by G minimum. Within the *ab initio* approach one considers most often conditions corresponding to fixed parameters T and V, and the equilibrium is described by the Helmholtz free energy:

$$F = E - TS. \tag{2}$$

Comparing equations (1) and (2), we see that the PV term is most often left off the theory.

A justification for this approach comes from the fact that for equilibria involving only solid and liquid phases, the phase boundaries are typically shifted only by the order of a few hundredths of a degree per bar change in *P*, and therefore the effect of *PV*-term upon the phase transitions temperature is often assumed to be negligible [4]. However, the rapid advances of modern technology as well as developments in high-pressure experiments invalidate the above argument. For instance, in modern cutting tools applications the force against the work piece, together with the minimal contact area, gives rise to stress or pressure levels of few GPa [5]. Such extreme compressions may affect structure and stability of construction materials.

To elucidate the effect of uniform stress or pressure on the structure and phase stability of Ti and Zr-based alloys, the temperature effects are neglected, thus putting the last term in the right-hand side of equations (1) and (2) to zero, and focus on the calculation of the enthalpy:

$$H = E + PV. \tag{3}$$

The lattice stability of phase α with respect to phase β is therefore defined by the entropy difference H_α-H_β at fixed composition, and its negative value corresponds to a stability of the phase α. On the other hand, a stability of the alloy with respect to its decomposition into pure components is given by the enthalpy of mixing:

$$\Delta H^{mix}_{A_{1-x}B_x} = H_{A_{1-x}B_x} - (1-x)H_A - xH_B \, , \tag{4}$$

where $H_{A(B)}$ is the enthalpy of the alloy pure components. In the following the isostructural mixing enthalpy is distinguish form the total one. In the former case the pure components are simulated in the same crystal structure as the alloy. In the latter case they are considered in their respective lowest energy a state that is in the ground states. The isostructural mixing enthalpy characterizes the tendency towards alloying (negative values) or segregation (positive values) at the fixed crystal lattice. It is important, e.g. for mutiscale modeling [3]. The total mixing enthalpy characterizes phase stability more directly, because it includes both, the effects of the lattice stability and the mixing.

2.2. Computational techniques

The calculations of the total energy E in equation (3) were performed in the framework of the Density Functional Theory (DFT) [1] using the exact muffin-tin orbital (EMTO) method combined with the coherent potential approximation (CPA) for the electronic structure of random alloys [6]. The self-consistent electron densities were obtained within the Generalized Gradient Approximation (GGA) [1] using full charge density formalism [6]. The applicability of DFT and EMTO for the pure alloy components has been demonstrated recently in [7], while CPA has been successfully applied for calculations of lattice parameters and bulk moduli of Zr-Nb and Ti-Al-Mo-V alloy in [8].

The energy integration has been carried out in the complex plane using a semi-elliptic contour comprising 24 energy points.The calculations were performed for a basis set including valence *s,p,d,* and *f* orbitals, whereas the core states were recalculated at each iteration of the self-consistency loop. The integration over the irreducible part of the Brillouin zone was performed using a 29×29×29 grid of *k*-points in the reciprocal space for bcc and face centred cubic (fcc) lattices. For the hcp it is used 21×21×17 grids of *k*-points. The convergence of the energy with respect to the calculation parameters was 10^{-8} Ry.

The equation of state (EOS) was determined by calculating the total energy of the alloy $E(V_{at})$ for a set of volumes (per atom) V_{at}. In this work $E(V_{at})$ dependence was fitted by the modified Morse function [9]. Note that for hcp crystals the ratio of the lattice constants c/a for each volume V_{at} is optimized by minimizing the energy $E(V_{at}, c/a)$ with respect to c/a. The equations of state are constructed with the help of the so-obtained minima of the function $E(V_{at}, c/a)$. The pressure in equation (3) was calculated as $P(V_{at}) = -\partial E(V_{at})/\partial V_{at}$ by analytical differentiation of the Morse function. The accuracy of the calculated EOS for the pure components using the computational tools employed in this work has been demonstrated in [7].

3. Results

Calculated total mixing enthalpies for bcc, fcc and hcp Mo-Ti and Nb-Zr alloys at two values of pressure, 0 GPa and P=10 GPa are shown in Figures 1 and 2, respectively. At first the reliability of these results is analyzed. The main tool for the experimental determination of the mixing enthalpy, the calorimetry, is getting rare, probably due to its relatively high cost. In particular, the authors could not find any experimental information for the two systems considered in this work. This underlines the importance of theoretical simulations. General accuracy of the adopted theoretical scheme is well established in the literature [3,5,6].

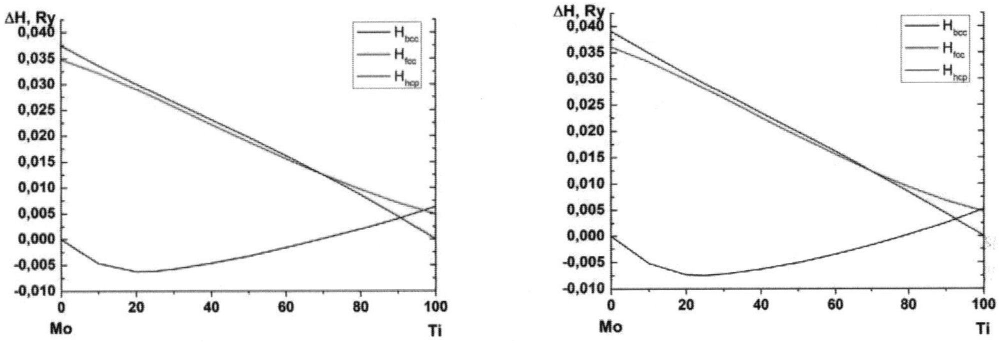

Figure 1. Calculated total mixing enthalpies for bcc (black lines), fcc (red lines) and hcp (blue lines) Mo-Ti alloys at P=0 GPa (left panel) and P=10 GPa (right panel) as a function of Ti concentration (in at. %).

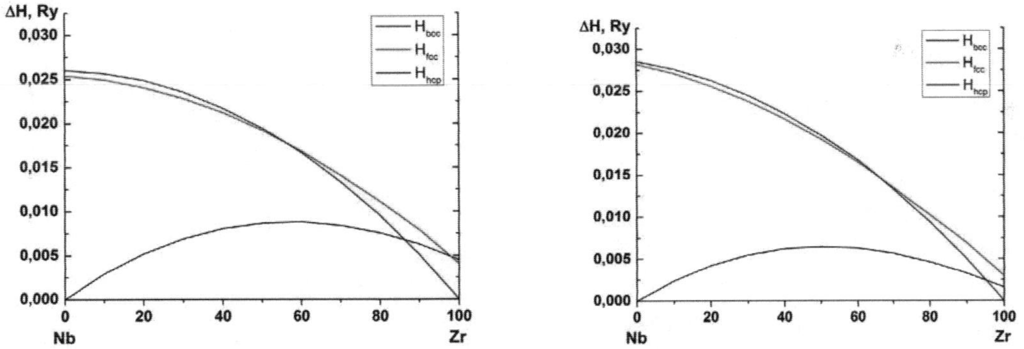

Figure 2. Calculated total mixing enthalpies for bcc (black lines), fcc (red lines) and hcp (blue lines) Nb-Zr alloys at P=0 GPa (left panel) and P=10 GPa (right panel) as a function of Zr concentration (in at. %).

Calculations reproduce known experimental trends in the two systems. Indeed, fcc alloys are nether stabilized in these two systems, in agreement with experiment. Therefore it is not discussed in the remaining part of this paper. At low temperature Nb-Zr system is characterized by a tendency towards phase separation into hcp Zr-rich and bcc Nb-rich phases, in complete agreement with the behavior of the mixing enthalpy shown in Figure 2. Next, in the calculations the enthalpy difference between bcc and hcp Mo-Ti alloys decreases rapidly when Mo is added to Ti (Figure 1). This is also in agreement with experiment, as Mo is known as β-Ti stabilizer (it stabilizes the bcc phase). Experimental situation on the stability of bcc solid solution in alloys with higher fraction of Mo in this system is somewhat more controversial. From the one hand, there are experimental studies that claim the presence of isostructural decomposition in bcc Mo-Ti system, in addition to the decomposition into bcc Mo-rich and hcp Ti-rich phases. On the other hand, SGTE and NIST phase diagrams claim the stability of the bcc Mo-rich solid solution. The isostructural mixing enthalpy of bcc Mo-Ti alloys is shown in Figure 3. It is negative in the whole interval of concentrations, indicating the stability of the solid solutions and supporting SGTE and NIST phase diagrams. Thus, the calculated results give reliable trends for the mixing enthalpies in Mo-Ti and Nb-Zr systems. At the same time, the absolute values should be taken with some care because of the neglect of the contribution from the local distortions of the crystal lattice in the CPA [3]. Moreover, recently it has been demonstrated that bcc phases of Zr and Ti are strongly anharmonic [10,11]. Therefore, one may expect an influence of temperature on the calculated properties of these alloys. However, neither of these effects should influence the calculated trends, which are discussed in the next section.

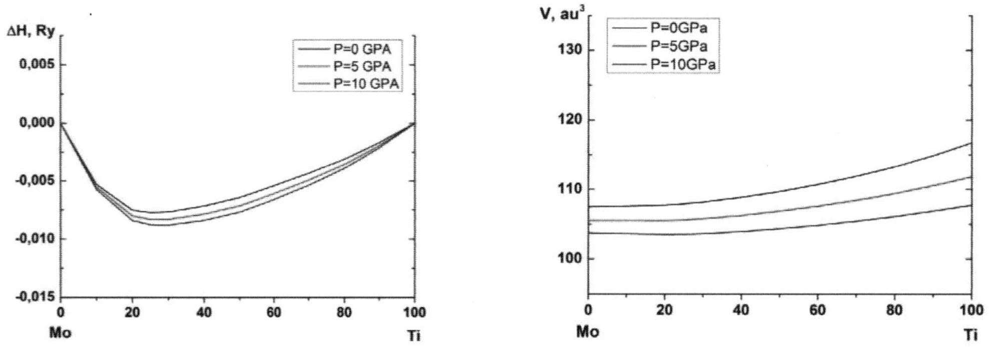

Figure 3. Calculated isostructural mixing enthalpies (left panel) and volume per atom *V* (right panel) for bcc Mo-Ti alloys at P=0 (black lines), 5 GPa (red lines) and 10 GPa (blue lines) as a function of Ti concentration (in at. %).

4. Discussion

Discussing the influence of pressure on the phase stability in the considered systems, in Mo-Ti system the bcc phase is getting more favourable energetically with respect to the hcp phase at about 10 at. % of Mo at ambient pressure. At 10 GPa the crossover occurs at even smaller concentration of 5 at. % Mo (Figure 1). The similar effect is seen in Nb-Zr system, that is both, the pressure and alloying with Nb stabilize the bcc phase with respect to the hcp one (Figure 2). Indeed, the line corresponding to calculated total mixing enthalpy for the bcc phase crosses the one for the hcp phase at ~12 at. % of Nb in Zr at ambient pressure, while the crossing occurs at ~5 at% at P=10 GPa.

The observed stabilization of the bcc phase can be explained by the electronic structure effects in the framework of the so-called canonical *d*-band model [12]. According to it the energy difference among different closed-packed crystal structures can be determined from a sum of one-electron, or band energies up to the highest occupied electronic level, the so-called Fermi energy. For transition metals the Fermi energy is defined by the number of *d*-electrons, and therefore the lattice stability can be written as a function of the occupation of the *d*-band [12]. The canonical band theory allows us to explain trends of the lattice stability upon the change of the band filling due to alloying of Ti and Zr with Mo and Nb, respectively. Isoelectronic elements, Ti and Zr, belong to the stability field of the hcp crystal structure. But both of them are close to the hcp-bcc boundary, and increasing their *d*-band filling by adding more electrons into the system should stabilize the bcc phase. Alloying with Nb/Mo increases the number of electrons because they have more d-electrons than Ti/Zr. Pressure increases the occupation of the d-band due to s- to d-charge transfer mechanism. Thus, both, pressure and alloying should stabilize the bcc phases in these systems, in agreement with our *ab initio* calculations.

Considering the effect of the uniform stress on the stability of solid solution phases against isostructural decomposition, one sees in Figure 3 that pressure decreases the mixing enthalpy, thus making the alloy more stable. This effect can be explained from the model proposed by Alling *et al.* in

IOP Conf. Series: Materials Science and Engineering **71** (2015) 012078 doi:10.1088/1757-899X/71/1/012078

[5]. Using the thermodynamic relations between the pressure derivative of the Gibbs free energy and volume per atom, Alling *et al.* showed that negative deviations of the concentration dependence of the latter should lead to more negative values of the former. In the right panel of Figure 3 we show calculated concentration dependence of the volume per atom in Mo-Ti alloys, which clearly have negative deviations from the linear concentration dependence. Thus, the predicted pressure-induced stabilization of bcc solid solutions in Mo-Ti alloys is in agreement with Alling's model.

5. Conclusion

The influence of stresses on structure and stability of Ti and Zr-based alloys have been investigated by means of state-of-the-art *ab initio* simulations. The exact muffin-tin orbital method combined with the coherent potential approximation was used for calculations of the electronic structure and total energy of random Mo-Ti and Nb-Zr alloys. The results observed that the application of the uniform stress and alloying with Mo/Nb stabilize bcc phases of Ti/Zr-based alloys relative to the hcp phase. In Ti-Mo system the pressure also increases the stability of the bcc solid solution against the isostructural decomposition. The observation is explained by the filling of Ti and Zr *d*-band by electrons upon the alloying and compression. In addition, our calculations provide support to a thermodynamic model of Alling *et al.* [5] that relates the pressure dependence of the stability of solid solutions to deviations of atomic volumes in alloys from linear concentration dependence.

6. Acknowledgements

This work was supported by the Grant of Russian Federation Ministry for Science and Education (grant no. 14.Y26.31.0005), Russian Foundation for Basic Researches (Grants No. 13-02-00606a, 12-08-00960-a) and the Program of Fundamental Research of State Academies of Sciences for the 2013-2020.

References

[1] Martin R M 2004 *Electronic structure Basic Theory and Practical Methods* (Cambridge: Cambridge University Press)
[2] Hammerschmidt T, Abrikosov I A, Alfe D, Fries S G, Höglund L, Jacobs M H G, Koßmann J, Lu X-G and Paul G 2014 *Phys. Stat. Sol.* (b) **251** 81
[3] Ruban A V and Abrikosov I A 2008 *Reports Progress Physics* **71** 046501:1-30
[4] Pelton A D 1996 *Phase Diagrams. Physical Metallurgy* (Amsterdam: North-Holland) p 471-533
[5] Alling B, Oden M, Hultman L and Abrikosov I A 2009 *Appl. Phys. Lett.* **95** 181906
[6] Vitos L, Abrikosov I A and Johansson B 2001 *Phys. Rev. Lett.* **87** 156401
[7] Abrikosov I A, Nikonov A Yu, Ponomareva A V, Dmitriev A I and Barannikova S A 2014 *Russian Physics Journal* **56**(9) 1030
[8] Barannikova S A, Ponomareva A V, Zuev L B, Vekilov Yu Kh and Abrikosov I A 2012 *Solid State Commun.* **152** 784
[9] Morruzi V L, Janak J F and Schwartz K 1988 *Phys. Rev.* B **37** 790-799
[10] Souvatzis P, Eriksson O, Katsnelson M I and Rudin S P 2008 *Phys. Rev. Lett.* **100** 095901
[11] Hellman O, Abrikosov I A and Simak S I 2011 *Phys. Rev.* B **84** 180301(R)
[12] Skriver H L *Phys. Rev.* B **31** 1909

IOP Conf. Series: Materials Science and Engineering **71** (2015) 012079 doi:10.1088/1757-899X/71/1/012079

Molecular dynamics study of diffusion along Ni-Al interphase boundary in the conditions of deformation

G Poletaev[1] and A Sannikov[1]
[1]Altai State Technical University, Department of Mathematics and Mathematical Modeling, Barnaul, 656038, Russia

E-mail: gmpoletaev@mail.ru

Abstract. The structural changes on Ni-Al boundary with orientations (100) and (111) and diffusion along it were studied by the method of molecular dynamics in the conditions of uniaxial compression-tension deformation. It was shown that during deformation the misfit dislocations network is distorted. Tension leads to a decrease of diffusion activation energy and to intensification of diffusion along the interphase boundary. During compression, on the contrary, the activation energy increases. Anisotropy of the diffusion along the interphase boundary under uniaxial deformation was not observed.

1. Introduction

At the interphase boundaries a situation can occur, that is qualitatively similar to that taking place at one crystal phase boundary, i.e. at the grain boundaries. As a result, in certain conditions, the acceleration of the matter diffusion transportation along the interphase boundaries is possible, that is proved experimentally [1, 2]. The fundamental difference between the phase boundaries and grain boundaries is that the second phase exists in many cases as an isolated inclusion in the matrix, i.e. phase boundaries do not form a united ramified network like grain boundaries. Dislocations that are typical for coherent interphase boundaries are formed not only as a result of contacting crystals disorientation (as in the case of grain boundaries), but as a result of mismatch of lattice parameters of the phases (so called misfit dislocations) [2, 3]. In [1, 2] it is highlighted because of structure imperfection the phase boundaries can detect high diffusion permeability, in some cases even higher than the grain boundaries. It is believed that the reverse case is also possible, and when the diffusion near the phase boundary is slower than in the grain volume.

In this paper the Ni-Al coherent boundary with two different orientations is considered: in (100) and (111) planes. The lattice parameters of contacting fcc metals are different (3.52 Å for Ni and 4.05 Å for Al), that leads to the formation of a high density of misfit dislocations at the Ni-Al interphase boundary. Earlier, in [4], using the method of molecular dynamics, we have shown that at the Ni-Al interphase boundary forms the network of edge misfit dislocations (with square cells of two systems of dislocations for the (100) boundary and with triangular cells of three systems for the (111) boundary), which play a key role in the diffusion along the interphase boundary at solid phase contact. It was found that the atoms migrate preferentially along the dislocation cores. Herewith Al atoms diffuse along the boundary more intense than Ni atoms. Activation energies of diffusion along the interphase boundary of Al and Ni atoms were found: 0.22 and 0.48 eV for the (100) boundary, 0.32 and 0.52 eV for the (111) boundary, respectively. Diffusion along the (111) boundary proceeded less intense than along the (100) boundary.

The present work is devoted to the study of the structural changes on Ni-Al boundary with orientations (100) and (111) and diffusion along it by the method of molecular dynamics in the conditions of uniaxial compression-tension deformation.

Previous research was carried out for a similar study for low-angle tilt [5, 6] and twist [7] boundaries in fcc metals. It was found that tension leads to an increase of diffusion intensity along grain boundaries, and on the contrary to decrease of diffusion intensity under compression. In [5-7] it was explained by a corresponding change of the free volume in the boundaries during deformation.

2. Description of the model

The interphase boundary was being created in the center of the simulation block (Figure 1). The boundary had orientation (100) or (111) in both contacting crystals. The block dimensions were chosen so that periodic boundary conditions could be used along the interphase boundary, i.e. along the X and Y axes in Figure 1 the endless repetition of the simulation block was imitated. Along the Z axis rigid conditions were imposed - atoms near the boundaries of the calculation block was parallel to the plane of the interphase boundary in the computer experiment remained motionless (in Figure 1 fixed atoms are dark grey). The number of atoms in the calculation block was about 30000.

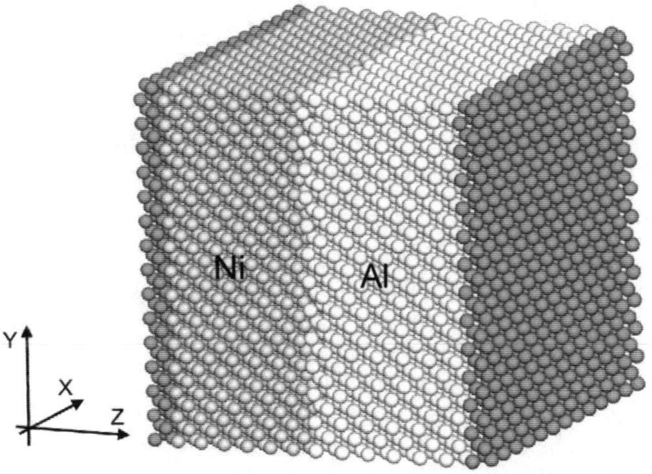

Figure 1. Simulation block containing Ni-Al (100) interphase boundary. The periodic boundary conditions were imposed along X and Y axes. Dark grey atoms remained motionless in the computer experiment (rigid boundary conditions along the axis Z).

To bring the structure of simulation block in the equilibrium state (at these conditions) structural relaxation was carried. As a result of the relaxation, the temperature of the block increased, so after its completion the block was cooled to 0 K. Integration time step in the molecular dynamics method was equal 5 fs. Interatomic interactions were described by Morse pair potentials, parameters of which were taken from [8].

When creating the interphase boundary disorientation contacting crystals was not made because it was found for Ni-Al boundary, that at any disorientation aluminum near the boundary aim to imitate the crystal orientation of nickel. As a result, when Al was initially disoriented relatively Ni, in Al the grain boundary parallel to the interphase boundary was formed. This phenomenon is apparently due to the relatively high energy of Ni-Al bonds in comparison with the Ni-Ni and Al-Al bonds.

Uniaxial compression or tension deformation was set by a corresponding change of the interatomic distances along the X or Y axis (Figure 1), i.e. deformation along the interphase boundary was considered. After the deformation of the calculation block structural relaxation and the subsequent cooling to 0 K was carried out again.

Investigating the diffusion and the mechanism of atom migration at the interphase boundary computer experiments 100-300 ps duration were held at different temperatures of the calculation block from 300 K to the melting point of aluminum. Temperature of the calculation block was set through the initial velocity of the atoms according to the Maxwell-Boltzmann distribution. Herewith total impulse of atoms in the calculation block was zero.

3. Results and discussion

In the relaxation process, during which conjugation of Ni and Al phases occurred, at the interphase boundary edge misfit dislocations formed (Figure 2). The dislocations formed, as already mentioned above, in the result of lattice parameters mismatch.

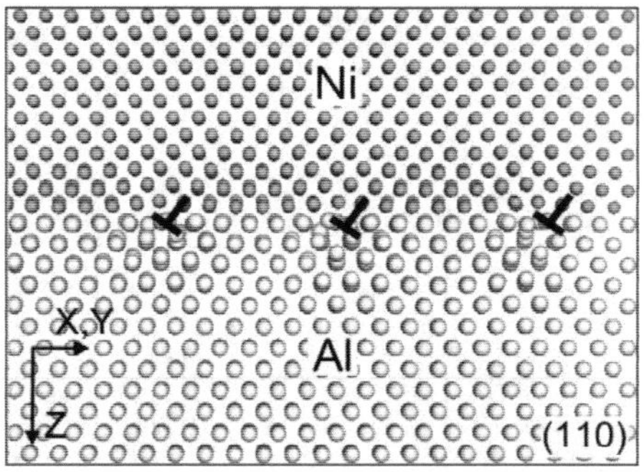

Figure 2. Misfit dislocations projected onto the plane (110) in the simulation block containing Ni-Al (100) boundary.

Figure 3 shows the distribution of free volume in the plane of Ni-Al (100) (Figure 3a) and (111) (Figure 3b) interphase boundaries at the absence of deformation. The visualization of free volume was carried out by calculating the average distance between nearest atoms. If the average distance was not significantly different from the distance corresponding to the ideal crystal, the atom was not depicted. Otherwise the atom was painted in one or another shade of grey.

In Figure 3 cores of misfit dislocations are seen, which in the case of Ni-Al (100) boundary form a network with square cells, consisting of two systems of edge dislocations, and in the case of Ni-Al (111) boundary - network with triangular cells, consisting of three systems of edge dislocations.

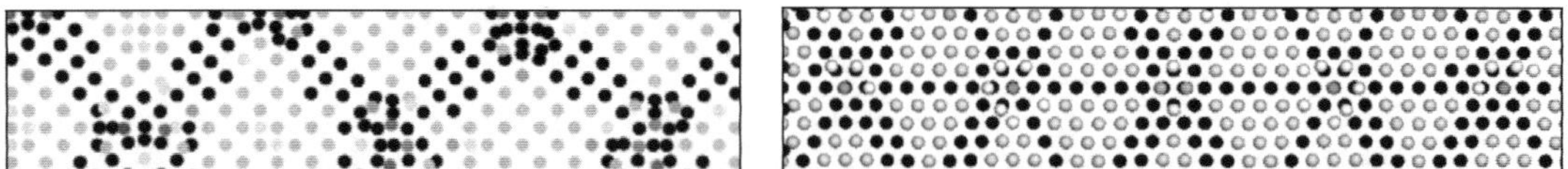

Figure 3. The image of misfit dislocations at the Ni-Al (100) (a) and (111) (b) boundaries using the free volume visualizer at the absence of deformation.

The dislocation network structure changed during the deformation. Figure 4 shows examples of misfit dislocation networks in the (100) (Figure 4a) and (111) (Figure 4b) boundaries at 3% compression along the X-axis. It was distortion of the networks and in the case of the (100) boundary there was a significant breach of the ordered arrangement of dislocation cores. For the (111) boundary the disorders of dislocations arrangement also observed, but to a lesser extent compared with the (100) boundary.

A measure of order of misfit dislocations arrangement at the interphase boundary depends largely on two factors: mismatch of lattice parameters of the contacting crystals (and hence the average distance between dislocations) and the bond energy of dissimilar atoms. The higher these values the higher the probability of formation of ordered dislocation network. In this case, in the result of the same deformation of the contacting phases, in Ni and Al different in magnitude tensions appeared (due to the difference of elastic moduli), which led to a curvature of the dislocation network.

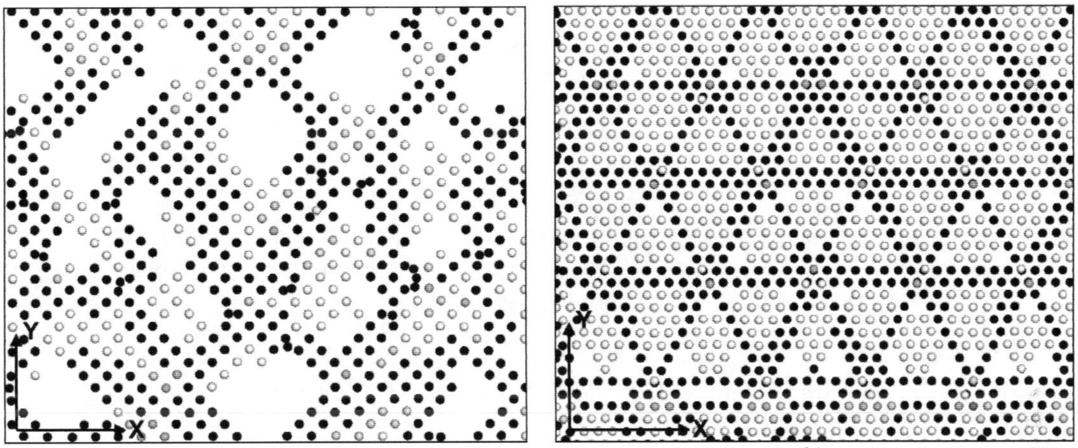

a) b)

Figure 4. The image of misfit dislocations at the Ni-Al (100) (a) and (111) (b) boundaries using the free volume visualizer at 3% stress deformation along the X-axis.

In the study of diffusion along the considered Ni-Al boundary, as in [4], a more intensive diffusion of Al atoms in comparison with the atoms Ni was observed. This is due to the fact that Ni-Ni and Ni-Al bonds are considerably stronger then Al-Al bond, whereby the defective areas (in the form of misfit dislocations cores) preferably contain Al atoms.

Atomic diffusion mechanism was investigated by a visualizer of atomic displacements relative to the initial positions. Displacements of atoms visualized by the software directly during the computer experiment allowed observing the atomic displacements in the dynamics.

Figure 5 shows examples of paintings of atomic displacements along the (100) and (111) interphase boundaries. The trajectories of the atomic displacements have directions as the cores of misfit dislocations. However, as can be seen from the figures, these displacements often did not coincide with the start positions of the cores corresponding to the absence of deformation, which indicates displacement of the dislocations, i.e. curvature of the dislocation network as a result of deformation.

For different values of the temperature at the end of the molecular dynamic experiment, the diffusion coefficient was calculated separately for the atoms of Ni and Al. The temperature during the computer experiment was kept constant. The diffusion coefficients in the XY plane (plane of the boundary), to determine the presence of diffusion anisotropy, were calculated separately along the X and Y axis:

$$D_x = \frac{\sum_{i=1}^{N} (x_{0i} - x_i)^2}{2tN}, \qquad D_y = \frac{\sum_{i=1}^{N} (y_{0i} - y_i)^2}{2tN}, \tag{1}$$

where x_{0i}, y_{0i} – coordinates of the initial position of i-th atom; x_i, y_i – coordinates of i-th atom in the time t; N - number of atoms in the simulation block.

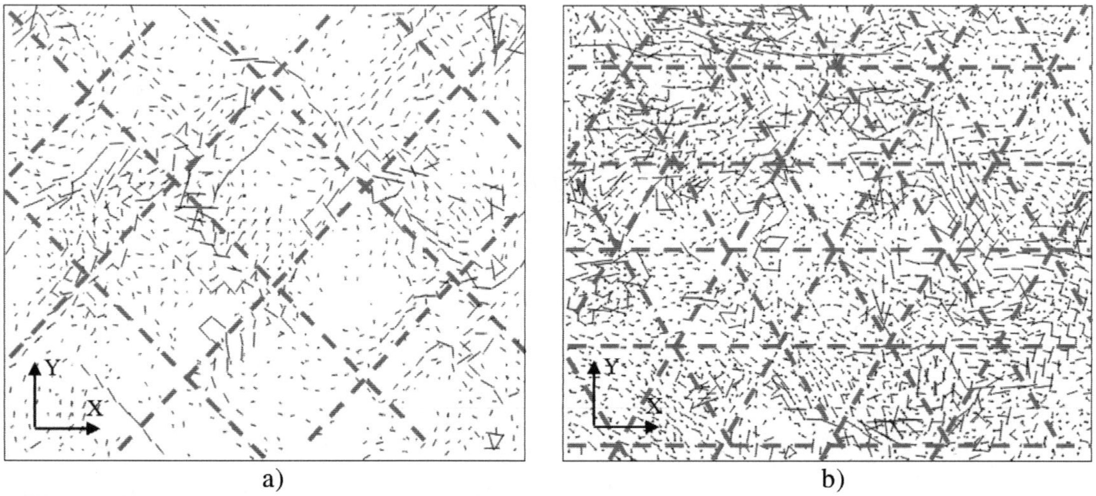

Figure 5. Migration of atoms in the plane of Ni-Al interphase boundary: a) (100) boundary at temperature 700 K for 250 ps; b) (111) boundary at temperature 900 K for 300 ps. Displacements of atoms are depicted by black segments. Gray bold dashed lines show the positions of misfit dislocations at the absence of deformation.

In the calculation of the diffusion coefficients the width of the interphase boundary was assumed equal 5 Å. The same width was assumed in the study of grain boundary diffusion in [5-7].

From the slope of dependences ln(D) on the T^{1} the activation energies of diffusion were found. The obtained values are shown in Table 1. This study was conducted for the values of compression and tension deformation of 3%.

Table 1. Activation energy of diffusion of Ni and Al atoms along Ni-Al (100) and (111) boundaries under conditions of uniaxial deformation, eV.

Type of deformation	(100)				(111)			
	Ni		Al		Ni		Al	
	E_x	E_y	E_x	E_y	E_x	E_y	E_x	E_y
normal state	0.48		0.22		0.52		0.32	
3% X tension	0.24	0.32	0.28	0.19	0.40	0.31	0.24	0.27
3% X compression	0.52	0.42	0.32	0.29	0.48	0.58	0.36	0.33
3% Y tension	0.36	0.25	0.21	0.26	0.45	0.38	0.30	0.28
3% Y compression	0.50	0.53	0.26	0.31	0.53	0.47	0.32	0.42

The scatter of the data (caused, in particular, the different distortion of the dislocation network at different temperatures) is quite large, making it difficult to assess the qualitative regularities. Nevertheless, the table shows that tension leads to a decrease of diffusion activation energy and to intensification of diffusion, compression - conversely (referring, of course, not talking about the deformations corresponding to the formation of intense plastic shears). In the papers [5-7], devoted to the study of diffusion along grain boundaries, this was explained by a corresponding change of the free volume within the boundaries during deformation. In addition, the table shows that the activation energy of Al atoms is smaller than for Ni atoms. Diffusion anisotropy was not detected, which may be explained by a large spread of the obtained values. However, atomic displacements (for example, Figure 5) also indicate the absence of a pronounced anisotropy of diffusion along the interphase boundary under uniaxial deformation.

4. Conclusion

Using the method of molecular dynamics in the present study it is shown that at Ni-Al interphase boundary network of edge misfit dislocations forms (with square cells of two systems of dislocations for the (100) boundary and with triangular cells of three systems for the (111) boundary), which play a key role in the diffusion along the interphase boundary at solid phase contact. The atoms migrate preferentially along the misfit dislocation cores. It was found that Al atoms diffuse along the boundary more intense than Ni atoms.

During deformation the misfit dislocations network is distorted. Tension leads to a decrease of diffusion activation energy and to intensification of diffusion along the interphase boundary. During compression, on the contrary, the activation energy increases. This is related, apparently, with the change of the free volume in the boundary. Anisotropy of the diffusion along the interphase boundary under uniaxial deformation was not observed.

It should be noted that the studies are devoted to the structurally correct (equilibrium) interphase boundaries. In fact contacting phases at the boundary contains various imperfections. Their role in the diffusion will be considered in subsequent studies.

Acknowledgment

The reported study was partially supported by RFBR, research projects No. 13-02-00301_a, No. 14-02-98000-r_sibir_a and No. 14-08-90416-Ukr_a.

References

[1] Bokshtain B S 1978 *Diffusion in metals* (Moscow, Russia: Metallurgiya)
[2] Bokshtain S Z, Bolberova E V, Ignatova I A, Kishkin S T and Razumovsky I M 1985 *Fizika metallov i metallovedenie (Rus)* **59** 938
[3] Poletaev G M and Starostenkov M D 2003 *Technical Physics Letters* **29** 454
[4] Poletaev G M, Sannikov A V and Mikrukov V R 2013 *Fundamental'nye Problemy Sovremennogo Materialovedenia (Rus)* **10** 112
[5] Rakitin R Y, Poletaev G M, Aksenov M S and Starostenkov M D 2005 *Fundamental'nye Problemy Sovremennogo Materialovedenia (Rus)* **2** 46
[6] Starostenkov M, Poletaev G, Rakitin R, Sinyaev D 2008 *Materials Science Forum* **567-568** 161
[7] Poletaev G M, Martynov A M, Dmitrienko D V and Starostenkov M D 2012 *Izvestiya Vysshikh Uchebnykh Zavedenij. Chernaya Metallurgiya (Rus)* **6** 64
[8] Tsaregorodtsev A I, Gorlov N V, Demyanov B F and Starostenkov M D 1984 *Fizika Metallov i Metallovedenie (Rus)* **58** 336

IOP Conf. Series: Materials Science and Engineering **71** (2015) 012080 doi:10.1088/1757-899X/71/1/012080

Estimation of the shear zone size in FCC – single crystals based on the model of dislocation interactions

V A Starenchenko[1], R I Kurinnaya[1], D N Cherepanov[1], M V Zgolich[1] and O V Selivanikova[2]
[1]Tomsk State University of Architecture and Building, Department of Higher Mathematics, Tomsk, 634003, Russia
[2]National Research Tomsk Polytechnic University, Department of Physical and Power Plants, Tomsk, 634050, Russia

E-mail: riklaz@mail.ru

Abstract. The main objective of the mathematical modeling of a shear plastic strain is estimation of the shear zone size. The paper is devoted to refinement of estimations of the shear zone size based on the model of dislocation interactions proposed by the authors. It is assumed that shear-forming dislocation loops, emitted by the dislocation source, are stopped by strong dislocation barriers that arise from the interaction of segments of expanding dislocation loops with reacting dislocations of non-coplanar slip systems. The formed barriers hinder further expansion of dislocation loops and, as a result, form the boundary of the shear zone. A refined estimation of the shear zone size has been obtained.

1. Introduction

Physical understanding of plastic strain of crystalline materials had arisen as a result of overcoming the contradictions between theoretical concepts and experimental results. For example, the crystal strength calculated based on the understanding of simultaneous breaking of atomic bonds in a glide plane was a thousand times greater than the actual strength. This problem was solved by introducing the notions of the dislocation, the displacement of which leads to breaking of atomic bonds along the dislocation line only. However, this case also includes a contradiction connected to the fact that the density of dislocations in crystals may increase by several orders, while the displacement of the dislocation from one edge to another does not increase the length of the dislocation, and the emergence of the dislocation to the surface even reduces their overall density. The assumption of the existence in a crystal of Frank-Read dislocation sources and the experimental observation of the sources, solved the problem of dislocation multiplication, but it turned out that slip lines (steps formed at the emergence of dislocations to the surface) are significantly shorter in length than the size of strained crystalline samples. To solve this problem, it is usually assumed that shear-forming dislocation loops, emitted by the dislocation source, are stopped by strong dislocation barriers that arise due to the interaction of segments of expanding dislocation loops with reactive dislocations of non-coplanar glide systems, such as Lomer-Cottrell barriers. The formed barriers hinder further expansion of dislocation loops and, thereby, form the boundary of the shear zone. The length of slip lines observed on the surface of the strained crystal should correlate with the distance between the barriers.

Thus, there are problems needed to be solved: 1) to identify the types of potential barriers; 2) to determine conditions under which the barriers do not break under the influence of stress on dislocation segments in the glide plane; 3) to determine conditions under which the length of forming barriers is large enough to prevent bending of barriers by neighboring segments of the dislocation loop; 4) to estimate the size of the shear zone as the distance between barriers.

In [1–4] these problems were solved within the framework of certain assumptions, however a number of proposed solutions have significant drawbacks. The model of dislocation interactions, proposed by the authors, allows refining the parameters used in estimation of the shear zone size and, thus, to obtain more substantiated average sizes of the shear zone in FCC-crystals.

2. Various methods for estimating the shear zone size

The simplest method for estimating the shear zone size is processing of experimental data on the length of slip lines. The dependencies of the average length of slip lines L on the shear strain for copper single crystals with different orientation of the strain axis (Figure 1) were obtained in [5–7].

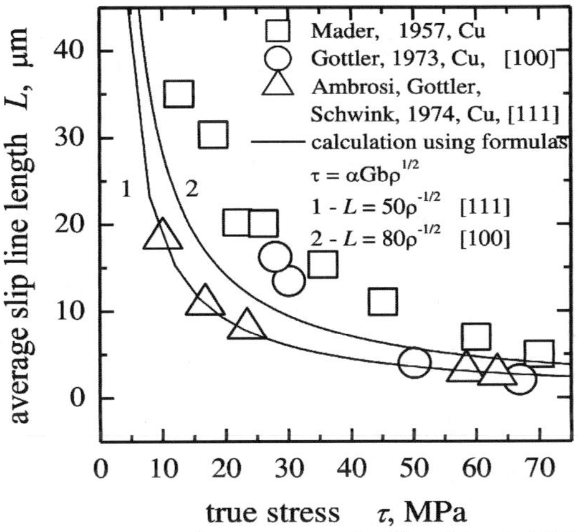

Figure 1. The average slip line length [5-7]

On the basis of these studies the empirical dependencies L on the degree of plastic strain have been proposed, such as $L = \Lambda\left(a - a^*\right)^{-1}$, where Λ is the coefficient, a is the shear strain, a^* is the degree of shear strain in the beginning of the second stage of the strain [5, 1], or $L = L_f + \left(L_i - L_f\right)e^{-k\varepsilon}$, where L_i and L_f are the initial and the limiting length of a free path of the dislocation, k is the constant, ε is the strain degree [8]. These dependencies contain parameters that depend on strain conditions and cannot serve as the basis for mathematical modeling of plastic strain.

According to the known theoretical estimates, L is directly proportional to $\rho^{-1/2}$ with coefficients varying within the limits of a single order, where ρ is the total dislocation density. Therefore, often the main parameters of models (the coefficient of the proportionality between L and the value τ^{-1} ~ $\rho^{-1/2}$) are found based on the experimental data. It is known that this parameter depends on the crystal orientation, the orientation of moving dislocation segments, the temperature (L increases approximately by 3 times at a reduction in the temperature from 573 K up to 4,2 K [9]), the presence of impurities, and other factors. For screw segments in [100] – copper single crystals such dependence cannot be obtained from the experimental data [6], and in [111] – single crystals the dependence

$L = 50\rho^{-1/2}$ has been obtained (Figure 1). Furthermore, at the same value L the intensity of dislocation accumulation can vary significantly depending on the nature of dislocation distribution on the boundary of the shear zone and the geometry of the shear zone [2].

Thus, common to all materials and strain conditions the dependence L on the density of dislocations, stress, or strain has not yet been proposed.

Emergence of simplest dislocation models of plastic strain, in which the intensity of dislocation accumulation depends on the average size D of the shear zone diameter, has allowed estimating the shear zone size based on its connection $\theta_{II} = \Delta\tau/\Delta a = 0{,}5\alpha GF\rho^{-1/2}D^{-1}$ with the hardening coefficient in the second stage of strain. However, for example, for [100] – single crystals it became impossible to accurately determine the end of the second stage of the experimental dependence. As a consequence, it is possible to obtain the values of the proportionality coefficient $0{,}5\alpha GF\theta_{II}^{-1}$ in the equality $D = 0{,}5\alpha GF\theta_{II}^{-1}\rho^{-1/2}$ differing by the order of magnitude.

In [1, 3] the methods for theoretical estimation of the shear zone size have been proposed based on the assumption that the length of the dislocation path is determined by dislocations of non-coplanar slip systems. To form a barrier it is necessary for the stress of the interaction of reacting dislocations to be higher than the internal long-range stress τ_2 in the secondary glide plane. Based on this condition, an effective capture radius $r_e = 0{,}5\pi^{-1}\alpha_e Gb\tau_{II}^{-1}$ is determined, where α_e is the parameter considering the geometry of the dislocation distribution. The parameter α_e has the value of an order $0{,}5$, and, thus, the capture radius may vary within the limits of the same order of the magnitude, which leads to the same spread of the diameter size of the shear zone.

In [3] the expression $D = B_r G^{-1}b^{-1}\tau\rho^{-1}$ for the diameter of the shear zone was obtained as follows. Assuming that the loop has a rectangular shape with the sides D and D_1, and the number of dislocations which interact with one of the sides during its path from the source at a distance of $0{,}5D$ equals to $0{,}25DD_1\xi\rho$, where $\xi\rho$ is the number of dislocations intersecting the unit area, $\xi = \rho_f\rho^{-1} \approx 0{,}5$ is the share of "forest" dislocations, ρ_f is the density of the "forest" dislocation. Barriers with the length D_1, capable to limit the length of the dislocation path, among all the reacting dislocations can form only β_φ of "forest" dislocations which are inclined at a small angle to the glide plane of the loop. During the path of the glide dislocation at a distance of $0{,}5D$ such an event can occur only once, therefore $0{,}25DD_1\xi\rho\beta_r\beta_\varphi = 1$. Taken into account the fact that the reaction takes place under the condition $\tau_2 \le \alpha_e Gb/2\pi r_e$, we obtain the limiting value $\varphi_m = arctg(2r_e D_1^{-1}) \approx \alpha_e Gb\pi^{-2}D_1^{-1}\tau_2^{-1}$ of the inclination angle φ of the forest dislocation to the junction line, at which the dislocation junction with the length D_1 can be formed. With a uniform distribution of values φ the probability of formation of a compound with the length D_1 equals to $\beta_\varphi = \varphi_m\pi^{-1} = \alpha_e Gb\pi^{-2}D_1^{-1}\tau_2^{-1}$.

As a result we obtain $D = 4\pi^2\tau_2\alpha_e^{-1}\beta_r^{-1}\xi^{-1}G^{-1}b^{-1}\rho^{-1}$. In the case of symmetric orientations of strain axes, when a slip is carried out under several equivalent systems, it can be assumed that $\tau = \tau_2$, then we obtain an expression for the diameter of the shear zone in the form $D = B_r\tau G^{-1}b^{-1}\rho^{-1} \approx \alpha B_r\rho^{-1/2}$, where $B_r = 4\pi^2\alpha_e^{-1}\beta_r^{-1}\xi^{-1} = 16\pi^2\beta_r^{-1}$. Given that $\beta_r = 0{,}14\ldots0{,}25, \alpha \approx 0{,}25$ we obtain $B_r \approx 1128$, $D \approx 282\rho^{-1/2}$, which is several times higher than the experimentally observed values.

The most accurate method for estimation of the shear zone size, in our opinion, can be based on the assumption that in the area S_D, swept out by the dislocation loop, there are 4 dislocations at different distances from the source that can form insurmountable barriers. Segments of the expanding dislocation loop are alternately stopped by these barriers, so that the final configuration of the shear zone boundary is a convex figure. Suppose β_D is the share of dislocations that form barriers, then $S_D\beta_D\beta_r\xi\rho = 4$ is the number of barriers. If the value $D = S_D^{1/2}$ is taken as the average size of the shear zone, we obtain $D = 2(\beta_D\beta_r\xi\rho)^{-1/2}$. The value β_D can be determined theoretically based on the model of dislocation interactions.

3. Estimation of parameters determining the diameter of the shear zone

To estimate the value D it is necessary to determine parameters β_r and β_D.

3.1. Estimation of the parameter β_r

As a result of dislocation reactions among dislocations of non-coplanar slip systems, dislocation junctions, which are obstacles for further motion of a slip dislocation, are formed at the intersection line of glide planes of reacting dislocations. To estimate the strength of dislocation junctions Schoeck and Frydman [10] proposed to determine the balance of a ternary dislocation node E on the basis of the virtual displacement principle, according to which the displacement of the node E (Figure 2) along the junction line is considered, and changes in the energy of dislocation segments are taken into account. In virtue of the task complexity a number [10] of simplifying assumptions have been introduced: 1) reacting dislocations intersect each other in the middle; 2) dislocation junction is motionless; 3) elastic interaction among dislocation segments is minor; 4) reacting dislocations are unextended; 5) forest dislocations are inflexible; 6) material is isotropic; 7) slip dislocations are screw or edgy only.

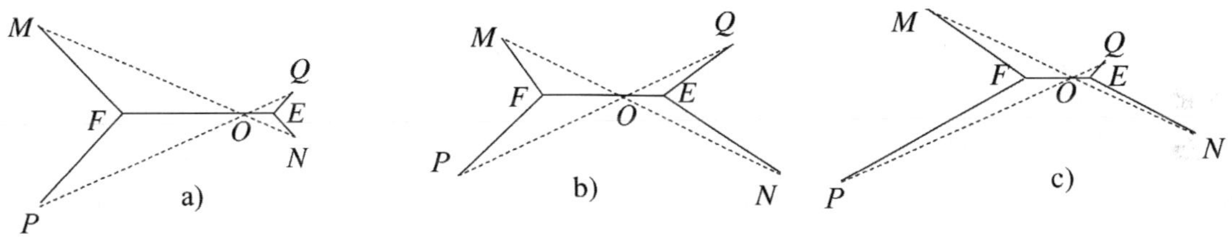

Figure 2. Examples of arbitrary intersections of reacting dislocations. The parameter $\gamma_g = QO : QP = 0{,}1 \div 0{,}9$ reflects the ratio of segment lengths of a slip dislocation. The parameter $\gamma_f = NO : NM = 0{,}1 \div 0{,}9$ reflects the ratio of segment lengths of forest dislocation. The point O – is the intersection point of reacting dislocations. Intersections at different values of parameters are presented: a) – $\gamma_g = 0{,}1$ and $\gamma_f = 0{,}1$; b) – $\gamma_g = 0{,}4$ and $\gamma_f = 0{,}6$; c) – $\gamma_g = 0{,}1$ and $\gamma_f = 0{,}5$.

This simplest model has been widely used in calculation of the strength of dislocation junctions in a number of papers by a group of authors with the weakening of certain simplifying assumptions. In particular, the authors of this paper have investigated the flexibility of forest dislocations and changes in the orientation of the axis of a slip dislocation with respect to the Burgers vector (change in the angle ψ_1).

As a result, for materials with a FCC structure the values of the parameter β_r were obtained [11, 12], characterizing the probability of formation of a dislocation junction (Figure 2) for different orientations of the strain axis of a crystal. The value of the parameter β_r was determined according to [13] using the relation $\beta_r = \rho_{rf}\rho_r^{-1} = 36^{-1}\Sigma_{i=1}^{8}(1-\cos\hat{\varphi}_i)$. The average value of the probability of formation of the junction β_r strongly depends [11, 12] on the type of a slip dislocation (on the angle ψ_1). The maximum value of the probability $\beta_r = 0{,}25$ corresponds to the 70^0-th slip dislocation, the minimum – $\beta_r = 0{,}145$ corresponds to a screw dislocation, to an edge dislocation – $\beta_r = 0{,}215$.

It should be noted that it was impossible to overcome some of the abovementioned simplifying assumptions within the framework of the Schoeck-Frydman model. Using the principle of virtual displacements [10, 13] it is impossible to directly determine the stress of a junction strain, to track changes in the geometry of the dislocation configuration under the influence of stress, as well as to overcome one of the strongest simplifying assumptions – the intersection of reacting dislocations in the middle, which is far from the actual intersection of dislocations in a crystal.

In this regard, the authors of this paper proposed a new model [14] enabling to describe the dislocation configuration, formed as a result of the reaction, in the real three-dimensional picture of the intersection of dislocations in non-coplanar slip systems.

3.2. Estimation of the parameter β_D

As part of the previously proposed model [14-16] by the authors of this paper, dislocation junctions at the intersection of arbitrary segments of reacting dislocations have been calculated. Intersections of reacting dislocations QP and NM (Figure 2) have been considered depending on the ratio of segment lengths $\gamma_f = NO : NM$ of the slip dislocation and the ratio of segment lengths of the forest dislocation (O is the intersection point of dislocations). Length spectra of dislocation junctions have been calculated for the following values of parameters $\gamma_g = QO : QP = 0{,}1 \div 0{,}9$ and $\gamma_f = NO : NM = 0{,}1 \div 0{,}9$.

As a result of computational experiments, in the given model [14-16] the dynamics of changes in the whole dislocation system formed as a result of the reaction of dislocations of non-coplanar slip systems under the influence of the applied stress τ, including its destruction, has been investigated. The change in the length of the dislocation junction has been studied directly under the influence of stress.

In the study of the destruction process of dislocation junctions the mechanism of formation of long dislocation junctions has been revealed [15, 16]. The essence of which consists in the following: under the applied stress a dislocation junction does not break but, on the contrary, its length increases. This occurs in the case when the position of dislocation nodes E and F begins to change under the influence of stress. At that, both nodes are moving along the junction line, but one of them, for example, the node E is more mobile than the node F. As a result, under the influence of stress, the length of the dislocation junction increases exceeding the original length. It has been revealed that rather high stresses of nearly $100 \div 500$MPa are needed for breaking of such junctions [16]. Long strong dislocation junctions are, practically, insurmountable barriers for motion of dislocations.

In this paper, a share of long insurmountable dislocation junctions, which were obtained at different intersections ($\gamma_g = 0{,}1 \div 0{,}9$ and $\gamma_f = 0{,}1 \div 0{,}9$) of reacting dislocations for all dislocation reactions implemented at orientation of the strain axis of the crystal [100], has been calculated. Each of these intersections has been considered for all types of slip reacting dislocations, from the screw orientation up to the edge orientation ($\psi_1 = 0^0 \div 90^0$), in which the given reaction is implemented. Calculations have been performed for the whole range of angles of forest dislocations in the case of the initial inclination angle to the junction line $\varphi = 10^0$.

The lengths of junctions under the above listed conditions have been determined, and the quantity of long junctions has been obtained for different values of the dislocation density $\rho = 10^{10} \div 10^{14}\,m^{-2}$ and for various dislocation reactions. According to the obtained results, a share of long junctions has been calculated for each implemented dislocation reaction. A share of strong (insurmountable) dislocation junctions from the total number of the obtained junctions has been determined [17].

Based on the obtained results, the probability of formation of a long strong dislocation junction for each value of the dislocation density has been determined using the total probability formula. From the analysis of the obtained results the average value of the parameter $\beta_D = 0{,}001$, characterizing the probability of occurrence of a long insurmountable dislocation junction in the case when the strain axis of the crystal is oriented towards [100], has been determined.

The presence of long strong dislocation barriers allowing blocking the motion of dislocations can be one of the mechanisms for determining the size of the shear zone.

4. Conclusion

The obtained estimates of the parameters β_r and β_D allow to obtain the value $2(\beta_D\beta_r\xi)^{-1/2} \approx 80$ of the coefficient in the formula $D = 2(\beta_D\beta_r\xi\rho)^{-1/2}$ of the dependence of the shear zone diameter on the dislocation density. This value agrees well with a variety of experimental data on slip lines and the hardening coefficient in the second stage, which allows using this value in estimation of values characterizing the defect structure of the shear zone.

References

[1] Berner R and Kronmüller H 1965 *Plastische Verformung von Einkristallen* (Berlin: Springer) p 272

[2] Popov L E, Sharkeev Yu P, Koneva N A, Esipenko V F, Popov V L and Kozlov Ed V 1977 *Phys. M. M.* **43** 1281-1288

[3] Popov L E, Kobytev V S and Kovalevskaya T A 1982 *Russ. Phys. J.* **52** 398-410

[4] Popov L E, Kolupaeva S N, Vihor N A and Puspesheva S I 2000 Dislocation dynamics of elementary crystallographic shear *Computational Materials Science* **19** 267–274

[5] Mader S 1957 *Zeitschrift für Physik* **149** 1 73-102

[6] Göttler E 1973 Versetzungsstruktur und Verfestigung von [100]-Kupfereinkristallen, I. Versetzungsanordnung und Zellstruktur zugverformter Kristalle *Phys. Stat. Sol.* **28** 1057-1076

[7] Ambrosi P and Schwink C 1978 *Scripta Metallurgica* **12** 303–308

[8] Roberts W, Bergström Y 1973 *Acta. Met.* **21** 457

[9] Garner A and Alden T H 1974 *Phil. Mag.* **29** 323-336

[10] Schoeck J and Frydman R, 1972 The contribution of the dislocation forest to the flow stress *Phys. Stat. Sol.* (B) **53** 661–674

[11] Kurinnaya R I, Ganzya L V, Popov L E 1982 Resistance to expansion of dislocation loop in FCC metal *Russ. Phys. J.* **25** 8 710-713

[12] Kurinnaya R I, Starenchenko V A, Kovalevskaya T A and Popov L E 1989 Interactions of the reactive superdislocations of the non-coplanar cubic slip systems *Metallofizika*. **11** 1 62–66

[13] Puschl W, Frydman F and Schoeck J 1982 The strength of the dislocation forest 30^0 u 60^0 dislocations *Phys. Stat. Sol.* (A) **74** 211–216

[14] Kurinnaya R I, Zgolich M V, Starenchenko V A 2004 Calculations of lengths of dislocation combinations in FCC crystals *Russ. Phys. J.* **47** 705 – 712

[15] Starenchenko V A, Zgolich M V and Kurinnaya R I 2009 Formation of extended compound dislocations and barriers in the course of interdislocation reactions in FCC crystals *Russ. Phys. J.* **52** 245 – 251.

[16] Zgolich M V, Kurinnaya R I and Starenchenko V A 2013 The influence of applied stress on the length of the dislocation junction formed at interaction of reacting dislocations in FCC–crystals *Bulletin of Tambov University Ser. Natural and Technological Sciences* **18** 1554–1555

[17] Starenchenko V, Cherepanov D, Kurinnaya R, Zgolich M and Selivanikova O 2014 The influence of dislocation junctions on accumulation of dislocations in strained FCC – single crystals *Advanced Materials Research* **1013** 272–279

IOP Conf. Series: Materials Science and Engineering **71** (2015) 012081 doi:10.1088/1757-899X/71/1/012081

Three-level modeling of fcc polycrystalline inelastic deformation: grain boundary sliding description

P Trusov[1], E Sharifullina[1] and A Shveykin[1]
[1]Perm National Research Polytechnic University, Mathematical Modeling Systems and Processes Department, Perm, 614990, Russia

E-mail: elvira16_90@mail.ru

Abstract. This paper considers three-level model of polycrystalline inelastic deformation based on crystal plasticity that allow describing the internal structure evolution of the material. It is assumed that crystallite inelastic deformation is realized by intragranular dislocation sliding with lattice rotation. The paper focuses on the problem of grain boundary sliding description. To solve this problem the additional scale level is introduced into the model, evolution and closing equations are formulated on the basis of physical analysis. To connect different scale levels the consistency conditions of constitutive relations are used. The results of computational experiments under uniaxial tension of a polycrystalline pure copper representative volume are obtained.

1. Introduction

Recently different methods of severe plastic deformation (for example, equal channel angular pressing, extrusion, rolling and their various combinations or modifications) are widely used in practice to obtain fine grain and ultra-fine grain metals with improved physical and mechanical properties. The possibility of material switching into superplastic state, producing of details with a complex shape by corresponding treatment and minimum cost, attracts to fine grain and ultra-fine grain materials [5]. Today there is increasingly more need for constructing mathematical models that allow describing integral characteristics of material processes mentioned above and internal structure evolution, which has an influence on physical and mechanical properties. Such mathematical models will define efficient process conditions to produce materials and details of the required quality. The multilevel models of inelastic deformation based on crystal plasticity [6,9] suggested to be the most suitable to achieve the goals indicated above (taking into account that influence on material in different parts of details and changing of internal structure caused by this can be essentially various). These models allow analyzing deformation processes at the macrolevel, to describe explicitly inelastic deformation mechanisms, material internal structure and it's evolution at lower scale levels. Up to this point it would seem that mentioned models have a high generality, allow description of both plastic (viscoplastic) and superplastic regimes within a common model. Brief review of multilevel models based on crystal plasticity is given in [1,6,9,10].

Three-level model of polycrystalline inelastic deformation is used in the paper. The model regards the temperature and includes the most important mechanisms: intragranular dislocation sliding, crystallite lattice rotations and grain boundary sliding, which are significant at elevated temperatures and fine grain structure of the material [2]. Special attention is devoted to description of critical stress changes for grain boundary sliding realizing due to different mechanisms of interaction between defects.

2. Description of three-level inelastic deformation model for polycrystalline

The three-level model is constructed on the base of two-level constitutive model previously developed [9]. It is a statistical model considered the deformation of single crystallites with different orientations, generalized Voigt's hypothesis is used to describe the influence on elements of lower scale levels, averaging procedure is applied to define the response of a representative volume [9]. Statistical models are computationally more efficient than the direct models and intensively used in the inelastic deformation modeling of real materials [7,8].

The scale level hierarchy is defined in this way (by descending scale): macroscale level, mesolevel I and mesolevel II. Mesolevel I is introduced for grain boundary sliding description. Crystallite inelastic deformation in mesolevel II model is assumed to be realized by intragranular dislocation sliding with crystallite lattice rotation. All macrolevel quantities are given by uppercase letters, quantities of mesolevel I are denoted by lowercase letters with the index I, mesolevel II quantities are denoted by lowercase letters with the index II.

The equation system of macroscale level is the following:

$$\begin{cases} \dot{\mathbf{\Sigma}} + \mathbf{\Omega}^{\mathrm{T}} \cdot \mathbf{\Sigma} + \mathbf{\Sigma} \cdot \mathbf{\Omega} = \mathbf{\Pi} : (\mathbf{D} - \mathbf{D}^{\mathrm{in}} - \mathbf{D}^{\mathrm{th}}), \\ \mathbf{\Pi} = \langle \mathbf{\pi}_I \rangle, \\ \mathbf{\Omega} = < \mathbf{\omega}_I >, \\ \mathbf{D}^{\mathrm{in}} = < \mathbf{d}_I^{in} > + \mathbf{\Pi}^{-1} :< \mathbf{\pi}'_I : \left(\mathbf{d}_I'^{in} \right) > + \mathbf{\Pi}^{-1} : \left(< \mathbf{\sigma}'_I \cdot \mathbf{\omega}'_I > - < \mathbf{\omega}'_I \cdot \mathbf{\sigma}'_I > \right), \\ \mathbf{D}^{\mathrm{th}} = < \mathbf{d}_I^{th} > + \mathbf{\Pi}^{-1} :< \mathbf{\pi}'_I : \left(\mathbf{d}_I'^{th} \right) >, \\ \dot{\Theta} = \langle \dot{\theta}_I \rangle. \end{cases} \tag{1}$$

Hooke's law in rate form $(1)_1$ is the constitutive relation, $\mathbf{\Sigma}$ is Cauchy stress tensor. The explicit internal variables of macrolevel model are spin of movable coordinate system $\mathbf{\Omega}$, effective anisotropic elastic properties $\mathbf{\Pi}$, inelastic $\mathbf{D}^{\mathrm{in}}$ and thermal $\mathbf{D}^{\mathrm{th}}$ components of the strain rate tensor. These variables depend on behavior of lower scale levels at each time moment. They are determined from consistency conditions of scale levels $(1)_2$, $(1)_3$, $(1)_4$ and $(1)_5$ [9]. The hypothesis of process adiabaticity is adopted at macrolevel: the rate of temperature change at macrolevel $\dot{\Theta}$ depends only on the rate of temperature change at mesoscale level due to grain boundary and intragranular shears; the hypothesis of temperature homogeneity is assumed at the initial moment at all scale levels.

Constitutive model at mesolevel I is presented by the following relation system:

$$\begin{cases}
\dot{\boldsymbol{\sigma}}_I = \boldsymbol{\pi}_I : \left(\mathbf{d}_I - \mathbf{d}_I^{in} - \mathbf{d}_I^{th}\right) + \boldsymbol{\omega}_I \cdot \boldsymbol{\sigma}_I - \boldsymbol{\sigma}_I \cdot \boldsymbol{\omega}_I, \\
\boldsymbol{\pi}_I = < \boldsymbol{\pi}_{II} >, \\
\boldsymbol{\omega}_I = < \boldsymbol{\omega}_{II} >, \\
\hat{\nabla}\mathbf{v}_I = \hat{\nabla}\mathbf{V}, \\
\mathbf{d}_I = \frac{1}{2}(\hat{\nabla}\mathbf{v}_I + (\hat{\nabla}\mathbf{v}_I)^{\mathrm{T}}), \\
\mathbf{d}_I^{in} = \mathbf{d}_{gb}^{in} + \mathbf{d}_{I_from_II}^{in}, \\
\mathbf{d}_I^{th} = \mathbf{d}_{gb}^{th} + \mathbf{d}_{I_from_II}^{th}, \\
\dot{\gamma}_{gb}^{(i)} = H\left(\tau_{gb}^{(i)} - \tau_{Cgb}^{(i)}\right)\dot{\gamma}_{0gb}\left(\frac{\tau_{gb}^{(i)}}{\tau_{Cgb}^{(i)}}\right)^{1/m} \exp\left(-\frac{U_{gb}}{k\theta_I}\right), \; i = 1,2,\ldots,K_{gb}, \\
\dot{\tau}_{cgb}^{(i)} = f\left(\dot{\gamma}_{gb}^{(i)}, \gamma_{gb}^{(i)}, \theta_I\right), \\
\hat{\nabla}\mathbf{v}_{gb} = \sum_{k=1}^{K_{gb}} \alpha_{vf}\dot{\gamma}_{gb}^{(k)}\mathbf{b}_{gb}^{(k)}\mathbf{n}_{gb}^{(k)}, \\
\mathbf{d}_{gb}^{in} = \frac{1}{2}(\hat{\nabla}\mathbf{v}_{gb} + (\hat{\nabla}\mathbf{v}_{gb})^{\mathrm{T}}), \\
\mathbf{d}_{gb}^{th} = \boldsymbol{\alpha}\dot{\theta}_I, \; \dot{\theta}_I = \alpha_I \frac{\boldsymbol{\sigma}_I : \mathbf{d}_{gb}^{in}}{C_m \rho} + < \dot{\theta}_{II} >, \\
\mathbf{d}_{I_from_II}^{in} = < \mathbf{d}_{II}^{in} > + \boldsymbol{\pi}_I^{-1} : < \boldsymbol{\pi}'_{II} : \left(\mathbf{d}'^{in}_{II}\right) > + \boldsymbol{\pi}_I^{-1} : \left(< \boldsymbol{\sigma}'_{II} \cdot \boldsymbol{\omega}'_{II} > - < \boldsymbol{\omega}'_{II} \cdot \boldsymbol{\sigma}'_{II} >\right), \\
\mathbf{d}_{I_from_II}^{th} = < \mathbf{d}_{II}^{th} > + \boldsymbol{\pi}_I^{-1} : < \boldsymbol{\pi}'_{II} : \left(\mathbf{d}'^{th}_{II}\right) >.
\end{cases} \qquad (2)$$

Hooke's law in rate form $(2)_1$ is the constitutive relation of mesolevel I model. The explicit internal variables of mesolevel I model are elastic modulus tensor $\boldsymbol{\pi}_I$, inelastic $\mathbf{d}_I^{\mathrm{in}}$ and thermal $\mathbf{d}_I^{\mathrm{th}}$ components of the strain rate tensor, spin of movable coordinate system $\boldsymbol{\omega}_I$ defined quasi-solid motion. These variables depend on behavior of lower scale level at each time moment. They are determined from consistency conditions of scale levels $(2)_2$, $(2)_3$, $(2)_{13}$ и $(2)_{14}$ [9]. Inelastic strain rate is defined by sum of two components: the first one is determined by grain boundary sliding realizing (equations $(2)_8$,$(2)_9$, $(2)_{10}$, $(2)_{11}$), the second component is defined by inelastic deformation at mesolevel II (consistency condition of constitutive relations at scale levels $(2)_{13}$). Thermal component of the strain rate tensor is also defined by sum of two elements: temperature change occurs by grain boundary sliding (relations $(2)_{12}$) and inelastic deformation at mesolevel II (relation $(2)_{14}$). Here $\boldsymbol{\alpha}$ is thermal expansion tensor, α_I is coefficient of heat eduction, C_m is specific heat, ρ is substance density. Velocity gradient $(2)_4$ is transmitted as forcing from macrolevel. In (2) notations are following: $< \beta >$ is the average value of β in a representative volume of level considered, β' is deviation from the average value for element considered (the element index is omitted), α_{vf} is the volume fraction of boundaries.

Shears on slip systems of grain boundary sliding are described by elastoviscoplastic relation with regard to thermally activated motion of grain boundary dislocations $(2)_8$, in which $\dot{\gamma}_{0gb}$ is grain boundary shear rate under shear stress $\tau_{gb}^{(k)}$ equal to critical shear stress $\tau_{Cgb}^{(k)}$, U_{gb} is the energy barrier (for grain boundary shear), $\tilde{k}$ is Boltzmann constant, θ_I is a temperature, $H(\cdot)$ is Heaviside function, K_{gb} is total slip systems of grain boundary sliding.

The evolution equation for critical stress is proposed in the following form:

$$\dot{\tau}_{cgb}^{(k)} = h\left(1-\exp\left(-\frac{q}{\gamma_{gb}^{(k)}}\right)\right)\dot{\gamma}_{gb}^{(k)} - \left(\sum_{j=1}^{2}\sum_{i=1}^{K_1}(g_i^k\,\dot{\gamma}^{(i)(j)}(\mathbf{b}^{(i)(j)}\cdot\mathbf{N}^{(i)(j)})\,H(\mathbf{b}^{(i)(j)}\cdot\mathbf{N}^{(i)(j)})\right) - \tau_{cgb}^{(k)}\frac{c}{S_j}\exp\left(-\frac{U_d}{\tilde{k}\theta_I}\right).$$

The first item describes increasing of critical stress by intrusion during grain boundary sliding, q is nondimensional model parameter defined due to model identification; $\gamma_{gb}^{(k)} = \int_0^t \dot{\gamma}_{gb}^{(k)} d\tau$ is cumulative shear on grain boundary sliding system, h is hardening coefficient. The second item describes the decreasing of critical stress by increasing boundary energy due to inflow of intragranular dislocations (the summation is assumed only for slip systems of two adjacent grains, dislocations from which are really going out to boundary), g_i^k is hardening (softening) matrix due to intragranular dislocations which are out to boundary facet, $\dot{\gamma}^{(i)(j)}$ is shear rate on i slip system in grain j, $\mathbf{b}^{(i)(j)}$ is unit vector in the direction of Burgers vector for i slip system in grain j, $\mathbf{N}^{(i)(j)}$ is outer normal to the boundary for grain j, $H(\cdot)$ is Heaviside function. The third item describes diffusion processes [3] on the base of Arrhenius law, U_d is activation energy of thermally activated (diffusion) processes, c is the diffusion coefficient, S_k is square of boundary facet j.

Constitutive model at mesolevel II has the form:

$$\begin{cases}
\dot{\boldsymbol{\sigma}}_{II} = \boldsymbol{\Pi}_{II} : \left(\mathbf{d}_{II} - \mathbf{d}_{II}^{in} - \mathbf{d}_{II}^{th}\right) + \boldsymbol{\omega}_{II}\cdot\boldsymbol{\sigma}_{II} - \boldsymbol{\sigma}_{II}\cdot\boldsymbol{\omega}_{II}, \\
\mathbf{d}_{II}^{th} = \boldsymbol{\alpha}\dot{\theta}_{II}, \\
\dot{\theta}_{II} = \alpha_{II}\dfrac{\boldsymbol{\sigma}_{II} : \mathbf{d}_{II}^{in}}{C_m\rho}, \\
\hat{\nabla}\mathbf{v}_{II} = \hat{\nabla}\mathbf{V} - \hat{\nabla}\mathbf{v}_{gb}, \\
\mathbf{d}_{II} = \dfrac{1}{2}(\hat{\nabla}\mathbf{v}_{II} + (\hat{\nabla}\mathbf{v}_{II})^{\mathrm{T}}), \\
\dot{\gamma}^i = \dot{\gamma}_0\left(\dfrac{\tau^i}{\tau_C^i}\right)^{1/n} H(\tau^i - \tau_C^i)\exp\left(-\dfrac{U}{k\theta_{II}}\right),\ i = 1,\dots,K, \\
\dot{\tau}_c^i = f(\gamma^j, \dot{\gamma}^j, \theta_{II}),\ i,j = 1,\dots,K, \\
\mathbf{d}_{II}^{in} = \dfrac{1}{2}\sum_{k=1}^{K_s}\beta_{vf}\dot{\gamma}^i\left(\mathbf{n}_s^{(k)}\mathbf{b}_s^{(k)} + \mathbf{b}_s^{(k)}\mathbf{n}_s^{(k)}\right), \\
\textit{relations for lattice spin } \boldsymbol{\omega}_{II},\ \dot{\mathbf{o}}\cdot\mathbf{o}^{\mathrm{T}} = \boldsymbol{\omega}_{II}.
\end{cases} \quad (3)$$

Hooke's law in rate form $(3)_1$ is the constitutive relation of mesolevel II model. The explicit internal variables of mesolevel II model are elastic modulus tensor $\boldsymbol{\Pi}_{II}$, inelastic $\mathbf{d}_{II}^{in}$ and thermal $\mathbf{d}_{II}^{th}$ components of the strain rate tensor, spin of movable coordinate system $\boldsymbol{\omega}_{II}$ defined quasi-solid motion. Thermal component $\mathbf{d}_{II}^{th}$ is defined by relation $(3)_2$, where $\boldsymbol{\alpha}$ is thermal expansion tensor. Relation $(3)_3$ describes temperature change due to inelasticity, α_{II} is coefficient of heat eduction, C_m is specific heat, ρ is substance density. The implicit internal variables are shear rates of intragranular dislocation sliding $\dot{\gamma}^i$ and critical shear stress rates $\dot{\tau}_c^i$: shear rate is determined by elastoviscoplastic relation $(3)_6$ with regard to activation energy of inelastic deformation U, hardening law $(3)_7$ is used for describing the critical shear stresses [9]. Macrolevel velocity gradient $\hat{\nabla}\mathbf{V}$ is transmitted as forcing to mesolevel II $(3)_4$ with taking into account that some forcing is dedicated to grain boundary sliding

IOP Conf. Series: Materials Science and Engineering **71** (2015) 012081 doi:10.1088/1757-899X/71/1/012081

realization $\hat{\nabla}\mathbf{v}_{gb}$ at mesolevel I. Schmid law as a criterion of shear activity (dislocation motion) is used to describe intragranular dislocation sliding [4]. Rotation model with regard to incompatibility of dislocation sliding in neighboring crystallites is suggested [9].

Introduction the tensor variable $\mathbf{z}$ for each crystallite as a characteristic of it's shape and boundary is proposed for grain structure description. The equation for this parameter evolution will be added to the relation structure at mesolevel II (3) on the base of forming, breaking and fragmentation consideration. In the future it is planned to connect the volume fractions of grain boundaries α_{vf} and grain interior β_{vf} included in the kinetic equations of mesoscale levels with the representative volume averaging of parameter mentioned above.

3. Results of modelling

The computational experiment under uniaxial tension of polycrystalline pure copper (1000 crystallites) is conducted. Parameters for shear description are taken as in [9]. Parameters for grain boundary sliding description are as follows: $\theta = 500\,K$, $\dot{\gamma}_{0gb}\exp\left(-\frac{U_{gb}}{\tilde{k}\theta}\right) = 10^{-4}\,s^{-1}$, $q = 1$, $c\exp\left(-\frac{U_d}{\tilde{k}\theta}\right) = 0.1\,m^2\cdot s^{-1}$, $g_i^k = 0.33\ \ \forall k,i$]. The results are obtained for four different choices of parameters (Figure 1): (1) – $h = 1$, $\tau_{cgb}^{(i)}(0) = 15\,MPa\ \ \forall i$, (2) – $h = 0.8$, $\tau_{cgb}^{(i)}(0) = 14\,MPa\ \ \forall i$, (3) – $h = 0.6$, $\tau_{cgb}^{(i)}(0) = 14\,MPa\ \ \forall i$, (4) – $h = 0.4$, $\tau_{cgb}^{(i)}(0) = 13\,MPa\ \ \forall i$. The first set of parameters corresponds to coarse grain structure, the last one corresponds to fine grain structure, when grain boundary sliding is relieved.

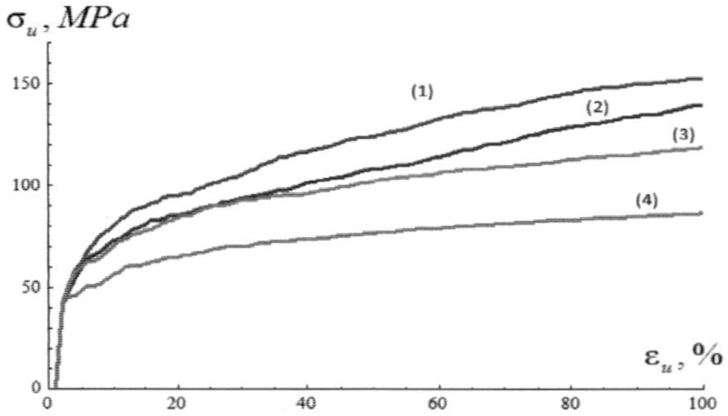

Figure 1. Dependence of the stress intensity from the strain intensity for four different types of grain structure ((1) – coarse grain structure, (4) – fine grain structure).

The results obtained show the increasing role of grain boundary sliding during deformation process, fine grain structure is characterized by a lower flow stress in comparison with coarse grain structure.

4. Conclusion

Three-level constitutive polycrystalline model is developed taking into account main mechanisms of inelastic deformation: intragranular dislocation sliding, lattice rotation and grain boundary sliding. The model includes temperature and consistency conditions of constitutive relations at different scale levels. Special attention is devoted to grain boundary sliding description due to included various mechanisms of interaction between defects into evolution relations: it is assumed on the base of existing theoretical and experimental data analysis that grain boundary sliding is occurred by inflow of intragranular dislocations, change in the boundary structure under realization of grain boundary sliding and diffusion processes. The results of computational experiments under uniaxial tension of a polycrystalline pure copper are obtained. They show model resources in description of increasing role the grain boundary sliding during deformation process.

Acknowledgements
The work was supported by the Russian Federation Ministry of Education and Science (basic part of the state task PNRPU), RFBR (grants Nos. 14-01-00069-a, 14-01-96008 r_ural_a), President Grant MK-4485.2014.1.

References

[1] Habraken A 2004 Modelling the plastic anisotropy of metals *Arch. Comput. Meth. Eng.* **11** 3-96
[2] Kaibushev O 1975 Plasticity and superplasticity of metals (Moscow: Metallurgy) 280
[3] Panin V, Egorushkin V and Elsukova T 2013 Physical mesomechanics of grain boundary sliding in a deformable polycrystal *Phys Mesomech* **16** 1-8
[4] Sharifullina E, Shveykin A and Volegov P 2012 Analysis of mesolevel constitutive relations used in two-level models of polycrystalline inelastic deformation **4** 186-192
[5] Shorshorov M, Tikhonov A, Bulat S, Gurov K, Nadirashvili N and Antipov V 1973 Superplasticity of metal materials (Moscow: Science) 220
[6] Trusov P, Ashikhmin V, Volegov P and Shveykin A 2010 Constitutive relations and their application to the description of microstructure evolution *Phys Mesomech* **13** 38-46
[7] Trusov P and Shveykin A 2013 Multilevel crystal plasticity models of single- and polycrystals. Statistical models *Phys Mesomech* **16** 23-33
[8] Trusov P and Shveykin A 2013 Multilevel crystal plasticity models of single- and polycrystals. Direct models *Phys Mesomech* **16** 99-124
[9] Trusov P, Shveykin A, Nechaeva E and Volegov P 2012 Multilevel models of inelastic deformation of materials and their application for description of internal structure evolution *Phys Mesomech* **15** 155-175
[10] Van Houtte P 2009 Crystal plasticity based modelling of deformation textures *Microstructure and texture in steels* (Springer) pp 209-224